Elrick

15 Oct., 1987

Numerical Simulation of Reactive Flow

Numerical Simulation of Reactive Flow

Elaine S. Oran, Ph.D.
Jay P. Boris, Ph.D.

Naval Research Laboratory
Washington, D.C.

Elsevier
New York • Amsterdam • London

Elsevier Science Publishing Co., Inc.
52 Vanderbilt Avenue, New York, New York 10017

Distributors outside the United States and Canada:
Elsevier Applied Science Publishers Ltd.
Crown House, Linton Road, Barking, Essex IG11 8JU, England

Library of Congress Cataloging in Publication Data

Oran, Elaine S.
Numerical simulation of reactive flow.

Includes bibliographies and index.
1. Fluid dynamics–Mathematical models. 2. Chemical reaction, Rate of–Mathematical models. 3. Transport theory–Mathematical models. I. Boris, Jay P.
II. Title. III. Title: Reactive flow.
QA911.066 1987 530.1'5 87-15588
ISBN 0-444-01251-6

Current printing (last digit)
10 9 8 7 6 5 4 3 2 1

Manufactured in the United States of America

To Elizabeth and Daniel

The mathematicians and the physics men have their mythology; they work alongside the truth never touching it; their equations are false but the things work.

Robinson Jeffers

Contents

Prologue

Numerical simulations can be used to predict and study the behavior of complex physical systems. To give quantitatively correct predictions, the simulation model must describe the individual contributing processes in the systems as well as their interactions. Once a model has been developed and tested, it can be used to interpret measurements and observations, evaluate new ideas, extend theoretical models into new parameter regimes, help in engineering design processes, and quantitatively test existing theories.

Perhaps the most important use of a computer model is to test the accuracy and completeness of our understanding of a physical system. We do this by comparing model predictions to experimental data. If both the simulation and the experiment are accurate, discrepancies between them point to previously unconsidered effects. A simulation can sometimes replace an experiment, but most often simulations and experiments are complementary. Experiments must often be interpreted by simulations, and simulations must often be calibrated by experiments.

The combined use of theory, simulations, and experiments to study a complicated system is a relatively new research approach. Experimental observations and approximate theoretical models suggest laws that we expect a physical system to obey. We test these laws against reality by incorporating a description of the contributing processes into a quantitative model, making quantitative predictions, and comparing these predictions to a definitive series of experiments.

Because even the simplest experiments are often quite complicated, the models describing them cannot be solved analytically and must be solved numerically. Today, a major part of a number of scientific programs, such as fusion, aerodynamics, and turbulence research, involves coordinating the results of experiments and simulations. The experiments provide input for the simulations as well as data against which the predictions, and the models on which they are based, can be tested.

The purpose of this book is to show how to construct, use, and interpret numerical simulations of chemically reactive flows. Constructing numerical simulations means making decisions at each step. These decisions affect the subsequent form and use of the model. For example, to formulate the mathematical models for processes that occur in a real system, we must decide which algorithms to use, how to couple the algorithms together, and how to organize and interpret the results. To understand the important trade-offs these decisions imply, we must understand the limitations and uses of the models being developed.

The computational techniques described here are useful for multidimensional problems with realistic boundary conditions and many interacting chemical species. These methods thus apply to reactive flow problems in combustion, aeronomy, partially ionized plasmas, aerodynamics, gas dynamic lasers, astrophysics, and general multiphase and magnetohydrodynamic flows. Because many of the underlying physical processes are similar, this text draws on numerical developments in all of these fields.

The specific examples we use to illustrate methods are drawn from combustion dynamics and involve interactions among a number of simultaneous processes occurring over a broad range of time and space scales. In combustion, chemical energy release generates pressure, temperature, and density gradients. These gradients, in turn, drive fluid processes, which transport mass, momentum, and energy throughout the system. The strong interplay between chemistry and fluid dynamics makes modeling reactive flow problems difficult. This book does not teach the physics and chemistry of combustion. It does describe how to develop detailed numerical simulations that can be used as tools for studying combustion.

This book is written for readers with advanced undergraduate or beginning graduate school knowledge of the basic principles of chemistry and fluid dynamics and their potential interactions. Some familiarity with the rudiments of numerical methods is helpful. Readers who have written and debugged simple programs to integrate differential equations will find this book relatively easy to follow.

The chapters in this book can be divided into three groups:

- Introductory Course on Modeling and Numerical Simulation — Chapters 1 through 4.
- Advanced Topics in Numerical Simulation of Reactive Flow Processes — Chapters 5 through 11.
- Simulating Complex Reactive Flows — Chapters 12 and 13.

Introductory Short Course

Chapters 1 through 4 are the foundation for material presented in the remainder of the book. These chapters

• Discuss the concepts and terminology used throughout the book. Chapter 1 describes modeling and simulation. This chapter gives an overview of numerical simulation and introduces the major themes of the book.

• Define the reactive flow problem and place it within the general context of numerical simulation. Chapter 2 describes the conservation equations for mass, momentum, energy, and chemical species, as well as some of the problems involved in solving them. Chapter 2 also describes individual processes in reactive flows and relates them to specific terms in the differential equations. The physical and mathematical assumptions implicit in this model are discussed. This chapter concludes with a general discussion of the types of hardware and resolution limitations we encounter when we try to solve these equations.

• Analyze limitations and trade-offs that arise in constructing and using computer models. Chapter 3 lists the sources of errors in numerical simulation and evaluates the costs and compromises needed to overcome them. The trade-offs that must be made at every stage of developing and using a simulation are determined as much by the computational and human resources available as by the specific requirements of the problem being tackled. Chapter 3 describes the elements of an efficient approach to simulation.

• Lead the reader into the numerical analysis and computational physics issues that are important for modeling reactive flows. Chapter 4 explains basic finite-difference approaches used to integrate the reactive flow equations numerically. Simple test problems lead to ordinary differential equations representing local processes, such as chemical kinetics, equilibration processes, and coupling terms, and partial differential equations for diffusion, convection, and wave phenomena.

Advanced Topics in Numerical Simulation

The short course chapters introduce a number of important topics and show the relations among them. The advanced topic chapters (5 through 11) discuss these topics in greater depth and at a higher level of sophistication. These chapters consider processes that are nonlinear, multidimensional, require variably spaced grids that may move in time, and involve many coupled equations.

Chapter 5 describes the spectrum of algorithms for solving ordinary differential equations as well as specific applications to reactive flows. Chapter 6 describes the types of representations and spatial and geometry problems. Chapter 7 describes solutions of diffusion equations which represent multi-species diffusion, thermal conduction, and radiation transport. Chapter 8 describes solution methods for a single continuity equation, and Chapters 9 and 10 describe Eulerian and Lagrangian solution methods for coupled continuity equations. Finally, Chapter 11 discusses the numerical implementation of boundary conditions and algorithms for matrix algebra.

Simulating Complex Reactive Flows

Chapters 12 and 13 tie the ideas and themes of the book together. They draw from the introductory material in Chapters 1–4 and from the separate advanced topic chapters. Chapter 12 surveys turbulence modeling, multi-phase flows, and radiation transport. Each of these topics alone deserves many volumes, and the material given is only an introduction that shows how the numerical methods described earlier in the book can be applied. In Chapter 13 we describe how to construct several different kinds of reactive flow computer programs, and present examples and calculations performed with them.

References and Related Texts

This book is complementary to a number of other books on combustion and reactive flow. In particular, the behavior of flames and detonations are treated in:

Combustion, Flames and Explosions of Gases
B. Lewis and G. von Elbe, Academic Press, New York, 1961.

Combustion Theory
F.A. Williams, Second Edition, Benjamin Cummings, Menlo Park, CA, 1985.

Combustion
I. Glassman, Academic Press, New York, 1977.

Detonation
W. Fickett and W.C. Davis, University of California Press, Berkeley, 1979.

Flames
A.G. Gaydon and H.G. Wolfhard, Chapman and Hall, London, 1979.

Combustion Dynamics: Dynamics of Chemically Reacting Fluids
T.-Y. Toong, McGraw-Hill, New York, 1983.

Combustion Fundamentals
R.A. Strehlow, McGraw-Hill, New York, 1984.

Computer Modeling of Gas Lasers
K. Smith and R.M. Thomson, Plenum, New York, 1978.

Transport Processes in Chemically Reacting Flow Systems
D.E. Rosner, Butterworths, Boston, 1986.

Principles of Combustion
K.K. Kuo, Wiley, New York, 1986.

The material in this book also complements books on computational fluid dynamics and computational physics, for example,

Difference Methods for Initial-Value Problems
R.D. Richtmyer and K.W. Morton, Interscience, New York, 1967.

Computational Physics
D. Potter, Wiley, New York, 1973.

Computational Fluid Dynamics
P.J. Roache, Hermosa Publishers, Albuquerque, New Mexico, 1982.

An Introduction to Computational Fluid Mechanics
C.-Y. Chow, Wiley, New York, 1979.

Finite-Difference Techniques for Vectorized Fluid Dynamic Calculations
Edited by D.L. Book, Springer-Verlag, New York, 1981.

Computational Fluid Mechanics and Heat Transfer
D.A. Anderson, J.C. Tannehill, and R.H. Pletcher, McGraw-Hill, New York, 1984.

References are included at the end of each chapter. A subsequent volume of software is planned in the form of simple modules, which implement many of the algorithms discussed and referenced here.

Acknowledgements

Many have helped us with various aspects of this book.

We would like to thank Paul Libby, Roger Strehlow, Howard Palmer, Steven Zalesak, David Fyfe, K. Kailasanath, Jill Dahlburg, Peyman Givi, C. Richard Devore, Howard Ross, Kenneth Laskey, Rainald Löhner, John Gardner, Gopal Patnaik, David Book, Theodore Young, Martin Fritts, Mel Baer, Jace Nunciato, Teman Burks, Shmuel Eidelman, Raafat Guirguis, Fernando Grinstein, Mark Emery, J. Michael Picone, and Sam Lambrakos for their technical contributions and comments.

We thank Daniel Oran, Michael Frenklach, Keith Ellingsworth, Judith Karpen, Douglas Ladouceur, Martin Rabinowitz, and Walter Shaub for their detailed comments on portions of the book.

We thank Anthony Leonard for permission to use material from his reviews on vortex dynamics, Robert May for permission to use material from his review on ordinary differential equations, Kenneth Kuo, Mel Baer, and Jace Nunziato for permission to use their materials on multiphase flows.

We thank Martin Fritts, Mark Emery, Pamela Barr, Robert Gelinas, Raafat Guirguis, Fernando Grinstein, James Hyman, Robert Noh, Ahmed Ghoneim, Ralph Metcalfe, Robert Siegel, and David Garvin for permission to use their illustrations.

We thank Fred Rettenmaier and Maureen Long for their editorial assistance, and Agnes Green for her artistic contributions.

There are a number of sponsors and sponsoring agencies whose support made this work possible. We would like to thank Timothy Coffey and William Ellis of the Naval Research Laboratory, Richard Miller, Robert Junker, Keith Ellingsworth, and Robert Whitehead of the Office of Naval Research, Julian Tishkoff of the Air Force Office of Scientific Research, Cecil Marek, Kurt Sackstedter, Dixon Butler, and Howard Ross of the National Aeronautics and Space Administration, and George Ulrich and Carl Fitz of the Defense Nuclear Agency.

In addition, we would like to thank Herbert Rabin, Albert Schindler, Fred Saalfeld, Homer Carhart, Fred Williams, and Denis Bogan for their continuing encouragement and help in the formative years of our program.

We would especially like to thank Darlene Miller, Francine Rosenberg, and Nancy Ciatti for running interference while we wrote this book. We thank Francine Rosenberg and Elizabeth Gold for their help in programming the equations in a number of chapters. We especially thank Francine Rosenberg, Theodore Young, and Robert Scott for providing and maintaining the tools we used.

We thank Norman Chigier and Vladimir Hlavacek for their efforts to convince us to write this book.

We thank Donald Knuth who crafted TeX, though neither of us know him personally. TeX allowed us to actually see the book as we wrote it, gave us control of its final appearance, and allowed us to bypass dealing with galley proofs.

We would like to thank Paul A. Boris for his scientific encouragement through the years.

We would like to thank Daniel Oran, Elizabeth Boris, David Boris, Paul Boris, Karen Clark, Minerva Oran, Herman Surick, Doris Luterman, Marcy

Docter, and Jonah Brown for their support, encouragement, and forbearance during the difficult years of this project.

Finally, we would like to thank Marjan Baco of Elsevier, who continually encouraged us and with whom it was delightful to work throughout process of writing this book.

Chapter 1
AN OVERVIEW OF NUMERICAL SIMULATION

1–1. SOME COMMENTS ON TERMINOLOGY

The words *model* and *simulation* are used repeatedly in this book. These words seem easy enough to understand and use in ordinary conversation, but their technical use is confused and imprecise. To different individuals and different research communities, these words mean different things.

To attack this technical Tower of Babel by proposing rigid definitions would be a disservice and not very productive. Instead, we discuss some of the nuances of key terms used in this book and try to use them consistently.

What is generally meant by modeling is much broader than what is generally meant by simulation. A simulation attempts to imitate the dynamic behavior of a system and to predict or calculate subsequent events. Modeling is generally used in a broader, more static sense: a model can represent a dynamic phenomenon in its entirety without specifying its evolution. A model can be a formula, a quantitative relation, a collection of empirical data, an equation, or even its equivalent analog circuit.

The objective of modeling is not to produce an exact copy of a system. Instead, a model tries to reproduce certain salient features of the system. Because approximations are made to derive the model, it is always understood to be an imperfect representation of the system. Consequently a model is invalid in some regimes and imprecise by some amount almost everywhere. A simulation exercises models for a particular choice of physical parameters, initial conditions, and boundary conditions. Throughout the literature, however, the use of the terms *modeling* and *simulation* overlap, and they are sometimes used interchangeably.

The terms *numerical* and *computational* are often used interchangeably, although they have somewhat different meanings. Numerical analysis, generally a theoretical subject whose main practical application is to solve problems on computers, is not all computational. Computational work in simulations is not all numerical. For example, a reactive flow computation may involve interactive systems programming, text manipulation, data management, graphics, etc. Here we consider models and algorithms as computational when they describe how the problem is broken down, represented, manipulated, and stored in the computer. We call techniques, methods, and algorithms numerical when they concern the calculation of numbers or the quantitative evaluation of formulas and equations.

An *algorithm* is not a model, and it is not necessarily numerical. An algorithm is a solution procedure used to implement a model. Not every model has an algorithm associated with it. There simply might not be an algorithm to implement a model. This book is an operational algorithm for implementing the equations of chemically reactive flow on a computer. It also presents many numerical algorithms for solving the model equations.

A simulation does not solve the equations that make up the mathematical model directly. The words *advance* and *integrate* are more appropriate within the context of a simulation than the word *solve*. We solve a set of discretized algebraic equations, the computational model, to advance the simulation model in time. This algebraic set of equations is derived from the original set of partial differential equations. Accurate algorithms are substituted for the various mathematical terms, for their interactions, and thus for the underlying physical phenomena.

As numerical and computational methods improve, simulations become more accurate. Computer memory and speed are increasing rapidly, and there is a corresponding increase in what can easily be calculated. Numerical simulation is no longer an esoteric art confined to a few laboratories with large computers. It is a tool available to most scientists and engineers. This book explains how to use this widespread and rapidly growing computational technology to solve complex reactive flow problems.

1–2. THREE TYPES OF PHYSICAL MODELS

Detailed Models

A *detailed model* or *fundamental model* describes the properties or behavior of a system starting with as many basic physical assumptions, or *first principles*, as possible. The problems that a detailed model can treat are limited in scope. Such models can, however, be extremely detailed and exact in what they do cover. They are often used to provide constants or other information to be used in more general but less rigorous calculations. The focus is usually one isolated type of process or interaction. For example, quantum mechanical calculations of the potential energy surface for a molecule or for cross sections of a chemical reaction are fundamental models providing information on chemical reaction pathways. Where to draw the line on what to call first principles is flexible and depends on the particular problem.

An important objective of detailed modeling is developing computational models with well understood ranges of validity. In principle, the broader the range of validity, the more generally useful the model is and the more expensive the model is to use.

Phenomenological Models

Phenomenological models and *empirical models* must be used when the scales of the physical processes are too disparate to resolve consistently in one calculation. In these situations, the usual process is to propose or derive macroscopic or averaged models of the small-scale processes. These are the phenomenological models that appear in the governing equations. For example, in the detailed models of flames considered later in this book, the chemical rate constants, diffusive transport coefficients, and equations of state cannot be determined simultaneously with the convective flow. These three types of models represent atomic-scale processes that have been averaged over distributions of particles to derive the continuous, fluid description. These quantities are phenomenologies representing processes that occur on a scale too small to be resolved in the fluid computation itself.

Often the exact forms of terms in the governing equations are not known. Alternatively, these terms may be too complex to evaluate, or we might not be able to find a numerical form for them. In this case a simpler, approximate form must be used. This approximate form is usually motivated physically and contains input data obtained from fits to experimental data or more fundamental theories or simulations. Several of the turbulence models discussed in Chapter 12 are examples of phenomenologies in which complex interactions are modeled by equations with physically reasonable forms. These models are then calibrated by adjusting coefficients to fit data derived from experiments or from specialized detailed simulations. Other examples include an energy release model for a chemical system, or a parametric model for soot formation. A phenomenological model is based on a prescription or formula that represents our intuitive, qualitative understanding of a particular physical situation, and thus must be calibrated with a more basic theory or an experiment.

Phenomenological models usually start with a simple theory for the phenomenon to be modeled. This theory is simple because it is only approximate. Often global constraints and conservation conditions are used to derive the phenomenological model. These conditions do not require detailed information about physical or chemical mechanisms. For example, the Riemann solutions of the Rankine-Hugoniot equations provide simple relationships describing gas dynamic discontinuities. These solutions require only the upstream and downstream conditions of the flow, and nothing about the structure of the discontinuity is contained or implied in the model. Obtaining information about the discontinuity requires a much more expensive, highly resolved, local integration of the fundamental equations.

Empirical Models

Empirical models are either direct fits of data to a given mathematical formula or are data used directly in tabular form. The important point is that the data are usually derived from experiments. Often the data include extraneous effects that we do not expect or want, such as measurement interference or equipment calibration errors. The caveat that goes along with such models is that they can be used for interpolation but not for extrapolation. This is true with respect to both the physical parameter range over which the model is valid and the particular physical environment in which it is used. Chemical rate constants are usually empirical models. Equations of state are usually taken from experimental data in tabular form. Other examples of empirical models are tables of the rate at which smoke diffuses through a chamber, the rate a fire spreads through a building, or the rate of chemical energy release in a turbulent diffusion flame.

1–3. LEVELS OF SIMULATION

The expression *chunking* (Hofstadter, 1979) means that microscopic complexity becomes statistically simplified, or chunked at a macroscopic level. Chunking reduces the number of degrees of freedom needed to effectively describe the behavior of a system. The chunked behavior of a system follows relatively simple laws that approximate the statistical behavior of the microscopically complicated physical system. For example, the collision dynamics of millions of molecules at a microscopic level may be used to derive the continuum fluid equations. Because chunking applies to physical systems, phenomenological models of small-scale processes can be fairly accurate representations of the larger scales. Chunking in physics is discussed further in Chapter 2.

In a detailed numerical simulation model, the individual governing processes are represented and calculated as accurately as needed to represent the behavior of systems in which several processes occur simultaneously. The submodels that make up the parts of the detailed model may be fundamental, phenomenological, or empirical. A detailed model often combines these different types of components in order to represent the wide range of time and space scales of the individual physical processes. It is important to understand that what qualifies as a detailed model is a matter of degree. A model is more or less detailed according to the degree of phenomenology or empiricism that is included.

In this book we are interested in detailed numerical simulations of reactive flows in which fluid dynamics, chemistry, and diffusive transport play important, interacting roles. Consider, for example, a detailed one-dimensional reactive flow model of a propagating laminar flame. The convection may be solved by a very accurate numerical method and can be considered a fundamental model of one-dimensional convective transport. The chemical reaction rate coefficients are phenomenological models if they are taken from fundamental calculations that have been fit into an analytic form or empirical models if they are fits to experimental data. The equation of state may be a table compiled from experiments. Diffusion constants could be estimates based on intelligent guesses, or they could come from experiments or theoretical studies. This detailed model of a flame would become more phenomenological and less detailed if the set of chemical rate equations were replaced by a shortened, less complete set of equations. It is important to consider the level of detail desired for the model being constructed and then to decide what is necessary to integrate this approximation into the exact problem.

1–4. A BRIDGE BETWEEN THEORY AND EXPERIMENT

Detailed numerical simulation is a way to produce solutions of the mathematical model of a particular physical system. It is a tool for studying physical systems that bridges theoretical analysis and laboratory experiments. As such, detailed simulations have some of the advantages and the disadvantages of both forms of research.

Simulations as Computer Experiments

A detailed computer simulation is not a theory. It is not even a single theoretical solution. Unlike an analytic theory, it does not give equations relating physical variables to each other and to the parameters of the problem. Instead, each simulation is a unique computer experiment that has been performed with one set of geometric, physical, initial, and boundary conditions. The simulation can tell us about something new and unexpected when the model is complex enough, much as a laboratory experiment can teach us something new about the physical environment. A major attractive feature of simulations is that theoretical insight is not needed to make discoveries.

Simulations and experiments contain similar types of errors. They both incorporate complicated interactions. Some are related to the problem under study, but some of the interactions are not. In simulations, some of the spurious interactions are bugs in the program and are analogous to experimental errors such as leaks or undetected sources of external energy. Calibration errors in an experiment are similar to invalid input constants, parameters, or submodels in a detailed simulation.

If the calibrations are incorrect, an experimental apparatus may still be working well. The interpretation of the data, however, will be wrong. By the same reasoning, if the chosen values for the controlling chemical rate constants in a simulation are incorrect, the results of the simulation are wrong even though the simulation may be quantitatively accurate in the context of the problem actually specified.

For example, consider an experiment in which hydrogen and oxygen are mixed and react. Various reactions dominate, depending on the pressure and density regimes, additives, etc. To give physically accurate results, a simulation must include a correct chemical mechanism to mimic this behavior in a natural way. It is usually difficult to know what is missing or what is incorrect when a simulation does not agree with an experiment.

Both simulations and laboratory experiments benefit greatly from focusing on specific mechanisms and interactions. For example, geometric complications, such as multidimensional effects and wall boundary effects, make both simulations and experiments more difficult. Much can be learned about fundamental interactions by idealizing and simplifying the problem as much as possible. The insight to construct systems for computational study that accomplish this simplification requires a different set of skills from those required for mathematical analysis.

Simulations as Extensions of Theory

Although a simulation may not provide the types of analytic relationships that a theory gives, it provides a similar flexibility. This flexibility is its ability to evaluate the importance of a physical effect by turning the effect on or off, changing its strength, or changing its functional form. This straightforward way of isolating interactions is also an important advantage that a simulation has over an experiment. In fact, an experiment is not always a better probe of our physical environment than a simulation.

Simulations may be used to test the range of validity of theoretical approximations. When a linear theory breaks down, the manner of breakdown can be studied by simulations. Simulations have been used to study the nonlinear evolution of the Kelvin-Helmholtz and Rayleigh-Taylor instabilities.

Theories of transition to turbulence, turbulent boundary layers, and chaotic nonlinear dynamics are other examples where simulations have been used to test and extend theory.

The reverse case is also true: theory plays a crucial role in validating a numerical model. Exact comparisons with closed-form solutions provide the most valuable benchmarks of the computational model. By turning off the terms that the theoretical solution does not include, the accuracy and failure modes of parts of the computer model can be evaluated before the program is used to study a more physically complex configuration in which errors go undetected.

Simulation Can Bridge the Gap between Theory and Experiment

Because results obtained from a detailed model may be more comprehensive than those from an analytic theory, detailed models can be used to bridge the gap between theory and experiment. In particular, an effective bootstrap approach to solving problems is to use detailed simulation models to calibrate phenomenological models of specific processes and interactions. These phenomenological models can then be used in the detailed model to extend its range of validity or its effective resolution. Proposed physical laws can be tested by including the major processes they describe in a simulation and then comparing the results of the simulations and experiments. The use of detailed simulation models to calibrate quantitative understanding of the controlling physical processes is perhaps its most important and fundamental use.

Diagnostics in Simulations and Experiments

The most time-consuming part of numerical simulation experiments, as with laboratory experiments, is often constructing and testing the diagnostics. Long calculations invariably require a lot of graphic and printed output which must be analyzed. The diagnostics usually involve processing the answers in some way, for example, by taking integrals of quantities, adding and differentiating quantities, etc. These diagnostics themselves are computer programs that help us relate the actual discretized simulation variables to quantities more easily interpreted or compared to theory or experiment.

Whereas simulation diagnostics are possible sources of error, the errors associated with experimental diagnostics can cause greater problems. Simulation diagnostics can be performed without interfering in the basic computations, but laboratory measurements usually perturb the experiment itself. Probes can change the local flow, provide sites for unwanted surface reactions, and absorb heat from the medium. This problem does not exist in the computer experiment. In a simulation, on the other hand, it is possible to drown in numbers.

1–5. THE TYRANNY OF NUMBERS

A detailed numerical simulation model is a program that is executed, or "run," on a computer. The simulations we are concerned with "integrate," or "advance," the approximate time-evolution equations of the model from one time t to another time $t+\Delta t$. The quantity Δt is called the *timestep* and is generally smaller than the shortest characteristic times we need to resolve in the calculation. The advancing simulation thus defines the changing state of the system at a sequence of discrete times. The usual assumption is that the state of the system at any time during a timestep can be inferred by interpolating between the state of the system at the beginning and at the end of the timestep.

Typically, tens or hundreds of thousands of numbers are needed to specify the state of a system at any particular time. Consider a detonation propagating in a gas phase reactive medium. If the problem is two-dimensional, roughly 100×100 discrete points in space are required to give a 1% spatial resolution of the flow field and of all the chemical species present. If a detailed chemical rate scheme for a diluted hydrogen-oxygen mixture is used, approximately 10 species have to be considered in addition to the gasdynamic mass, momentum, and energy density variables. Thus, at least 140,000 numbers are required to specify any state of this system at a given time.

A major problem with simulations is how to deal with so many numbers. We cannot afford to store all the data from every timestep, nor do we generally want to. For this reason, algorithms are preferred that advance the system one timestep with minimal information required from previous timesteps. In this book, we highlight single-step algorithms which only require information from the previous timestep. Even if the cost of computer memory continues to drop, the time needed to store, compute, and transmit all these numbers makes the sheer size of a large simulation an obstacle.

The tyranny of numbers that plagues most detailed reactive flow simulations also hampers diagnosing the physics and chemistry that goes on within them. To look at the actual numbers associated with one of the many variables describing a single timestep can require printing out approximately 10,000 numbers, about 10 single-spaced pages of output. If the comparison solution is printed out at the same time, 20 pages are needed. Graphics and more sophisticated diagnostic software, necessary tools for interpreting simulations, are themselves perpetual headaches. Graphics and diagnostics can be as time-consuming and expensive to develop, apply or use, as the simulation model itself. The way certain types of graphics are used, however, should become evident from the examples presented in later chapters.

In addition to producing large amounts of data, detailed numerical simulations usually require large amounts of computer time. For this reason, simulations are run in segments of generally a few hundred or a few thousand timesteps. Thus a software facility must be provided to "dump" out and save the information describing a particular state of the system at selected timesteps. The calculation can be restarted from this data at a later time. At each restart one has the option of changing the parameters or the diagnostics of the simulation model.

1–6. THEMES OF THIS BOOK

Several themes run through this book. These themes consist of guidelines for carrying out reactive flow simulations, some points of view that are helpful, and some useful rules of thumb which we have developed after doing many detailed reactive flow simulations.

Choosing the Type and Level of Simulation

We have already differentiated between modeling based on phenomenologies and empirical laws and modeling from first principles. We have also introduced the idea of chunking, which helps determine the level for each component process in the overall simulation. The optimal combination of modeling approaches depends as much on the resources available and the type of answer required as on the parameters of the system being simulated.

Building Modular Simulation Models

In a modular approach, the problem is divided into a number of individual physical processes. Models can be built so that each of these processes is calculated accurately and calibrated separately. The resulting algorithms are then coupled together to study the resulting reactive flow mechanisms and interactions. In addition to providing a reasonable organization for attacking the overall problem, this approach allows use of the best algorithm for each aspect of the problem.

Evaluating Numerical Methods for Accuracy and Efficiency

Throughout this text, we evaluate many numerical methods with respect to what they can and cannot do in reactive flow calculations. We try to explain their strong and weak points, emphasizing accuracy, efficiency, ease of use, and flexibility. Numerical simulation is a developing field. The power opened up by the availability of new and diverse computing capabilities is enormous. Currently, the rate at which good algorithms are developed is comparable to the rate of increase of computational capability and its availability. When the goal is to solve a particular kind of reactive flow problem, the best methods to use are those which have been extensively tested and debugged. However, we also try to introduce more advanced methods and point out some areas to watch for important developments in computational methods.

Calibrating the Simulation Model

It is crucial to compare simplified analytic solutions with the output of the simulations when developing numerical simulation models. This check should be done with each process as it is programmed and with combinations of processes whenever possible. It is extremely important to do these tests when developing the model, and to repeat them after any major changes have been made.

Computational Rules of Thumb

A number of useful rules of thumb for simulating reactive flows emerge from experiences in a number of different fields of physics, chemistry, and engineering. These evolve from common sense and from working through the trade-offs required to find optimum algorithms for each of the relevant physical processes and for coupling them together. These rules provide invaluable guides to rapid progress as well as warnings of the problems likely to arise. We discuss these in the text and summarize them in the last chapter.

References

Anderssen, R.S., and F.R. de Hoog, 1983, The Nature of Numerical Processes, *Math. Sci.* 8: 115–141.

Box, G.E.P., W.G. Hunter, and J.S. Hunter, 1978, *Statistics for Experimenters*, Wiley, New York.

Hofstadter, D. R., 1979, *Gödel, Escher, Bach: An Eternal Golden Braid*, Basic Books, New York.

Chapter 2
THE REACTIVE FLOW MODELING PROBLEM

Modeling gas phase reactive flows is based on a generally accepted set of time-dependent coupled partial differential equations maintaining conservation of density, momentum, and energy. These equations describe the convective motion of the fluid, the chemical reactions among the constituent species, and the diffusive transport processes such as thermal conduction and molecular diffusion. Many kinds of problems are described by such apparently simple equations when they are combined with various initial and boundary conditions.

The material presented in the chapter is very condensed, and it will not be instructive to those totally unfamiliar with the individual topics. The purpose is first to review these equations to establish the notation used throughout this book, and then to relate each term in the equations to a physical process important in reactive flows. The chapter can then be used as a reference for the more detailed discussions of numerical methods in subsequent chapters. It would be most valuable to skim the chapter the first time through the book, and then to read it again before reading Chapter 13.

2–1. REACTIVE FLOW CONSERVATION EQUATIONS

2–1.1 Time-Dependent Conservation Equations

The equations used to model a gas phase reactive flow are the continuum time-dependent equations for conservation of the mass density ρ, the individual chemical species number densities, $\{n_i\}$, the momentum density $\rho\mathbf{v}$, and the energy density E. These equations may be written as

$$\frac{\partial \rho}{\partial t} = -\nabla \cdot (\rho \mathbf{v}), \qquad (2-1.1)$$

$$\frac{\partial n_i}{\partial t} = -\nabla \cdot (n_i \mathbf{v}) - \nabla \cdot (n_i \mathbf{v}_{di}) + Q_i - L_i n_i, \qquad i = 1, ..., N_s, \quad (2-1.2)$$

$$\frac{\partial \rho \mathbf{v}}{\partial t} = -\nabla \cdot (\rho \mathbf{v}\mathbf{v}) - \nabla \cdot \mathbf{P} + \sum_i \rho_i \mathbf{a}_i, \qquad (2-1.3)$$

and

$$\frac{\partial E}{\partial t} = -\nabla \cdot (E\mathbf{v}) - \nabla \cdot (\mathbf{v} \cdot \mathbf{P}) - \nabla \cdot (\mathbf{q} + \mathbf{q}_r) + \mathbf{v} \cdot \sum_i m_i \mathbf{a}_i + \sum_i \mathbf{v}_{di} \cdot m_i \mathbf{a}_i \ . \tag{2-1.4}$$

Table 2–1 contains a glossary of the quantities used in these equations and in the additional equations given in this chapter.

The first term on the right hand side of each of Eqs. (2–1.1) – (2–1.4) describes the convective fluid dynamics effects. The remaining terms contain the source, sink, coupling, external force, and transport terms that drive the fluid dynamics. The pressure tensor, heat flux, total number density, and energy density used in these equations are defined by

$$\mathbf{P} \equiv P(N,T)\mathbf{I} + \left(\frac{2}{3}\mu_m - \kappa\right)(\nabla \cdot \mathbf{v})\mathbf{I} - \mu_m[(\nabla\mathbf{v}) + (\nabla\mathbf{v})^T], \tag{2-1.5}$$

$$\mathbf{q}(N,T) \equiv -\lambda_m \nabla T + \sum_i n_i h_i \mathbf{v}_{di} + P \sum_i K_i^T \mathbf{v}_{di}, \tag{2-1.6}$$

$$N \equiv \sum_i n_i, \tag{2-1.7}$$

and

$$E \equiv \frac{1}{2}\rho\mathbf{v} \cdot \mathbf{v} + \rho\epsilon. \tag{2-1.8}$$

In addition, expressions are required for the thermal equation of state and the caloric equation of state, which have generic forms

$$\begin{aligned} P &= f_t(N,T) \\ \epsilon &= f_c(N,T) \ , \end{aligned} \tag{2-1.9}$$

which indicates that f_t and f_c are functions of N and T. In most of this book we consider gas phase reactive flows at densities where the thermal equation of state is the ideal gas equation of state,

$$P = Nk_BT = \rho RT. \tag{2-1.10}$$

The caloric equation of state can be considered as the relation

$$h_i = h_{io} + \int_{T^o}^{T} c_{pi}\, dT \tag{2-1.11}$$

which relates the enthalpies $\{h_i\}$ to the heat capacities, $\{c_{pi}\}$, and the enthalpies are defined through the equation

$$\rho\epsilon = \sum_i \rho_i h_i - P \ . \tag{2-1.12}$$

Table 2–1. Symbols in Chapter 2

Symbol	Definition
$\mathbf{a}_i$	External force per unit mass (cm/s ^{2})
c_o	Speed of sound (cm/s)
c_{pi}	Heat capacity of species i (cm^2/s^2-K)
D_{ik}	Binary diffusion coefficient for species i and k (cm^2/s)
E	Total energy density (erg/cm^3)
g	Gravitational acceleration constant (980.67 cm/s^2)
h_i	Specific enthalpy for species i (erg/g)
h_{io}	Specific heat of formation for species i (erg/g) evaluated at temperature T_o
$\mathbf{I}$	Unit tensor (nondimensional)
k^f	Forward chemical rate constant (see Section 2–2.3)
k^r	Reverse chemical rate constant (see Section 2–2.3)
k_B	Boltzmann constant (1.3805×10^{-16} erg/K)
K_i^T	Thermal diffusion coefficient of species i (cm^{-1})
L_i	Chemical loss rate of species i (s^{-1})
m	Mass of a species (g)
M	Mach number, see Eq. (2–2.1)
n_i	Number density of species i (cm^{-3})
N	Total number density (cm^{-3})
N_s	Number of chemical species present
P	Scalar pressure (dynes/cm^2)
$\mathbf{P}$	Pressure tensor (dynes/cm^2)
$\mathbf{q}$	Heat flux (erg/cm^2-s)
$\mathbf{q}_r$	Radiative heat flux (erg/cm^2-s)
Q_i	Chemical production rate of species i (cm^{-3}s^{-1})
R	Gas constant (8.3144×10^7 erg/deg-mol)
s	Specific entropy (erg/K)
S	Surface (cm^2)
T	Temperature (K)
$\mathbf{v}$	Fluid velocity (cm/s)
$\mathbf{v}_{di}$	Diffusion velocities of species i (cm/s)
V	Volume (cm^3)
Greek	
ϵ	Specific internal energy (erg/g)
γ	Ratio of specific heats, c_p/c_v
Γ	Circulation, (cm^2/s)
κ	Bulk viscosity coefficient (poise, g/cm-s)
λ	Thermal conductivity coefficient (g-cm/K-s^3), also Wavelength
μ	Coefficient of shear viscosity (poise, g/cm-s)

Table 2–1. Symbols in Chapter 2 *(continued)*

Symbol	Definition
ν	Frequency (Hz)
ϕ	Velocity potential (cm^2/s)
ψ	Steam function (cm^2/s)
ρ	Mass density (g/cm^3)
τ	Viscous stress tensor (dynes/cm^2)
ω	Vorticity (s^{-1})
Superscripts	
T	Transpose operation on a matrix
Subscripts	
m	Quantity defined for a mixture of species
i, j, k, or l	Individual species
r	Quantity due to the presence of radiation

The external forces $\{m_i \mathbf{a}_i\}$ in Eqs. (2–1.3) and (2–1.4) can be different for each individual species. For gravity, the accelerations are all the same,

$$\mathbf{a}_i = \mathbf{g}, \tag{2 – 1.13}$$

where $\mathbf{g}$ is the acceleration due to gravity. Then if Eq. (2–1.4) is rewritten as an equation for the internal energy, $\partial\epsilon/\partial t$, the external force term drops out (see, for example, Williams, 1985; Brodkey, 1967). The force terms remain, however, in the momentum Eq. (2–1.3). The additional complications of electric and magnetic forces are neglected here because we are considering uncharged gases.

The chemical production terms and loss rates, $\{Q_i\}$ and $\{L_i\}$ in Eqs. (2–1.2), can be written as

$$Q_i = \sum_j k^f_{i,j} n_j + \sum_{j,k} k^f_{i,jk} n_j n_k + \sum_{j,k,l} k^f_{i,jkl} n_j n_k n_l \tag{2 – 1.14}$$

$$L_i = k^r_i + \sum_j k^r_{i,j} n_j + \sum_{j,k} k^r_{i,jk} n_j n_k, \tag{2 – 1.15}$$

where k^f and k^r are forward and reverse chemical rate constants. The chemical rate constants can be functions of temperature and pressure.

The set of species diffusion velocities $\{\mathbf{v}_{di}\}$ are found by inverting the matrix equation

$$\mathbf{G}_i = \sum_k \frac{n_i n_k}{N^2 D_{ik}} (\mathbf{v}_{dk} - \mathbf{v}_{di}). \tag{2 – 1.16}$$

The source terms $\{\mathbf{G}_i\}$ in Eq. (2–1.16) are defined as

$$\mathbf{G}_i \equiv \nabla\left(\frac{n_i}{N}\right) - \left(\frac{\rho_i}{\rho} - \frac{n_i}{N}\right)\frac{\nabla P}{P} - K_i^T \frac{\nabla T}{T}. \qquad (2-1.17)$$

These diffusion velocities are also subject to the constraint

$$\sum_i \rho_i \mathbf{v}_{di} = 0. \qquad (2-1.18)$$

There is no net mass flux arising from the relative interspecies diffusion because the total mass flux is defined as $\rho\mathbf{v}$.

Equations (2–1.1) – (2–1.18) are the reactive flow equations for neutral gases. In the discussions in Chapter 12 on turbulence and multiphase flow, other forms of the reactive flow equations are defined.

2–1.2 The Continuum Representation

A reactive medium is composed of molecules and atoms, and these are composed of even smaller particles. The interactions of these particles result in the processes we want to simulate. The reactive flow equations presented above do not solve for the behavior of individual particles, but assume that the medium is a continuous, macroscopic fluid. This continuum approximation incorporates statistical averaging processes performed over the small-scale particle phenomena.

This natural process in which the small-scale phenomena are averaged to produce a global quantity or interaction law is called *chunking* by Hofstadter (1979). For example, complete, detailed knowledge of quarks and charmed chromodynamics is (thankfully) not necessary for understanding many aspects of the composition and dynamics of elementary particles. Nuclear physicists often work on an even more macroscopic level by considering protons and neutrons because subelementary particle physics can be chunked at the nuclear physics level for all but extremely high energy collisions. Taking this several steps further, physical chemists can consider composite nuclei and electrons, chunking the proton and neutron systems that make up nuclei. In reactive flow problems, we usually use an even more macroscopically chunked fluid or continuum pictures of the nuclei and electrons, which give us distinct chemical species. Hofstadter describes the relations among various levels of chunking:

> Although there is always some "leakage" between the levels of science, so that a chemist cannot afford to ignore lower level physics totally, or a biologist to ignore chemistry totally, there is almost

> no leakage from one level to a distant level. That is why people can have an intuitive understanding of other people without necessarily understanding the quark model, the structure of the nuclei, the nature of electron orbits, the chemical bond, the structure of proteins, the organelles in a cell, the methods of intercellular communication, the physiology of the various organs within the human body, or the complex interactions among organs. All that a person needs is a chunked model of how the highest level acts; and as we all know, such models are very realistic and successful....

A chunked model no longer has the ability to predict exactly.

> In short, in using chunked high level models, we sacrifice determinism for simplicity.... A chunked model defines a "space" within which behavior is expected to fall, and specifies probabilities of its falling in different parts of that space....

Consider the reactive flow Eqs. (2–1.1) – (2–1.18). It is not possible to resolve aspects of the microscopic interactions using a continuum approximation. However, it is possible to include some of the macroscopic consequences of atomic and molecular phenomena within a fluid model. Chemical reactions, interspecies molecular diffusion, thermal conductivity, temperature-dependent enthalpies and equations of state are all examples of macroscopic models of microscopic processes that result from chunking the microscopic physics.

In Chapter 4 we introduce some fundamental aspects of finite-difference methods for solving Eqs. (2–1.1) – (2–1.18). These methods divide time and space into discrete intervals, and then define discrete variables that approximate the continuous functions. This process puts the equations in a form suitable for numerical computation. We expect the corresponding numerical solutions to converge and become better representations of the continuous fluid variables as the size of the discrete intervals becomes smaller and smaller. The problems that arise due to limits on the attainable resolution are discussed extensively in Chapters 3 and 6.

Even if practical computational restrictions were removed and arbitrarily good resolution were possible, the continuous variables would still have to be represented in a computer by a discrete set of distinct real numbers of finite precision. It is amusing to note that fluid equations were developed originally to simplify the discrete equations of individual particle dynamics. Now we must reformulate the continuum problem to solve equations for finite volumes of material. However, the new set of equations is considerably smaller and simpler to solve than the original equations of particle dynamics.

2–1.3 Physical Phenomena Represented in the Equations

There are basically four types of physical processes represented in reactive flow equations. The first two, chemical reactions and diffusive transport, originate in the atomic and molecular nature of matter. The third and fourth, convection and wavelike properties, are collective phenomena.

1. Chemical reactions (or chemical kinetics), represented here by the production and loss terms Q_i and $L_i n_i$.

This is an example of a "local" phenomenon, one which does not depend on spatial gradients. Other examples of local phenomena are phase changes, external source terms such as laser or spark heating, and sinks such as optically thin radiation loss. The macroscopic form of these effects used in the continuum equations arises from processes that average over microscopic effects.

2. Diffusion (or diffusive transport), represented generally as $\nabla \cdot Y$ where Y may be $n_i \mathbf{v}_{di}$, $\mu \nabla \mathbf{v}$, or $\mathbf{q}$.

The last two terms in Eq. (2–1.5) describe the diffusive effects of viscosity. The last terms in Eq. (2–1.6) describe the change in energy due to molecular diffusion and chemical reactions. Note that the species diffusion velocities, $\{\mathbf{v}_{di}\}$ in Eqs. (2–1.2), also appear in the energy Eq. (2–1.6) representing part of the heat diffusion process.

3. Convection (or convective transport), describing the motion of fluid quantities in space, how fluid parcels interpenetrate, and how compression changes density.

These effects are represented in the equations by fluxes of conserved quantities through volumes, e.g., $\nabla \cdot \mathbf{v}X$ where X is $\rho, n_i, E, \rho\mathbf{v}$ or P. Convection is a continuum concept, which assumes that quantities such as density, velocity, and energy are smoothly varying functions of position. The fluid variables are defined as averages over the actual particle distributions, so that only a few degrees of freedom are necessary to describe the local state of the material.

4. Wavelike and oscillatory behavior, which are described implicitly in the reactive flow equations by coupled continuity equations.

The important point about wavelike motion is that energy is not carried throughout the system by fluid or particle motions. It is transferred from one element of the fluid to others by waves that can travel much faster than the fluid velocity. The main type of waves considered are shock waves, which move as discontinuities through the system, and sound waves, in which there are alternating compressions and rarefactions in density and pressure of the fluid. Other types of waves included in these reactive flow equations are gravity waves and chemical reaction waves.

The complete set of reactive flow equations is so difficult to solve that entire disciplines and scientific communities flourish solving subsets of these equations for particular applications. For example, hydrodynamics applications eliminate the chemical reactions and even the energy equation, and assume the flow is incompressible ($\nabla \cdot \mathbf{v} = 0$). Aerodynamics problems often deal with time-independent, nonreacting flows over curved surfaces. Chemical kinetics research often considers the chemical reactions without any of the convection and diffusive transport effects. Chemical processing problems deal with solutions of a combination of the chemical kinetics and diffusive transport equations. The coupled set of equations also can be used to describe laminar and turbulent flows, and flame and detonation phenomena. The equations and solution techniques described in this book are applicable to these systems and to many other kinds of reactive flows.

2–1.4 Boundary and Initial Conditions

A specific solution of the reactive flow equations is determined by the initial conditions, the set of chemical species and their fluid dynamic, thermophysical and chemical properties, and by the boundary conditions that describe the geometry of the system and exchange of mass, momentum, and energy occurring between the system and the rest of the physical world. Boundary and initial conditions select those solutions which apply to the particular problem under study from the many families of possible solutions. The physical and numerical implications of these conditions also influence the solution techniques used.

It is often difficult to specify boundary conditions that are simple, mathematically tractable and also physically meaningful. Inflow, outflow, free surfaces, fluid interfaces, and flows about solid objects all require different analytic representations and the corresponding numerical models are different. Sometimes imposing physically accurate boundary conditions often means that certain solution methods cannot be used. We return to this discussion on boundary conditions in Chapter 11.

Specifying the initial conditions should be easier than specifying boundary conditions because initial conditions are input at the beginning of the calculation, rather than appended as constraints on the calculation at every timestep. In practice, however, specifying all the necessary chemical kinetic, equation of state, molecular, and thermal transport data for the particular materials in the system can be an extremely time-consuming process and provides the opportunity for introducing large errors.

2–2. DISCUSSION OF THE CONSERVATION EQUATIONS

In this section we consider important concepts and quantities related to the four main types of processes occurring in the reactive flow conservation equations. This section does not contain a derivation of these equations, but rather introduces the terminology and notation used when referring to these processes throughout the book. In addition, we try to expose the approximations assumed in formulating these equations and to mention the limits of their applicability.

2–2.1 Equations of Fluid Dynamics

Consider a subset of Eqs. (2–1.1) – (2–1.18) describing a flow in which there is only one species, no chemical reactions, and no external forces. These equations are

$$\frac{\partial \rho}{\partial t} = -\nabla \cdot (\rho \mathbf{v}), \qquad (2-2.1)$$

$$\frac{\partial \rho \mathbf{v}}{\partial t} = -\nabla \cdot (\rho \mathbf{v}\mathbf{v}) - \nabla \cdot \mathbf{P}, \qquad (2-2.2)$$

$$\rho \frac{\partial \epsilon}{\partial t} = -\rho \nabla \cdot (\epsilon \mathbf{v}) - \mathbf{P} \cdot (\nabla \mathbf{v}) - \nabla \cdot \mathbf{q}, \qquad (2-2.3)$$

where

$$\mathbf{q} = -\lambda \nabla T. \qquad (2-2.4)$$

We have replaced Eq. (2–1.4) with the equivalent equation for the rate of change of internal energy density. The pressure tensor is defined by

$$\mathbf{P} = P(N,T)\, \mathbf{I} + \boldsymbol{\tau}\,, \qquad (2-2.5)$$

where we have defined the *viscous stress tensor* as

$$\boldsymbol{\tau} \equiv -\mu[\nabla \mathbf{v} + (\nabla \mathbf{v})^T] + \left(\frac{2}{3}\mu - \kappa\right)(\nabla \cdot \mathbf{v})\mathbf{I}\,. \qquad (2-2.6)$$

Equations (2–2.1) – (2–2.6) are the standard equations discussed in most textbooks on fluid dynamics. They consist of two parts: the convective transport terms, and simplified source and diffusive transport terms.

Equations (2–2.1) – (2–2.3) describe the behavior of the mass density, vector momentum density, and internal energy density as functions of time and position. The fluxes of these properties through the surface of a volume element are usually represented as gradients. Equation (2–2.1) is the mass density continuity equation and Eq. (2–2.2) is the *Navier-Stokes equation*.

Sometimes the set of Eqs. (2–2.1) – (2–2.3) are referred to as the Navier-Stokes equations. The *Euler equation* is obtained from Eq. (2–2.2) by setting κ and μ equal to zero. Equations (2–1.1) – (2–1.3) are also called the Euler equations when all of the diffusive transport terms are set to zero.

The pressure tensor, Eq. (2–2.5), has two parts, one associated with compression and another associated with viscous dissipation. In general, κ is negligible except in the study of shock-wave structure and absorption and attenuation of acoustic waves. As it is otherwise generally very small, it is often dropped from consideration.

When the mean free path of a particle (that is, the average distance traveled by a typical particle between collisions) is small compared to the macroscopic dimensions of the system, the medium behaves as a continuous fluid. The properties such as density, velocity, and energy are well defined at each point. Using the short mean free path assumption, the fluid equations can be rigorously derived from the Boltzmann equation. These fluid equations reduce to the Navier-Stokes equations which contain expressions for the vector fluxes of mass, momentum, and energy. Transport coefficients are then defined in terms of these flux vectors. Cases where the equations are not valid include very dilute gases for which the container size is the same order as the mean free path, and in the region near a shock wave where there are abrupt changes in very short distances. For turbulent flows, the equations are considered valid because the characteristic length of the smallest scale size of interest is generally greater than the mean free path.

The *vorticity*, $\boldsymbol{\omega}$, is defined by

$$\boldsymbol{\omega} \equiv \nabla \times \mathbf{v}. \qquad (2-2.7)$$

Vorticity is a vector quantity describing the rotation of a fluid parcel. It satisfies an evolution equation which can be derived from Eqs. (2–2.1) – (2–2.2) using Eq. (2–2.7),

$$\frac{\partial \boldsymbol{\omega}}{\partial t} + \boldsymbol{\omega} \nabla \cdot \mathbf{v} = \boldsymbol{\omega} \cdot \nabla \mathbf{v} + \frac{(\nabla \rho \times \nabla \cdot \mathbf{P})}{\rho^2}. \qquad (2-2.8)$$

A net *circulation* is defined for an area as the integral of the fluid velocity

$$\Gamma = \int \mathbf{v} \cdot d\mathbf{r} = \oint_{\mathbf{S}} \boldsymbol{\omega} \cdot d\mathbf{S} \qquad (2-2.9)$$

where $\mathbf{S}$ is a closed surface. The evolution of the net circulation is determined by

$$\frac{d\Gamma}{dt} = \int_{\mathbf{S}} \left[\frac{(\nabla \rho \times \nabla \cdot \mathbf{P})}{\rho^2} \right] d\mathbf{S}\ , \qquad (2-2.10)$$

Table 2–2. Mach Numbers Characterizing Flow Regimes

Flow Regime	Mach Range	Comment
Incompressible	$M<0.4$	No compression; density and pressure are constant moving with flow
Subsonic	$0.4<M<1$	Compression becomes important
Transonic	$M\sim 1$	Weak shocks occur
Supersonic	$1<M\leq 3$	Strong shocks occur
Hypersonic	$M\geq 3$	Very strong shocks occur, most of energy is kinetic energy

an integration of the vorticity source term over the area S.

When $\omega = 0$, there is no rotation in the fluid. In this case the flow can be described entirely by a *velocity potential*, ϕ, where

$$\mathbf{v} = \nabla\phi \,. \tag{2-2.11}$$

When the flow described by Eq. (2–2.8) is also *incompressible*, that is, $\nabla\cdot\mathbf{v} = 0$, then

$$\nabla\cdot\mathbf{v} = \nabla^2\phi = 0, \tag{2-2.12}$$

which is the Laplace equation for determining ϕ. This type of flow is called *potential flow*.

In the general case, the vector velocity flow field $\mathbf{v}(\mathbf{r},t)$ can be expanded as the sum of three terms,

$$\mathbf{v} = \nabla\times\boldsymbol{\psi} + \nabla\phi + \nabla\phi_p \,. \tag{2-2.13}$$

The vector *stream function* $\boldsymbol{\psi}$ for the rotational component satisfies

$$\nabla\times(\nabla\times\boldsymbol{\psi}) = \boldsymbol{\omega} \tag{2-2.14}$$

and the compressional component

$$\nabla\cdot\nabla\phi = \nabla\cdot\mathbf{v} \tag{2-2.15}$$

is no longer zero. The third term, $\nabla\phi_p$, is a particular potential solution, which can be added to the rotational and compressional contributions in Eq. (2–2.12) as both the curl and its divergence are zero. This particular solution is usually chosen to adapt the composite solution for $\mathbf{v}$, given by Eq. (2–2.13), to prescribed boundary conditions.

Regimes of fluid flow are usually characterized in terms of the *sound speed*, the velocity at which a small pressure disturbance moves in the fluid relative to the flow speed. An expression for the sound speed can be derived from the continuity and momentum equations by considering steady motion with no viscous terms. The sound speed, c_s, is

$$c_s = \sqrt{\left.\frac{\partial P}{\partial \rho}\right|_s} , \qquad (2-2.16)$$

where the derivative is taken at constant entropy (see, for example, Landau and Lifshitz, 1959, Chapter VIII). For an ideal gas, this reduces to

$$c_s = \sqrt{\frac{\gamma P}{\rho}} . \qquad (2-2.17)$$

The *Mach number* of the flow is defined using Eq. (2–2.17),

$$M \equiv \frac{|\mathbf{v}|}{c_s} . \qquad (2-2.18)$$

Different flow regimes are characterized by the value of the flow Mach number, as in Table 2–2.

2–2.2 Equations of State

The ideal gas equation of state, Eq. (2–1.10), is strictly valid for a collection of vanishingly small particles with no forces acting between them, and it works best for a dilute gas. However, it breaks down even for gases under atmospheric conditions. More complex equations of state have been proposed that consist of expansions of pV/Nk_B, and the expansion coefficients are fit either theoretically or from experimental data. If the intermolecular forces are known, it is possible to derive the equation of state theoretically.

In general descriptions of fluid dynamics (and thus of reactive flows), two equations are state are needed: the thermal equation of state,

$$P = P(V,T) , \qquad (2-2.19)$$

and the caloric equation of state,

$$\epsilon = \epsilon(V,T) . \qquad (2-2.20)$$

Extensive statistical mechanical derivations and discussions of various forms of equations of state are given in Hirschfelder, Curtiss, and Bird (1964). Another practical reference is Reid, Prausnitz, and Sherwood (1977). Most analytic forms for the equations of state can be incorporated in the fluid dynamic equations and solved numerically along with them. Often, however, and usually for condensed phases, the equations of state are tables of numbers derived from experiments or complex, expensive calculations.

2–2.3 Equations of Chemical Reaction Kinetics

In combustion kinetics, we generally consider unimolecular, bimolecular, and termolecular reactions among the chemical species. These three cases can usually be written as

$$a \underset{k^r}{\overset{k^f}{\rightleftharpoons}} b + c \qquad (2-2.21a)$$

$$a + b \underset{k^r}{\overset{k^f}{\rightleftharpoons}} c + d \qquad (2-2.21b)$$

$$a + b + m \underset{k^r}{\overset{k^f}{\rightleftharpoons}} c + m \qquad (2-2.21c)$$

where the a, b, c, d, and m indicate generic chemical species, and k^f and k^r are forward and reverse chemical rates, respectively.

Equations (2–1.2) are the time-dependent conservation equations for the N_s chemical species number densities. These terms are conservative in the sense that change in the total number of particles in a region occurs only as a result of convection into or out of that region. The chemical production and loss rates, $\{Q_i\}$ and $\{L_i\}$ in Eqs. (2–1.2), are not individually conservative as the individual species change through chemical reactions.

The terms in Eqs. (2–1.2) describing changes due to chemical reactions are

$$\frac{\partial n_i}{\partial t} = Q_i - n_i L_i, \qquad i = 1, ..., N_s. \qquad (2-2.22)$$

Equation (2–2.22) describes a set of nonlinear, first order, ordinary differential equations. Both the $\{Q_i\}$ and $\{L_i\}$ are linear combinations of products of the number densities $\{n_i\}$ multiplied by rate constants. Their generic forms are given in Eqs. (2–1.4) – (2–1.15). The individual forward and reverse reaction rate constants k^f and k^r may be complicated functions of temperature and sometimes pressure. The expressions for these rates commonly encountered in combustion kinetics are an *Arrhenius form* or *modified Arrhenius form*,

$$Ae^{-C/T} \qquad \text{or} \qquad AT^B e^{-C/T}, \qquad (2-2.23)$$

respectively. The validity of these forms depends on the existence of a Boltzmann distribution in the vibrational and electronic states of the participating molecules. When these forms are not applicable, other forms or a table of values as a function of temperature and pressure may be more appropriate. Sometimes the individual vibrational or electronic states must be considered separately, as occurs in atmospheric, solar, and chemical laser applications. Basic problems in the numerical solution of first-order ordinary differential equations are discussed in Chapter 4. Solving coupled sets of temperature-dependent nonlinear ordinary differential equations is discussed in Chapter 5.

When Eq. (2–2.22) is coupled to the equation of state, Eqs. (2–1.9) – (2–1.12), the temperature can be determined from an algebraic equation expressing the energy conservation constraint. When these thermophysical quantities are not known, it is often possible to solve an ordinary differential equation that describes the change in temperature due to chemical reactions. This case often occurs in atmospheric chemistry problems in which the vibrational or electronic states are not in equilibrium.

When the chemical reactions are approximately thermoneutral, that is, when little energy is absorbed or emitted by the chemical reactions or when the reactants are highly dilute, there is limited feedback between the fluid dynamics and the chemistry. Then the major effect of chemistry on the fluid dynamics may be through a change in the number density which causes changes in pressure. The thermoneutral assumption is correct to first order when we relate the ion chemistry of the Earth's ionosphere to neutral wind motion, or the density of the heavy ions to the motion of the bulk plasma in the solar atmosphere. However, when the reactions are endothermic or exothermic, as in combustion, or when ionization is significant, as in laser-produced plasmas, the strong coupling between chemistry and fluid dynamics cannot be ignored.

2–2.4 Equations of Diffusive Transport

The pressure tensor, P, and heat flux vector, $\mathbf{q}$, represent fluxes of momentum and energy in a region of space. There is also a mass flux vector $\rho_i \mathbf{v}_{di}$ constrained by Eq. (2–1.7). These fluxes are written in terms of the transport coefficients and gradients of mass, velocity, and temperature. Such relationships are valid when conditions deviate only slightly from equilibrium, so that the underlying particle distribution functions are nearly Maxwellian and variations of average fluid properties over a mean free path are small. When the gradients are small enough, the fluxes can be approximated as first derivatives of density, velocity, or temperature.

Diffusive transport processes express the transfer of mass, momentum, or energy due to gradients in concentration, velocity, or temperature. The transport coefficients and the gradients with which they are associated are given in Table 2–3. Molecular diffusion represents a transfer of mass and energy due to a concentration gradient. Viscosity represents a transfer of momentum due to a velocity gradient. Thermal conduction and thermal diffusion represent a transfer of energy due to a temperature gradient. Extensive descriptions of diffusive transport properties are given in Hirschfelder,

Table 2–3. Transport Coefficient and Gradients

Diffusion Coefficient	Symbol	Associated Gradients
Molecular (ordinary, or binary) Diffusion	D_{ij}	$\nabla \cdot (n_i/N), \nabla T, \nabla P$
Viscosity	μ	$\nabla \mathbf{v}$
Thermal conduction	λ_m	$\nabla T, \nabla P$
Thermal diffusion	K_j^T	∇T

Curtiss, and Bird (1964), Chapman and Cowling (1970), and Reid, Prausnitz, and Sherwood (1977). Discussions related to combustion are given by Strehlow (1984).

In Chapter 7 we give expressions for transport coefficients for gases in combustion applications, and a new method for solving the matrix of equations for multispecies diffusion velocities.

2–2.5 Equations of Radiation Transport

Equation (2–1.4) includes the effects of energy transport by radiation. Here radiation transport is encompassed in the term $\nabla \cdot \mathbf{q}_r$. In addition to a contribution to the heat flux vector $\mathbf{q}$, there can also be contributions from radiation to the pressure tensor and the internal energy. For gas phase flames and detonations, these additional contributions are usually small and so are neglected in Eqs. (2–1.3) and (2–1.4).

In general, the term representing the radiation flux in Eq. (2–1.4) can be written

$$\nabla \cdot \mathbf{q}_r = R_E - R_A \tag{2-2.24}$$

where R_E is the rate of emission of radiation per unit volume and R_A is the rate of absorption of radiation per unit volume. Usually the absorption rates are proportional to the densities of the absorbing species. Often radiation transport must be described as a function of wavelength, such as when particular atomic transitions and hence particular spectral lines are important. In these cases the flux has to be treated as an integral over wavelength, λ,

$$\mathbf{q}_r(\mathbf{r}) = \int_0^\infty \mathbf{q}_\lambda(\mathbf{r}) d\lambda \, . \tag{2-2.25}$$

Generally $\mathbf{q}_\lambda$ is derived in terms of the coefficients of absorption of a material, $\alpha_\nu(T, \rho)$.

There are two cases which can be treated rather directly. The first is the case of an opaque material, such as might occur in the hot gases in a car engine. This situation is referred to as *optically thick*. Here the absorption is so intense that the photons travel only a very short distance before being absorbed. The radiation is essentially in local thermodynamic equilibrium with the material. For this application, a mean free path for the photons making up the radiation field can be defined and the process can then be treated as a form of diffusion. The other limit occurs in some relatively nonluminous flames when the material is essentially transparent to the radiation. There is little self-absorption so that the term R_A in Eq. (2–2.24) is very small and the emission term, R_E, dominates. This situation is often referred to as the *optically thin* limit.

The problems of treating radiation transport in a numerical model and a more detailed explanation of the terms involved are given in Chapter 12. References to the physics of these processes are Zel'dovich and Raizer (1966), Rybicki and Lightman (1979), Siegel and Howell (1981), and Mihalas and Mihalas (1984).

2–2.6 Characteristic Trajectories

Consider an ideal, one-dimensional, planar, constant entropy compressible flow, with fluid velocity v and sound speed c_s defined in Eq. (2–2.16). Information from any point in the flow propagates according to the equations

$$\frac{dx}{dt} = v \qquad (2-2.26)$$

and

$$\frac{dx}{dt} = v \pm c_s \,. \qquad (2-2.27)$$

Equation (2–2.26) defines a trajectory C_o, along which the entropy s is constant. In addition, when s is constant throughout the flow, the *Riemann invariants*,

$$R_{\pm} = v \pm \int \frac{dP}{\rho c_s} \,, \qquad (2-2.28)$$

are constant along the trajectories $C_{\pm}$ defined by Eq. (2–2.28) (Landau and Lifshitz, 1959; Courant and Friedrichs, 1948). The $C_{\pm}$ trajectories follow forward and backward sound waves moving relative to the flow. The C_o trajectory follows a particle path. Variations in the entropy are convected according to the adiabatic equation

$$\frac{\partial s}{\partial t} + v\frac{\partial s}{\partial x} = 0 \,. \qquad (2-2.29)$$

Equations (2–2.16) and (2–2.26) – (2–2.29) completely describe flows in which entropy is constant following the fluid motion. The flows are described in terms of motions along the *characteristic trajectories*, or *characteristics*, denoted above by $C_\pm$ and C_o. An arbitrarily small perturbation can also be written as a superposition of disturbances propagating according to these three laws. The basic ideas of the methods of characteristics are treated in some detail in Courant and Friedrichs (1948) and Courant and Hilbert (1962), and reviewed in Liepmann and Roshko (1957).

In flows for which the entropy is not constant, two types of discontinuities, or interfaces between different states of the same medium, can occur. One type is the *contact discontinuity*, across which density, energy, and entropy are discontinuous, but pressure and normal velocity are continuous. The other type is the *shock*, across which all dependent variables change discontinuously. Contact discontinuities propagate at the fluid velocity, $\mathbf{v}$, that is, along C_o. Shock waves move with a velocity v_s which uniquely determines the discontinuous changes in ρ, $\mathbf{v}$ and P. In the limit of very weak shocks, the shock trajectory reduces to an acoustic disturbance, one of the two trajectories $C_\pm$.

2–2.7 Classification of Partial Differential Equations

The general one-dimensional second-order differential equations discussed so far can be written in a general form

$$a\frac{\partial^2 \rho}{\partial t^2} + b\frac{\partial^2 \rho}{\partial t \partial x} + c\frac{\partial^2 \rho}{\partial x^2} + d\frac{\partial \rho}{\partial t} + e\frac{\partial \rho}{\partial x} + f\rho + g \;=\; 0 \;, \qquad (2-2.30)$$

where now ρ may represent any one of the variables for which we are solving. This equation can be classified as *hyperbolic*, *parabolic*, or *elliptic* depending on whether the discriminant

$$\mathcal{D} \;\equiv\; b^2(x,t) - 4\; a(x,t)\; c(x,t) \qquad (2-2.31)$$

is positive, zero, or negative (see, for example, Potter, 1973, Richtmyer and Morton, 1967, Liepmann and Roshko, 1957, Carnahan et al., 1969). Specifically,

$$\begin{aligned} \text{hyperbolic} \quad b^2 - 4ac \;&>\; 0\;, \\ \text{parabolic} \quad b^2 - 4ac \;&=\; 0\;, \\ \text{elliptic} \quad b^2 - 4ac \;&<\; 0\;. \end{aligned} \qquad (2-2.32)$$

When a, b, c, d, e, f, or g are functions of the dependent variable ρ as well as space and time, the equation is nonlinear. Because the coefficients can vary in time and space, the class of an equation may vary according to the

Table 2–4. Classification of Equations

Type	Example	
hyperbolic	$\nabla^2 \rho - \frac{1}{c}\frac{\partial^2 \rho}{\partial t^2} = 0$	Wave equations
elliptic	$\nabla^2 \rho = 0$	Laplace equation
	$\nabla^2 \psi = \xi$	Poisson equation
parabolic	$\nabla^2 \rho - \frac{1}{\kappa}\frac{\partial \rho}{\partial t} = 0$	Diffusion equation
	$\nabla^2 \rho + \frac{2m}{\hbar^2}(E - V)\rho = 0$	Schrödinger equation

particular time or location in space. These definitions are generalized to multidimensions in Carnahan et al. (1969).

Table 2–4 shows several examples of this classification drawn from the reactive flow equations and related partial differential equations of physics. In these examples, the cross-derivative term is absent, and thus $b(x,t)$ is zero. Then for hyperbolic equations, the functions $a(x,t)$ and $c(x,t)$ must have the opposite sign. This condition describes wave equations. In elliptic equations, there are usually two spatial second derivatives. Here the $a(x,y)$ and $c(x,y)$ have the same sign.

For hyperbolic equations, the characteristics are real (see, for example, Mathews and Walker, 1965) and the speed of propagation of a signal is finite. In parabolic equations, such as diffusive equations, a speed of propagation is not meaningful, and the characteristics are complex. In elliptic equations, there is effectively an infinite propagation speed. The solution at any time is related to the solution at the boundaries at the same time. The Poisson equation is an example of an elliptic equation.

2–2.8 Dimensionless Numbers

For most combustion problems, it is difficult if not impossible to solve the governing equations analytically. When theory cannot provide a solution and we wish to study general properties of the system, it is often useful to study geometrically similar systems and make comparisons. For example, consider two systems in which the size, fluid conditions, or even the type of fluid may differ. In some cases, it appears that the behavior of similar systems should scale with each other. Methods such as *dimensional analysis* are used when the variables are known but the equations are not. Methods such as *inspection analysis* of the governing equations are used when the equations are known. These methods lead to a host of dimensionless numbers which are ratios of parameters of the system. With care, and an eye on the underlying assumptions involved, systems may be characterized by the values of these dimensionless numbers. A more detailed discussion is given in Brodkey (1967).

As an example, consider the momentum equation, Eq. (2–1.3). A component or structure in the flow can be characterized by a length in the fluid l, a time of the flow t, and a velocity of the flow v, and an external force ρg. Now consider another flow with geometrically similar boundaries, so that this flow is characterized by parameters indicated by primes: l', t', and v'. If these flows are similar, we can write down a set of relations of the form $l = c_1 l'$, $t = c_2 t'$, and so on, and replace the variables in Eq. (2–1.3) by the primed variables. Then equating the coefficients in the regular and primed equations, a number of relations emerge. For example,

$$\frac{lv\rho}{\mu} = \frac{l'v'\rho'}{\mu'} \equiv \text{Re} = \text{constant}. \qquad (2-2.33)$$

Thus quantity $lv\rho/\mu$ must be the same in both systems. This dimensionless ratio is the *Reynolds number*, the ratio of inertial to viscous forces. We also have

$$\frac{v^2\rho}{P} \equiv \frac{1}{\text{Eu}} \left(= \frac{v^2\gamma}{c_s^2} \right). \qquad (2-2.34)$$

This relation defines the *Euler number*, the ratio of pressure to inertial forces. The term in parentheses on the right hand side of Eq. (2–2.34) holds if the gas obeys the ideal gas law. This definition shows that Eu is the gas constant times the square of the Mach number. If the body force is gravity, so that $f = \rho g$, we obtain

$$\frac{v^2}{lg} \equiv \text{Fr} = \text{Froude number}, \qquad (2-2.35)$$

which defines the *Froude number*. The Froude number is the ratio of kinetic to gravitational potential energy.

The equations of motion for two flows will be the same if their Reynolds numbers, Mach numbers, Froude numbers, and the ratio of their specific heats are equal. Inspection of the continuity equation does not produce any new dimensionless numbers, but inspection of the energy equation does. These include the *Peclet number*, which is the ratio of heat transported by convection to heat transported by conduction,

$$\frac{vl\rho c_p}{\lambda} = \text{Peclet number} = \text{Pe}, \qquad (2-2.36)$$

Table 2–5. Dimensionless Numbers

Name	Symbol	Definition	Significance
Atwood	A	$\Delta\rho/(\rho_1+\rho_2)$	Density difference/Density sum
Brinkman	Br	$\mu v^2/T\lambda$	Heat from viscous dissipation/ Heat from thermal conduction
Capillary	Cp	$\mu v/\sigma$	Viscous force/ Surface tension force
Crispation	Cr	$\mu\lambda/\sigma l\rho c_p$	Diffusion/ Surface tension
Damköhler			
	Da_1	$R_i l/v$	Chemical reaction rate/ Convective transport rate
	Da_2	$R_i l^2/D$	Chemical reaction rate/ Diffusive transport rate
	Da_3	$Ql/\rho hv$	Heat from chemistry/ Heat from convection
	Da_4	$Ql/\lambda T$	Heat from chemistry/ Heat from conduction
	Da_5	$Ql/\rho Dh$	Heat from chemistry/ Heat from diffusion
Euler	Eu	$P/\rho v^2$	Pressure force/Inertial force
Drag Coefficient			
	C_D	$(\rho'-\rho)/lg\rho' v^2$	Drag force/Inertial force
Froude	Fr	v^2/lg	Kinetic energy/ Gravitational energy
Grashof	Gr	$(g\Delta T l^3)/T\mu^3$	Buoyancy forces/Viscous forces
Karlovitz	K	$(l/v)(dv/dy)$	Stretch factor: criterion for area increase in flame front
Knudsen	Kn	δ/l	Convection time/Collision time
Lewis	Le	$\lambda/\rho D_{ij}c_p$ = Sc/Pr	Thermal conduction transport/ Diffusive energy transport (Sometimes defined as inverse)
Mach	M	v/c_s	Magnitude of compressibility effects
Newton	Nt	$F/\rho l^2 v^2$	Imposed force/Inertial force
Nusselt	N	$\alpha l/\lambda$	Total heat transfer/ Thermal conduction
Peclet	Pe	$vl\rho c_p/\lambda$	Convective heat transport/ Heat transport by conduction
	Pe,s	lv/D_{ij}	Convective mass transfer/ Molecular mass transfer
Poisseuille	Po	$l\Delta p/\mu v$	Pressure force/Viscous force
Prandtl	Pr	$c_p\mu/\lambda$	Momentum transport/ Thermal conduction

Table 2–5. Dimensionless Numbers *(continued)*

Name	Symbol	Definition	Significance
Reynolds	Re	$l v \rho / \mu$	Inertial forces/Viscous forces
Richardson	Ri	$gl/(\Delta v)^2$	Buoyancy effects/ Vertical shear effects
Rossby	Ro	$v/2\Omega l \sin\Lambda$	Inertial Force/Coriolis force
Schmidt	Sc	$\mu/\rho D_{ij}$	Viscous momentum transport/ Diffusive mass transport
Stanton	St	$\alpha/\rho v c_p$	Thermal conduction loss/ Heat capacity
Stokes	S	$\mu/\rho \nu l^2$	Viscous damping rate/ Vibrational frequency
Strouhal	Sr	$\nu l/v$	Vibrational rate/ Convective flow rate
Weber	W	$\rho l v^2/\sigma$	Inertial force/ Surface tension forces

Legend

c_p	Specific heat at constant pressure (cm ^{2}/K-s^2)
F	Imposed force (g-cm/s^2)
g	Gravitational constant (cm-s^2)
l	Scale length, radius (cm)
Q	$\sum h_{oj} R_i \rho_i$ (g/s^{-3}), where
R_i	$=P_i/n_i - L_i$, (s^{-1}) see Table 2–1.
T	Temperature (K)
v	Characteristic flow velocity (cm/s)
α	Newton's-law heat coefficient (g/s^{-3}-K), $\lambda(dt/dx)=\alpha\Delta T$
δ	Collisional mean free path (cm)
Δ	Difference between two quantities
Λ	Latitude of position on Earth's surface
μ	Viscosity coefficient (poise, g/cm-s)
ν	Frequency (Hz)
Ω	Solid-body rotational angular velocity (s^{-1})
ρ	Mass density (g-cm^3)
σ	Surface tension coefficient (g/s^2)

the *Prandtl number*, the ratio of momentum diffusivity (kinematic viscosity) to thermal conduction,

$$\frac{c_p \mu}{\lambda} = \text{Prandtl number} = \text{Pr}, \qquad (2-2.37)$$

the *Brinkman number*, the ratio of heat produced locally by viscous dissipation to that transported by thermal conduction,

$$\frac{\mu v^2}{\lambda T} = \text{Brinkman number} = \text{Br}. \qquad (2-2.38)$$

By considering molecular diffusion, we can define the *mass transfer Peclet number*, the ratio of convective to molecular mass transfer,

$$\frac{lv}{D_{ik}} = \text{Peclet number} = \text{Pe,s}, \qquad (2-2.39)$$

and also the *Schmidt number*, the ratio of momentum diffusion to mass diffusion,

$$\frac{\mu}{\rho D_{ik}} = \text{Schmidt number} = \text{Sc}. \qquad (2-2.40)$$

These and other dimensionless ratios are defined in Table 2–5.

Characterizing a system or a set of apparently similar systems by a dimensionless number is often useful. For example, it can be estimated from the Reynolds number whether or not a flow is unstable to a shear instability. Whether a droplet will break up can sometimes be estimated by evaluating the Weber number. But these nondimensional numbers can also be misleading, because a flow is generally characterized by many such numbers which may be changing in time. For example, the Reynolds number evaluated at some location and at some time might be the same for two systems. The flow stability based on Reynolds number might be wrong due to the effects of time-dependent chemical reactions and subsequent energy release. Even though the Weber number for breakup of a droplet for an idealized case could be one value, the presence of strong shears in the flow could cause droplets of much smaller radius than predicted to break up. These nondimensional numbers must be used with caution, and as general guides to system behavior.

2–2.9 Finding Input Data

Table 2–6 is a summary of Eqs. (2–1.1) – (2–1.18) written in terms of the physical processes and the required input information. Obtaining good values for the inputs to the equations is as important as using good numerical algorithms to solve them. Often there is as much difficulty in determining these values as in building the simulation model. These input data are usually the least accurate part of the computation. Both convection and diffusion processes can usually be computed to a few percent accuracy. The transport coefficients and reaction rate coefficients are seldom known to ten percent accuracy.

Consider, for example, the forward and reverse chemical kinetics rates mentioned in Eqs. (2–1.14) – (2–1.15). Figure 2–1 is a summary of the predicted values of one chemical rate constant as a function of the year it

Table 2–6. Processes in Conservation Equations

Process	Inputs	Chapter
Chemical kinetics	Species and reactions, Enthalpies, Heats of formation	5
Molecular diffusion	Molecular diffusion coefficients	7
Thermal diffusion	Thermal diffusion coefficients	7
Viscosity	Viscosity coefficients	7
Thermal conduction	Thermal conduction coefficients	7
Convective transport	Boundary and initial conditions	8–11

was measured. It shows that the numbers that go into the model must be scrutinized carefully and tests must be performed to see how sensitive the answers are to the input data. In each pertinent chapter we discuss sources for obtaining input data. In Chapter 5 we discuss the use of sensitivity analyses to determine how the selection of a particular input parameter affects the calculated results.

2–3. FLUID DYNAMICS WITH CHEMICAL REACTIONS

2–3.1 Feedback and Interaction Mechanisms

Fluid dynamics and chemical kinetics interact in several fundamental ways that become crucial aspects of reactive flows. Convective flow causes a small volume element to move from one location to another, to penetrate other elements of fluid, and to change density in response to changing flow conditions along its path. Each of these three convection effects interact differently with ongoing chemical reactions in the system.

Equation (2–2.13) shows that the velocity field can be broken up into three separate components. The potential flow component of the velocity field is the part with no curl and no divergence. Thus it represents only advective flows which are translations free of both rotation and compression. In one-dimensional problems, advection involves motion at a constant velocity. In multidimensions, the velocity potential arises by solving the Laplace equation, Eq. (2–2.11), with the suitable boundary conditions. There are no sound waves, shocks, vortices, or turbulence in potential flow.

Because nearby fluid elements remain nearby in potential flow, the main effect of the flows on chemical reactions is to move the reactions with the

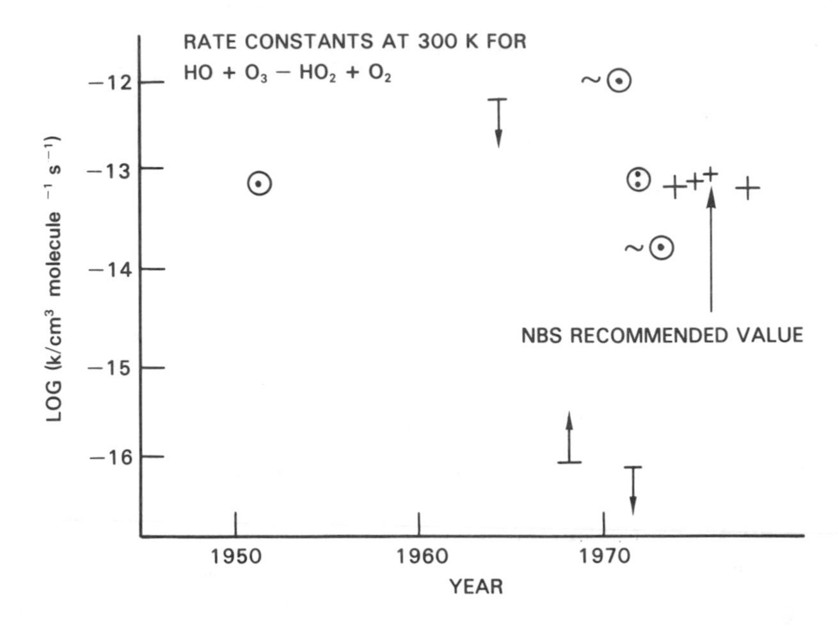

Figure 2–1. Rate constant for the reaction of hydroxyl radical and ozone, as a function of the year of its measurement. Measurements are indicated by points, circles, daggers, and horizontal bars. Arrows above or below the bars indicate measurements which give upper and lower bounds. (Figure provided by Dr. David Garvin of the National Bureau of Standards.)

associated fluid element from one place to another. Although this sounds simple, the numerical problems of trying to represent continuous fluid motion through a discretized spatial domain have dominated computational fluid dynamics for decades. Chapters 6, 8, 9, and 10 discuss methods of accurately representing fluid motion.

When the vorticity is nonzero, there are gradients transverse to the local direction of fluid motion. These rotational, or shearing, motions allow fluid from one place to penetrate and convectively mix with other fluid because these motions can separate two initially close fluid elements. Shearing in the local flow also enhances species gradients and causes faster molecular mixing. A natural consequence of large-scale rotation in flows is turbulent mixing, which brings widely separated fluid elements close together where they can mix and chemically react. In this way vorticity enhances chemical reactivity. Vorticity can also affect chemical reactions in an initially premixed medium by mixing hot reacted fluid with cold, unreacted fluid. Then the temperature rise as well as the introduction of reactive intermediate species increases the rate of chemical reactions. Numerical methods designed to consider highly rotational flows are discussed in Chapters 8, 9, and 10.

When the flow is curl free but not divergence free, that is, when $\nabla \times \mathbf{v} = 0$ but $\nabla \cdot \mathbf{v} \neq 0$, there are velocity gradients in the direction of motion leading to compressions and rarefactions. Compression generally accelerates chemical reactions and expansion generally retards them. A runaway effect is possible, in which higher densities increase the rate of chemical reactions, which in turn increase pressure even faster.

Fluid motions arising directly from local expansions can be violent, but they decrease in strength with distance from the source and they stop when the energy release is complete. The indirect effects of expansion are longer lived and can be more important. Once expansion is complete, it leaves pockets of hot reacted gas in the midst of unreacted fuel or oxidizer. Gradients in temperature and density become frozen into the fluid and move with it. Embedded density gradients interact with external pressure gradients to generate additional vorticity. Vorticity is generated slowly by this passive influence of reaction kinetics on the flow, but its integrated effect can be much larger than even the long-lived vortices generated directly during the expansion. Although the vortices have flow velocities which are only a few percent of the expansion velocities that produced them, they persist for long times after the expansion has stopped. Unlike direct expansion, these rotational flows cause mixing.

Each of the four types of processes in Eqs. (2–1.1) – (2–1.18), chemical kinetics, diffusive transport, convection, and wave phenomena can interact with every other. So far we have explicitly considered the interactions of convection and energy release from chemical reactions. Perhaps the most important of the remaining interactions are between molecular diffusion and chemical kinetics and between molecular diffusion and convection. The rotational component of the flow generates vorticity and causes fluid elements to interpenetrate. This process increases the surface area of the interfaces between different materials and, by shearing the fluid and stretching the vortices, increases the size of gradients. Thus diffusive transport of all types is enhanced by the rotational component of fluid flow. The strongest effects of molecular diffusion on convection occur indirectly through effects such as heat release resulting from mixing fuel and oxidizer. The direct interactions between molecular diffusion and chemical kinetics occur because each changes the nature of the species present in a fluid element. Molecular diffusion does this through moving new material into the volume, and chemical kinetics does this through chemical transformations. There is also a change in background temperature associated with each: molecular diffusion by moving hot or cold material from one region to another, and chemical kinetics by releasing or absorbing energy. The interactions of these processes are particularly important in flames.

2–3.2 Three Reactive Flow Problems

We now describe three reactive flow problems that contain most of the interactions and processes we will be considering: (1) a propagating gas phase detonation, (2) a laminar, premixed-gas flame, and (3) a multiphase reacting flow with fuel droplets. These examples of very idealized reactive flows are used to show how the physical processes contained in Eqs. (2–1.1) – (2–1.18) interact in combustion problems. As we describe pertinent numerical methods in later chapters, we use these types of problems as examples of their application.

Gas Phase Detonation Propagation

Consider a planar, one-dimensional detonation propagating in an open space filled with a combustible material. Before the detonation passes, the material is homogeneous in composition, pressure, and temperature. A detonation is a supersonic wave structure consisting of a leading shock wave, followed by a reaction zone, followed by a region of burned gas. The shock wave heats and compresses the undisturbed reactive mixture as it passes through it. The raised temperature triggers chemical reactions, and energy release eventually occurs some distance behind the shock. The energy release generates pressure waves, some of which propagate forward and accelerate the shock wave. A propagating dynamic equilibrium configuration results from a balance between shock heating, compression, and the energy release behind the shock. This simple one-dimensional picture is illustrated qualitatively in Figure 2–2a. In a detonation, energy release occurs so quickly that molecular diffusion and thermal conductivity are usually not important energy transport mechanisms. The velocity of the detonation depends only on the energy content of the material and the background temperature and pressure of the gas. For now we ignore the effects of walls or heat sources or sinks. This highly idealized picture of a detonation contains the basic interactions of convection and chemical reactions with energy release.

In gas-phase multidimensional detonations, the flow is much more complicated. The detonation front often consists of intersecting shocks which change their relative orientation as a function of time. These shocks are still followed by reaction zones and regions of burned material, but the reaction zones also vary in size and shape as a function of time. An idealized

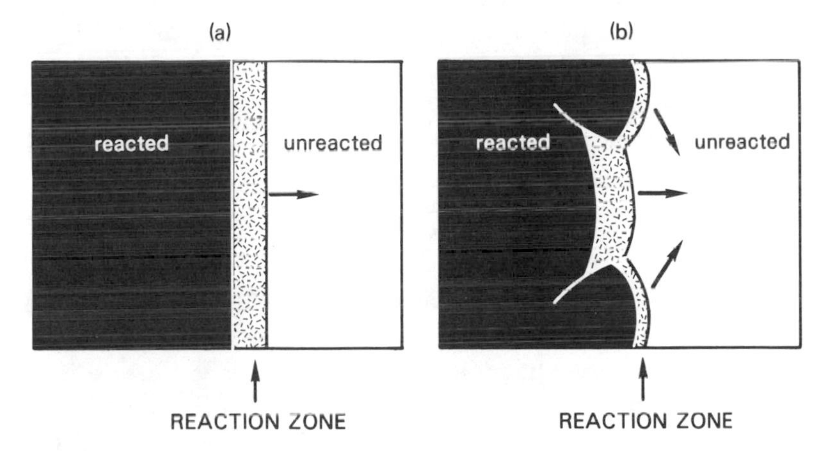

Figure 2–2. Schematic diagram of the structure of a propagating detonation. (a) A planar one-dimensional detonation consisting of a shock, followed by a reaction zone, followed by totally reacted material. (b) A multidimensional detonation, consisting of interacting shock waves.

two-dimensional configuration is shown in Figure 2–2b. Rotational flows, excluded in the one-dimensional picture, occur behind the leading shock fronts because shear flows and vorticity are generated at the slip lines behind the multiple-shock interactions. As they move through space, the intersections of the leading shock waves, called triple points, trace out patterns that are often called detonation cells. In real detonations, effects of inhomogeneities in the reactive material, and the presence of walls, obstacles, and boundary layers can be important. Excellent reviews of the structure and theory of these cells are given in Strehlow (1984) and Fickett and Davis (1979).

The basic processes which have to be modeled numerically to simulate a propagating detonation are convection including compressibility effects and chemical reactions leading to energy release. An accurate two-dimensional calculation combining these two processes can produce a numerical model of a detonation structure. Techniques for modeling supersonic flows are discussed in Chapters 8 and 9. A particular difficulty encountered in this problem is avoiding numerical errors that tend to smear out the structure we want to resolve. One approach to avoiding these errors is to use algorithms that directly incorporate conservation properties of the continuity equation. Techniques for modeling energy release from chemical kinetics are discussed in Chapter 5. In addition to stand-alone methods for integrating coupled sets of ordinary differential equations, we discuss more phenomenological methods that can be efficiently incorporated into a reactive flow calculation. In general, following the chemical kinetics is the most expensive part of the calculation. Research in numerical methods for ordinary differential equations is still needed to find faster, more efficient algorithms to couple with fluid dynamics. In comparison, the accurate convection algorithms are relatively inexpensive. Descriptions of both one-dimensional and two-dimensional detonation models are given in Chapter 13.

Premixed Laminar Flame Propagation

Laminar flame propagation in a premixed gas combines the effects of convective transport, thermal conductivity, and molecular diffusion with chemical kinetics and heat release. In addition, processes such as losses to the walls and radiation can be of major importance. Because of the number of physical processes required to describe a flame, there are many more potentially important interactions to be considered.

Consider an idealized one-dimensional flame front moving through a combustible, premixed gas, and assume that wall and radiation losses are negligible. The flame front consists of gradients in temperature and species densities separating a low density, high temperature region from a high density, low temperature region. Unlike the shock front, which is characterized by pressure, temperature, and density jumps, the pressure through a flame front is essentially constant. This situation is illustrated qualitatively in Figure 2–3. Energy is released in the vacinity of the flame front causing convective expansion. The chemical reactions create a source of unstable intermediate species that may diffuse ahead of the front. Thermal conduction heats the cold material ahead of the flame. The flame velocity, the structure of the flame front, and the physical mechanisms by which a flame propagates, are controlled by the relative magnitudes of the interactions among all of the important, controlling processes.

In one-dimensional flow, convection consists of compression and advection. In two or three dimensions, rotation induced by vorticity generation is also important. If rotational flows interact with a flame front, the surface area of the interface increases, and this affects the diffusive transport. The additional interactions among processes discussed in Section 2–3.1 occur when the flow has a rotational component, and these interactions can be extremely important in a multidimensional flame.

Modeling a flame numerically requires algorithms for all of the important processes as well as algorithms for coupling the processes together. There are two major numerical problems. One is resolving the convective flow without excess numerical errors which could easily mask the physical

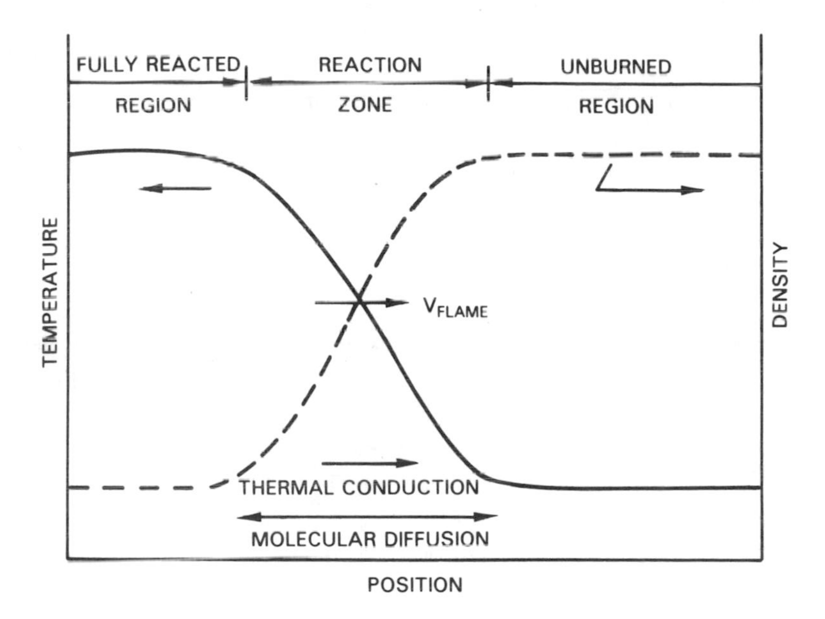

Figure 2–3. Schematic diagram of the structure of a propagating laminar flame. The flame is moving from the left to the right, leaving behind it burned, reacted gases.

diffusion processes. Techniques for doing this are described in Chapters 10 and 11. The other problem is inverting a matrix equation to find the diffusion velocity of each species. This problem is treated in Chapter 7. A detailed example of a one-dimensional numerical flame model is given in Chapter 13.

Multiphase Flows with Droplets

When liquid fuel droplets are injected into a background of hot oxidizing gas, multiphase processes become important that were not important in the idealized gas-phase flame and detonation. The droplets move with the background flow and interact with it, so both the background and the droplets change in time. In general, the droplets are stressed by all of the changing conditions in the background fluid: shear flows, turbulence, and gradients in temperature or density. The background gas flow perturbs the shape and temperature of the droplet and changes the flow patterns within the droplet through viscous effects. When the background is hot, the droplets evaporate, becoming smaller as they shed new gases which diffuse into the background. As fuel gases mix with a background oxidizer, flames may form around the droplet in the flowing gas. The droplets cause flow perturbations in the background gas through distorting the flow patterns and changing the background composition as they evaporate. This extremely complicated multiphase flow problem is important in applications ranging from automobiles to rocket engines.

The basic equations of multiphase reactive flow are more complicated than Eqs. (2–1.1) – (2–1.18). The fluids exist in different phases and the physics and chemistry at the interfaces play an important role. Resolving the physical interactions of one or two droplets themselves in a background flow is a difficult computational problem that is just now being addressed. Models for physical effects such as evaporation, interphase drag, surface tension, and boundary layers must be considered. If the individual droplets are not resolved, it is not satisfactory to assume that everything within a volume element is premixed and homogeneous: a certain fraction might be in a condensed phase that is not molecularly mixed with the background gas. The computational models are complex and many phenomenologies and submodels are required to represent all of the processes. The particular problems of modeling multiphase flows are discussed in Chapter 12.

References

Brodkey, R.S., 1967, *The Phenomena of Fluid Motions*, Addison-Wesley, Reading, MA.

Carnahan, B., H.A. Luther, and J.O. Wilkes, 1969, *Applied Numerical Methods*, Wiley, New York.

Chapman, S., and T.G. Cowling, 1970, *The Mathematical Theory of Non-Uniform Gases*, 3rd Edition, Cambridge University Press, Cambridge, England.

Courant, R., and K.O. Friedrichs, 1948, *Supersonic Flow and Shock Waves*, Interscience, New York.

Courant, R., and D. Hilbert, 1962, *Methods of Mathematical Physics*, Vol. II, Interscience, New York.

Fickett, W., and W.C. Davis, 1979, *Detonation*, University of California Press, Berkeley, CA.

Hirschfelder, J.O., C.F. Curtiss, and R.B. Bird, 1964, *Molecular Theory of Gases and Liquids*, Wiley, New York.

Hofstadter, D. R., 1979, *Gödel, Escher, Bach: An Eternal Golden Braid*, Basic Books, New York.

ation

nd precisely what information is needed to solve a
planning the simulations much easier. Such a focal
an be used to organize the development of the sim-
determine what model, algorithmic, and diagnostic
ded. Too many computational projects are started
eloping a general purpose model capable of accurate
de range of parameter space but without a particular
When the focus of a simulation project is too diffuse,
generally gained compared to the effort expended.

s of the real system and the important technical issues
select a representative standard case to simulate. The
ould be as idealized as possible and yet still be related
fic problem. For example, a standard case for the gas safety
be: How is the minimum ignition energy changed as air is
xture of 97% methane and 3% ethane? Specifically, what
is required if the mixture is stoichiometric methane in air at
, and heat is added at a fixed rate and distributed uniformly
f fixed radius? This standard case should produce answers in
s corresponding to the most relevant experimental or analytical
able.

time the standard-case simulation is completed, the numerical
udes the relevant physical processes and the numerics have been
The results of the standard case can be compared to results of sub-
imulations in which different parameters of the system are varied.
nple, pressure sensitivity of the ignition energy could be checked by
g the initial pressure and repeating the simulations. The sensitivity
type of natural gas could be checked by evaluating how the ignition
changes as the percentage of higher hydrocarbons is increased. When
r modified algorithms are incorporated in the simulation, the standard
provides a familiar basis for evaluating potential improvements or gen-
zations. For example, a new integration algorithm might be used, and
results compared with the standard case as a benchmark.

The specific problem and the standard case chosen from it do not have
model the real system exactly to be useful. It is usually best to solve a
htly different problem well rather than do a poor job on exactly the right
blem. By leaving out many of the components of a complex system and
sing on only a few interactions, we can often formulate a computational
lem whose predictions help our understanding of the real system. For
ple, the simulation probably does not have to reproduce the compli-
geometry of the area into which the natural gas tank vents. We might

Landau, L.D., and E.M. Lifshitz, 1959, *Fluid Mechanics*, Pergamon Press, New York.

Liepmann, H.W., and A. Roshko, 1957, *Elements of Gasdynamics*, Wiley, New York.

Mathews, J., and R.L. Walker, 1965, *Mathematical Methods of Physics*, Benjamin, New York.

Mihalas, D., and B.W. Mihalas, 1984, *Foundations of Radiation Hydrodynamics*, Oxford University Press, New York.

Reid, R.C., J.M. Prausnitz, and T.K. Sherwood, 1977, *The Properties of Gases and Liquids*, McGraw-Hill, New York.

Rybicki, G.B., and A.P. Lightman, 1979, *Radiative Processes in Astrophysics*, Wiley, New York.

Siegel, R., and J.R. Howell, 1981, *Thermal Radiation Heat Transfer*, Hemisphere, Washington, D.C.

Strehlow, R.A., 1984, *Combustion Fundamentals*, McGraw-Hill, New York.

Williams, F.A., 1985, *Combustion Theory*, Benjamin Cummings, Menlo Park, CA.

Zel'dovich, Ya. B., and Yu. P. Raizer, 1966–1967, *Physics of Shock Waves and High-Temperature Hydrodynamic Phenomena*, Academic, New York.

Chapter 3

MODELS AND SIMULATION

From Chapter 4 on, this book is primarily concerned with numerical algorithms. We describe numerical representations for various terms in the reactive flow equations, the algorithms used to implement these numerical representations, how the algorithms are optimized, and how they are coupled together to form numerical simulations models. The most profitable use of a simulation model is as strongly affected by the research and software environments as by particular models or algorithms. Therefore it is reasonable to consider the process of developing and using simulation methods before describing the details of the algorithms.

There are three main stages in developing and using a numerical simulation:

1. Specifying the Mathematical and Computational Model
   - Selecting and posing the problem for simulation
   - Choosing the computational representation
2. Constructing the Numerical Simulation Model
   - Choosing solution algorithms
   - Optimizing the model algorithms
   - Installing and testing the model on the computer
3. Using the Numerical Simulation Model
   - Performing detailed simulations
   - Analyzing and understanding the results
   - Optimizing the overall process by using time efficiently

This chapter gives a general overview of the first two stages and a number of suggestions about the third. Much of the material suggested here can only be learned by trying things and coping with the consequences. Our purpose here is to explain some of the considerations and options.

3–1.1 Posing S

The first step in d
entific questions in
For example, we may
niter near the vent of a
is it is for combustible
answer this question, we
minimum ignition energies,
flame speeds of gas mixtures
various amounts of air. Simu
cific questions, and thus provid
problem of the safety of a natur

The difficult underlying ques
the real system from a simulatio
question.

First, what we *intend to learn* cl
gresses. Because we begin the problem
edge, we must expect to ask the wrong
the right questions but from the wrong p
always bias the process of choosing the pro

Second, the constraints imposed by the
fect on the problem we can study. The best wa
quickly is to recognize and work within the con
ulation capability. Otherwise, it is necessary to
and resources to undertake an extensive developm

Often finding answers to the questions posed in
teraction among several physical processes as a para
varied. For example, we might ask how the minimum
mixture of 10% air and 90% natural gas changes as sma
are added. To answer this question, the computer prog
ments the model must represent the interacting effects (c
and fluid dynamics) at the right level of accuracy, in all th
sures and temperatures) where they are important, and for a
parameters (for example, the percentage of ethane in the mixtur
ing on detailed physical interactions suspected to be of primary i
less relevant aspects of the real system can be treated in an ideali
the model or perhaps not included at all.

calculate the ignition energy in planar and spherical geometries to bound the answers. If a main effect is controlled by the geometry, however, the calculation has to be multidimensional.

3–1.2 Choosing the Computational Representation

Constructing Mathematical and Computational Submodels

In Chapter 2, we described the importance of chunking in theoretical models of physical systems. The statistical nature of many of the usual physical interactions gives meaning to relatively large-scale averages such as density and temperature. Chunked models are useful because it is not practical to solve complicated problems at the most microscopic level of detail. A more averaged level may be adequate, depending on the accuracy of the experimental data or the type of information we need.

Because the simulation codes solve sets of equations describing chunked models, they should be built to reflect this natural chunking. The most natural way of chunking the computer program is into software modules that mirror the natural mathematical and physical chunkings. We thus recommend developing computational algorithms which separately describe chunked portions of the underlying physics, and then coupling these separate algorithms to create the complete simulation model. In this way it is easier to replace a submodel with a more detailed submodel, if it is eventually needed.

The Representation Affects the Answers

The *representation* of a physical system is the mathematical form of the equations, the discretization used, and any auxiliary conditions needed to describe the problem. Because the model equations must be discretized for digital simulation, the representation is the framework within which space and time are described. This could mean, for example, considering moving fluid parcels or treating volume elements fixed in space.

The ways in which simulation results can be prejudiced by the representation are often surprising. These surprises come when we automatically adopt a mathematical or computational representation and then forget the assumptions required to justify its use. Because the representation establishes the framework of the model, it affects the answers both directly and indirectly. For example, a reactive flow model might predict a temperature field with a physically reasonable value for each computational cell. However, if the representation does not allow time-variation, the results may be misleading when there are very rapid transients. Determining the representation of the computational model for the complete set of equations and for the various physical submodels is the next important step after choosing the problem to solve.

The Navier-Stokes equations, for example, are a reasonable starting point for calculating the ignition energy of the natural gas mixtures discussed above. Adapting this model means that changes in physically averaged variables over a mean free path must be small so that the mass, momentum, and energy fluxes through a volume element can be related by first derivatives to the density, velocity, and temperature. Once this representation is chosen, the possibility of fundamental descriptions of processes with large gradients over a mean free path is excluded. This chunked fluid model cannot predict the detailed structure of a shock, whose width is determined by the viscosity term in the Navier-Stokes equations or by nonphysical smoothing effects in the numerical algorithm chosen.

It is possible to include most of the macroscopic consequences of the atomic and molecular phenomena in a fluid model. The presence of terms describing microscopic processes in these macroscopic representations can lull us into thinking that we have a detailed enough representation of the model to make a prediction.

In calculating ignition energies for the natural gas mixture, molecular diffusion is one of the macroscopic continuum effects that must be included (Chapters 4 and 7). On the microscopic level, a random walk transports the lighter, hotter species ahead of the flame front and thus prepares the mixture for ignition. If we use the standard second-derivative form to represent molecular diffusion, we are assuming a continuum representation is valid. If the random walk is not well represented by a diffusion equation, the ignition process is not properly represented.

An important part of the fluid description of this process is the value of the effective molecular diffusion coefficients. Because it is hard to measure diffusion coefficients over the wide pressure and temperature ranges required in combustion systems, calculated and extrapolated values are often used. If, for example, we have a polar molecule and the assumed force law does not include this, the diffusion coefficient can be wrong and the diffusion process is not properly represented.

As another example of how the representation can prejudice the answer, consider a statistical turbulence model describing the time-averaged flow properties of a turbulent jet. Such models usually contain coefficients which are fit to experimental data. The model might predict the global quantities well for experiments which are close to those whose results were

Landau, L.D., and E.M. Lifshitz, 1959, *Fluid Mechanics*, Pergamon Press, New York.

Liepmann, H.W., and A. Roshko, 1957, *Elements of Gasdynamics*, Wiley, New York.

Mathews, J., and R.L. Walker, 1965, *Mathematical Methods of Physics*, Benjamin, New York.

Mihalas, D., and B.W. Mihalas, 1984, *Foundations of Radiation Hydrodynamics*, Oxford University Press, New York.

Reid, R.C., J.M. Prausnitz, and T.K. Sherwood, 1977, *The Properties of Gases and Liquids*, McGraw-Hill, New York.

Rybicki, G.B., and A.P. Lightman, 1979, *Radiative Processes in Astrophysics*, Wiley, New York.

Siegel, R., and J.R. Howell, 1981, *Thermal Radiation Heat Transfer*, Hemisphere, Washington, D.C.

Strehlow, R.A., 1984, *Combustion Fundamentals*, McGraw-Hill, New York.

Williams, F.A., 1985, *Combustion Theory*, Benjamin Cummings, Menlo Park, CA.

Zel'dovich, Ya. B., and Yu. P. Raizer, 1966–1967, *Physics of Shock Waves and High-Temperature Hydrodynamic Phenomena*, Academic, New York.

Chapter 3

MODELS AND SIMULATION

From Chapter 4 on, this book is primarily concerned with numerical algorithms. We describe numerical representations for various terms in the reactive flow equations, the algorithms used to implement these numerical representations, how the algorithms are optimized, and how they are coupled together to form numerical simulations models. The most profitable use of a simulation model is as strongly affected by the research and software environments as by particular models or algorithms. Therefore it is reasonable to consider the process of developing and using simulation methods before describing the details of the algorithms.

There are three main stages in developing and using a numerical simulation:

1. Specifying the Mathematical and Computational Model
   - Selecting and posing the problem for simulation
   - Choosing the computational representation
2. Constructing the Numerical Simulation Model
   - Choosing solution algorithms
   - Optimizing the model algorithms
   - Installing and testing the model on the computer
3. Using the Numerical Simulation Model
   - Performing detailed simulations
   - Analyzing and understanding the results
   - Optimizing the overall process by using time efficiently

This chapter gives a general overview of the first two stages and a number of suggestions about the third. Much of the material suggested here can only be learned by trying things and coping with the consequences. Our purpose here is to explain some of the considerations and options.

3–1. DEVELOPING A NUMERICAL SIMULATION MODEL

3–1.1 Posing Scientific Problems for Modeling

The first step in developing a numerical simulation model is posing the scientific questions in terms that can be answered effectively by a simulation. For example, we may want to know how dangerous it is to use a spark igniter near the vent of a natural gas tank. In more scientific terms, how likely is it is for combustible gases from the vent to ignite or detonate? To help answer this question, we ask related but more specific questions about the minimum ignition energies, the flammability and detonation limits, and the flame speeds of gas mixtures which are composed of natural gas diluted with various amounts of air. Simulations can now provide answers to these specific questions, and thus provide information that helps evaluate the general problem of the safety of a natural gas tank.

The difficult underlying question is: *What do we intend to learn about the real system from a simulation?* There are two aspects of this basic question.

First, what we *intend to learn* changes dynamically as the study progresses. Because we begin the problem-solving process with limited knowledge, we must expect to ask the wrong questions initially, or perhaps ask the right questions but from the wrong perspective. Preconceived notions always bias the process of choosing the problem.

Second, the constraints imposed by the computational tools have an effect on the problem we can study. The best way to get answers to problems quickly is to recognize and work within the constraints of the existing simulation capability. Otherwise, it is necessary to have the time, experience, and resources to undertake an extensive development project.

Often finding answers to the questions posed involves simulating an interaction among several physical processes as a parameter in the model is varied. For example, we might ask how the minimum ignition energy of a mixture of 10% air and 90% natural gas changes as small amounts of ethane are added. To answer this question, the computer program which implements the model must represent the interacting effects (chemical reactions and fluid dynamics) at the right level of accuracy, in all the regimes (pressures and temperatures) where they are important, and for a range of model parameters (for example, the percentage of ethane in the mixture). By focusing on detailed physical interactions suspected to be of primary importance, less relevant aspects of the real system can be treated in an idealized way in the model or perhaps not included at all.

Specifying clearly and precisely what information is needed to solve a specific problem makes planning the simulations much easier. Such a focal problem for research can be used to organize the development of the simulation model and help determine what model, algorithmic, and diagnostic developments are needed. Too many computational projects are started with the goal of developing a general purpose model capable of accurate prediction over a wide range of parameter space but without a particular application in mind. When the focus of a simulation project is too diffuse, little information is generally gained compared to the effort expended.

The properties of the real system and the important technical issues should be used to select a representative standard case to simulate. The standard case should be as idealized as possible and yet still be related closely to a specific problem. For example, a standard case for the gas safety problem would be: How is the minimum ignition energy changed as air is added to a mixture of 97% methane and 3% ethane? Specifically, what ignition energy is required if the mixture is stoichiometric methane in air at 1 atm, 298 K, and heat is added at a fixed rate and distributed uniformly in a sphere of fixed radius? This standard case should produce answers in physical units corresponding to the most relevant experimental or analytical results available.

By the time the standard-case simulation is completed, the numerical model includes the relevant physical processes and the numerics have been checked. The results of the standard case can be compared to results of subsequent simulations in which different parameters of the system are varied. For example, pressure sensitivity of the ignition energy could be checked by changing the initial pressure and repeating the simulations. The sensitivity to the type of natural gas could be checked by evaluating how the ignition energy changes as the percentage of higher hydrocarbons is increased. When new or modified algorithms are incorporated in the simulation, the standard case provides a familiar basis for evaluating potential improvements or generalizations. For example, a new integration algorithm might be used, and the results compared with the standard case as a benchmark.

The specific problem and the standard case chosen from it do not have to model the real system exactly to be useful. It is usually best to solve a slightly different problem well rather than do a poor job on exactly the right problem. By leaving out many of the components of a complex system and focusing on only a few interactions, we can often formulate a computational problem whose predictions help our understanding of the real system. For example, the simulation probably does not have to reproduce the complicated geometry of the area into which the natural gas tank vents. We might

calculate the ignition energy in planar and spherical geometries to bound the answers. If a main effect is controlled by the geometry, however, the calculation has to be multidimensional.

3–1.2 Choosing the Computational Representation

Constructing Mathematical and Computational Submodels

In Chapter 2, we described the importance of chunking in theoretical models of physical systems. The statistical nature of many of the usual physical interactions gives meaning to relatively large-scale averages such as density and temperature. Chunked models are useful because it is not practical to solve complicated problems at the most microscopic level of detail. A more averaged level may be adequate, depending on the accuracy of the experimental data or the type of information we need.

Because the simulation codes solve sets of equations describing chunked models, they should be built to reflect this natural chunking. The most natural way of chunking the computer program is into software modules that mirror the natural mathematical and physical chunkings. We thus recommend developing computational algorithms which separately describe chunked portions of the underlying physics, and then coupling these separate algorithms to create the complete simulation model. In this way it is easier to replace a submodel with a more detailed submodel, if it is eventually needed.

The Representation Affects the Answers

The *representation* of a physical system is the mathematical form of the equations, the discretization used, and any auxiliary conditions needed to describe the problem. Because the model equations must be discretized for digital simulation, the representation is the framework within which space and time are described. This could mean, for example, considering moving fluid parcels or treating volume elements fixed in space.

The ways in which simulation results can be prejudiced by the representation are often surprising. These surprises come when we automatically adopt a mathematical or computational representation and then forget the assumptions required to justify its use. Because the representation establishes the framework of the model, it affects the answers both directly and indirectly. For example, a reactive flow model might predict a temperature field with a physically reasonable value for each computational cell. However, if the representation does not allow time-variation, the results may be misleading when there are very rapid transients. Determining the representation of the computational model for the complete set of equations and for the various physical submodels is the next important step after choosing the problem to solve.

The Navier-Stokes equations, for example, are a reasonable starting point for calculating the ignition energy of the natural gas mixtures discussed above. Adapting this model means that changes in physically averaged variables over a mean free path must be small so that the mass, momentum, and energy fluxes through a volume element can be related by first derivatives to the density, velocity, and temperature. Once this representation is chosen, the possibility of fundamental descriptions of processes with large gradients over a mean free path is excluded. This chunked fluid model cannot predict the detailed structure of a shock, whose width is determined by the viscosity term in the Navier-Stokes equations or by nonphysical smoothing effects in the numerical algorithm chosen.

It is possible to include most of the macroscopic consequences of the atomic and molecular phenomena in a fluid model. The presence of terms describing microscopic processes in these macroscopic representations can lull us into thinking that we have a detailed enough representation of the model to make a prediction.

In calculating ignition energies for the natural gas mixture, molecular diffusion is one of the macroscopic continuum effects that must be included (Chapters 4 and 7). On the microscopic level, a random walk transports the lighter, hotter species ahead of the flame front and thus prepares the mixture for ignition. If we use the standard second-derivative form to represent molecular diffusion, we are assuming a continuum representation is valid. If the random walk is not well represented by a diffusion equation, the ignition process is not properly represented.

An important part of the fluid description of this process is the value of the effective molecular diffusion coefficients. Because it is hard to measure diffusion coefficients over the wide pressure and temperature ranges required in combustion systems, calculated and extrapolated values are often used. If, for example, we have a polar molecule and the assumed force law does not include this, the diffusion coefficient can be wrong and the diffusion process is not properly represented.

As another example of how the representation can prejudice the answer, consider a statistical turbulence model describing the time-averaged flow properties of a turbulent jet. Such models usually contain coefficients which are fit to experimental data. The model might predict the global quantities well for experiments which are close to those whose results were used to determine model parameters. This model, however, cannot predict deterministic local values or phenomena such as the formation and breakup of coherent structures (see Chapter 12). The model could not tell us something new and fundamental about turbulence. It is also dangerous to use the model outside the calibrated parameter regime or to use it coupled to models of other physical effects. The limitations are clear, yet there is a tendency to expect the models to be better than they are.

Finally, a lumped-parameter chemistry representation might be derived from a detailed chemical kinetics mechanism involving many elementary reactions among many chemical species. This is a calibrated phenomenology that can calculate, but not really predict, chemical induction times, energy release rates, and equilibrium conditions, given initial temperatures and pressures. It interpolates the known results in regions where it has been calibrated. In regions where it has not been calibrated, the answers are essentially unknown, even though the values given by the model may be reasonable.

These examples illustrate how the representation affects the validity of the answers. When developing a model and deciding whether it can be used to solve a particular problem, it is important to keep in mind that *a computer simulation cannot correctly describe an effect if the controlling physical processes are not adequately represented.* Remember which physical processes and interactions have been approximated or removed from the equations, so that a crucial process is not left out. This is obvious advice, but the results of not considering it are insidious.

3–2. LIMITING FACTORS AND CONSTRAINTS

The reactive flow conservation equations include representations of chemical kinetics, thermal and molecular diffusion, convective transport, and several other physical processes. To derive this model, a fair amount of chunking of the physics and chemistry has occurred in a macroscopic, continuum representation. There are a number of additional limitations and constraints beyond those implicit in the model chosen.

To integrate the mathematical model numerically, we must discretize the equations in both time and space. This gives a consistent, computable representation. Space is divided into finite volume elements called *computational cells*. Time is divided into finite intervals call *timesteps*. The approximate numerical solution in each of these computational cells is advanced through each timestep. With shorter timesteps and smaller computational cells, the solution obtained numerically should approach the solution of the original continuum equations. It is important to make sure that the computation actually implemented on the computer goes to the correct continuum limits.

In practice we can seldom approach the continuum limit as closely as we would like. The accuracy and range of validity of the computational simulations are determined by limitations on how many computational cells and timesteps can be used and by the accuracy and stability of the numerical methods. These resolution-limiting and accuracy-limiting factors are imposed primarily by computer hardware constraints such as finite computer memory, finite processing speeds, finite precision data formats, and data storage facilities.

Another type of computational limitation is the high cost associated with solving complex problems. These costs often appear as software constraints. The limitations arising from complexity and from accuracy are intertwined and often hard to distinguish. The physical and geometric complexity of a problem can make the cost of constructing and carrying out simulations prohibitive. These costs of complexity are based in personnel and computer costs and the time involved in developing and maintaining reliable software, rather than the performance of the computer hardware. Complexity leads to bigger models with more special cases to be programmed and tested. It means more chance for error and more initial data to be found and cross-checked. It means less efficient use of the computational hardware and slower running programs than we would encounter simulating simple, idealized problems.

We can identify three types of accuracy limitations brought on by hardware constraints: temporal, spatial, and computational. We can also identify three types of costs due to the complexity of the problem: physical, geometric, and computational. Below we discuss these in some detail, and show how phenomenologies are used to try to circumvent some of the limitations.

3–2.1 Limitations in Accuracy

Temporal Accuracy Limitations

Limitations in temporal accuracy arise from trying to represent different, highly disparate time scales simultaneously. The important time scales can easily range over ten or more orders of magnitude in flame and detonation problems. As long as these important scales can be resolved by the numerical method, obtaining the desired simulation accuracy is not necessarily a limitation. Computing for more timesteps takes longer, but does not cause the problem to exceed the computer memory. Cost, however, can quickly

become prohibitive. What starts out as a cost becomes a limitation due to finite computer speed.

When we wish to model processes with characteristic times of variation shorter than the affordable timestep, the equations that describe these phenomena are usually called "stiff." Sound waves are stiff with respect to timesteps well suited for modeling a flame. The equations describing many chemical reactions are stiff with respect to convection, diffusion, or even the characteristic timestep based on the speed of sound in the material. Stiffness is sometimes a matter of practical timestep limitations, but can also be an inherent property of certain types of equations. Two distinct modeling approaches, the global-implicit approach and the timestep-split asymptotic approach, have been developed to treat temporally stiff phenomena. These two approaches are discussed again in more detail in Chapter 4 and in subsequent chapters.

Spatial Accuracy Limitations

There are also limitations that arise from the need to resolve a wide range of space scales. In continuum representations, greater accuracy comes with finer resolution in space as well as in time. If we want to calculate accurately the behavior of a small structure in the flow, there must be a minimum of three or four computational cells spanning the structure. If steep gradients are involved, more cells are usually necessary. Structures that are not resolved well enough are "numerically diffused," as will be explained in Chapter 4. Because the computational cells have a finite size, it is uncertain exactly where in the cell some feature of the fluid actually is. This uncertainty decreases as the spatial resolution increases. A given accuracy requires a certain cell spacing.

To resolve steep gradients in a one-dimensional propagating flame front, computational cells of 10^{-3} to 10^{-2} cm might be required. On the other hand, cells of 1 to 10 cm size might be adequate to model convective transport. This relative factor of 10^3 means that the spatial resolution problem will not disappear just by making the data arrays bigger; the computer itself is simply not adequate for more than a one-dimensional calculation with 10^{-3} cm resolution everywhere.

The requirements of adequate spatial resolution can be more severe than those of temporal resolution because to maintain a stable, accurate solution, the choice of timestep is tied to the size of the computational cells. Doubling the spatial resolution in a one-dimensional calculation usually requires a fourfold increase in computation to integrate a given length of time. One factor of two arises for the additional cells and another factor of two for the halved timestep required to integrate the equations using these smaller cells. Doubling the resolution in two dimensions requires a factor of eight times as much work to advance the calculation the same physical time. In three dimensions, a factor of sixteen times as much work is required. Algorithms which give accurate answers on coarse grids are extremely valuable.

When the space scales to be resolved are highly disparate, adaptive gridding methods may be used to put spatial resolution only where it is needed (Chapter 6). A still more difficult problem is encountered in turbulent flows, in which a very large range of spatial scales may be important throughout the flow field. In general, we seek numerical methods which maximize accuracy with a minimum number of grid points. Even with the best algorithms, approximations and phenomenologies are required to represent phenomena which occur at subgrid scales.

Computational Accuracy Limitations

Computational accuracy limits arise from trying to represent continuous functions as discrete elements in time and space. In doing this on a computer, we are forced to assign finite-precision, real numbers to describe the properties of these fluid elements. Modern digital computers operate in terms of finite-precision real numbers. Computational acccuracy limitations are due both directly and indirectly to the precision of the computer hardware. Even though the representation for each number may be valid to six, ten, or twelve digits, there will still be accumulated roundoff errors. For example, a 32-bit floating point number is accurate to one part in a few million. Calculations with many thousands of timesteps often show accumulated roundoff errors in the third decimal place with 32-bit numbers. Other types of hardware problems can arise in the arithmetic calculations, such as "underflows" and "overflows," in which the product of two numbers is too small or too large to be represented in the computer.

The algorithms used to implement the mathematical equations are also limited in accuracy. Evaluating the accuracy and efficiency of these algorithms is a major theme of this book. The *order* of an algorithm refers to the highest power of the expansion parameter that is retained in the solution method (Chapter 4). When the error made per timestep is without bound, the algorithm is *unstable*. The computational accuracy limitations are due to truncation errors and the associated problems of numerical stability. When these truncation and instability properties of an algorithm are combined with the precision limitations of the hardware, the accuracy can be

reduced even further. For example, when an algorithm is said to be "second-order accurate," it usually means that increasing the spatial and temporal resolution by a factor of two decreases the error by a factor of four. However, the effective order of accuracy will sometimes be less due to finite precision.

Interactions of the stability properties of algorithms with computer limitations on precision are not easily predictable. Although it is customary to view numerical stability as an algorithmic issue, numerical stability also interacts with errors introduced through precision limitations. Roundoff errors can adversely affect stability, especially if, in the name of economy, the choice of the timesteps and cell size puts the algorithm close to the stability limit. In practice, algorithms must be chosen which are stable regardless of whether the computer hardware precision is 32-bit or 64-bit.

As computers become faster and more precise, slowly growing numerical errors, that may not have appeared previously in shorter or coarser calculations, appear with new and bizarre properties. These relatively weak effects of solution errors, algorithm instabilities, precision errors, and interactions among these, were either damped previously in coarser calculations or did not appear because the calculations were not run long enough. An example of this is roundoff errors in very long calculations using 32-bit floating point numerical representations. Single-bit errors are of little or no consequence after 1000 timesteps, but many dominate the solution after 100,000 timesteps. The general use of supercomputers allowing a factor of 100 between the largest and smallest resolved spatial scales makes these types of modes a potential issue.

3–2.2 The Costs of Complexity

Physical Complexity Costs

One way that physical complexity costs is in computer time. For example, combustion systems usually have many interacting species. The sets of coupled equations that describe such systems must be solved simultaneously to follow the system evolution. Complicated sets of stiff ordinary differential equations describing reactions and large matrices describing the multispecies diffusion process are time-consuming to solve on the computer. To simulate this level of detail requires calculation times orders of magnitude greater than needed for idealized, empirical, or phenomenological models.

There is another cost of physical complexity: the increase in personal time required to gather input data and then to debug the more complicated numerical algorithms. The software development time can grow as the square of the number of important physical processes, since interactions among submodels can generate problems even though the separate modules may appear to be working correctly. This complexity can be considered as a software constraint.

Often the accuracy in the input data is not sufficient for the level of accuracy desired from the simulation. Figure 2–1 shows a typical example in which the rate of the chemical reaction $OH + O_3 \rightarrow HO_2 + O_2$ at 300 K is given as a function of the year of the measurement. We note with amusement and chagrin that using this reaction before 1970 in a kinetics mechanism gives a rate uncertain by five orders of magnitude! Similar tales of horror also exist for thermochemical data, even though these data are generally more accurate (private assurances, David Garvin, National Bureau of Standards).

There are several lessons from Figure 2–1. First, if all the input data have large inaccuracies, the accuracy and level of detail asked of the simulation may be unreasonable. A much less expensive calculation with lower resolution might do as well. If only some of the data are poorly known, the simulations could be used in conjunction with relatively more accurate empirical data to help determine the uncertain input parameters.

Geometric Complexity Costs

Real systems often have complicated geometries. The spatial, temporal, and computational problems associated with modeling this more complicated geometry can, in principle, be solved by larger, faster computers. There are again software and personal time costs which arise from the need to represent, program, and diagnose the geometric complexity. These costs are not decreased by improvements in the computer system.

It takes more time to program a system with complex geometry than one with a simple geometry, and there is more computer code that can go wrong. Even if the program runs properly the first time, it will take longer to run, more time to get to the point where there were any results, require more computer storage, and the output will be harder to interpret. When enormous software structures are built, they tumble down very easily.

Geometric complexity often requires increased dimensionality and more complex boundary conditions. For example, experimental probes in a reactive flow act as obstacles, and their effects must often be modeled. In the absence of a probe, port, or exhaust valve, a one- or two-dimensional model of the problem might be adequate. Realistic geometric complications force a reactive flow problem to be represented in two or three dimensions. Thus a major part of experimental and computational ingenuity is often directed toward finding a meaningful problem of lower dimensionality.

One-dimensional models are often used, even though ignoring the other two spatial directions gives a limited picture of the physical process. Idealized combustion systems such as a premixed laminar flame or a planar detonation wave can be modeled accurately in one dimension. Practical combustion systems, including flames and detonations, generally involve multidimensional effects such as boundary-layer growth, formation of vortices, separating flows, or internal sources and sinks. These systems require at least two-dimensional fluid dynamics. Even with sixth generation parallel processing supercomputers, what can be achieved with a two-dimensional detailed model is limited by computer time and storage requirements. Here again, costs transform into accuracy limitations.

A problem which often accompanies increased boundary complexity and higher dimensionality is that a favorite algorithm no longer works. When we need to include the effects of holes, bumps, and knobs on a surface bounding a flow, certain otherwise ideal algorithms are not flexible enough to treat unusual boundary conditions or do not generalize to multidimensions. Abandoning good algorithms for less accurate but more general ones is a major cost of geometric complexity. Making these trade-offs is a fundamental consideration in building simulation models.

The Costs of Computational Complexity

Computational complexity is the result of implementing physically and geometrically complex problems in a simulation model. The individual algorithms grow in complexity and the simulation model grows in size. Not only the computer running time but also personnel costs for developing, maintaining, and running the simulation model become large. In addition, numerical problems with the computation occur more often because there are more places where the errors might be and more reasons why they might occur. As a simulation model becomes larger and more comprehensive, it becomes more difficult to verify and modify. The software then becomes a serious constraint. A careful, modular approach to model development with many test problems is the only reasonable way to proceed.

3–2.3 Using Phenomenological Models

The list of costs and limitations encountered in developing and applying simulations is formidable. To deal with them we can improve our numerical algorithms and other software, and we can take advantage of larger, faster, more accurate, and friendlier computers. Over the last few decades, improved algorithms have contributed to simulation capability at least as much as hardware development. Together, hardware and software advances have increased our capability by orders of magnitude. Early one-dimensional gas dynamics calculations were resolved with forty or fifty computational cells. Such problems can now be performed in three dimensions, about four or five orders of magnitude more computing, in the same time as the old one-dimensional calculations.

Another way to deal with resolution limitations and costs is to isolate the expensive processes and to replace them in the simulation with simplified phenomenological models. As discussed in Chapter 1, these phenomenologies are ad hoc, physically reasonable, approximate models with parameters that are guesses, fits to experiments, or calibrated from detailed but more restricted numerical models. For example, a large chemical reaction mechanism is composed of many chemical species and perhaps hundreds of reaction rates linking them. Integrating the ordinary differential equations gives the time history of the individual chemical species and bulk properties such as the temperature and pressure. Using this chemical mechanism in a multidimensional fluid dynamics model is possible in theory, but is extremely expensive in reality. An alternate approach is to use the detailed mechanism to calculate bulk properties, such as final temperatures and pressures as a function of a range of initial temperatures and pressures. This table of numbers forms the basis of a computationally inexpensive phenomenology for use in multidimensional fluid models. The more accurate the underlying theoretical framework for the phenomenology, the smaller the tables of results needed and the broader the validity of the model.

There is always a tendency to overestimate how much a phenomenology can actually predict. Phenomena cannot be predicted when the controlling physical processes are not resolved accurately enough in time, in space, or by the computational model. If one of the controlling physical or chemical processes in a simulation is being treated by a phenomenological model, the entire simulation might not be more accurate than the phenomenology, even if the other effects are treated by more detailed models. For example, representing an elementary chemical reaction rate with an Arrhenius form seems fundamental, but assumes that there is a Boltzmann distribution among the electronic and vibrational states of the molecule. However,

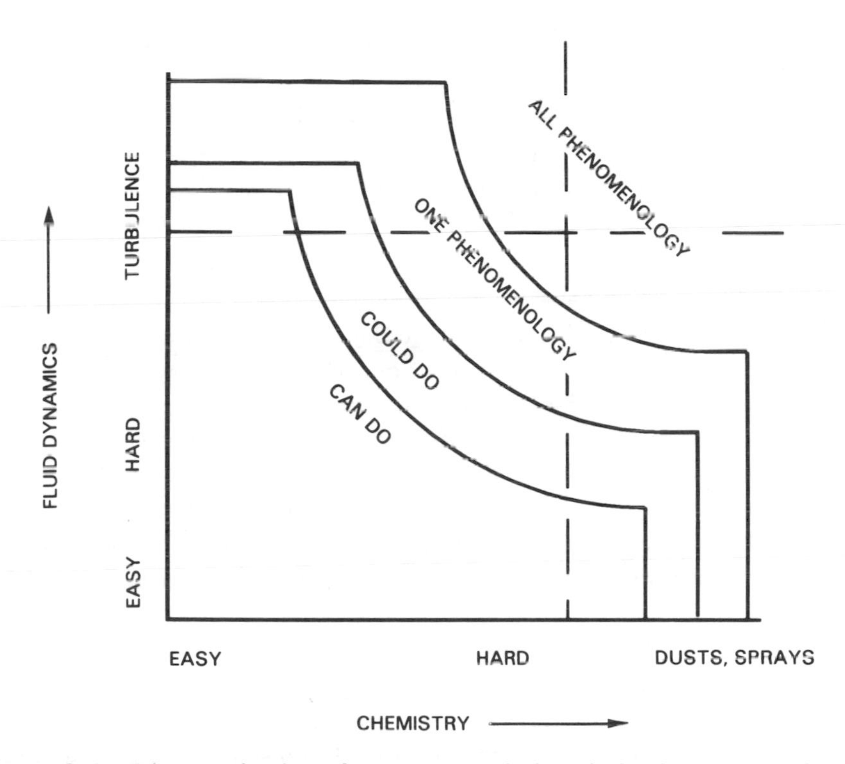

Figure 3–1. Diagram showing what can currently be calculated as more and more phenomenologies are added to the model.

because nonequilibrium reactions are not represented in this formalism, a computational model based on Maxwellian distributions is questionable for slow equilibration effects in gases and high density reactions which occur in explosive materials.

Figure 3–1 shows schematically what levels of detailed simulations are practical for complex, multidimensional, reactive-flow problems in fuel-oxidizer systems. Two "dimensions" of difficulty are considered: the complex physics of chemical kinetics and local multiphase material effects along the horizontal axis, and the resolution-bound processes of fluid dynamics on the vertical axis. Each axis ranges in difficulty from EASY to HARD.

For example, an easy problem on the horizontal axis might involve several coupled chemical reactions among a few species. A hard problem might involve hundreds of chemical reactions among dozens of chemical species. Even more difficult problems have been indicated to the right of the vertical dashed line. In this diagram, the local multiphase properties of dusts, sprays, droplets, and other heterogeneous phenomena have been indicated as an extension of local chemical kinetics. These types of processes and interactions usually need a phenomenology to be included in a fluid simulation.

Along the vertical axis, an easy problem might be a one-dimensional shock calculation. A hard problem might involve transient multidimensional fluid dynamics where both divergence and curl components of the flow field interact. Detailed modeling of turbulence is a very hard fluid dynamic problem. Generally phenomenologies are needed to represent turbulence, and this is indicated high on the vertical axis, above the dashed horizontal line.

The region indicated CAN DO in the figure includes problems with either an easy fluid dynamic component or an easy chemistry component, so that the computational effort can be concentrated on the harder aspect of the problem. Today detailed modeling of chemical and fluid processes is possible for certain problems with one difficult aspect, provided that the other aspects are easy enough. The COULD DO region is an extension of the CAN DO region. If ample computational resources are available and the computations are directed at answering a few, specific, well-focused questions, the COULD DO region can be simulated. No new technology is needed, but ten to forty hours of supercomputer time are required, compared to the one half to two hours for a CAN DO calculation.

The next region is marked ONE PHENOMENOLOGY to indicate that the overall problem difficulty is so great that full resolution of all the important physics, fluid dynamics, and chemistry cannot be done simultaneously. At least one of the major aspects has to be treated phenomenologically. This phenomenological component may involve quantities such as induction times, energy release rates, or flame speeds which have to be interfaced with the rest of the simulation.

The outer region of Figure 3–1 is labeled ALL PHENOMENOLOGY. Both the fluid dynamics and the chemical kinetics of the problem are so difficult that detailed solution of neither is considered practical. Thus interacting phenomenologies must be constructed. Turbulent reacting flow models fall into this category. As a warning, note that if one phenomenology is a suspect representation, the interactions between two are at least doubly suspect.

Consider, for example, building a simulation of the interacting shock wave structures at the front of a propagating gas phase detonation. The physics of this process was discussed briefly in Section 2–3.2. We currently have detailed chemical kinetic mechanisms for some gases, such as hydrogen-oxygen mixtures and methane-oxygen mixtures, which are good enough to qualify as practical, calibrated reaction-kinetics models. Using the detailed chemical kinetics mechanism and a compressible fluid dynamics model, a three-dimensional calculation for hydrogen-oxygen would qualify as a COULD DO category, probably taking up the maximum amount of computer time because so much information would have to be transferred in and out of the computer memory in the course of the calculation. A two-dimensional hydrogen-oxygen simulation falls in the CAN DO category. An analogous two-dimensional calculation for a methane-oxygen system is a COULD DO calculation because the kinetics reaction mechanism for methane oxidation is more complicated and expensive to use. Using less detailed models of the chemistry allows better resolution of the fluid dynamics and moves the calculations into the ONE PHENOMENOLOGY range. In this example, the phenomenology is no more complicated for hydrogen than methane, and, once the phenomenology is developed, the two-dimensional calculations are in the CAN DO range.

One advantage of the chemical kinetics phenomenologies is that we know, in principle, how to incorporate these processes into calculations. Though not all of the input chemical rates or specific heats necessary for particular reactions or species are known, we have some confidence in the equations and how they fit into the simulation model. This helps in developing calibrated phenomenologies and in evaluating analytic models with unknown input parameters. In turbulence, we are in a much weaker position with respect to what we know and how to use that knowledge. We assume that the effects of turbulence are already in the conservation equations, but various limitations and costs keep us from being able to resolve the effects in detailed simulations. This problem has been dealt with by proposing phenomenologies and fitting the constants in them to experiments and other calculations. Turbulence, dusts, and sprays are areas which are more than just difficult to incorporate from first principles. New physical insights as well as numerical methods have to be devised before the complete spectrum of multiscale, multiphase reactive-flow phenomena can be simulated nearly as well as phenomena in laminar, gas-phase combustion.

3–3. CONSTRUCTING AND TESTING THE MODEL

Chapters 1 and 2 described many of the processes involved in developing and carrying out numerical simulations of complicated, time-dependent reactive flow systems. We indicated that implementing these physical processes computationally is an evolving procedure with a number of trade-offs to be made. These trade-offs involve all aspects of the simulation process and require early choices with significant consequences. The knowledge required to optimize the simulation requires foresight into the entire process, and this foresight is only gained through experience.

Almost every decision in constructing a numerical model involves evaluating trade-offs. For example, the computational representation should be chosen keeping in mind the various available algorithms. In turn, algorithms should be chosen keeping in mind what kind of programming techniques would be best for a particular kind of computer. Knowing that a good algorithm exists for treating certain processes or interactions makes choosing the corresponding representation straightforward. The availability of a computer which allows vectorization would be of little benefit if the representation or algorithms chosen involve many operations that cannot be vectorized. In this an algorithm which vectorizes easily might save a factor of five in execution time, even though it requires twice as many statements and takes twice as long to program as a competing scalar algorithm.

3–3.1 Trade-Offs in Selecting and Optimizing Models

To understand the trade-offs in constructing a numerical simulation model, we consider some of the desirable qualities that we want a reactive flow simulation model to have. The representations, algorithms, and testing procedures should be chosen to optimize this set of quantities.

An ideal numerical model is:

1. Computationally fast and efficient,
2. Frugal with computer memory and input and output requirements,
3. Accurate, both numerically and physically,
4. Robust and reliable over a known range of conditions,
5. Flexible and easy to modify,
6. Quick to produce useful results,
7. Well documented.

Unfortunately, these desirable qualities usually conflict with each other. If an algorithm is very fast, it is seldom the most accurate. If an algorithm is both fast and accurate, its range of applicability is often inadequate for the problem at hand. If an algorithm is valid over a wide range, it may require large amounts of computer memory or be inflexible with respect to changes in boundary conditions. The most accurate algorithms for modeling important physical effects often border on numerical instability, so that computer models using them are not robust and reliable. These constraints and costs are all expressions of a cosmic conservation theorem which says "there is no free lunch, " or "you can't get something for nothing."

Making the right choices to optimize the model and its performance involves assigning relative importance to the qualities listed above. These weights depend on the charactersitics of the particular problem and on the abilities and knowledge of those performing the computations. As a specific example of how these properties conflict, consider trade-offs involving computational speed versus computer memory. In fluid simulations, it is possible to recompute the fluid pressure from the equation of state every time it is used during a timestep, or to compute the pressure once each timestep and store it. The best thing to do depends on the circumstances. The answer may differ for a one-dimensional or a three-dimensional model, because of the different relative weights on speed and memory for the two cases. Storage is often at a premium in a three-dimensional calculation. Storage limitations are usually of little concern in a one-dimensional problem. On the other hand, if the simulations do not fit into the memory, the calculated numbers are stored on backup devices and brought into the main computer memory by sections. Computing time or information transfer time is then the limiting factor. In this case, it may again pay to store the information rather than recompute it for a three-dimensional model.

Consider some more examples of algorithm trade-offs. Often good one-dimensional fluid dynamic algorithms do not generalize easily to two or three dimensions. If the focal problem is two-dimensional, a slower, more general one-dimensional algorithm might be used in a preliminary one-dimensional model, with the eventual two-dimensional model in mind. The choice of algorithm also depends on whether the flows are subsonic or supersonic. The best algorithms for convective transport in flames allow considerably longer timesteps than those required for detonations, even though each timestep requires more computer time. Thus, it might be best to develop the model initially with the shorter-timestep, more expensive method appropriate for shocked flow if we ultimately want to model the transition of a flame to a detonation. Finally, if we are interested in solving a set of equations for a large chemical reaction mechanism, the algorithm we choose depends on whether the calculation will be coupled to fluid dynamics. As discussed in Chapter 5, the stand-alone case often requires more expensive, more accurate methods. Trade-offs in choosing of algorithms are a major part of the discussions in Chapters 5–11.

3–3.2 Steps in Constructing a Simulation Model

Before programming a simulation model, major decisions have to be made to

– Select a level of modeling commensurate with the problem, computational resources, and the expertise available.

– Choose a suitably idealized standard case that stresses the most important physical effects and allows subsequent generalizations.

– Develop a mathematical model which includes the standard case and reasonable deviations.

Then, the program development can begin. The following steps form a check list for constructing a detailed reactive flow simulation model.

Step 1. Write a control program for the simulation which transfers control to, or *calls*, subprograms to calculate each physical process independently. To begin with, the calls are calls to *dummy* subprograms, which simply return to the main control program. Each dummy subprogram will eventually solve the equations for one of the physical processes modeled. The control program with the calls to dummy subprograms becomes the outline of the numerical model. This outline is filled in as each dummy routine is replaced by a subroutine which actually performs a calculation. The logic of program control, storing data, restarting a calculation, and diagnostics of results can be developed at this step.

Step 2. Create a working list of all physical and chemical variables, parameters, control variables, and indices. Determine the computer memory they will require as a function of the dimensions of variable arrays. Physical variables include quantities such as mass momentum, energy, and species densities, temperatures, and pressures, as a function of position. Control variables include logicals that, for example, turn on processes in the model and control input and output. This list can be used to estimate the memory requirements of the program. The estimate, however, increases as dummy subprograms and subroutines are filled in.

Step 3. Choose a data representation for storing all the the problem variables. This process builds the *common blocks* for the model and lays out the data structure for the numerical implementation. A schematic diagram of the computational grid should be developed. This diagram is one of the primary references for developing and later explaining the model.

Step 4. Choose some reasonable initial conditions for a test problem, either for the standard case or for a preliminary benchmark. These data are ideal for testing diagnostics if the values to be printed, plotted, or analyzed are known exactly. It usually pays to have the dummy modules do something artificial to test and exercise the initial suite of diagnostics.

Step 5. Write the initializing subroutine that will be called from the control program. This subroutine fills all the variable and data arrays with the initial conditions for the first standard case. The design criteria here should be simplicity and flexibility. Because the initialization is executed only once per run, its efficiency is not an important issue.

Step 6. Run the program repeatedly with the dummies, the initializer, and the diagnostics until the logical ordering of the processes within the timestep is correct and the diagnostics print the correct numbers in the correct places. During this phase it is essential to use known data and to vary these data. Nothing is more disconcerting and misleading than incorrect diagnostic routines whose shortcomings become interpreted as bugs in the simulation algorithms. This is a good stage to develop data output facilities as well as the capability to restart the calculation from output data.

Step 7. Begin implementing the physics and chemistry algorithms, each one alone with all of the others turned off. Do not allow two processes to interact before each has been tested alone. This is the stage when many of the algorithm trade-offs discussed above become apparent. Each process simulated should be checked in as many limits as possible to bound the behavior. Using the simulation to reproduce analytic solutions is the best way to check the numerical model, because these solutions give detailed profiles for comparison.

Step 8. After the individual modules have been tested, test the interactions among them by doing calculations where they are turned on in pairs. When there are many possible interactions, a lot of tests are necessary. Again, limiting cases should be tested and comparisons with theoretical models should be conducted. During Steps 7 and 8, it is helpful to modify the model in simple ways, for example, by temporarily changing a diffusion coefficient or gas constant so that comparisons to theoretical test problems are exactly valid.

Step 9. Turn everything on ... and hope.

3–3.3 An Example: Simulating a Propagating Laminar Flame

We now describe the optimizations and main trade-offs involved in developing a usable simulation model for a typical laminar premixed flame. Resolution limitations and problem complexity translate into cost, in this case impractically high computer-time requirements. This idealized analysis spotlights the fact that brute-force numerical approaches often require too much computer time for even this idealized problem. Clever numerical methods are essential to make the simulation possible, not to mention practical.

A 1 m long closed tube contains a homogeneous, combustible gas mixture. When ignited at one end, a flame propagates along this tube at an average velocity of 100 cm/s until the reactants are consumed. This entire process takes about one second. If the flame is laminar and losses to the walls and other boundary-layer effects are negligible, a one-dimensional description of variation along the tube may be adequate. A burning gas may be described by a chemical reaction rate scheme which involves some tens of species and perhaps a hundred chemical rates. The simulation determines how properties such as temperature, species densities, and position of the flame front change as a function of time after the mixture is ignited.

We want to know these quantities to about one percent accuracy. Thus 100 cells of width 1 cm and 100 timesteps of length 10^{-3} s should nominally be enough to resolve the problem. Unfortunately, this expectation is rather naive. The physical and chemical processes cover time scales ranging over nine orders of magnitude and space scales ranging over five orders of magnitude, as shown in Table 3–1. Figure 3–2 recasts the information in Table 3–1 graphically in a space scale versus time scale diagram for the simulation. Since the space and time scales are logarithmic, the exponential of the area in this diagram is proportional to the cost of the calculation.

A sound wave crosses the system in about 10^{-3} s, about three orders of magnitude faster than the flame or the fluid velocities. Thus timesteps of 10^{-6} s are required to resolve the sound waves with the 1 cm cells. Assuming that the flame zone is about 10^{-2} cm wide and that a grid spacing of 10^{-3} cm is needed to resolve the steep density and temperature gradients, a timestep of about 10^{-9} s is required to resolve the sound waves. Timesteps to resolve the chemistry, typically 10^{-6} s or so, would not pose a problem until the limitation on the timestep due to the sound speed is removed.

To estimate the time needed for this calculation, we need to multiply three numbers together:

1. The number of grid points in the calculation,
2. The number of timesteps,
3. The average time needed to integrate each of the grid points for each timestep on the particular computer used.

The third number is typically about 10^{-3} s of supercomputer time per grid point per timestep. This estimate includes solving the chemical and fluid dynamic equations with appropriate diffusive transport and is based on a hydrogen-oxygen flame problem optimized on a computer capable of 50 million floating point operations per second (megaflops). In Figure 3–2, the area spanned by the simulation is the number of point-steps in the calculation.

The brute-force calculation with 10^{-9} s timesteps and 10^{5} cells (which are 10^{-3} cm wide), requires 10^{14} point-steps to complete the simulation. It would take about 3000 years of computer time to simulate the one-dimensional flame and resolve all the space and time scales in Table 3–1! This is unacceptable. Ideally such a simple calculation should take about 100 seconds of computer time, 10^{3} timesteps for 100 grid points. More than a faster computer is required to make this problem computationally feasible. We need algorithms with fine resolution in time and space only where it is

Table 3–1. Important Scales in the Flame Problem

Time Scales	(s)	Space Scales	(cm)
(a) Sound Speed (Courant limit)	10^{-9}	(a) Flame Resolution	10^{-3}
(b) Chemical Reactions	10^{-6}	(b) Flame Zone	10^{-2}
(c) Sound Transit Time	10^{-3}	(c) Diffusion Scale	10^{-1}
(d) Flame Transit Time	1	(d) Convective Scales	10
		(e) System Size	100
Average Flame Velocity 100 cm/s			

Figure 3–2. The time required for the direct computation of a flame propagating down a closed one-meter tube may be calculated by taking the exponential of the area in the larger square box on this figure. When sound waves are resolved and a unform grid is used, the calculation could take 3000 years. We would like the calculation to take 100 seconds. The letters (a) – (e) across the axes are keyed to the scales listed in Table 3–1.

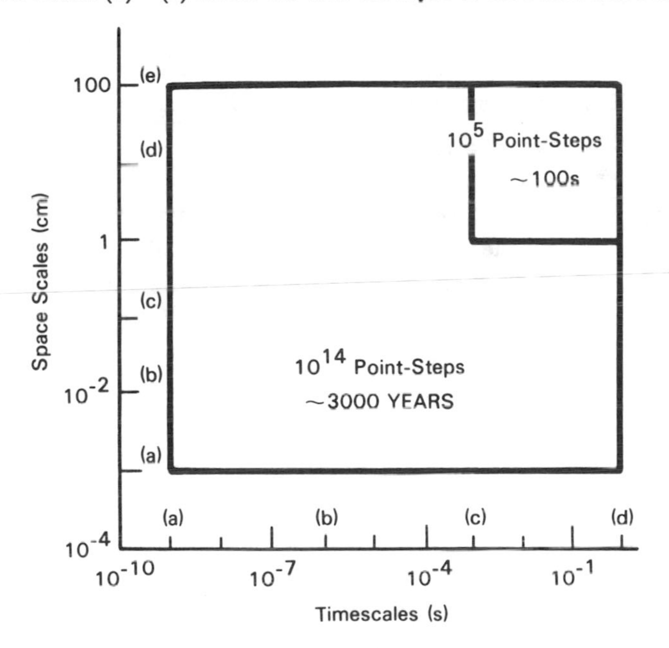

required. Further, the representation of the problem has to be optimized to take advantage of what is known about the physics and chemistry.

The first step in making this problem manageable is to recognize that the flame front moves much slower than the speed of sound. This means that the pressure in the tube is fairly constant in space. Because the tube is closed, the pressure increases monotonically in time as the reactants are consumed. The chemistry at the flame front determines the speed at which the flame propagates. However, the timestep of 10^{-9} s is determined by the sound speed. Because the pressure must increase in a closed tube, we cannot eliminate the integrated macroscopic effects of the sound waves which transmit pressure information throughout the system. The characteristic velocities of interest, the flame velocity and the fluid velocity that result from the flame, are subsonic. If a suitable algorithm can be found, we would not have to resolve the system on such short time scales. To do this, we solve a pressure equation instead of an energy equation by an *implicit* method which overcomes the sound speed restriction on the timestep. Details of implicit methods are discussed in Chapters 4, 9, and 10.

Using an implicit pressure algorithm with uniform grid spacing, the timestep for convection and flame propagation can be increased about three orders of magnitude because the velocities that must be resolved are three orders of magnitude below the sound speed. The number of point-steps in the integration is decreased accordingly from 10^{14} to 10^{11}, and the computational time decreases from 3000 to 3 years.

Three years is still a ridiculously long time. The next trick for reducing the cost further is to eliminate unnecessary grid points by using algorithms that adapt the grid to the solution. *Adaptive gridding* is an area of computational physics that has developed rapidly in the last few years. It is based on putting finely resolved cells only in regions of the grid where they are needed. Adaptive gridding has paid particular dividends in one-dimensional modeling, and work is progressing on the multidimensional adaptive gridding problem which is clearly much more difficult. This subject is treated in Chapter 6.

Orders of magnitude reduction in the computer memory requirements and the running time of this test problem can be obtained by putting the small cells only in and around the flame front. Suppose 100 cells of 1 cm length are used to represent the spatial variations of the problem away from the flame, and the region surrounding the flame front is finely gridded with 100 additional cells of 10^{-3} cm length. The timestep is now governed by the flame speed in the smallest cells, but a total of only 200 cells are needed rather than 10^5. This savings, about a factor of 500, reduces the time to

2×10^8 point-steps, or about two days. The trade-off for the reduced cost is additional program complexity associated with treating variably spaced cells, lost accuracy from the reduced accuracy of the finite-difference algorithms using variably spaced cells, and the need to develop algorithms which move the fine grid region to encompass the flame at every instant.

Finally, we consider the computational savings obtained by using an adaptive gridding technique called *intermittent subgrid embedding*. The idea here is to use the same adaptive gridding trick with the timesteps that we just used with space. Intermittently, a detailed numerical model to determine the flame speed and the structure of the flame front is embedded in the macroscopic calculation which otherwise needs only 1 cm resolution. The background flow parameters (such as the pressure and temperature of the unburned gases) determine the boundary conditions for the 10^{-1} cm flame front region. The response of the detailed flame front simulation to these changing boundary conditions determines the energy released during each of the larger, 10^{-1} s timesteps. If particular attention is given to determining the slowly varying, macroscopically relevant proportion of the thin flame zone every now and then, the low resolution calculation determines the evolution of the flame location to the one percent accuracy required.

Assume that 100 cells are enough to resolve changes in the flame zone during the short period of time when the detailed flame front simulation is embedded in the calculation. During this embedded calculation, the flame front need move only a fraction of one of the coarse 1 cm cells, so that it takes 100 timesteps to cross 10 of the fine zones. This 10 second calculation is sufficient to determine the flame speed and burned-gas conditions used in the more coarsely spaced calculation. The embedded calculation, we assume, should be done only once as the flame passes through each of the large cells. Thus a total of $[100+(100\times 10)]$ s $= 1100$ s of computational time is required for the overall simulation.

This example shows the importance of finding the right set of algorithms for a specific type of problem. The penalty for misjudgment in this phase of model building may be severe. This example also emphasizes the value of adaptive gridding.

Most of the cost of detailed reactive flow simulations arises from integrating the stiff ordinary differential equations that describe the chemical kinetics. In detonation and flame simulations with detailed chemical mechanisms, we generally find that the fluid dynamics and diffusive transport consume 10 to 20% of the running time. Chemical kinetics for a reasonably realistic mechanism consumes the rest. A large fraction of this time is spent evaluating the temperature dependence of the reaction rates.

3–3.4 Testing Programs and Models

In summary, we recommend choosing a focal problem to help select the computational representation, then choosing a specific standard case to implement the model, and then making variations of this standard case to test the model. It is important to go through a systematic testing process. It provides a way to evaluate parts of the model before the entire simulation is put together. It can also give answers for direct comparison with either experimental data or analytic theories. When comparing calculations to experimental data, such tests are good ways to exercise modules in the simulation model. When comparing calculations to analytic theories, it is often possible to predict the results of a theory and then computationally carry the answer into a regime where the theory is no longer valid. This process teaches how the computer model works and what the costs are to get answers.

Once the model is constructed, it is necessary to make sure that the physical processes are represented correctly and that the model is as bug-free as possible. At this point, knowledge and intuition of the behavior of the important physical processes become important. The material below is not an organized methodology, but a collection of hints and suggestions for testing the model.

Avoid Programming in Dimensionless Numbers

This sounds contrary to everything a student is taught in theory classes. Picking a set of standard units and keeping the program variables in these units, however, can save a large amount of time in writing, debugging, and interpreting complex reactive flow simulations with many interacting physical processes. It is inordinately difficult to debug a program when all of the variables are of order one. Many "factor of two" errors do not occur, or are much easier to find, if the units in the program relate directly to a real system. It is also questionable how much can be gained from scaling when there are many interacting processes on many time scales. Finally, working in real units that are relevant to the focal problem helps develop intuition about the system and makes it easier to communicate with experimenters.

Intermediate Benchmarks

The primary objective is an accurate, informative simulation of the standard case. To reach this, a number of smaller benchmark calculations, each addressing a physical or numerical question, should be carried out, progressively leading up to the standard case. Sometimes it is useful to choose intermediate benchmarks with characteristic dimensions or ratios of system parameters that are different from the standard case. For example, when characteristic time and space scales vary widely, a benchmark computation on a closely related problem with a smaller range of scales requires fewer timesteps to run and check. This reduces calculation costs during the development and testing phases and more quickly gets the project to the point where highly resolved or complicated problems can be simulated reliably.

It is also possible to develop and test the simulation model on problems with lower dimensionality than the standard case. This requires less computer time and memory and the answers can be scrutinized more thoroughly than equivalent two- or three-dimensional problems. However, it is important to ensure that the methods being tested will generalize easily and efficiently to more than one dimension. The test problems should be meaningful both from the point of view of solving the specific problem and testing the computational implementation.

When a benchmark is successfully completed, a copy of the whole computer program that produced it along with the results should be filed safely away. This provides a description of the model as it evolves. Hints for documentation are discussed in Section 3–4 below.

Keep the Computer Program Clean

There is a tendency not to remove parts of computer programs which are no longer used, but to turn them off internally in the model by "commenting" them out or branching around them. Keeping these parts in the program slows down the the whole computational process from editing through printing. The superfluous statements can also cause problems if they are accidently turned on or parts are not turned off. It is simpler and safer to discard the rejected or incorrect statements, and to reprogram or reinsert them again when they are needed.

Change the Model Slowly and Systematically

It is extremely important not to make too many changes at once when the model is being upgraded or modified. It is better to perform several independent tests which can be compared with each other and with the standard case. Even when new ideas or components seem to be logically independent, they often are not. Thus separate tests not only reduce the overall development time, they also increase confidence in the model. Putting three or four tests into a single run, unless they are serial and truly independent, invites confusion and possibly trouble.

Document the Model Internally

Internal model documentation is an important aspect of both changing and developing models. Section 3–4.2 is dedicated to this subject. Everyone has trouble remembering exactly what was done six months or a year ago. Although it may seem like a waste of time to go back and add comments to a program, you will be grateful for having done this. Even if you write the computer program and do not expect others to use it, you will profit from putting in prolific internal documentation.

Interactive versus Batch Mode

Whether to use interactive or batch mode for computing is a part of the question of how to use your personal and computer time best. Much of the preliminary model construction and testing described above is best done interactively. As the project proceeds, however, there comes a point when the tests take more computer time to complete. Interactive debugging can be unnecessarily time consuming and does not leave you with an unambiguous record to document progress. Colleagues may also become irate over the amount of high priority interactive computing required as the calculations get bigger and longer.

From here on, the interactive session is best for editing and submitting different batch tests and for checking the results when they are complete, not for doing the calculations. Debugging calculations for an eventual one-hour simulation requires five or ten minutes to simulate enough to test the changes being considered. In five minutes, four different batch tests can be devised and submitted. Then the next thirty minutes can be used for thinking, writing, or programming while these batch runs are executing.

Maintain a Healthy Skepticism

Never believe the results of your calculations. Always look for errors, limitations, unnatural constraints, inaccurate inputs, and places where algorithms break down. Nothing works until it is tested, and it probably does not really work even then. This skepticism should help avoid embarrassing situations, such as occur when a neat physical explanation is painstakingly developed for a computational blunder.

3-4. USING TIME EFFICIENTLY

Developing and testing computer programs, looking at the results, thinking about the output, and planning the next stage of the project, all take time. It makes sense to do all of this as quickly and efficiently as possible. We now discuss four aspects of using time efficiently:

1. Doing a number of tasks in parallel,
2. Documenting the entire process,
3. Using what others do,
4. Starting over without wasting more time.

3-4.1 Partitioning Time among Tasks

Formulating the simulation in a completely serial way, without having the next and the previous steps in mind, can be costly. This is also true when testing and applying the evolving numerical model. Time is best spent divided among several aspects of the project, some requiring deep thought and others which are mundane. The trick is to find a good use for what would otherwise be dead time. To accomplish this, it helps to do several things in parallel so that progress is always being made on some aspect of the problem.

For example, while programs are running or graphical output is being processed, it might be useful to plan the next stages of the project. Perhaps simple analytic estimates are needed to determine exact parameters for the next case. Perhaps a book or journal article needs to be read. Algorithms can be modified, developed, or tested that add the next piece of physics or chemistry to the model while the production calculations are executing. Even when the computer is slow, understanding the results takes time and can occur in the background while more mundane tasks are done.

3–4.2 Documenting and Retrieving Results of Simulations

Recording steps in the research and writing accounts of the work as it proceeds is not universally adopted in practice, though most of us agree that research should be done this way. Under time pressure, it is easy to forget to document what is being done. Curiosity to see the result of the next program improvement is also a good excuse to delay documentation efforts. We tend to rationalize not documenting using the notion that progress will actually be slowed down by stopping to record or analyze. We tell ourselves that we will remember the important aspects of what was done. We sometimes do. We also tell ourselves that we can write with more perspective when we are farther along. This is also sometimes true.

These comments apply to preparing documentation on any level, either in the program itself or writing up the results of a calculation. Improper or sparse documentation can cost a lot of time for yourself and for others. Conscientiously documenting the analysis, algorithms, programs, and results can save hours or days in the future at the cost of minutes or hours while you still understand what is going on. Documenting focuses your thoughts and makes it easier to understand what has been done. It is especially important when trying to repeat previous results and to solve follow-on problems which build on current work.

A notebook, kept faithfully through the model construction and testing stage, keeps track of what worked, where problems were, and what was fixed. Various tests of individual modules and sets of modules should be documented here. Output files from these tests should be included, labeled with dates, and titled so that they are easily identified. If one algorithm is replaced by another, the changes should be noted and explained.

An important part of documentation during the model construction and testing phases is documentation in the simulation program itself. This should be done liberally while you still remember what the program is doing. When program comments are omitted, it becomes difficult to find otherwise trivial errors. This internal documentation also brings an added bonus. When the model is completed, gathering the internal documentation provides a rough first draft of the material needed to document the computer program itself.

The input data and output data from a simulation should be documented. Each output *dump* file, which contains all of the variables needed to restart the computation at an intermediate timestep, should be labeled internally so that reading the file tells what the data are. This labeling should be generated automatically when the model is executed on the computer. Thus an up to date record is produced which labels all of the output files.

When to keep something and when to throw it away is a difficult decision. With the gradual replacement of paper output and graphical hardcopy by files on an electronic storage system, it is possible to keep more worthless output and to forget what is stored more easily. Without a good filing system and a good notebook, it is difficult to find things quickly and reliably. Careful choice of mneumonic names for files and ways to inspect files quickly are extremely useful.

Finally, the project notebook should include the standard case and simulations following it. Output must be labeled with dates and descriptions coupling it to the corresponding input data. A good way to document results is to generate figures or graphs as the problem proceeds. Each figure should be chosen to summarize a certain result, aspect, or concept in the project. For each of these summary figures, an extended figure caption should be written, typically a few hundred words. In addition to having presentations and progress reports ready instantly, the figures and extended captions can be assembled into technical reports. This procedure is similar to using collected internal documentation from the computer program to produce a document describing the computer model.

One of the very first figures produced in a simulation project should be a schematic figure explaining the computational grid. It should show the cell interface locations, problem geometry, indices, and computational region boundaries. It can also show geometric locations where averaged physical variables are defined and indicate what kind of boundary conditions are used. This figure sets the stage for results presented in tabular or graphical form and will be used repeatedly to explain the model and the results to others.

3–4.3 Working with Others

The basic simulation model is usually developed and programmed early in the course of a project. Its physical and chemical basis is sporadically changed or upgraded every few months or years. In numerical simulations, just as in laboratory experiments, a significant and sometimes dominant fraction of the technical effort is spent on developing new diagnostics, using them, and analyzing the results they produce. The basic simulation model becomes surrounded by a suite of special programs more extensive than the basic model itself. These special utilities use the primary simulation variables to create, for example, graphics for interpreting the output and integrals over or derivatives of various quantities that serve as diagnostics.

Joining forces with other users of the same or similar computers can reduce the effort of producing and debugging these utility programs. Software obtained this way is seldom exactly what is required, but it can often be changed easily at little cost. Sometimes it is only necessary to write an interface program to hide inconsistencies between the numerical model and the diagnostic software.

If you are developing software for yourself, but especially if you are developing software that someone else might use, it is important to make the program that implements the simulation model:

Clear and well-documented — A clear program can be understood and modified accurately and quickly. It is important in a continuing research environment for the original programmer to be able to quickly understand a program that he has not seen in a year or two.

Simple — A simple program is much less prone to error and is more easily modified than a complicated program. Simplicity and clarity also help make the program flexible. Flexibility is not a necessary quality because a program need only do one kind of calculation well. However, flexibility is highly desirable.

Fast — Making the program fast has economic benefits. It also allows flexibility because some of the computational efficiency can be traded off for greater accuracy. Making the program fast also has the benefit of allowing longer calculations with larger grids and better resolution.

Accurate — The compromise between speed and accuracy has always posed one of the most troublesome and provocative challenges for the scientific programmer. Almost all advances in numerical analysis have come about trying to reach these twin goals. Changes in the basic algorithms often give greater improvements in accuracy and speed than using special numerical tricks or changing programming languages.

Robust — A robust program works adequately over a broad spectrum of input data and has few special cases that it cannot handle. A robust program can be trusted to give correct answers reliably.

Much has been written about structured programming and programming style. Simple and clear discussions are found in the books by Ledgard (1975) and by Kernighan and Plauger (1974). In addition, we recommend the books by Higgins (1979), Hughes et al. (1978), and Van Tassel (1978). We next present some general programming guidelines taken from these works, but mostly from our own experience (see, for example, Boris and Winsor, 1982). Scientific programming is an art, done in a difficult medium and requiring a special talent. Guidelines are therefore more appropriate than rigidly enforced standards such as might be developed for documentation.

3–4.4 Some General Programming Guidelines

Program clearly. Programming tricks should be avoided to maintain clarity. If tricks are used, they should be well documented internally. The computer program which implements the numerical model should be easy to read.

> Claritymeansinpartreadability.Readabilitycancomeinmanyform
sandshapessothebestsingleguidelinewecancomeupwithistoforcethepro
gramtoresemblewhatmostofusseemorethananyothertypeofvisualcom
munication,Englishprose.Inthiswayitiseasiesttoassimilateprogr
aminformation.Thereareundoubtedlyavastnumberofwaystoforcethe
programstatementsintoamuchclearerandmorereadableform.Unfortu
natelymanyprogrammersdothisaboutasmuchasthisrunonbutratherfor
tranlikeparagraphdoes.Legibilityisreallyonlyanextensionofthe
goldenrule.

Put blanks in the program statements. Put blanks after commas, between variable names, after subroutine names, around + and − signs, and between words, such as "GOTO." Blank lines make the program neater and more easily read. These are courtesies to the next programmer (most often yourself).

Proper commenting helps to make the program clearer. The comments and the program statements should agree, and the comments should add overview information or a new interpretation that is not obvious from reading the program statements. The comments should also remind the reader of the overall goal of that particular part of the program, as well as any hidden conditions which it assumes. For easy reading, indent the program statements from the comments, so that the program looks like an outline with the comments as topic headings.

The layout of the program can be almost as useful as the comments in helping to explain what is being done. It is possible to format and structure a program to help the reader find:

a. The structure of program execution — including loops, IF tests, and common blocks.
b. The data structure — statements that give dimensions to variables, variable declarations, data statements that initialize variables, and assignment statements that change their values.
c. The other parts of compound structures — the ends of loops, the labels that are targets of GO TOs, and the format statements that go with READ and WRITE instructions.

There are many different concepts of good program layout. Whatever style is chosen, it should make these structures easy to find and scan. Programs are clearer and bugs are easier to find if the program follows a top-to-bottom logical path. This means that branches and tests should not transfer control to earlier statements unless absolutely necessary. When a back transfer is useful, a comment explaining it is usually in order. Some authors suggest eliminating the GO TO statement altogether, an extreme but tenable position. No amount of patching can fix a badly structured program. Do not be afraid to rewrite sections of the program completely. The time lost in doing this is usually saved many times over by avoiding later bugs and confusion.

Label the data input, output, and defaults, and use variable names and grammatical constructions that mean something in English as well as in the chosen programming language. Choose variable names that are difficult to confuse with each other and choose a data represenatation that makes the program simple. Do not avoid using long descriptive names. Using subroutines and library functions cuts down the scope of what must be comprehended simultaneously.

Break program statements in logical places. Try to use a convenient operation at an outer level to begin each continuation line. Use short constructions, if they suffice, but do not abbreviate comments or variable identifiers. Do not sacrifice readability to save a few lines or spaces. Just as in English, do not mix several ideas into a single line or statement of the program. It is best to keep the program in clear English and simple algebra.

There are some places in programs where it pays to be wordy. It never hurts to use parentheses to avoid ambiguity or to avoid mixed-mode arithmetic that can cause misunderstandings between the programmer and the computer. If a logical expression in the program is hard to understand, it is worth the time to transform it, even if the more legible format is longer.

There are also some places where it pays to be terse. You do not have to spell everything out to a computer. It will often do a better job than you can. For example, it often pays to avoid temporary variables and unnecessary branches. Avoid IF statements with multiple destinations. A logical expression will often substitute for a conditional branch, will usually compile as efficiently, and can often shorten the program by eliminating the need for an explanatory comment.

The most tedious and hence poorly written aspects of programs are the input and output data structures. Test input for plausibility and validity as it is read into the program so that bad input can be identified and either fixed or the program execution stopped. In particular, if the algorithm has a range of validity, the input should be checked to make sure this range is not exceeded. Print out explanations when the data are suspect. Uniform input formats make data easy to read. The program should echo all the input data in a neat, well-documented format that can be detached and used to document the simulation.

Debugging is easier when good programming techniques are used. For example, all variables should be initialized before they are used. Constants can be initialized by statements in the program, while variables should be initialized by input or executable statements. When free form input is used to overwrite default values (for example, the NAMELIST in FORTRAN), both the current value and the default values should be printed in the output and identified there. Take care to branch the correct way and be careful when a loop exits to the same statement from both the side and the bottom. Do not compare floating point numbers solely for equality, since 10.0×0.1 is often not exactly 1.0 in a computer.

Go through the program by hand, using reasonable input numbers. This can be an extremely useful way to find logical and typographical errors. It sometimes works when nothing else does. The trick here is to try to do what the computer has actually been programmed to do, which is not necessarily what you want it to do. We call this "playing computer." A particular advantage of playing computer is that it forces an analysis of what is actually written in the program and exactly what the algorithm does.

After the numerical model is properly understood by both the computer and the programmer and is running correctly, it might be useful to consider ways to speed it up. It may be possible to vectorize parts of the program if the computer has this feature. It may also be possible to find ways to arrange blocks of program so they execute more efficiently.

Developing good code structure and internal documentation habits is as important when developing software for others to use as it is when you are programming for yourself. There are, however, additional considerations when someone else is going to use the program. One example is naming conventions for variables. What may be obvious for one person can be inscrutable for someone else. Therefore, glossaries of variables, control parameters, subroutine arguments, and logical segments of the program are necessary for a new user and can be extremely helpful for the original author of the program.

3–4.5 Consider Starting Over

In the process of developing, testing, and using the numerical simulation model, changes occur in our perceptions of the problem and thus in what we wish the model to do. The simulations are being performed to extend our understanding of a particular kind of problem. As our understanding evolves, the specification of the job which must be done naturally changes. It is important to be flexible and not to be afraid to backtrack and rethink what you are doing. It is better to change the algorithms or structures as early in the game as possible to save much larger efforts later.

There is another level to the question of starting over. It occurs much later in the life of the numerical model. After being used for a while, the large simulation model becomes a hodgepodge, partly written by the person who originally wanted to solve a problem, partly by those who took over the model and made changes for their applications, and partly by others unrelated to the project who wrote software that is being adapted and interfaced. The model grows as new pieces are added. Eventually it grows to the point where it is no longer clear, simple, or fast, and there are faster, more accurate, and more robust algorithms to use for modeling the various processes.

There eventually comes a decision point. Do you try to replace old algorithms with new ones, or do you start completely over and rewrite the program? The answer depends on just how unwieldy the program has become. If it is fairly new, perhaps a few years old, replacing old algorithms and cleaning house may be the best route. But, *never be afraid to start over.* This is a good opportunity to get rid of unused parts of the program, upgrade the algorithms and data structures, and fix the model to use the configuration and special capabilities of the newest available computer properly. The general guideline here is to adapt and clean up until some major change is required, and then to rewrite.

References

Boris, J.P., and N.K. Winsor, 1982, Vectorized Computation of Reactive Flow, in G. Rodrigue (Ed.), *Parallel Computation*, pp. 173–215, Academic Press, New York.

Higgins, D.A., 1979, *Program Design and Construction*, Prentice-Hall, Englewood Cliffs, NJ.

Hughes, C.E., C.P. Pfleeger, and L.L. Rose, 1978, *Advanced Programming Techniques*, Wiley, New York.

Kernighan, B.W., and P.J. Plauger, 1974, *Elements of Programming Style*, McGraw-Hill, New York.

Ledgard, H.F., 1975, *Programming Proverbs for FORTRAN Programmers*, Hayden, Rochelle Park, NJ.

Van Tassel, D., 1978, *Program Style, Design, Efficiency and Debugging and Testing*, 2nd ed., Prentice-Hall, Englewood Cliffs, NJ.

Chapter 4
SOME GENERAL NUMERICAL CONSIDERATIONS

This chapter presents and analyzes some simple finite-difference methods for simulating the four major processes in reactive flows: chemical reactions, diffusion, convection, and wave motion. The material presented here is an overview and short course on solving idealized forms of the equations representing these processes and introduces the more advanced discussions in Chapters 5–11.

Table 4–1 shows the mathematical representation of these four processes and indicates where their numerical solution is discussed in this book. In this chapter we analyze these four linear, constant-coefficient equations separately, expanding and clarifying the analysis originally presented by Oran and Boris (1981). For each process, we choose several simple methods and derive overall numerical accuracy and stability criteria. This brings out many of the numerical difficulties which also appear in the coupled problems where several of the processes interact, and also highlights the strengths and weaknesses of some of the classical solution methods. Throughout the presentation, we evaluate the major computational and algorithmic trade-offs that arise in simulating each process numerically.

4–1. INTRODUCTION TO FINITE DIFFERENCES

4–1.1 Discretizing Time and Space

Conventional computers have finite-sized memories segmented into floating-point words of data. This structure forces us to represent the continuous fluid variables in the computational domain by a finite number of discrete real values. Typically the spatial domain is broken up into *computational cells*, also called *fluid elements* or *zones*. We generally use the term *cells*, although we occasionally use the three terms interchangeably. The cells are the volumes defined by a lattice of points, often called the *computational mesh*, the *grid*, or the *cell interfaces*. Usually the discrete variables in these cells are evaluated at a specific time. Because the computer representation uses discrete numbers rather than continuous variables, the resolution in time

Table 4–1. Terms in the Reactive Flow Equations

Equation	Processes	Reference
$\frac{\partial \rho}{\partial t} = \gamma \rho + S$	Local Processes: Source, Sink, Coupling; Chemical Reactions	Section 4–1; Chapter 5
$\frac{\partial \rho}{\partial t} = \nu \frac{\partial^2 \rho}{\partial x^2}$	Diffusive Processes: Molecular Diffusion; Thermal Conduction; Thermal Diffusion	Section 4–3; Chapter 7
$\frac{\partial \rho}{\partial t} = \frac{\partial (\rho v)}{\partial x}$	Convective Processes: Advection; Compression; Rotation	Section 4–4; Chapters 8–10
$\frac{\partial^2 \rho}{\partial t^2} = v_w^2 \frac{\partial^2 \rho}{\partial x^2}$	Waves and Oscillations: Sound Waves; Gravity Waves; Other Oscillations	Section 4–5; Chapter 10

is also quantized into discrete intervals. These intervals, called *timesteps*, provide convenient increments over which to advance the numerical solution.

We want to compute solutions for a continuous dependent variable $\rho(x,t)$, where x is a spatial variable and t represents time. The function $\rho(x,t)$ will be approximated by updating the discrete cell values at the times defined by the sequence of discrete timesteps. The choice of variable name ρ is made purposely to remind the reader that, in general, our objective is to simulate numerically the properties of a spatially varying density function, such as mass density, momentum density, energy density, or chemical species concentration. The variables ρ and x can be scalar or vector.

Throughout this book, the subscript j indicates the jth computational cell. A sequence of adjacent computational cells is denoted by the indices $j-1$, j, $j+1$, ..., and the interfaces separating these cells are labeled $j-\frac{1}{2}, j+\frac{1}{2}, \ldots$. The quantity ρ_j^n is the cell-centered quantity ρ at location x_j at time t^n. This quantity ρ_j is the cell value of the continuous variable ρ, typically the average of $\rho(x,t)$ over the computational cell. The coordinate x_j is the location of cell j. The values $\{x_{j+\frac{1}{2}}\}$ label the cell-interface locations between the corresponding cell locations x_j and x_{j+1}. The cell width is given by

$$\Delta x_j = x_{j+\frac{1}{2}} - x_{j-\frac{1}{2}}. \qquad (4-1.1)$$

The index j is used instead of i to avoid confusion between the cell index and the imaginary number $i = \sqrt{-1}$. In this chapter, the imaginary unit i appears in Fourier expansions used to analyze the finite-difference algorithms. We also use k (units of cm^{-1}) to represent the wavenumber of a Fourier mode with spatial dependence $e^{i\mathbf{k}\cdot\mathbf{x}}$. In multidimensional applications, the cells are often labeled i, j, k to represent three spatial directions. Because we rarely use Fourier analysis in multidimensional contexts, this should not cause confusion.

We use the superscript n, for example ρ^n, to indicate a particular variable evaluated at the time t^n. Thus a sequence of times are denoted by superscripts $\ldots\ n-1,\ n,\ n+1\ \ldots$. Superscripts $n-\frac{1}{2}$ or $n+\frac{1}{2}$ indicate times halfway through the timestep. When we write down algorithms that advance an equation by a timestep, we sometimes use superscript o to denote the value of the variable at the "old" timestep, $n-1$, and then n denotes the "new" timestep. The timestep is usually Δt. In the more general algorithms described in later chapters, Δt can vary with time, so that

$$\Delta t^n \equiv t^n - t^{n-1} = t^n - t^o . \qquad (4-1.2)$$

Often, as in Chapter 5, the symbol h is adopted to denote the integration interval. This notation is common in analyzing ordinary differential equations where h can represent an incremental step in any independent variable.

There are a number of ways to interpret the discretized variables in terms of the continuous physical quantities being approximated. Consider x_j, the location of cell j. Depending on how we choose to interpret x_j, it could be

- the location of the cell center,
- the location of the the cell center-of-mass, or
- a particular point in the cell where ρ_j happens to be known.

Consider ρ_j^n, the cell value of ρ at x_j. This can be

- the value of $\rho(x,t)$ at a particular position x_j and at a time t^n,
- a characteristic value of $\rho(x,t)$ near the specific location x_j at time t^n,
- an average of the continuous solution in the volume enclosed by the computational cell interfaces at a specific time t^n,
- the average value in time of ρ between $t^{n-\frac{1}{2}}$ and $t^{n+\frac{1}{2}}$ in the cell x_j,
- the average in space and time, or
- the coefficients of a set of basis functions in a linear expansion of the dependent variable $\rho(x,t)$.

Each of these interpretations has different properties, each approximates some situations better than others, and each corresponds to different solution algorithms.

In each case, there is only a finite number of discrete values in the representation, and each value is only specified to finite precision. This means that information is inevitably lost in the computational solution relative to the continuous problem being solved. The result is uncertainty in the computation arising from the discretization. The source of the uncertainty is the missing information about the detailed solution structure within the discrete spatial cells and timesteps. All approaches which use a finite number of values to represent a continuous profile have this problem.

Consider a continuous independent variable z defined in the bounded region

$$z_1 \leq z \leq z_2 , \tag{4-1.3}$$

where z represents either a space or a time variable. Since ρ is defined everywhere on the continuous region $[z_1, z_2]$, the set of values $\{\rho_j\}$ is an incomplete description of $\rho(z)$. Figure 4–1 shows two possible profiles of $\rho(z)$, each of which has the same average in the computational cells. From this figure, we see that including more values in the sequence $\{\rho_j\}$ gives a better representation of the function. When fewer points are used, more information about ρ is lost. By choosing a finite, discrete representation of ρ, we only describe the long-wavelength properties of ρ. If z represents time, we see that choosing a particular timestep is the same as choosing a filter which wipes out high-frequency noise. If z is a space variable, we effectively filter out the short wavelengths by choosing a particular spatial mesh.

4–1.2 Advancing Equations in Time and Space

The elementary definition of the derivative of a continuous variable ρ with respect to z is

$$\begin{aligned} \frac{d\rho(z)}{dz} &= \lim_{\Delta z \to 0} \frac{\rho(z+\Delta z) - \rho(z)}{\Delta z} \\ &= \lim_{\Delta z \to 0} \frac{\rho(z) - \rho(z-\Delta z)}{\Delta z} . \end{aligned} \tag{4-1.4}$$

The calculus of ordinary and partial differential equations treats the limit as the interval Δz goes to zero. In finite-difference expressions for the derivative, the limit is not taken because there is no continuum of values on which to shrink the increment to zero. The step size Δz must remain finite. The derivative is approximated on the computational mesh using the discrete values $\{\rho_j\}$ and $\{\Delta z_j\}$. For example, one finite-difference approximation to the derivative, the first-order forward difference, is

$$\lim_{\Delta z \to 0} \frac{\rho(z+\Delta z) - \rho(z)}{\Delta z} \approx \frac{\rho_{j+1} - \rho_j}{z_{j+1} - z_j} . \tag{4-1.5}$$

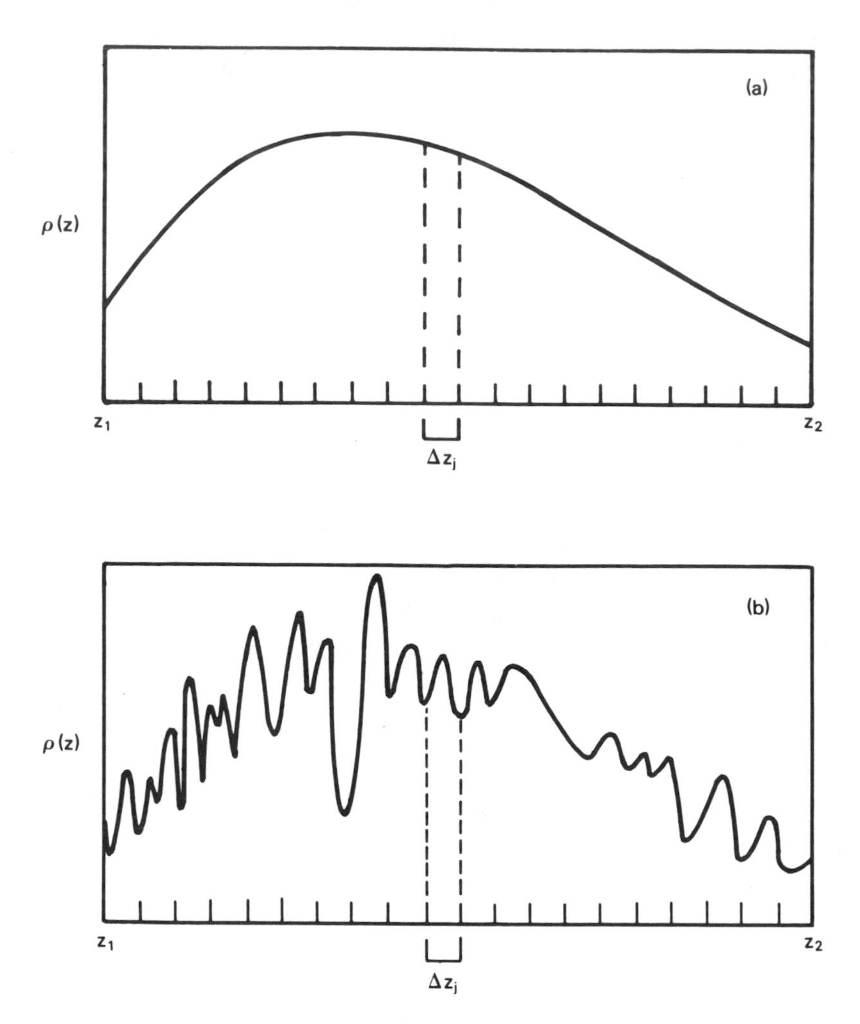

Figure 4–1. Two functions $\rho(z)$. In (a), ρ varies smoothly and does not change much in the discretization interval Δz_j. In (b), ρ has substantial structure in this interval and the discrete values carry little information.

It is usually assumed that the numerical solution approaches the exact answer as the values of $\{\Delta z_j\}$ become smaller.

Table 4–2 lists several of the simpler finite-difference forms for derivatives that can be used for the terms in Table 4–1. When the solutions are smooth, it is generally true that a higher-order solution means a more accurate solution. Note that increasing the order of accuracy means increasing

Table 4–2. Simple Finite-Difference Forms for Derivatives

Derivative	Form	Order of Accuracy
$\frac{\partial \rho}{\partial z}$	$\frac{\rho_{j+1} - \rho_j}{\Delta z}$	First (forward difference)
	$\frac{\rho_j - \rho_{j-1}}{\Delta z}$	First (backward difference)
	$\frac{\rho_{j+1} - \rho_{j-1}}{2\,\Delta z}$	Second (centered difference)
$\frac{\partial^2 \rho}{\partial z^2}$	$\frac{\rho_{j+1} - 2\rho_j + \rho_{j-1}}{\Delta z^2}$	Second (centered difference)

the range of the variables included in the finite-difference formulas. Thus higher order usually means more computational work. For example, the algorithms for the first-order derivatives involve ρ_j and either ρ_{j+1} or ρ_{j-1}. For the algorithm to be of second order, ρ_{j+1} and ρ_{j-1} are included, and so on. When the solutions are not smooth and have appreciable structure, using high-order difference methods can actually be less accurate than lower-order methods because approximation to higher-order derivatives requires information from further away on the grid.

In this chapter and generally throughout this book, we emphasize *finite-difference methods*, constructed using formulas similar to those given in Table 4–2. Other generic types of approaches, such as finite-element and spectral methods, are discussed in later chapters.

4–1.3 Qualitative Properties of Numerical Algorithms

Each of the four processes in Table 4–1 has important qualitative physical properties stemming from the form of the equations being solved. These properties are listed in Table 4–3. Numerical solutions of these equations should reflect these qualitative properties and much is gained by making the numerical algorithms do so. The first four qualitative physical properties (conservation, causality, positivity, and reversibility) are properties of the physical system and of the continuum mathematical representation. The fifth property, accuracy, is a requirement on the quantitative solution of the computational model equations, not on the physical system.

First consider *conservation*. Conservation laws are expressed as integrals over continuous fluid quantities. These conservation integrals remain constant, regardless of the details of the fluctuations and flow phenomena

Table 4–3. Desirable Properties of Numerical Algorithms

Property	Comment
Conservation	Applies to mass, momentum, energy, species number densities, and other conserved quantities
Causality	Due to finite propagation speed, or finite time for information transmission
Reversibility	Important for waves and convection
Positivity	Important for all but wave processes
Accuracy	Algorithm must give quantitatively correct answers, must converge, and must be stable

occurring. Usually a conservation law equates the rate of change of the integral over a finite volume with the integral of a flux quantity normal to the surface of that volume. The terms in Table 4–1 are all expressions of such *conservation laws*. The conservation laws can be written in this *flux conservation* form because they reflect local properties of the system — they are derived as integrals of the partial differential equations.

For example, let ρ represent a mass density. Local conservation of mass means that the density ρ changes due to the flow of material into and out of the surfaces surrounding a given volume of space. Because any flux leaving one cell through an interface enters an adjacent cell, this formulation conserves ρ globally as well as locally.

Local conservation laws can be used to formulate finite-difference algorithms which conserve sums of the simulated physical quantities. These sums are finite-difference analogues of the conservation integrals. Demanding that an algorithm conserve a specified quantity both locally and globally introduces a constraint that forces the algorithm to represent the physical laws correctly. When conservation laws are not enforced, truncation and round-off errors can grow without bound, making even very simple systems behave in a bizarre manner. Many algorithms discussed in this and later chapters are conservative. If we give up strict conservation in an algorithm, we demand something else as a trade-off. For example, in Section 4–2 we discuss asymptotic algorithms which do not conserve strictly, but which produce accurate answers for very long timesteps. When a conservation law of the physical system is not built into the algorithm, how well the algorithm actually conserves the specified quantity becomes a measure of its accuracy.

Physical constraints limit the rate at which material and energy can be transmitted in a physical system. Furthermore, material or energy going from one location to another must pass through the intermediate locations. This statement of *causality* has important implications for the choice of the solution method. For example, implicit algorithms, discussed throughout this chapter, tie the solution at one location in space to that at far distant locations. Thus they transmit numerical information across distances of many computational cells each timestep. Problems sometimes arise because this numerical transmission speed is too fast. For example, the material in front of a gas dynamic shock physically should not have any forewarning that the shock is coming. Implicit algorithms may generate unphysical precursors of the shock in the undisturbed, unshocked fluid.

Positivity means that quantities such as the mass density, which are initially positive everywhere, cannot become negative. This property holds for any quantity described by the convection or diffusion equations. Quantities described by the wave equation can change sign. Much is gained by simultaneously enforcing the constraints of positivity and conservation in the solution of convection processes, the third term in Table 4–1. The results are the monotone algorithms discussed extensively in Chapter 8.

Reversibility means that the equation is invariant under the transformation $t \rightarrow -t$. It is important to preserve reversibility for waves and convection, but not really relevant for diffusion or chemical kinetics. Computationally, reversibility means that if at time t^n the calculation is stopped and the timestep is made negative, the solution retraces its evolution back to time t^o. Reversibility is not easy to achieve in practice even when an algorithm is designed to be reversible because of nonlinear terms in the equations, a moving grid, or other complicating effects.

Accuracy involves computer precision, numerical convergence, and algorithm stability. An algorithm converges when successively reducing Δx and Δt produces successively more accurate answers. A method need not always be convergent to be useful. Asymptotic methods, used near equilibrium or for the solution of stiff equations, are an example of this. Asymptotic numerical approximations do not necessarily get better as the stepsize is made smaller. These methods are discussed in Section 4–2 below and in Chapter 5. An algorithm is stable if a small error at any stage in the calculation produces a smaller cumulative error. All useful numerical algorithms must be stable, at least in the regime where they are used.

4–1.4 Approaches to Analyzing Accuracy

To evaluate a numerical algorithm requires accuracy criteria. This means quantifying how much information is lost when a continuous function is represented by a discrete set of values. A practical approach to determining accuracy is to choose an idealized, soluble linear problem, usually one that has constant coefficients. We then compare the exact solution to the solution implied by the particular finite-difference algorithm being used.

Comparisons of the exact solution and the finite-difference solutions for ordinary differential equations are usually done by comparing coefficients of the successively higher-order terms in power series expansions of the two solutions. For example, assume that ρ is an exponential:

$$\rho(z) \;=\; e^{\gamma z} \;=\; 1 + (\gamma z) + \frac{(\gamma z)^2}{2} + \frac{(\gamma z)^3}{6} + \cdots \,, \qquad (4-1.6)$$

and the expansion for $\rho(z)$ obtained from a finite-difference algorithm can be written

$$\rho(j\Delta z) \;=\; 1 + (\gamma j\,\Delta z) + \frac{(\gamma j \Delta z)^2}{2} + \frac{(\gamma j \Delta z)^3}{4} + \cdots \,. \qquad (4-1.7)$$

The finite-difference solution is said to be *accurate to second order* in Δz and has a third-order error $(\gamma j \Delta z)^3/12$.

Because finite-difference algorithms for the nonlocal processes in Table 4–1 involve spatial as well as temporal derivatives, they are often evaluated by comparing the exact solution for a single Fourier harmonic with the same Fourier harmonic using a particular finite-difference form. When the problem is linear, each harmonic of the system can be treated independently and analyzed as a separate problem. Then the same analytical techniques can be used for the partial differential problem as for ordinary differential equations.

The infinite discrete Fourier expansion of the continuous function $\rho(z)$ with periodic boundary conditions is given by

$$\rho(z) \;=\; \sum_{k=-\infty}^{\infty} \rho(k) e^{i2\pi kz/(z_2 - z_1)} \,, \qquad (4-1.8)$$

where k is here a dimensionless integer. The discrete Fourier transform coefficients $\rho(k)$ used in Eq. (4–1.8) are defined by the Fourier integral

$$\rho(k) \;\equiv\; \frac{1}{z_2 - z_1} \int_{z_1}^{z_2} \rho(z) e^{-i2\pi kz/(z_2 - z_1)} dz \,. \qquad (4-1.9)$$

By expanding the discrete grid density values in a finite Fourier series for $\{\rho_j\}$, we obtain

$$\rho_j = \sum_{k=1}^{J} \rho(k) e^{-i2\pi kj/J}, \qquad (4-1.10)$$

where the discrete Fourier transform coefficient $\rho(k)$ are now defined by a sum rather than an integral,

$$\rho(k) \equiv \frac{1}{J} \sum_{j=1}^{J} \rho_j e^{-i2\pi kj/J} \ . \qquad (4-1.11)$$

As seen from Eq. (4–1.10), only a finite number of discrete wavelengths can be represented on the finite mesh. The representation $\{\rho_j\}$ is only a long-wavelength approximation to the continuous function $\rho(z)$, as there is a cutoff wavelength below which no structure in the solution can be resolved. Analyses of this type are carried out in this and subsequent chapters to indicate the severity of the finite-difference approximations on different wavelength phenomena.

When we use Fourier series to analyze a particular algorithm, it is useful to define an *amplification factor* for each wavenumber k in the solution,

$$A(k) \equiv \frac{\rho^n(k)}{\rho^{n-1}(k)} \ . \qquad (4-1.12)$$

The amplification factor is the ratio of the kth Fourier harmonic at the new timestep to its value at the previous timestep. We show in Section 4–3.4 on diffusion processes that $A(k)$ measures how much the amplitude of mode k decays during a timestep. To represent smoothing of a gradient, $|A|$ should be less than one. In general, $A(k)$ is a complex number which can be written as

$$A(k) = A_R(k) + iA_I(k) = |A(k)| e^{i\varphi(k)}, \qquad (4-1.13)$$

where $\varphi(k)$ is the *phase shift* associated with a single timestep of mode k.

Individual Fourier harmonics make convenient test functions because the discrete Fourier series expansions can be manipulated analytically almost as easily as the continuous Fourier integrals. Thus it is relatively straightforward to see the correspondence between the continuous variable and the discretized approximation. This correspondence is convenient, but it can also be misleading. Because of the global extent of the sinusoidal basis functions in a Fourier expansion, problems can arise in trying to represent physical causality and positivity. For example, fluid upstream from a shock in a gas is physically unable to respond until the shock actually arrives because the shock travels faster than any characteristic speed in the fluid. In a Fourier treatment of shocks, numerical precursors arrive ahead of the shock.

Table 4–4. Types of Errors on a Grid

Type of Error	Properties Affected	Sources of Error
Local	Conservation Reversibility Positivity Accuracy	Precision (roundoff) Truncation (time derivative) Hardware errors
Amplitude	Conservation Reversibility Positivity Causality Accuracy	Truncation (space derivative) Numerical instability Numerical diffusion Precision
Phase	Causality Positivity Accuracy	Truncation (space derivative) Numerical dispersion Precision
Gibbs	Causality Positivity Accuracy	Truncation of grid Grid uncertainty Precision

4–1.5 Types of Finite-Difference Errors

The worst errors of all are programming errors. Many hours of numerical analysis and reams of paper have been expended in a sometimes futile search for "bugs." To avoid the confusion of blaming a flawed solution on the algorithm being used rather than on a programming error, it is valuable to be able to recognize the types of computational errors which occur in algorithms and what can be done to minimize them. The four types of errors which occur as a result of using finite-difference algorithms are listed in Table 4–4 along with the source of each error and the desirable properties of the solution which are affected.

Local errors occur in all algorithms for any of the terms in Table 4–1. They originate from the finite precision of the computer and from hardware errors. They also arise as local discretization errors of the time derivative. Some effects of roundoff are discussed in Chapter 5. Discussion of these types of errors come up frequently throughout this book.

When spatial derivatives are present, errors appear in the phase and amplitude of the amplification factor, $A(k)$, for each Fourier mode. Amplitude errors in convection and wave equations occur when $|A| \neq 1$ and introduce too much smoothing, in which case they result in *numerical diffusion*, or too much steepening, which often causes *numerical instability*. When $|A| < 1$, the result is numerical diffusion or numerical damping. When $|A| > 1$, the result is numerical instability. Amplitude errors in the diffusion, convection, and wave equations affect all of the desirable properties in Table 4–4 and are illustrated in Figure 4–2a. When $t_1 < t_2 < t_3$ in the figure, an initially localized quantity spreads out and its peak value and gradients decrease in time. The numerical process in which the amplitudes of the short-wavelength modes decay faster than the larger wavelengths is often called numerical diffusion, and can be thought of as nonphysical damping. In the reverse case, suggested by taking the curves in the sequence $t_1 > t_2 > t_3$, the system is numerically unstable.

Dispersion is a physical phenomenon in which waves of different wavelengths travel at different speeds in a medium. For example, a prism breaks white light into a rainbow through dispersion. Many types of waves are dispersive, but convection is not. The dispersion arising from discrete approximations to the spatial derivatives is a numerical error, called *phase error* or *numerical dispersion*. These errors occur in convection and wave equations, but not in the local or diffusion equations. Phase errors can adversely affect causality, positivity, and accuracy. The spurious ripples they introduce are illustrated in Figure 4–2b.

A localized quantity of mass in a flow or a localized acoustic pulse should travel coherently for a long distance. Though the profile is composed of many modes, the phases of all of these modes, in the absence of numerical dispersion, are coupled together to keep the peaks and gradients in the profile well-defined and sharp. Numerical dispersion, like numerical diffusion, usually affects the short wavelengths most severely. In the shortest wavelength that can be represented (about two cells), phase changes appear as amplitude errors and thus these wavelengths cannot even propagate. Thus sharp gradients develop a train of ripples, as shown in Figure 4–2b for a steep density profile moving to the right.

Phase errors do not always appear as numerical dispersion. If all modes have the same phase error, for example, the wave or the localized profile simply moves at the wrong speed. No dispersion is involved. Unlike numerical diffusion, however, numerical dispersion is not self-limiting. In numerical diffusion, the errors are limited in size because short-wavelength modes disappear. Dispersion errors, however, can increase linearly with time as the phase maxima of different modes grow continually further apart. Thus dispersion errors are the most insidious type of errors. Because the amplitudes of the individual modes stay bounded, there is no clearly identifiable point beyond which the solution becomes unsatisfactory.

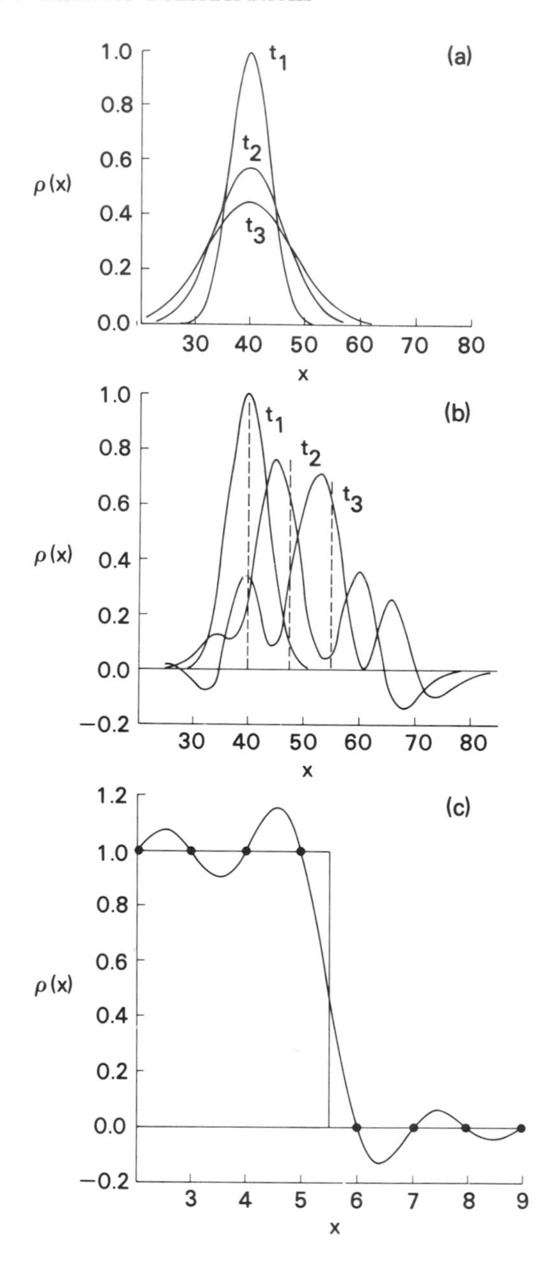

Figure 4–2. Schematic showing three types of numerical errors: (a) Amplitude errors. When $t_1 > t_2 > t_3$, the function ρ should not change in time from its shape at t_1. Numerical diffusion causes ρ to spread. When $t_3 > t_2 > t_1$, ρ should not change from its shape at t_3, but numerical instability amplifies it. (b) Phase (dispersion) errors, which cause the solution to deviate from the profile indicated by the dashed line. (c) Gibbs errors (numerical grid uncertainty) caused by using a finite representation of a continuous function.

The fourth type of finite-difference error, the *Gibbs phenomenon* or *Gibbs error*, looks very much like a dispersion error (compare Figures 4–2b and 4–2c). Gibbs errors occur in wave and convection equations because the higher harmonics of the continuous solution are not included in the numerical representation. Gibbs errors are limited in amplitude and are an inherent problem associated with representing a continuous function on a discrete grid. They cannot be eliminated even by ideal phase and amplitude properties in an algorithm. In calculations, they appear as undershoots and overshoots which are worse near large gradients and discontinuities. They appear when a steep-sided profile moves a fraction of a cell from its initial location. Sometimes it appears as if they are not present, and sometimes they seem to disappear temporarily in a calculation. One of the worst aspects of Gibbs errors is the bound they place on the possible accuracy of discrete methods for convection. This bound arises from intrinsic uncertainty in the representation rather than from the properties of the solution algorithm.

Figure 4–2c shows the Gibbs error for a calculation of the translation of a square wave whose height is $\rho = 1$ and whose width is 20 cells. The square wave is convected by a method which has no amplitude or phase errors, so only the Gibbs error remains. An example of such a method is the spectral method, discussed in Chapter 8. The dots are the known density values at the grid points. Using the lowest 100 Fourier harmonics, the solid curve is synthesized as the smoothest continuous curve passing through all the known values. When the profile moves a half cell, the Gibbs errors appear as wiggles between the grid points. The resulting unphysical fluctuations are typically of order 10–15% of the height of the discontinuity.

4–2. LOCAL PROCESSES AND CHEMICAL KINETICS

4–2.1 Explicit and Implicit Solutions

The first process in Table 4–1 is represented by equations of the form

$$\frac{\partial \rho}{\partial t} = \gamma \rho + S \, , \qquad (4-2.1)$$

where ρ is one of the fluid variables such as the mass density or the number density of a chemical species. This idealized ordinary differential equation represents spatially localized processes such as local mass, momentum, or energy source and sink terms, which may be a function of position as well as time,

$$S(\mathbf{x}, t) \, . \qquad (4-2.2)$$

Local processes can also have the linear form shown in the first term on the right hand side of Eq. (4–2.1),

$$\gamma(\mathbf{x},t)\rho(\mathbf{x},t) \ . \qquad (4-2.3)$$

Here γ is an exponential growth or damping coefficient when it is positive or negative, respectively.

Local processes include damping, equilibration in chemical reactions, dissipative effects, or coupling between equations. Generalizations of Eq. (4–2.1) also represent coupling between different evolution equations. Examples of coupled local processes are the drag terms in multiphase momentum equations. If ρ is a two-component vector consisting of an electron and an ion temperature, the generalization of Eq. (4–2.1) expresses thermal equilibration at the rate γ (here negative). If ρ is a vector of chemical reactants, Eq. (4–2.1) generalizes to the set of chemical kinetic rate equations and γ is a second-order tensor.

Now we consider integrating the simplest case, the single ordinary differential equation, Eq. (4–2.1), with γ and S constant. This equation has an analytic solution which we compare directly to numerical solutions derived from an explicit algorithm, an implicit algorithm, and an asymptotic algorithm. Each type of algorithm defined below can be solved in closed form. We show that the most accurate and cost-effective technique within this framework is a hybrid algorithm consisting of an explicit (or centered) method and an asymptotic method. The explicit method is used when the stable timestep restriction is not prohibitive (normal equations). The asymptotic method is used when the timestep required for stability of the explicit algorithm is too small for practical applications (stiff equations).

When γ and S are constant, Eq. (4–2.1) has the solution

$$\rho(t) \ = \ \left(\rho(0) + \frac{S}{\gamma}\right) e^{\gamma t} - \frac{S}{\gamma} \ , \qquad (4-2.4)$$

whether γ is positive or negative. When γ is positive, the solution eventually grows exponentially from its initial value $\rho(0)$, though it may initially decrease and change sign. When γ is negative, the solution relaxes exponentially toward the asymptotic value $-S/\gamma$.

Equation (4–2.1) is really an ordinary differential equation, though we have been writing it with a partial derivative notation to remind us of the larger multidimensional reactive flow context. We can approximate Eq. (4–2.1) by a simple finite-difference formula

$$\frac{\rho^n - \rho^{n-1}}{\Delta t} \ = \ S + \gamma[\theta\rho^n + (1-\theta)\rho^{n-1}] \ , \qquad (4-2.5)$$

where

$$\rho^n \equiv \rho(n\,\Delta t)\,, \tag{4-2.6}$$

and Δt is the fixed finite-difference timestep. The numerical values of ρ are produced only at discrete times $t^n = n\,\Delta t$. The implicitness parameter θ ranges from 0 to 1 and determines whether the right hand side of Eq. (4–2.5) is evaluated using $\rho(n\,\Delta t)$ at the new time ($\theta = 1$, fully implicit solution), at the old time using $\rho((n-1)\,\Delta t)$ ($\theta = 0$, fully explicit solution), or somewhere in between (semi-implicit solution).

The finite-difference equation, Eq. (4–2.5), can be solved formally for ρ^n in terms of the initial density $\rho(0)$. This solution,

$$\rho^n = \left(\rho(0) + \frac{S}{\gamma}\right) E(\gamma\,\Delta t) - \frac{S}{\gamma}\,, \tag{4-2.7}$$

is formally similar to Eq. (4–2.4) where $E(\gamma\,\Delta t)$ is the particular finite-difference approximation to the exponential function $e^{\gamma \Delta t}$. From the particular approximation Eq. (4–2.5), we find

$$E(\gamma\,\Delta t) \equiv \frac{1 + (1-\theta)\gamma\,\Delta t}{1 - \theta\gamma\,\Delta t}\,. \tag{4-2.8}$$

The highest-order approximation of the exponential $e^{\gamma t}$ occurs when $\theta = \frac{1}{2}$, though this is not the most accurate approximation. For any value of γ, a value of θ can be chosen which makes the approximation exact. This particular value, θ^*, is called an integrating factor, where

$$\theta^* = \frac{e^{\gamma\,\Delta t} - \gamma\,\Delta t - 1}{\gamma\,\Delta t(e^{\gamma\,\Delta t} - 1)}\,. \tag{4-2.9}$$

When $\theta = \frac{1}{2}$, Eq. (4–2.8) can be expanded as

$$E(\gamma\,\Delta t) \cong 1 + \gamma\,\Delta t + \frac{(\gamma\,\Delta t)^2}{2} + \frac{(\gamma\,\Delta t)^3}{4} + \cdots\,. \tag{4-2.10}$$

Here E is an accurate representation of the exponential to second order in $\gamma\,\Delta t$, with an error term of $-(\gamma\,\Delta t)^3/12$. When $\gamma\,\Delta t$ approaches zero, θ^* approaches $\frac{1}{2}$.

Figure 4–3 shows these solutions for the relaxing case where $\gamma < 0$ and $S = 0$. The solutions should approach zero asymptotically starting with $\rho(0)$. These solutions represent a stable relaxation which is characteristic of most coupling and chemical kinetic processes. The three panels show the results of numerical integrations using timesteps $-\gamma/2$, $-\gamma$, and $-2/\gamma$.

The characteristic relaxation time is $|\gamma|^{-1}$. The four solutions shown for each choice of timestep are labeled by the letters E for the explicit solution ($\theta = 0$), C for the centered solution ($\theta = \frac{1}{2}$), I for the fully implicit solution ($\theta = 1$), and T for the theoretical (analytical) solution.

When $-\gamma\,\Delta t$ is less than unity, as in panel (a), the explicit solution never changes sign. All of the numerical solutions replicate the qualitative behavior of a decaying exponential and they are all numerically stable. The centered solution (C) is nearly indistinguishable from the exact theoretical solution. The explicit solution (E) is lower than the exact solution (T) by about the same amount as the implicit solution (I) is too large. Unfortunately, ensuring $|\gamma|\,\Delta t \leq 1$ at each timestep can be expensive computationally when the equations are *stiff*. Here being stiff means that

$$\frac{-\gamma\rho}{\partial\rho/\partial t} \gg 1\,. \qquad (4-2.11)$$

When $\gamma\,\Delta t = -1$ in panel (b), the centered solution still approximates the exact solution very well, although the implicit solution is noticeably less accurate than with a smaller timestep. The explicit solution displays a rather singular behavior: it goes exactly to zero in one timestep and then stays there. The asymptotic state, $\rho = 0$, has been achieved far too early. All solutions are stable and behave in a physically reasonable way.

In panel (c), the timestep is increased further so that $\gamma\,\Delta t = -2$. The explicit solution behaves strangely, oscillating at fixed amplitude about the asymptotic $\rho = 0$ state. This situation marks the borderline of classical mathematical stability for the explicit solution. As shown in panel (b), the explicit solution is incorrect even when Δt is a factor of two smaller, that is, $\Delta t = -1/\gamma$. For Δt larger than $-1/\gamma$, the explicitly computed value of ρ goes below zero and changes sign each timestep. This occurs because the approximation to $e^{\gamma t}$, $E(\gamma\,\Delta t)$, becomes negative.

Solutions of coupled systems of equations for real problems generally cannot tolerate the mathematically stable but physically bizarre behavior shown in Figure 4–3c. Chemical kinetics calculations, for example, are explosively unstable when species number densities have negative values. An important rule of thumb in computational physics is that the numerical timestep should always be chosen at least a factor of two smaller than required by the most stringent mathematical stability condition. In other words, practical stability, which gives the qualitatively correct physical behavior, is at least a factor of two more expensive in terms of computer requirements than mathematical stability. Not keeping the timestep well below the stability estimate is a frequent source of error in computations.

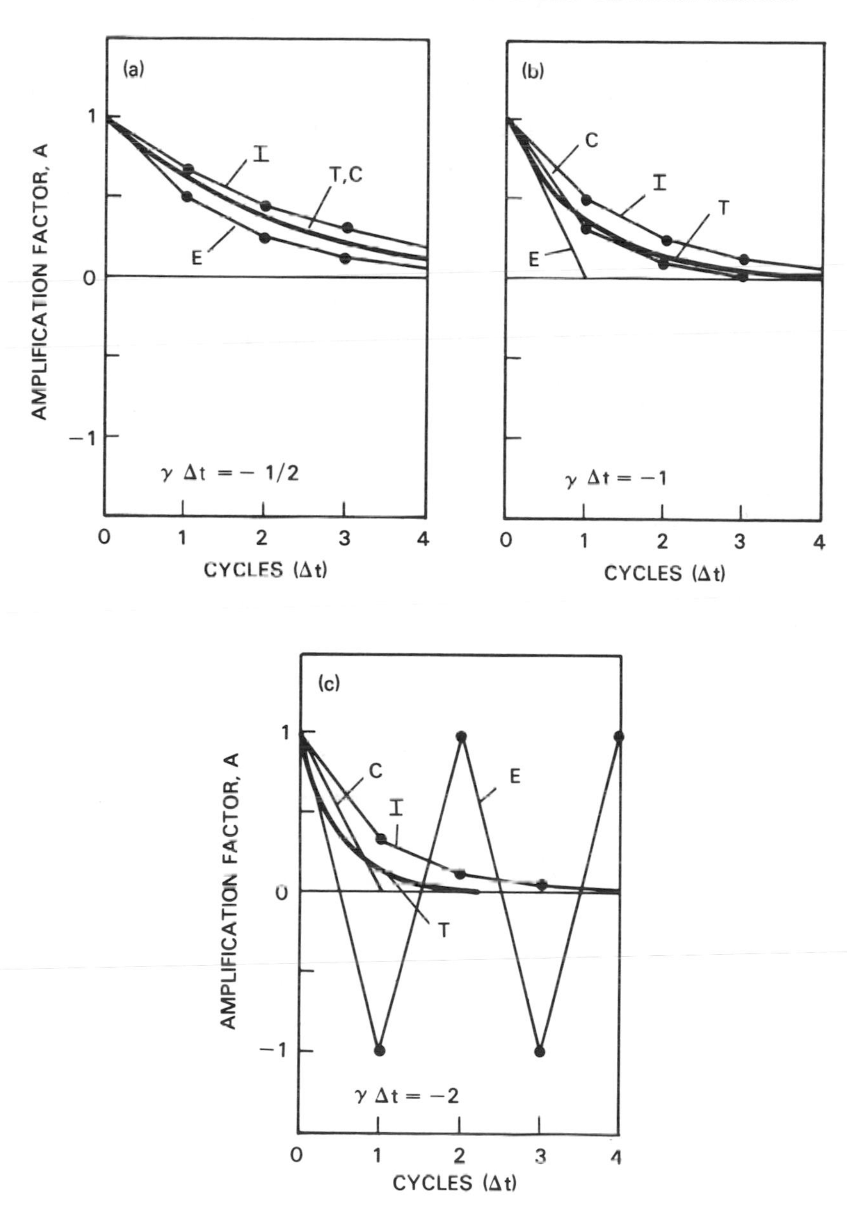

Figure 4–3. Solutions for a local scalar process for various values of the parameter $\gamma\,\Delta t$. The labels T, E, C, and I indicate the analytical (true), explicit, centered, and implicit solutions, respectively.

When $\gamma\,\Delta t = -2$, the centered solution still behaves acceptably but reaches the limiting case before instability occurs in the form of negative values of ρ. The implicit solution is behaving physically and stably, although it is not decaying quickly enough. This guaranteed stability is the lure of the implicit approach. Unfortunately the error in the implicit solution, relative to the rapidly decaying theoretical curve, has grown even larger than in the previous two panels. This potentially large error in the implicit relaxation rate of stiff equations is a hidden sting. The practical timestep stability condition for an explicit method is also the effective accuracy condition for the implicit method.

4–2.2 Asymptotic Solutions

There are many situations where the decaying nature of the exponential relaxation makes large relative errors progressively less important because the correct equilibrium is being approached. There are also physically unstable situations where these same errors are more important. The implicit algorithm also becomes considerably more expensive in complex systems of many coupled equations because the division sign in Eq. (4–2.8) then becomes a matrix inversion operation. Thus there are continuing efforts to develop less expensive methods to treat the case $|\gamma|\,\Delta t \gg 1$. In an asymptotic expansion, mathematical accuracy is obtained when $|\gamma|\,\Delta t$ is large. Thus it is distinguished clearly from the methods just described, which are based on a Taylor-series expansion and which converge best when $|\gamma|\,\Delta t$ is small.

Let both γ and S be time dependent. Variations in these quantities introduce nonlinear coupling and multiple time scales into the problem when several variables are involved. These are not major here, but they are stressed in the more detailed discussion in Chapter 5. We assume also that ρ relaxes rapidly towards its limiting value, but that the rate of approach keeps changing as $\gamma(t)$ and $S(t)$ evolve.

To derive a numerical asymptotic approximation to Eq. (4–2.1), the right-hand side is set to zero and the derivative $\partial\rho/\partial t$ is iteratively approximated using the previous approximation. For example, consider Eq. (4–2.1) at time n,

$$\gamma^n \rho^n = -S^n + \left.\frac{\partial \rho}{\partial t}\right|^n . \qquad (4-2.12)$$

The first-order asymptotic solution is

$$^{(1)}\rho^n \approx -\frac{S^n}{\gamma^n}\,, \qquad (4-2.13)$$

assuming $\partial\rho/\partial t$ is negligible compared to the other terms. Including the time derivative of ρ in Eq. (4–2.13) gives the next approximation. This derivative can be approximated using the lowest-order solution just obtained,

$$\left.\frac{\partial \rho}{\partial t}\right|^n \approx -\frac{1}{\Delta t}\left(\frac{S^n}{\gamma^n} - \frac{S^{n-1}}{\gamma^{n-1}}\right) \approx -\frac{1}{\Delta t}\left(\frac{S^n}{\gamma^n} - \rho^{n-1}\right) . \qquad (4-2.14)$$

Thus the next order of ρ can be written

$$^{(2)}\rho^n \approx -\frac{S^n}{\gamma^n} - \frac{\left(\frac{S^n}{\gamma^n} - \frac{S^{n-1}}{\gamma^{n-1}}\right)}{\gamma^n\,\Delta t} . \qquad (4-2.15)$$

Note in Eq. (4–2.12) that the derivative $\partial\rho/\partial t$ is strictly evaluated at time n but is at time $n - \frac{1}{2}$ in Eq. (4–2.14). This phase lag in the time at which derivatives can be evaluated seems to be characteristic of asymptotic numerical approximations. It often means that the lower-order approximations are more accurate than the higher-order asymptotic approximation.

An expansion such as Eq. (4–2.15) is adequate only when $|\gamma|\,\Delta t$ is large. For example, when $|\gamma|\,\Delta t = 2$,

$$^{(2)}\rho^n \approx -\frac{3}{2}\frac{S^n}{\gamma^n} + \frac{1}{2}\frac{S^{n-1}}{\gamma^{n-1}} . \qquad (4-2.16)$$

When $|\gamma|\,\Delta t = 1$,

$$^{(2)}\rho^n \approx -2\frac{S^n}{\gamma^n} + \frac{S^{n-1}}{\gamma^{n-1}} . \qquad (4-2.17)$$

When $|\gamma|\,\Delta t = \frac{1}{2}$,

$$^{(2)}\rho^n \approx -3\frac{S^n}{\gamma^n} + 2\frac{S^{n-1}}{\gamma^{n-1}} . \qquad (4-2.18)$$

As $|\gamma|\,\Delta t$ becomes small, the value of $^{(k)}\rho^n$, the kth approximation, is dominated by an improperly computed time derivative.

Major gains in solution speed and accuracy are achieved by using an asymptotic method combined with another type of method, switching between them at the timestep needed for optimal accuracy. In particular, consider an asymptotic method when the equations are stiff and extremely short timesteps would be otherwise required, and then switch to another method to get high accuracy when the equations are not stiff. In the simple problem illustrated in Figure 4–3, the asymptotic formula gives the final relaxed state $\rho = 0$. Since the explicit and centered formulas are unstable for

$\Delta t > 1/|\gamma|$ and $\Delta t > 2/|\gamma|$, respectively, the asymptotic formula should be used whenever Δt exceeds these values.

The maximum error made by using either the explicit or centered formulas for normal equations and the asymptotic formula for the stiff equations occurs at an intermediate value of Δt. With a combination of explicit and lowest-order asymptotic methods, the maximum error in Δt is e^{-1}. A centered-asymptotic combination has an error of e^{-2} at the transition value $\Delta t = 2/|\gamma|$.

Because the implicit formula is always stable, the transition point can be taken at any value of Δt. It is useful to determine where the two formulas, implicit and asymptotic, are most accurate. From Eq. (4–2.8), the error made in using the implicit formula for one timestep (taking $\theta = 1$) is

$$\begin{aligned} \text{Implicit error} &= \text{implicit solution} - \text{exact solution} \\ &= \frac{1}{1-\gamma\,\Delta t} - e^{\gamma \Delta t} . \end{aligned} \tag{4-2.19}$$

The corresponding error using the asymptotic formula is

$$\text{Asymptotic error} = -e^{\gamma\,\Delta t} , \tag{4-2.20}$$

because the lowest order asymptotic solution is zero. These errors are equal when $-\gamma\,\Delta t \approx 1.679$. For timesteps less than $\Delta t \approx -1.679/\gamma$, the implicit formula is more accurate. For longer timesteps, the asymptotic formula is more accurate. The implicit-asymptotic combination has a worst error of $e^{-1.679}$. Because the worst error of the centered-asymptotic combination is smaller than that of the implicit solution, there really is no reason to use a fully implicit technique.

The material presented above introduced the concepts of implicit, explicit, centered, and asymptotic methods. These concepts will be amplified and expanded substantially in Chapter 5, which surveys methods for solving coupled nonlinear ordinary differential equations. Asymptotic methods are good for stiff regions of the solution but not as good for following rapid variations. They are most useful when coupled to fluid dynamics equations. Asymptotic methods are also useful for treating sound waves when they are considered as stiff phenomena, a topic which is discussed in conjunction with the slow-flow algorithm in Chapter 9.

4–3. DIFFUSIVE TRANSPORT

4–3.1 The Diffusion Equation

Diffusion, the second process in Table 4–1, is a nonlocal process, one that has spatial as well as temporal variations. The general multidimensional diffusion equation

$$\frac{\partial}{\partial t}\boldsymbol{\rho}(\mathbf{x},t) = \nabla \cdot \mathbf{D}(\mathbf{x},t,\boldsymbol{\rho})\nabla\boldsymbol{\rho}(\mathbf{x},t) , \qquad (4-3.1)$$

where $\boldsymbol{\rho}$ and $\mathbf{x}$ are vectors, is simplified to a single, one-dimensional, constant-coefficient diffusion equation,

$$\frac{\partial}{\partial t}\rho(x,t) = D\frac{\partial^2}{\partial x^2}\rho(x,t) , \qquad (4-3.2)$$

where D is assumed constant in space and time. By analogy with Eq. (4–1.1), a spatially and temporally varying source term in the form of $S(x,t)$ could be added to the right hand side of Eqs. (4–3.1) and (4–3.2). This extra term is also set to zero here, so the steady-state solution of Eq. (4–3.2) has the constant value ρ^∞, determined by conservation throughout the computational domain.

The diffusive flux, defined as

$$F_x \equiv -D\frac{\partial \rho}{\partial x} , \qquad (4-3.3)$$

is proportional to the gradient of the fluid variable $\rho(x,t)$. The sign of the flux is such that large values of ρ tend to decrease as regions with lower values of ρ fill in.

There are two cases with very steep gradients for which Eq. (4–3.3) is not a good representation of the diffusion process. In the first case, the variations are appreciable over only a few mean free paths. These large variations occur when a discrete, random-walk process is represented as a continuous diffusion term.

The second case occurs at very low densities where the gradients may seem small, but are very large relative to the local density. If the flux of ρ, Eq. (4–3.3), is divided by the value of ρ at a location x, the characteristic diffusive transport velocity of ρ is

$$v_c = \frac{x}{2t} , \qquad (4-3.4)$$

a curious result considered further in Chapter 7. For any finite time, the diffusive transport velocity becomes arbitrarily large at distances far enough away from the initial source of ρ. This means that particles in the tails of the diffusion-equation profiles move too fast compared to the random walk of particles which they are trying to represent. These anomalies are considered again in Chapter 7 where techniques for correcting this, called *flux limiters*, are discussed.

4–3.2 An Analytic Solution: Decay of a Periodic Function

To evaluate numerical solution methods, we use an idealized problem for which there is an analytic solution to Eq. (4–3.2). As with the simplified ordinary differential equation methods discussed in Section 4–1, this problem allows us to compare the "exact" results with those from approximate finite-difference algorithms. This also permits phase and amplitude errors for various algorithms to be compared analytically.

Consider the decay in time of periodic, sinusoidal profiles $\rho(x,t)$, which initially have the amplitude $\rho(x,0)$. The spatial structure is fixed in shape because sines and cosines are eigenfunction of Eq. (4–3.2), but the amplitude changes in time. We use the complex exponential trial function

$$\rho(x,t) \;=\; \rho(t)(\cos kx + i \sin kx) \tag{4-3.5}$$

in Eq. (4–3.2). The part of ρ containing the time variation, designated here by $\rho(t)$, can be found assuming that the wavenumber k is constant.

The complete analytic solution

$$\rho(x,t) \;=\; \rho_0 e^{-Dk^2 t} e^{ikx} \;, \tag{4-3.6}$$

shows exponential decay in time of the sine and cosine spatial variations. Because the time-varying coefficient e^{-Dk^2t} is real, the phase of each harmonic is not affected by the diffusion operator. Sine and cosine components stay separate. Each Fourier harmonic of the profile decays in such a way that the total amount of ρ is conserved. The conservation integral is

$$\frac{d}{dt}\left\{ \int_0^{2\pi/k} \rho(x,t)dx \right\} \;=\; 0 \;, \tag{4-3.7}$$

where the function $\rho(x,t)$ is integrated over one wavelength. For Eq. (4–3.6), this conservation integral constraint also means that a solution, which initially has a real value of $\rho(0)$ everywhere, stays real. This zero phase shift property puts a symmetry requirement on the algorithm chosen to represent the spatial derivative. Because the time dependence has the form of a real exponential, the sign of the decaying wave cannot change and positivity must be preserved in a reasonable solution.

Thus any numerical algorithm for the diffusion equation should reproduce these fundamental qualitative properties:

1. The total integral of $\rho(x,t)$ should be conserved,
2. The amplitude, $|\rho(t)|$, should decay monotonically,
3. There should be no phase errors introduced by the algorithm (such as might arise from asymmetric forms of the derivative), and
4. Positivity should be preserved.

4–3.3 Explicit and Implicit Solutions

One way to write Eq. (4–3.2) in simple finite-difference form is

$$\begin{aligned}\frac{\rho_j^n - \rho_j^{n-1}}{\Delta t} &= \frac{D\theta}{(\Delta x)^2}\left[\rho_{j+1}^n - 2\rho_j^n + \rho_{j-1}^n\right] \\ &+ \frac{D(1-\theta)}{(\Delta x)^2}\left[\rho_{j+1}^{n-1} - 2\rho_j^{n-1} + \rho_{j-1}^{n-1}\right] .\end{aligned} \tag{4-3.8}$$

This formula is similar to Eq. (4–2.5) used to approximate local processes and the time derivative is differenced identically. The space derivative is evaluated using a fraction θ of the values of $\{\rho\}$ at the new time t^n and a fraction $(1-\theta)$ of the values at the beginning of the timestep, t^{n-1}. Again, we select three specific values of θ from the range $0 \le \theta \le 1$. When $\theta = 0$, the method is explicit, denoted E, and new values of ρ depend only on previous values of ρ. The explicit algorithm is computationally the least expensive to use because the new values are not coupled to each other. When $\theta = \frac{1}{2}$, the resulting algorithm is called centered, denoted by C. The centered algorithm has a higher mathematical order of accuracy than algorithms for which θ is not equal to $\frac{1}{2}$. The final choice, $\theta = 1$, is the most stable, fully implicit algorithm denoted by I.

As in the theoretical example given in Eqs. (4–3.5) and (4–3.6), we assume that ρ initially has a sinusoidal spatial variation which can be written as e^{ikx}. For each separate mode, Eq. (4–3.8) yields

$$\begin{aligned}\frac{\rho^n(k) - \rho^{n-1}(k)}{\Delta t} &= -\frac{2D\theta}{(\Delta x)^2}(1 - \cos k\,\Delta x)\rho^n(k) \\ &- \frac{2D(1-\theta)}{(\Delta x)^2}(1 - \cos k\,\Delta x)\rho^{n-1}(k) .\end{aligned} \tag{4-3.9}$$

Values of $\rho(k)$ at both the new timestep n and old timestep $n-1$ appear on both sides of the equation. However, Eq. (4–3.9) can be solved for $\rho^n(k)$

in terms of the value at the previous timestep $\rho^{n-1}(k)$ because we consider only one harmonic k at a time. In the general case, Eq. (4–3.8) can be solved by a tridiagonal matrix inversion algorithm.

In terms of A, the single-timestep amplification factor defined in Eq. (4–1.12), the finite-difference solution to Eq. (4–3.9) can be written as

$$A^{FD}(k) = \frac{[1 - \frac{2D\Delta t}{(\Delta x)^2}(1-\theta)(1-\cos k\,\Delta x)]}{[1 + \frac{2D\Delta t}{(\Delta x)^2}\theta(1-\cos k\,\Delta x)]} \,. \qquad (4-3.10)$$

Here the superscript "FD" indicates the finite-difference solution. Over the time interval Δt, the analytic solution given by Eq. (4–3.6) predicts the amplification factor

$$A^T(k) = \left(\frac{\rho^n}{\rho^{n-1}}\right)^T = e^{-Dk^2\Delta t} \,. \qquad (4-3.11)$$

The superscript "T" designates the theoretical or true solution to the sine wave test problem posed above. The amplitude of each mode decays exponentially at the rate Dk^2.

We now want to determine the range of wavenumbers and the range of timesteps Δt over which the individual numerical algorithms work well. To do this, we compare the solution obtained from three numerical algorithms ($\theta = 0$ (E), $\theta = \frac{1}{2}$ (C), and $\theta = 1$ (I)) with the theoretical amplification factor in Eq. (4–3.11) of the same problem. Two nondimensional parameters are useful,

$$\beta \equiv \frac{2D\Delta t}{(\Delta x)^2} \quad \text{and} \quad k\,\Delta x \,.$$

Here β is a measure of the size of the timestep relative to how fast gradients diffuse on the scale of a cell size Δx. The other parameter, $k\,\Delta x$, is the wavenumber of the harmonic being considered in units of the cell size. The values of $k\,x$ range from 0 to π (two cells per wavelength), and β generally ranges from zero to infinity.

The values of A for the three finite-difference algorithms, given by Eq. (4–3.10), and the theoretical exact solution, Eq. (4–3.11), may be rewritten as

$$A^{FD}(k) = \frac{[1 - \beta(1-\theta)(1-\cos k\,\Delta x)]^{1/\beta}}{[1 + \beta\theta(1-\cos k\,\Delta x)]^{1/\beta}} \,, \qquad (4-3.12)$$

and

$$A^T(k) = e^{-\frac{1}{2}(k\,\Delta x)^2} \,, \qquad (4-3.13)$$

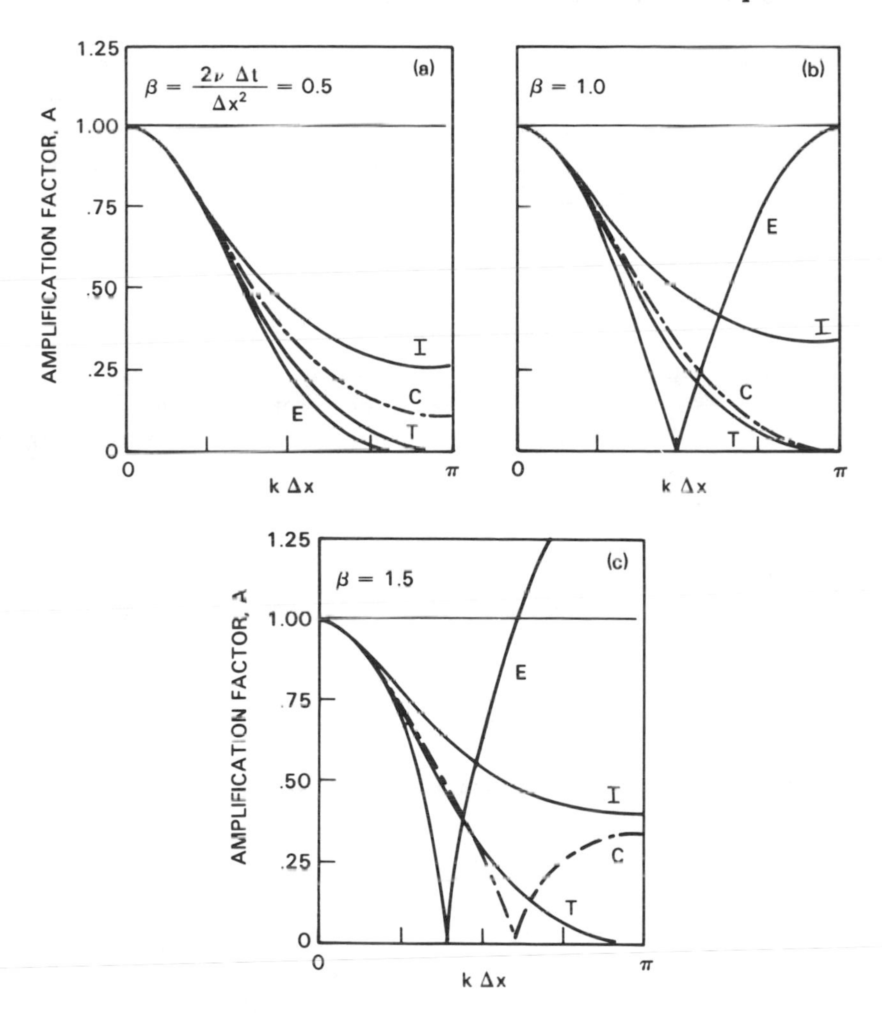

Figure 4–4. The amplification factor, A, for a diffusion equation, as a function of $k\,\Delta x$ for various values of the parameter β. The T, E, C, and I refer to the theoretical (analytical), explicit, centered, and implicit solutions, respectively.

respectively. The theoretical result in Eq. (4–3.13) is the $(1/\beta)$th root of Eq. (4–3.6) and expresses the amplification factor per unit time rather than per timestep. The exponential factor $1/\beta$ is included in the definition of Eq. (4–3.12) so that comparisons of various solutions may be made at fixed times even though different timesteps have been used. The exact theoretical solution is then compared with the approximate solutions in Figure 4–4.

Figure 4–4 shows A as a function of $k\,\Delta x$ for various values of β. The far right of each panel is the shortest wavelength that can be represented, two cells per wavelength. The far left is the infinite wavelength limit. The points where the solution reflects from the horizontal, where $A = 0$, indicate the part of the short-wavelength spectrum where the mode amplitude changes sign because the timestep is too long. This occurs first for solution E when $\beta = 1$ and subsequently for solution C at $\beta = 1.5$. For $k\,\Delta x$ larger than the value at reflection, the corresponding algorithms predict negative values of the positive quantity ρ. Thus these reflection points are the practical if not mathematical limit of numerical stability. For the panel in which $\beta = 1.5$, the short wavelength portion of the spectrum computed using the explicit algorithm is unstable: the amplitude is greater than unity and the sign is wrong. For $k\,\Delta x \geq 2\pi/3$, the centered algorithm is practically although not absolutely unstable. The amplitudes of these short wavelengths decrease rapidly from one timestep to the next, which is the correct trend, but their sign oscillates.

The best method for solving the diffusion equation depends on the diffusion coefficient and the affordable timestep for a given cell size. Equation (4–3.12) should be equal to the correct answer, Eq. (4–3.13), when θ is properly chosen. It is not possible to find one integrating factor as we could for Eq. (4–2.5) because each harmonic of the spatial variation requires a different value $\theta^*(\Delta t)$. In the long wavelength limit, however, the spatial mode number variation of θ^* drops out and

$$\theta^* \approx \frac{3\beta - 1}{6\beta} \,. \qquad (4-3.14)$$

When the nondimensional timestep β is small, θ^* is negative in Eq. (4–3.14). This means that the explicit solution is most accurate for the values $\beta \leq \frac{1}{3}$, a regime in which no sensible algorithm is exact because the correct answer is not allowed with a value of θ in the range 0 to 1. This means that there is *no* regime in which the fully implicit algorithm is the most accurate, and whenever the explicit algorithm is stable for all wavelengths, it is generally the most accurate at long wavelengths.

Clearly the explicit algorithm is the fastest and most accurate whenever a short enough timestep can be selected. When longer timesteps are required, a partially implicit algorithm with θ chosen according to Eq. (4–3.14) should be used. For wavelengths shorter than four cells, strong additional damping must be added to ensure the positivity of the composite algorithm. Chapter 7 addresses flux limiters for the explicit algorithm which also ensure positivity and stability of the solution.

4–3.4 The Cost of Implicit Methods

In all of these comparisons, the implicit algorithm is stable and behaves in a physically reasonable way at short wavelengths even though it is never the most accurate algorithm. Strict conservation of ρ is satisfied because Eq. (4–3.8) can be written in flux conservation form. In the regimes where the other algorithms are also stable, the implicit algorithm is somewhat less accurate but this should not be a major concern. Short-wavelength parts of the solution, where numerical errors are the worst, decay quickly and the solution becomes smoother as time progresses. All of these considerations indicate that implicit diffusion algorithms can be used wherever practical.

There are drawbacks to using partially or fully implicit algorithms, however. Complications arise when we have to deal with more than one spatial dimension, nonlinearity from coupled diffusion equations, and spatial dependence of the transport coefficients. Solving the multidimensional diffusion equation implicitly involves inverting large but sparse matrix equations of the type discussed in Chapter 11. When the equations are linear, this is straightforward although computationally expensive. When the equations are nonlinear, the functional form of the coefficients is complicated and iteration of linearized matrix solutions is required. Then the expense can be very high. It is possible to use an asymptotic solution in the limit of very fast diffusion, and this also involves solving a large matrix. An asymptotic solution might again be more accurate than the implicit formulation, as was the case for local processes discussed above, but may not maintain conservation.

4–3.5 Physical Diffusion and Numerical Diffusion

Numerical diffusion, a major source of error in finite-difference algorithms, has the same form and effect as physical diffusion. Figure 4–2a shows the loss in amplitude of a quantity $\rho(x,t)$ due to numerical diffusion. The three curves in panel (a) describe equally well the analytic solution of the evolution of a point-source profile at three successive times, as determined by Eq. (4–3.7). As the profile decreases in amplitude, it becomes broader. For numerical diffusion, as for physical diffusion, the amplitudes of the solution harmonics change but the phases do not.

To complete the analogy between numerical diffusion and Eq. (4–3.2) and Eq. (4–3.7), suppose that the direction of time is reversed in these equations. Graphically this is shown in Figure 4–2a with $t_1 > t_2 > t_3$. The solution evolves the way many strong numerical instabilities appear to behave in simulations. When these algorithms go unstable, their behavior can be analyzed to show that there has been a change of sign in the diffusion term. Taking too big a timestep is a simple way to induce this. A Gaussian profile run backward in time by an "antidiffusion" equation shows increasing spikiness. As the profile gets narrower, its height increases until it becomes singular at some finite time. Diffusion equations, when reversed in time, are both mathematically and physically unstable. The δ-function singularity which we have shown at $t = 0$ is a typical result of this property.

Because diffusion smooths profiles, an evolving solution loses track of details in initial conditions and any fluctuations subsequently added by truncation or roundoff errors. In a finite time, short wavelength errors in the numerical solution decay away. Thus the solution techniques with numerical diffusion become more accurate, in an absolute sense, as the integration proceeds. Inaccuracy and instability usually have their maximum effect in the short wavelength properties of the solution where large relative errors are effectively masked by the overall solution decay.

The phrases *diffusive effects* and *dissipative effects* are often used interchangeably. Diffusion and dissipation, in the thermodynamic sense, both imply entropy increases. However, "dissipative" has other connotations which do not apply to diffusion. Characterizing a system as dissipative implies that the system is irreversible and has losses. This type of process is usually not conservative. Diffusion, however, is thermodynamically irreversible (that is, entropy goes up during the diffusive mixing of two media), but is fully conservative in that the total amount of ρ stays the same while its distribution in space varies.

4–4. CONVECTIVE TRANSPORT

4–4.1 Fluid Dynamic Flows: Advection and Compression

The continuity equation, the third process in Table 4–1, describes the compression, rotation, and translation (advection) of a fluid. These three effects together constitute convection, that is, continuous fluid dynamic flow. In the ideal one-dimensional limit, the equation

$$\frac{\partial}{\partial t}\rho(x,t) \;=\; -\frac{\partial}{\partial x}\rho(x,t)v(x,t) \qquad (4-4.1)$$

describes the convection of a quantity $\rho(x,t)$ by a flow velocity $v(x,t)$. Limiting the problem to one dimension excludes rotations, but allows both advection and compression.

Solving continuity equations satisfactorily is one of the important, difficult problems in computational fluid dynamics. Fluid dynamics is governed by coupled continuity equations: one scalar continuity equation for the mass density ρ; one vector continuity equation with one, two, or three components for the momentum density ρv; and one scalar continuity equation for the energy density E. The discussion in this section is a starting point for the treatment of more sophisticated solution techniques for coupled and multidimensional continuity equations.

The one-dimensional, Cartesian continuity equation is written in *conservation form* in Eq. (4–4.1). It can also be written in *Lagrangian* form as well,

$$\frac{\partial \rho}{\partial t} + v\frac{\partial \rho}{\partial x} \equiv \frac{d\rho}{dt} = -\rho\frac{\partial v}{\partial x} . \qquad (4-4.2)$$

The two terms on the left hand side of Eq. (4–4.2) are the *Eulerian time derivative* and the *advective space derivative*, respectively. Together they comprise the *Lagrangian derivative, $d\rho/dt$,* which gives the rate of change of the density variable ρ in a frame of reference which moves with the local fluid velocity. This representation is useful because the Lagrangian derivative differs from zero only through the compression term, $v\partial\rho/\partial x$. Advection and rotation do not appear explicitly in the Lagrangian representation.

The compression term also includes the inverse effect, rarefaction. The compression term contains a spatial derivative of the velocity, $\partial v/\partial x$, but is linear in the convected quantity ρ. Therefore the net effect of compression is really local because the gradient of ρ does not appear. If we treat $\partial v/\partial x$ as a known quantity while solving Eq. (4–4.2), we can treat compression as one of the local effects discussed in Section 4–2. To simplify the discussion of advection in this section, we eliminate compressional effects by assuming the velocity is constant, so that $\partial v/\partial x = 0$. This allows use of Fourier analysis to study advection where the short-wavelength components of the phase error become extremely important.

Equation (4–4.2) suggests numerical algorithms for the continuity equation which take advantage of a Lagrangian frame of reference that moves with the fluid. In a Lagrangian representation, fluid parcels (cells) in the discrete model have time-varying boundaries that move with the local fluid velocity. These cell boundaries are defined so that the convective flux into or out of the cell is zero. Thus the mass of a Lagrangian fluid parcel is constant. In the Lagrangian frame of reference, the troublesome convective derivatives, the source of numerical diffusion, are no longer in the equations.

At first glance Lagrangian approaches seem to be the ideal way to eliminate major advection errors. Unfortunately, multidimensional Lagrangian fluid algorithms are more difficult than Eulerian algorithms because the moving grid becomes tangled as the flow evolves. Points which were originally close can move quite far from each other. To carry Lagrangian simulations more than a few timesteps, the grid must be restructured or reconnected.

Thus the geometry of the distorted Lagrangian grid enters into the problem and makes the calculations difficult and expensive. Another complication is that it is not possible to treat all of the compressible reactive flow equations in the same Lagrangian frame of reference. Because of the complications associated with Lagrangian representations, this book concentrates on Eulerian methods which we find the most robust for multidimensional problems. Lagrangian methods for computation have great potential, much of which is still untapped. They are an active frontier in computational fluid dynamics research and are discussed further in Chapter 10.

As with local processes and diffusion equations (Sections 4–2 and 4–3 above), the major errors in representing advective effects are caused by discretizing the continuous variables. In modeling convection, the discretization process introduces phase errors, amplitude errors, and Gibbs errors. Thus solving these continuity equations combines the difficulties of the diffusion processes described above and the wave processes considered in Section 4–5 below. A perfect method for convection would transport the quantity ρ at the correct velocity, maintain the correct amplitudes, and preserve the steep gradients at the correct locations.

4–4.2 The Square-Wave Test Problem

Figure 4–5 shows an application of the continuity equation which is very useful for testing numerical solution methods. A cloud of material is injected into a background gas and drifts along with it. The location and shape of the cloud are shown in panel (a) at two different times, t_1 and t_2. In one dimension, the density of the cloud can be modeled by a square wave, as shown in panels (b) and (c) of the figure, if

- the cloud is initially deposited over a finite region,
- the flow is incompressible, so that the density of the two materials is nearly constant, and
- there is essentially no diffusion present, so that the materials do not interpenetrate.

Panels (b) and (c) show one-dimensional density profiles of ρ, measured through the center of the cloud along the flow direction. The uniform velocity convects the cloud a distance equal to about twice its spatial extent after $t_2 - t_1$ seconds have elapsed. The sharp edge of the profile represents the narrow region at the edge of a cloud where the mixture varies rapidly from nearly pure cloud, to a thin region of equally mixed cloud and background, to a much larger region of nearly pure background fluid.

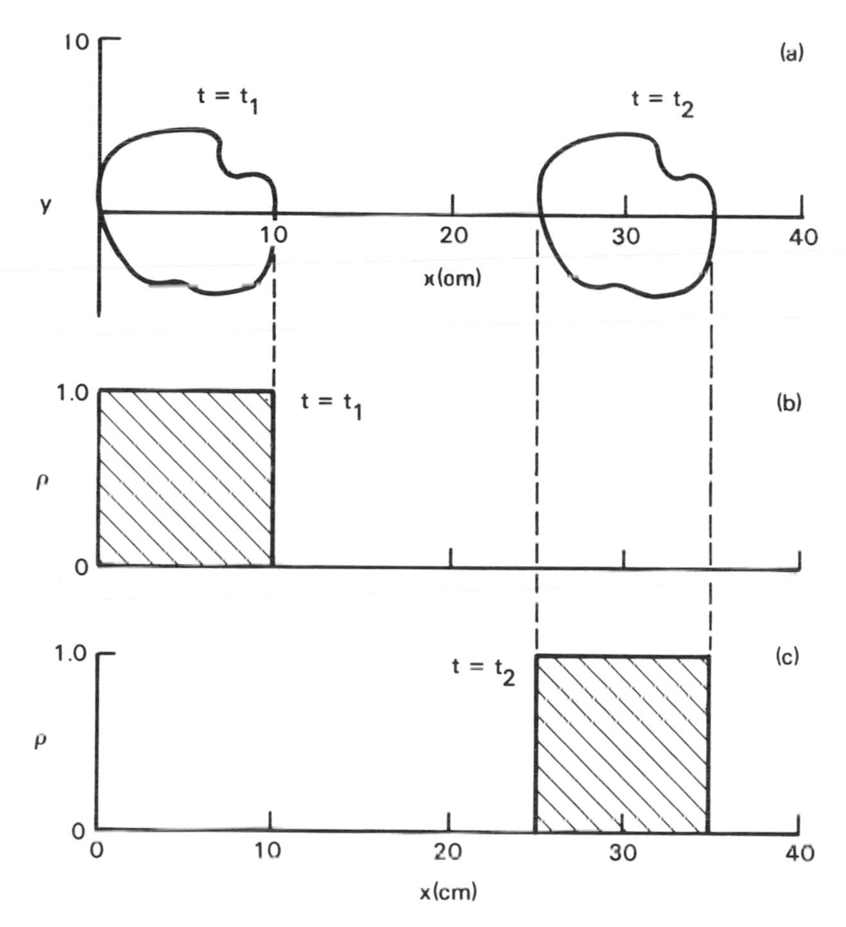

Figure 4–5. Schematic for the square wave test problem. (a) A cloud of gas moves at a constant velocity, unaffected by diffusion. (b) Density along the centerline of the cloud at the initial time t_1. (c) Density along the centerline of the cloud at a later time t_2.

The square density profile from $x = 0$ to $x = 10$ cm at time $t = 0$ is a good test of algorithms designed to calculate convection. It is a stringent test because all wavelengths of the Fourier expansion are involved in the solution and the correct phase relationship among all of the harmonics is important for maintaining the correct shape of the profile. Because v is constant, the true profile $\rho(x)$ moves through space without any relative distortion and the ρ "square wave" is centered at $x_2 = vt_2$, as shown in panel (c).

Many finite-difference methods have been explored to solve Eq. (4–4.2). These range from the explicit, first-order, "donor-cell" algorithm, to a number of second-order methods such as the Lax-Wendroff and "leapfrog" algorithms (see, for example, Roache, 1982; Potter, 1973; Anderson et al., 1984), to the higher-order, nonlinear monotonic methods. We consider several of these below applied to the square wave test problem, and substantially expand this topic in Chapter 8.

4–4.3 Explicit and Implicit Solutions

When $\partial v/\partial x = 0$, Eq. (4–4.2) reduces to the one-dimensional advection equation

$$\frac{\partial}{\partial t}\rho(x,t) = -v\frac{\partial}{\partial x}\rho(x,t) . \qquad (4-4.3)$$

This equation is linear in ρ with first-order derivatives in both space and time. Despite the obvious fact that the spatial derivative is first rather than second order, this equation looks rather similar to a diffusion equation. Similar finite-difference approximations are used now to study the advection equation.

As above, let Δx be the cell size and Δt be the fixed timestep interval. Then ρ_j^n is defined here as the average value of ρ in cell j at time $t^n = n\Delta t$. It can be represented over the interval Δx at time Δt as

$$\rho_j^n \equiv \int_{x_j-\Delta x/2}^{x_j+\Delta x/2} \rho(x, n\,\Delta t)\frac{dx}{\Delta x} . \qquad (4-4.4)$$

In this representation the conserved sum is

$$\sum_{j=1}^{J}\rho_j^n\,\Delta x = \sum_{j=1}^{J}\rho_j^{n-1}\,\Delta x + \int_{(n-1)\,\Delta t}^{n\,\Delta t} v\left[\rho\left(x_J+\frac{\Delta x}{2}\right) - \rho\left(x_1-\frac{\Delta x}{2}\right)\right] dt , \qquad (4-4.5)$$

which replaces the corresponding conservation integral. The last two terms on the right hand side of Eq. (4–4.5) are the fluxes of ρ through the boundaries of the computational domain, taken to be at $x_{\frac{1}{2}} \equiv \Delta x/2$ and $x_{J+\frac{1}{2}} \equiv (J+\frac{1}{2})\Delta x$.

We use centered time and space differences with the first-order derivative formula to represent the advection term. As for Eqs. (4–2.5) and (4–3.8), the finite-difference equation approximating Eq. (4–4.3) is

$$\frac{\rho_j^n - \rho_j^{n-1}}{\Delta t} = -v\left[\theta\frac{(\rho_{j+1}^n - \rho_{j-1}^n)}{2\,\Delta x} + (1-\theta)\frac{(\rho_{j+1}^{n-1} - \rho_{j-1}^{n-1})}{2\,\Delta x}\right] \qquad (4-4.6)$$

and θ is again the implicitness parameter. The explicit algorithms (E) have $\theta = 0$, the centered algorithms (C) have $\theta = \frac{1}{2}$, and the fully implicit algorithms (I) have $\theta = 1$. These choices are of interest because the explicit algorithms are again the easiest to implement and least expensive, the time-centered algorithms generally have the highest order of accuracy, and the implicit algorithms are the most stable numerically.

Equation (4–4.6) can be simplified by combining timestep, cell size, and flow velocity into a single nondimensional parameter,

$$\epsilon \equiv v \frac{\Delta t}{\Delta x} . \qquad (4-4.7)$$

Here ϵ is the number of cells of size Δx that the fluid traverses in a timestep Δt. The value of ϵ should usually be less than unity because the finite-difference formula uses information only from adjacent cells. Using this ϵ in Eq. (4–4.6) yields

$$\rho_j^n + \frac{\theta \epsilon}{2}(\rho_{j+1}^n - \rho_{j-1}^n) = \rho_j^{n-1} - \frac{(1-\theta)}{2} \epsilon (\rho_{j+1}^{n-1} - \rho_{j-1}^{n-1}) . \qquad (4-4.8)$$

This equation gives the values of ρ at the new timestep n (on the left hand side) in terms of the values at the previously calculated timestep $n-1$ (on the right hand side). Equation (4–4.8) contains values of ρ^n at three different locations for any $\theta > 0$, so that a tridiagonal matrix results that must be solved at each timestep. When $\theta = 0$, this reduces to a simple explicit algebraic expression for ρ^n in terms of ρ^{n-1}. When the matrix must be solved at each timestep, the cost can be substantial.

Figure 4–6 summarizes the results obtained using several methods to solve Eq. (4–4.3) for the square wave test problem introduced in Figure 4–5 with $\Delta x = 1$ cm. The top panel in this figure shows the exact solution of the test problem at four different times. Note that the edges of the square wave are not perfectly vertical. This reflects the fact that the variables $\{\rho_j\}$ are defined by only a discrete set of numbers.

The test problem has a uniformly spaced grid of 100 cells extending from $x = 0$ to $x = 100$ cm. Initially $\rho = 1$ in the first ten cells and $\rho = 0$ everywhere else. The boundary conditions are periodic, so material flowing out of cell 100 through the interface at $x = 100$ cm reenters the computational region at the same speed and density through the interface to cell 1 at $x = 0$ cm. The flow velocity is a constant, $v = 1$ cm/s. With the timestep fixed at $\Delta t = 0.2$ s, the fluid moves 2/10 of a cell per timestep, and $\epsilon = 0.2$. At $t = 0$, the computed and exact solutions are identical by virtue of the initial conditions. The computed solutions obtained from Eq. (4–4.8) for later times are indicated in Figure 4–6 by data points in panels (a)–(c).

Panel (a) shows the solution calculated with the explicit algorithm, Eq. (4–4.8) with $\theta = 0$, after 100 timesteps. Unfortunately this simple,

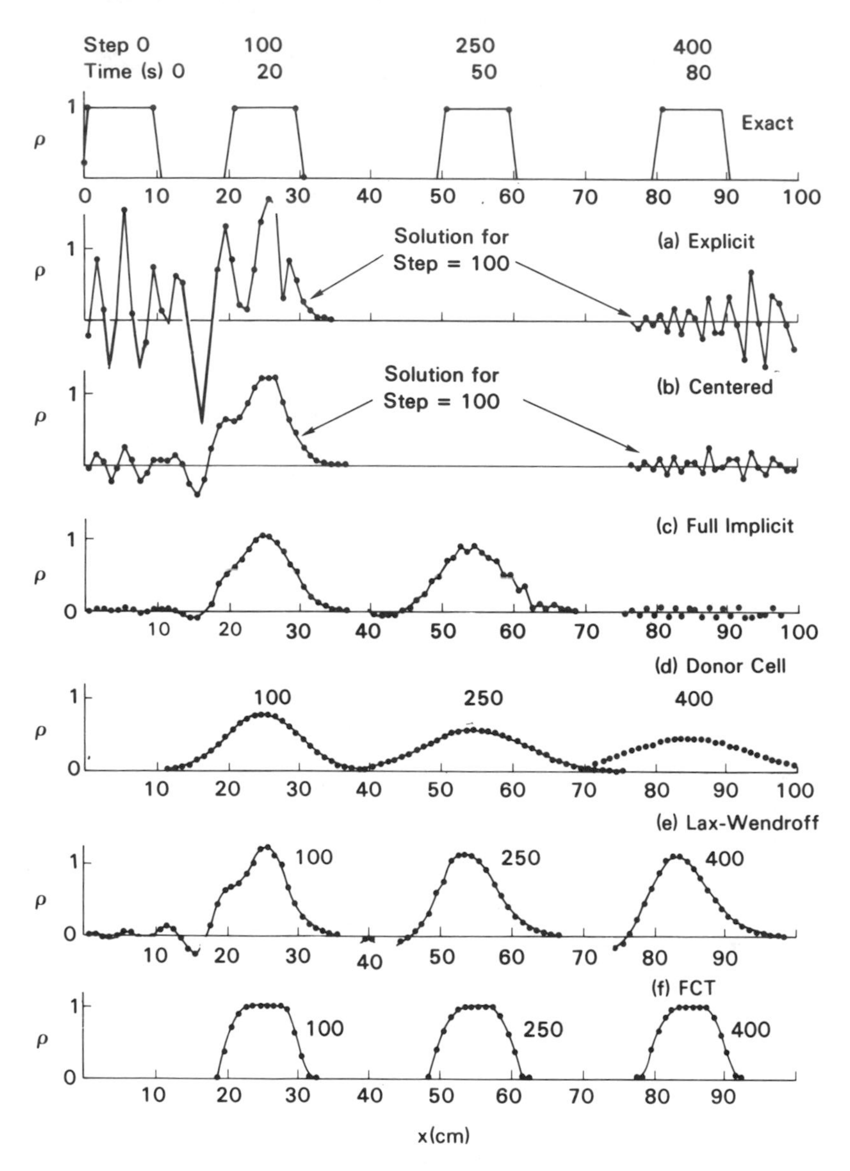

Figure 4–6. Solutions of the square wave test problem using six different numerical algorithms. The top panel is the square wave as interpreted on the numerical grid described in the text. (a) Explicit algorithm, at step 100. (b) Centered algorithm, at step 100. (c) Fully implicit algorithm, at steps 100 and 250. Note that the noise on the far right corresponds to part of the solution at step 100. By step 250, the noise which has come in through the periodic boundary interacts with the main part of the solution. This gives the ragged appearance. (d) Donor-cell, at steps 100, 250, 400. (e) Lax-Wendroff, at steps 100, 250, 400. (f) Flux-Corrected Transport, at steps 100, 250, 400 (discussed in Chapters 8 and 9).

inexpensive algorithm gives terrible results. It shows large oscillations which get worse in time, as well as errors which have propagated back onto the far right side of the grid due to the periodic boundary conditions. The results show little resemblance to the exact solution and are numerically unstable. No matter how small the timestep used, this kind of forward-differenced convection is unstable.

The centered solution is shown in panel (b). Because this algorithm is time reversible, it gives marginally better results than the explicit algorithm. Mode amplitudes neither increase nor decrease in time. Though only a linear result, this is a highly desirable property. The centered solution also has higher-order accuracy than the explicit or fully implicit solutions. Nevertheless, the profile is not even roughly correct and there are large nonphysical positive and negative values of density where ρ should be zero. The errors in the centered solution arise from a combination of Gibbs errors and phase errors which grow steadily in time.

Panel (c) was computed using the implicit algorithm. It has larger phase errors than the centered algorithm, but looks somewhat smoother. Including all the spatial difference evaluated at the new time t^n introduces numerical damping into the solution. This damping appears to compensate partially for the phase errors. Neither the explicit, centered, nor fully implicit algorithms are really useful, as can be seen by the large difference between the exact and computed solutions.

Panels (d) and (e) were computed using two other explicit, linear algorithms which are stable. Panel (d) is computed using the *one-sided* (also called *upwind* or *donor-cell*) algorithm which is only first-order accurate. Panel (e) shows the *Lax-Wendroff* algorithm, which is second-order accurate. Both of these algorithms are in common use, and both, as can be seen from the figures, have severe limitations.

For the donor-cell method, ρ at the new time t^n is calculated from

$$\rho_j^n = \rho_j^{n-1}(1-\epsilon) + \epsilon\, \rho_{j-1}^{n-1} . \qquad (4-4.9)$$

Notice that the spatial derivative is approximated using two values of ρ, one of which is located upstream of the grid point j. This is the origin of the name donor-cell: the space derivative is computed using values associated with the fluid parcel which is "donated" to cell j by the flow. When the velocity is negative and the flow is coming from the right, the donor-cell fluxes have to be calculated using the j and $j+1$ values of ρ. In this case the fluid arrives at cell center j at time t^n from the cell at $j+1$. Because the derivative is not centered and uses values from only twocells, the algorithm is accurate only to first order in space.

This linear interpolation in Eq. (4–4.9) introduces very severe numerical diffusion which overshadows the dispersive errors. The initially sharp edges of the density profile, which should propagate as sharp edges, smooth out rapidly. By step 500, more than half the material has numerically diffused out of the square wave. This occurs even though the square wave is resolved by 10 computational cells. In many practical three-dimensional calculations, a resolution of $20 \times 20 \times 20$ cells may be affordable on the computer available. Then the effect of numerical diffusion is even more severe than in the illustrative test case, panel (d).

Panel (e) shows the square wave test performed using the Lax-Wendroff algorithm. One way to derive this algorithm is to use a second-order Taylor series expansion in the time variable,

$$\rho(x,t^n) \cong \rho(x,t^{n-1}) + \Delta t \frac{\partial}{\partial t}\rho(x,t^{n-1}) + \frac{1}{2}(\Delta t)^2 \frac{\partial^2}{\partial t^2}\rho(x,t^{n-1}) \, , \quad (4-4.10)$$

to estimate ρ at time $t^n = t^{n-1} + \Delta t$, given the values of $\{\rho_j\}$ at t^{n-1}. The next step is to recast the first and second time derivatives in terms of space derivatives, and then in terms of finite differences using Eq. (4–4.3). The derivatives in Eq. (4–4.10) become

$$\frac{\partial}{\partial t}\rho_j^{n-1} \cong -v \frac{(\rho_{j+1}^{n-1} - \rho_{j-1}^{n-1})}{2\,\Delta x} \quad (4-4.11)$$

and

$$\frac{\partial^2}{\partial t^2}\rho_j = v^2 \frac{(\rho_{j+1}^{n-1} - 2\rho_j^{n-1} + \rho_{j-1}^{n-1})}{(\Delta x)^2} \, , \quad (4-4.12)$$

when v is a constant. Substituting Eqs. (4–4.11) and (4–4.12) into Eq. (4–4.10), we obtain

$$\rho_j^n = \rho_j^{n-1} - \frac{\epsilon}{2}(\rho_{j+1}^{n-1} - \rho_{j-1}^{n-1}) + \frac{1}{2}\epsilon^2(\rho_{j+1}^{n-1} - 2\rho_j^{n-1} + \rho_{j-1}^{n-1}) \, , \quad (4-4.13)$$

the simplest form of the Lax-Wendroff algorithm.

The results of the two first-order algorithms shown in panels (c) and (d) are dominated by numerical diffusion. In the Lax-Wendroff algorithm which is second order, much less of the material originally in the square wave at $t = 0$ has diffused out at later times. Moreover, the amplitude errors are small enough that we can see the dispersion errors, which initially appear at the shortest wavelength and look similar to Gibbs errors. These appear as numerical ripples, predominantly in the wake of the square wave, and produce unphysical regions of $\rho < 0$. The pronounced undershoots and overshoots pose problems which are just as serious as those caused by excess numerical diffusion.

Panel (f) shows the same problem computed by the Flux-Corrected Transport algorithm (FCT) discussed in Chapters 8 and 9. This is an intrinsically nonlinear algorithm designed to eliminate nonphysical ripples due to the Gibbs phenomenon and numerical dispersion. Nonlinear monotone methods, of which FCT is an example, are designed to minimize phase and amplitude errors and still maintain positivity and conservation.

4–4.4 Phase and Amplitude Errors for Common Algorithms

The phase and amplitude properties for the five linear convection algorithms described below can be analyzed in terms of the amplification factor $A(k\,\Delta x)$ defined in Eqs. (4–1.12) and (4–1.13). For a specific algorithm, A is found by substituting the test function

$$\rho_j^n = \rho^n(k)e^{ikj\delta x} \qquad (4-4.14)$$

into the equation defining the particular algorithm, collecting the $\rho^n(k)$ and $\rho^{n-1}(k)$ terms, and then forming the ratio $A(k)$.

For diffusion equations, the imaginary part of the amplification factor is zero, which means that the phase shift for a given harmonic k is zero. However, the existence of a prevailing flow direction in convection ensures that A_I cannot be zero. By substituting Eq. (4–4.14) into Eq. (4–4.3) and taking the spatial derivative analytically, we obtain the exact theoretical result,

$$A^T(k\,\Delta x) = e^{-i\epsilon k\,\Delta x} \,. \qquad (4-4.15)$$

This theoretical amplification factor A^T has unit magnitude, consistent with a profile that moves in space without growing or shrinking in size. We can also write the exact phase shift from Eq. (4–4.15), φ_T, as

$$\varphi^T = -\epsilon k\,\Delta x \,. \qquad (4-4.16)$$

The amplification factors for the five linear algorithms considered above are determined by substituting the test function, Eq. (4–4.14), into Eqs.(4–4.8), (4–4.9), and (4–4.13). The results are:

1. Explicit solution, Eq. (4–4.8) with $\theta = 0$,

$$A^E(k\,\Delta x) = 1 - i\epsilon\sin(k\,\Delta x) \,, \qquad (4-4.17)$$

2. Centered solution, Eq. (4–4.8) with $\theta = \frac{1}{2}$,

$$A^C(k\,\Delta x) \;=\; \frac{1 - i\frac{\epsilon}{2}\sin(k\,\Delta x)}{1 + i\frac{\epsilon}{2}\sin(k\,\Delta x)}\,, \qquad (4-4.18)$$

3. Fully implicit solution, Eq. (4–4.8) with $\theta = 1$,

$$A^I(k\,\Delta x) \;=\; \frac{1}{1 + i\epsilon\sin(k\,\Delta x)}\,, \qquad (4-4.19)$$

4. First-order donor-cell algorithm,

$$A^{DC}(k\,\Delta x) \;=\; 1 - \epsilon + \epsilon\cos(k\,\Delta x) - i\epsilon\sin(k\,\Delta x)\,, \qquad (4-4.20)$$

5. Second-order Lax-Wendroff algorithm,

$$A^{LW}(k\,\Delta x) \;=\; 1 - \epsilon^2(1 - \cos k\,\Delta x) - i\epsilon\sin(k\,\Delta x)\,. \qquad (4-4.21)$$

The amplification factor A for the explicit algorithm has unity in the denominator. Because the magnitude of the numerator is larger than unity, the algorithm is unstable. It is surprising how poor the result in Figure 4–6a is with only a small growth factor, $1.02 \approx \sqrt{1+\epsilon}$. Partially and fully implicit algorithms, which require solution of a tridiagonal matrix, generally have a denominator that is greater than unity which counteracts the instability arising from the numerator.

Figure 4–7 shows the magnitude of the amplification factor, $|A(k)|$, and $\varphi_R(k)$, the ratio of the phase of each algorithm to the phase of the exact solution, as a function of $k\,\Delta x$ for each of the five convection algorithms. Note the changes of scales on the graphs of $|A(k)|$. The horizontal axes in the graphs of $|A|$ are the exact solution. Also note how the errors in the phase increase with increasing values of $k\,\Delta x$.

As seen from Figures 4–2 and 4–3, the combination of phase errors and amplitude errors make it a challenging, tricky problem to represent convection numerically on an Eulerian grid. Phase errors grow linearly in time and cause growing deviations from the correct solution. Amplitude errors in convection, unlike those associated with diffusion, also result in growing rather than decreasing errors.

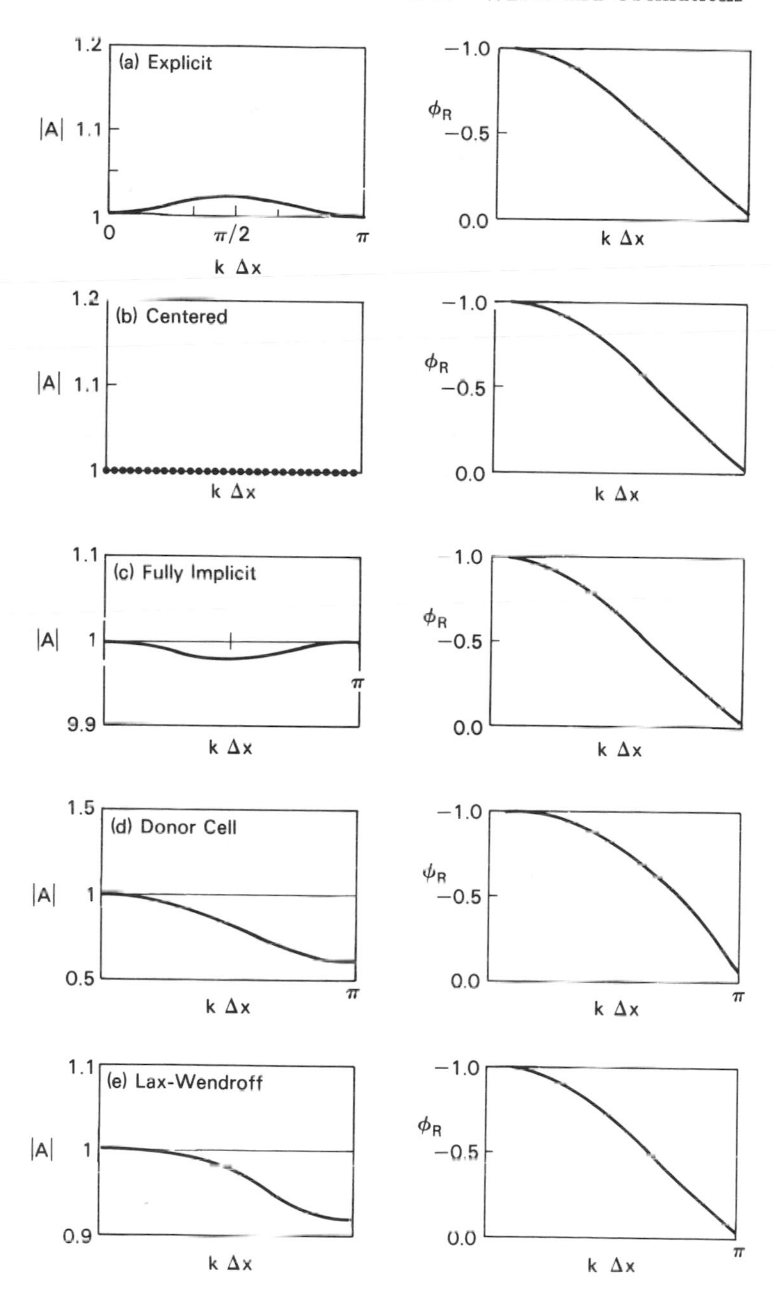

Figure 4–7. Amplitude and phase of the amplification factors for the five convective algorithms described in the text. (a) Explicit. (b) Centered. (c) Fully implicit. (d) Donor cell. (e) Lax-Wendroff. The dots on the horizontal axis for the centered solution mean that $|A|=1$ everywhere.

4–5. WAVES AND OSCILLATIONS

4–5.1 Waves from Coupling Continuity Equations

The fourth and last process in Table 4–1 is wave propagation. In the one-dimensional case, a linear wave in the scalar variable ρ is governed by the second-order wave equation

$$\frac{\partial^2 \rho}{\partial t^2} = v_w^2 \frac{\partial^2 \rho}{\partial x^2} \,. \tag{4 - 5.1}$$

Here v_w is the phase velocity of the wave, the velocity at which the wave profile moves.

Equation (4–5.1) does not appear explicitly in the full set of reactive flow equations, but it arises from the interaction of two first-order continuity equations through a coupling term (such as a pressure gradient or gravity). If $v(x,t)$ represents the fluid velocity, one possible set of first-order equations that generate a wave equation is

$$\frac{\partial \rho}{\partial t} = -v_w \frac{\partial v}{\partial x} \tag{4 - 5.2a}$$

$$\frac{\partial v}{\partial t} = -v_w \frac{\partial \rho}{\partial x} \,. \tag{4 - 5.2b}$$

Several kinds of waves are contained in the complete reactive flow conservation equations, depending which terms combine to provide the two space and time derivatives. Sound waves couple the energy and momentum equations. Gravity waves arise from coupling the mass and momentum equations. Other types of waves are propagated by convection or driven by body forces which arise from convection in curved flows. Each kind of wave can exist over a spectrum of wavelengths, can propagate in a number of directions, and is generally accompanied by an energy flux.

All waves have some common physical and mathematical properties. These properties provide a partial basis for defining useful solution algorithms. One important property of most waves is energy conservation, and another is that the net displacement of fluid is usually quite small. The amplitudes of the variables ρ and v either change sign or fluctuate about a mean value as the wave oscillates. Thus positivity is not an important property in wave equations, though it can be in the underlying continuity equations.

The second-order derivatives in space make wave equations appear similar to the diffusion equation discussed in Section 4–3. Because the wave equation is also second order in time, it is reversible unless a physical dissipation process is present. Thus there should be no amplitude dissipation in a wave equation algorithm. Implementing reversibility in the algorithm guarantees that modes do not damp when the algorithm is stable. Because waves often travel extremely long distances without much damping, the algorithm used to calculate their behavior should have minimal numerical damping. This implies that the best algorithms are nearly unstable and can be easily driven unstable by small truncation errors, additional uncentered terms, and nonlinear effects that do not appear in the linear stability analysis.

Equations (4–5.1) or (4–5.2a,b) have oscillatory solutions in space and time of the form

$$\begin{aligned} \rho(x,t) &= \rho(k)e^{ikx}e^{-i\omega t} \\ v(x,t) &= v(k)e^{ikx}e^{-i\omega t} \,, \end{aligned} \qquad (4-5.3)$$

where $\rho(k)$ and $v(k)$ are complex. Here k is the wavenumber in the x direction, a real quantity related to the wavelength of the mode by $\lambda = 2\pi/k$. The symbol ω denotes the angular frequency of the wave in the laboratory frame of reference. Inserting Eqs. (4–5.3) into Eqs. (4–5.2) yields the analytical solution,

$$\omega = v_w k \,. \qquad (4-5.4)$$

In this case, the frequency of the wave increases linearly with the wavenumber k.

In the case of sound waves, velocity fields in three dimensions can be reduced to a single scalar wave equation by defining a velocity potential,

$$\mathbf{v} = \nabla\phi \,, \qquad (4-5.5)$$

and by assuming that the fluctuations in pressure, velocity, and density are small enough that linear approximations are valid. In this case the variable ρ in Eq. (4–5.1) is replaced by ϕ, and v_w is the speed of sound $c_s \equiv \sqrt{(\gamma P/\rho)}$.

Only when v_w is constant can the wave equation be written as Eq. (4–5.1). More generally, v_w appears inside one or both of the spatial derivatives. This complication changes the character of the wave propagation appreciably in that the amplitude may now change when v_w varies in space and time. In this section, we analyze the special case in which the wave speed is independent of position and time and does not depend on the local values of $\rho(x,t)$ and $v(x,t)$. This situation is complicated enough. Both explicit and implicit reversible algorithms for waves are described below.

4–5.2 An Explicit Staggered Leapfrog Algorithm

Simple finite-difference formulas for Eqs. (4–5.2) are

$$\frac{\rho_j^n - \rho_j^{n-1}}{\Delta t} = -v_w \left(\frac{v_{j+\frac{1}{2}}^{n-\frac{1}{2}} - v_{j-\frac{1}{2}}^{n-\frac{1}{2}}}{\Delta x} \right) \qquad (4-5.6a)$$

$$\frac{(v_{j-\frac{1}{2}}^{n+\frac{1}{2}} - v_{j-\frac{1}{2}}^{n-\frac{1}{2}})}{\Delta t} = -v_w \frac{(\rho_j^n - \rho_{j-1}^n)}{\Delta x} \, . \qquad (4-5.6b)$$

The variable ρ is defined at the old and new times and at cell centers. The velocity variable v is defined at cell interfaces and at times which are midway between the times when the ρ is defined. Figure 4–8 shows a *grid portrait* for this algorithm. Because of the symmetry between the velocity and density grids, it is not entirely correct to think of either grid as cells, or the other grid as the interfaces. These are called *staggered grids.*

This formulation is called an explicit, *staggered leapfrog algorithm* because the two variables jump forward over each other to advance the system in time, as shown in Figure 4–8. Because the new values of ρ depend only on the old values of ρ and known values of v, not on each other, the algorithm is explicit and can be computed very efficiently. The large computational cost of solving implicit equations which couple the new values together is absent.

The staggered leapfrog algorithm is time reversible, as can be seen by inspecting Eqs. (4–5.6) and Figure 4–8. At any step in the integration, the sign of the timestep can be changed and the exact values of the variables at previous steps can be recovered by running the algorithm backward. This property of the leapfrog algorithm is advantageous because it mirrors the reversibility property of the continuum wave Eqs. (4–5.1) and (4–5.2). Because it is reversible, the leapfrog algorithm maintains accurate mode amplitudes as long as it is numerically stable.

Using the definitions

$$\begin{aligned} x_j &= j\,\Delta x \\ x_{j+\frac{1}{2}} &= (j + \frac{1}{2})\,\Delta x \\ t^n &= n\,\Delta t \\ t^{n+\frac{1}{2}} &= (n + \frac{1}{2})\,\Delta t \end{aligned} \qquad (4-5.7)$$

in Eqs. (4–5.3) and substituting into Eqs. (4–5.6) results in the following *numerical dispersion relation*

$$\sin\left(\frac{\omega_E\,\Delta t}{2}\right) = \epsilon \sin\left(\frac{k\,\Delta x}{2}\right) \, . \qquad (4-5.8)$$

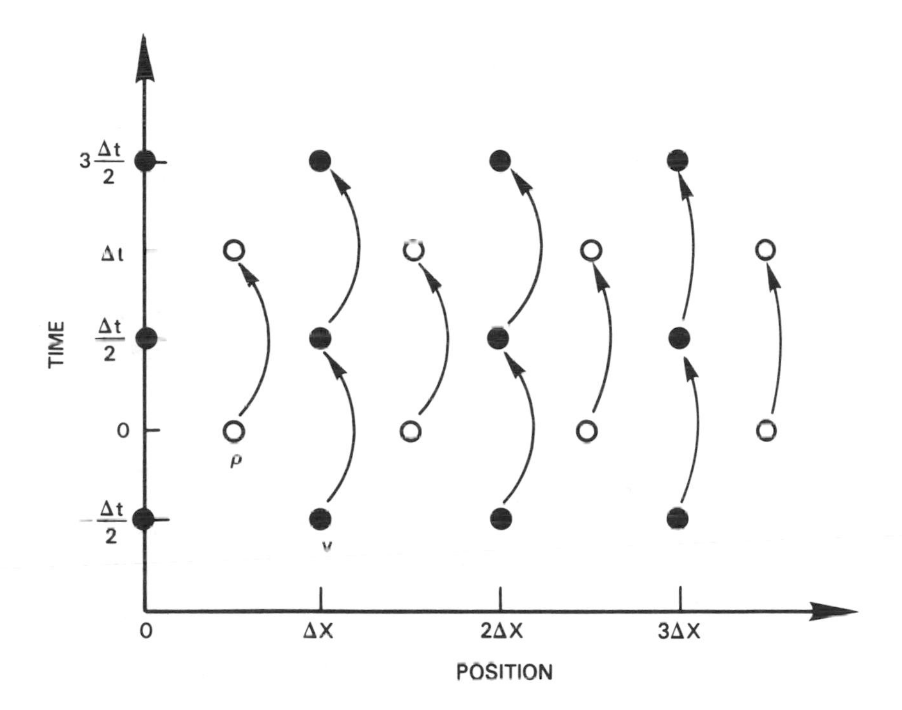

Figure 4–8. Diagram showing the relation between the variables ρ and v in the staggered leapfrog algorithm for solution of wave equations.

Equation (4–5.8) gives the numerical wave frequency ω_E as a function of the wavenumber k and other parameters of the numerical system. Here the nondimensional timestep ϵ is

$$\epsilon \equiv \frac{v_w \, \Delta t}{\Delta x} \,. \tag{4 – 5.9}$$

Equation (4–5.8) can be solved formally for the numerical frequency,

$$\omega_E = \frac{2}{\Delta t} \sin^{-1}\left[\epsilon \sin\left(\frac{k\,\Delta x}{2}\right)\right] \,. \tag{4 – 5.10}$$

Equation (4–5.8) or (4–5.10) gives stability and accuracy bounds for the leapfrog algorithm. The algorithm converges because ω_E approaches the theoretical value ω in Eq. (4–5.4) as Δx approaches zero for fixed $\epsilon < 1$. As long as ϵ is less than unity, that is, as long as the real wave travels less than a cell per timestep, the magnitude of the argument of $\sin^{-1}$ in Eq. (4–5.8) can never exceed unity. This means that a real value of the numerical frequency ω_E can always be found with zero imaginary part, implying that the algorithm is numerically stable even though this frequency is quantitatively in error.

The numerical dispersion relation approaches the analytic solution given by Eq. (4–5.4) when the wavelength is many computational cells long, that is, $k\,\Delta x \ll 1$. For finite values of $k\,\Delta x$, the computed and exact dispersion relations differ. The exact solution gives one phase velocity for all waves, but each Fourier harmonic propagates at a different phase velocity in the finite-difference solution. Although the numerical waves are undamped, they propagate at the wrong phase velocity. As with a single continuity equation, this error is called dispersion.

The numerical solution becomes inaccurate when ϵ and $(k\,\Delta x)/2$ become appreciable fractions of unity. The worst dispersion errors occur at short wavelengths. There the sine function on the right hand side of Eq. (4–5.8) causes the numerical dispersion relation to deviate from the theoretical expression Eq. (4–5.4). Inaccuracy has clearly set in by the time $k\,\Delta x$ approaches $\pi/4$, that is, eight computational cells per wavelength. Most of the modes in the system, however, have a shorter wavelength than eight cells. Thus the error in the computed frequency is appreciable even when the timestep (and hence ϵ) is very small. Because the dispersion comes from the finite-difference approximations to the spatial derivatives, it is not surprising that taking a smaller timestep cannot help.

Figure 4–9 shows the dispersion relations for the explicit leapfrog algorithm (E), the exact theoretical value (T), and an implicit algorithm (I) discussed below. The curves show $(\omega\,\Delta t)/\epsilon \equiv (\omega\,\Delta x)/v_w$ as a function of $k\,\Delta x$. The theoretical curve is now a straight line.

Four different values of ϵ are shown in the four panels, spanning the transition from numerical stability to numerical instability for the explicit algorithm. At the stability boundary for E ($\epsilon = 1$), the theoretical and explicit dispersion relations agree exactly. This is shown by the diagonal line labeled T, E in Figure 4–9b. This exact result is obtained at the stability limit because the wave speed is exactly one spatial cell per timestep, a special situation which the leapfrog algorithm handles well.

The two lower panels of Figure 4–9 show cases where ϵ is increased above unity by taking a timestep larger than the explicit stability limit for short wavelength modes. There is a maximum value of $k\,\Delta x/2$ for which the algorithm is numerically unstable. In the lower two panels at $k\,\Delta x/2 = \sin^{-1}(1/\epsilon)$, the quantity $\epsilon \sin(k\,\Delta x/2)$ equals unity. At shorter wavelengths, the modes are numerically unstable, as indicated by the termination of curve

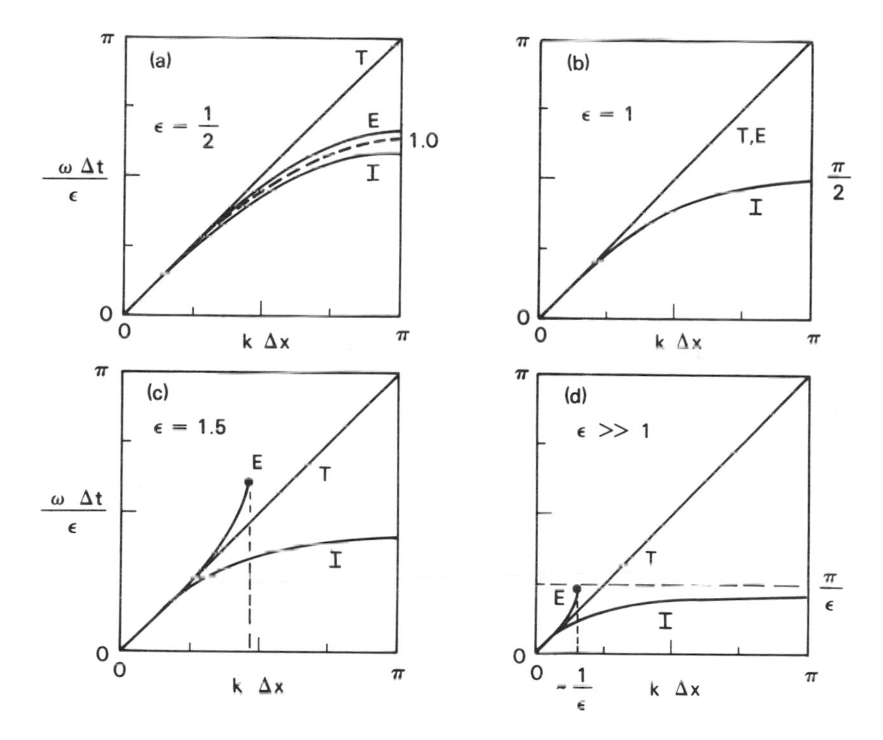

Figure 4–9. The dispersion relation for the solution of wave equations using an explicit (E) and an implicit (I) leapfrog algorithm, and the exact theoretical value (T). These are shown for four values of the parameter $\epsilon = v_w \, \Delta t / \Delta x$.

E. As the timestep is increased, fewer modes are numerically stable and those that are stable, have longer wavelengths.

4–5.3 A Reversible Implicit Algorithm

Figure 4–9 also presents dispersion curves for an implicit algorithm (I) based on the formulas

$$\rho_j^n - \rho_j^{n-1} = -\frac{\Delta t\, v_w}{2}\left[\frac{(v_{j+\frac{1}{2}}^n - v_{j-\frac{1}{2}}^n)}{\Delta x} + \frac{(v_{j+\frac{1}{2}}^{n-1} - v_{j-\frac{1}{2}}^{n-1})}{\Delta x}\right] \tag{4-5.11a}$$

$$v_{j+\frac{1}{2}}^n - v_{j+\frac{1}{2}}^{n-1} = -\frac{\Delta t\, v_w}{2}\left[\frac{(\rho_{j+1}^n - \rho_j^n)}{\Delta x} + \frac{(\rho_{j+1}^{n-1} - \rho_j^{n-1})}{\Delta x}\right] . \tag{4-5.11b}$$

The grid here is staggered in space but not in time. The new variables ρ and v are now determined simultaneously at a given timestep. This algorithm is reversible when the right sides of the equations are evaluated as a time-centered average of the spatial derivative. Reversibility, though computationally expensive, does guarantee that the individual mode amplitudes are preserved.

Because this implicit algorithm requires a matrix inversion, more computation is required at each timestep than with the explicit algorithm. To see the form of this matrix equation, we substitute Eq. (4–5.11b) into Eq. (4–5.11a) and eliminate the new values v. The result,

$$\begin{aligned} \rho_j^n - \frac{\epsilon^2}{4}(\rho_{j+1}^n - 2\rho_j^n + \rho_{j-1}^n) &= \rho_j^{n-1} + \frac{\epsilon^2}{4}(\rho_{j+1}^{n-1} - 2\rho_j^{n-1} + \rho_{j-1}^{n-1}) \\ &\quad - \frac{\Delta t\, v_w}{\Delta x}(v_{j+\frac{1}{2}}^{n-1} - v_{j-\frac{1}{2}}^{n-1}) \, , \end{aligned} \tag{4-5.12}$$

is a second-order finite-difference approximation to Eq. (4–5.1). The left side of Eq. (4–5.12) involves three adjacent values at the new time and is generally solved using a tridiagonal matrix algorithm. This extra work is similar to implicit solutions discussed for diffusion and convection equations. In two or three dimensions, the computational difficulties are far worse because the algebraic finite-difference equations are coupled in all directions. The tridiagonal equation, Eq. (4–5.12), becomes pentadiagonal in two dimensions, or septadiagonal in three dimensions.

Substituting the oscillatory trial solutions into Eqs. (4–5.11) or Eq. (4–5.12) gives the following numerical dispersion relation for the reversible implicit algorithm,

$$\omega_I = \frac{2}{\Delta t} \tan^{-1}\left[\epsilon \sin\left(\frac{k\,\Delta x}{2}\right)\right] \, . \tag{4-5.13}$$

Equation (4–5.13) is the same as Eq. (4–5.10) except that an inverse tangent function replaces the inverse sine. This change ensures that the implicit method is always stable numerically because a real value of ω can always be found regardless of the values of $k\,\Delta x$ and ϵ. Figure 4–9 shows that ω_I always falls below both the theoretical value and also below the explicit value when the explicit algorithm is stable. Thus the explicit algorithm is again everywhere more accurate than the implicit algorithm whenever the explicit algorithm is stable.

4–5.4 Reversibility, Stability, and Accuracy

Performing the actual calculations to verify that an algorithm is reversible is a crucial test of the computer program. Almost all programming errors, inconsistencies in definitions of variables, grid location errors, or incorrect finite-difference formulas destroy the exact reversibility property. One of the very best checks that can be made on the nominally reversible components of the model is to run the calculation for many timesteps, stop it, change the sign of Δt, and run the problem back to the initial conditions.

Just because an algorithm is fully reversible does not mean that it is necessarily accurate. As long as the explicit reversible algorithm is stable, it is more accurate than the corresponding implicit algorithm. Algorithms generally become inaccurate before the stability boundary is reached. As long as $\epsilon \leq 1$, that is, the wave moves less than a cell per timestep, we can always find a real value of ω for any choice of $k\,\Delta x$, and thus the solution is stable. When $\epsilon > 1$, then $\sin(\omega\,\Delta t/2) > 1$ for modes with $k\,\Delta x/2$ near $\pi/2$. These modes are unstable because complex values of ω are required to satisfy the dispersion relation. It is usually courting disaster to run an algorithm with the longest stable timestep. Nonlinear effects, spatial variations of parameters, and variable cell sizes all tend to enforce a practical stability limit which is a factor of two more stringent than the mathematical limit.

When ϵ is very large, almost all of the waves travel more than a cell per timestep and thus are unstable if integrated explicitly. This is shown in Figure 4–9d. If $\epsilon > 1$, the maximum stable wavenumber for the explicit algorithm k_{max} satisfies

$$\frac{k_{\max}\,\Delta x}{2} \sim \frac{1}{\epsilon} > 1\,, \qquad (4-5.14)$$

and the maximum value which the implicitly determined frequency ω_I can have is

$$\omega_{\max} = \frac{\pi}{\Delta t}\,. \qquad (4-5.15)$$

For wavelengths which are explicitly unstable, that is, for $k > k_{\max}$ from Eq. (4–5.14), the argument of the inverse tangent in the dispersion relation for the implicit algorithm, Eq. (4–5.13), is on the order of unity. Errors in the frequency of the waves, and hence the phase velocities of the modes, are thus also of order unity. In wave equations, as in convection and diffusion equations, stability conditions for the explicit method are generally the accuracy conditions for the stable, fully implicit methods. When the timestep is very large compared to what would be stable in an explicit algorithm, only the very longest waves in the system are properly calculated by the implicit algorithm. Thus we must be extremely careful using implicit algorithms. Just because an implicit calculation is stable does not mean that it is accurate, even at long wavelengths.

For diffusion equations, we did not have to consider the phases of the various modes. The amplitudes were the important quantities. In the case of waves, questions of accuracy focused on the phases. In the two algorithms discussed here, the mode amplitudes were held fixed by construction. The modes most in error, the short wavelengths, do not decay as they do in diffusion. Using the reversible, centered implicit algorithm, with timesteps much longer than the explicit stability condition requires, must be done cautiously. If some physical damping or diffusion is present in addition to the waves, the short wavelengths decay. There might not be a problem if the amplitudes become small enough before the growing phase errors become objectionable. If, however, waves are weakly damped, the short wavelengths may be troublesome enough to require the use of a Fourier or spectral method which minimizes both phase and amplitude errors (though Gibbs errors still require special consideration).

Asymptotic methods are more difficult to formulate for waves than for local effects. Examples of such methods are the WKB approximation and various multiple time-scale techniques. An asymptotic method might be considered to solve the wave equation if it is possible to average over fast oscillations and extract quadratic effects which grow in time. The mean values of the oscillating variables themselves are zero.

4–6. COUPLING

4–6.1 Problems in Coupling the Terms

Modeling reactive flows involves more difficulties than the idealized problems treated in this chapter. The models must correctly describe the interactions of the basic chemical and physical processes in Table 4–1 — even when such interactions are not described by the simplified forms given. Occasionally a problem reduces to solving coupled equations where all the terms are of the same form, such as when we solve the nonlinear ordinary differential equations of chemical kinetics. More generally, it is necessary to solve equations of several forms which couple together the various physical processes to simulate phenomena such as flames and detonations. Many of the interesting physical effects and many of the numerical difficulties arise from the two-process interactions listed in Table 4–5. These six interactions are a good starting point for considering how to couple algorithms representing several

Table 4–5 Interactions and Their Effects

Interaction Processes	Some Effects Caused
Chemistry / Diffusion	Laminar flames, corrosion
Chemistry / Convection	Bouyant flames, turbulent burning
Diffusion / Convection	Molecular mixing in turbulence, Double-diffusion instabilities
Chemistry / Waves	Chemical-acoustic instability, Deflagration-to-detonation transition
Diffusion / Waves	Damping of high frequency sound waves
Convection / Waves	Shocks, turbulent noise generation

processes. If there are N processes to be coupled, there are $N(N-1)/2$ interactions among them which should be tested and calibrated separately.

Several methods have evolved to solve sets of time-dependent, nonlinear, coupled partial differential equations. The major methods are the *global-implicit method*, also called the *block-implicit method*, and the *fractional-step method*, also called *timestep splitting*. Other general approaches to coupling, such as the *Method-of-Lines* and *finite-element methods*, are also used. Each of these has a somewhat different philosophy, although they do have common elements. The various approaches are often combined into "hybrid" algorithms in which different coupling techniques are used for different interactions. For general reading on different ways multiple time scales are treated, we recommend the book edited by Brackbill and Cohen (1985).

Small errors in simulating convection can cause large errors in the overall simulation. Consider, for example, a simulation of chemical reactions occurring behind a shock. The shock heats the combustible mixture, and the raised temperature and pressure initiate chemical reactions. In some mixtures, an error of even a few percent in the temperature calculated immediately behind a shock can cause an order-of-magnitude change in the very sensitive chemical reaction rates. This can subsequently cause large changes in the time delay for appreciable energy release to begin. More generally, each of the interactions listed in Table 4–5 has regions of parameter space where one variable is very sensitive to small errors in the calculation of other variables.

4–6.2 Global-Implicit Coupling

The nonlinear sets of coupled partial differential equations describing a reactive flow system can be written

$$\frac{d}{dt}\boldsymbol{\rho}(\mathbf{x},t) = \mathbf{G}(\boldsymbol{\rho},\nabla\boldsymbol{\rho},\nabla\nabla\boldsymbol{\rho},\mathbf{x},t) \qquad (4-6.1)$$

where $\boldsymbol{\rho}$ is now a vector, each component of which is a function of the time t and the vector position $\mathbf{x}$. A global-implicit method solves this equation using fully implicit finite-difference formulas because of their superior stability properties. This means that the right side of Eq. (4–6.1) is formally evaluated using information at the advanced time. To do so, the nonlinear terms in $\mathbf{G}$ have to be linearized locally about the known solutions at the previous timestep. This approach is the direct generalization of the implicit algorithms presented earlier in this chapter to the case where several types of terms are treated at once.

To see more clearly how a global-implicit method works, write Eq. (4–6.1) in the form

$$\frac{\boldsymbol{\rho}^n(\mathbf{x},t^n) - \boldsymbol{\rho}^{n-1}(\mathbf{x},t^{n-1})}{\Delta t} = \mathbf{G}^n(\boldsymbol{\rho}^n,\nabla\boldsymbol{\rho}^n,\nabla\nabla\boldsymbol{\rho}^n,\mathbf{x},t^n)\,. \qquad (4-6.2)$$

This can be rewritten more simply as

$$\boldsymbol{\rho}^n \approx \boldsymbol{\rho}^{n-1} + \Delta t\,\mathbf{G}^n \qquad (4-6.3)$$

where the arguments $\mathbf{x}$ and t are suppressed. We have assumed that changes in the values of $\boldsymbol{\rho}$ are small in one timestep, that is, that Δt is suitably small.

The difficulty with using Eq. (4–6.3) to find the new values of $\{\boldsymbol{\rho}\}$ is the implicit dependence of $\mathbf{G}$ on the new (unknown) values of $\boldsymbol{\rho}$. If the changes in $\boldsymbol{\rho}$ are small during a timestep, it is consistent to assume that the changes in $\mathbf{G}$ are also small. In this case we can approximate $\mathbf{G}^n$ by the first-order Taylor series expansion

$$\mathbf{G}^n = \mathbf{G}^{n-1} + \Delta\boldsymbol{\rho}\frac{\partial\mathbf{G}^{n-1}}{\partial\boldsymbol{\rho}}\,. \qquad (4-6.4)$$

The Jacobian, $\partial\mathbf{G}^{n-1}/\partial\boldsymbol{\rho}$, is a tensor when $\boldsymbol{\rho}$ is a vector of variables. It is evaluated using the known values of $\boldsymbol{\rho}$ and hence can be evaluated explicitly. Defining $\Delta\boldsymbol{\rho}$ as the change in $\boldsymbol{\rho}$ from the old time to the new time,

$$\Delta\boldsymbol{\rho}^n \equiv \boldsymbol{\rho}^n - \boldsymbol{\rho}^{n-1} = \Delta t\,\mathbf{G}^n\,, \qquad (4-6.5)$$

we can combine Eqs. (4–6.4) and (4–6.5) to obtain a linearized global-implicit approximation for the new values of $\boldsymbol{\rho}$,

$$\Delta\rho^n = \Delta t\, \mathbf{G}^{n-1}\left[\mathbf{1} - \Delta t\, \frac{\partial \mathbf{G}^{n-1}}{\partial \rho}\right]^{-1} . \qquad (4-6.6)$$

An implicit matrix equation results from this procedure, and a large evolution matrix, $1 - (\Delta t)\partial \mathbf{G}^{n-1}/\partial \boldsymbol{\rho}$, must generally be inverted.

In order to use this formalism, linear finite-difference approximations to the convective derivatives must be used. This generally prevents use of the more accurate nonlinear methods. A rigorously convergent treatment requires iterations of matrix inversions at each timestep. In one dimension the problem usually involves a block tridiagonal matrix with M physical variables at N_x grid points. In this case an $(MN_x \times MN_x)$ matrix must be inverted at each iteration of each timestep. The blocks on, or adjacent to, the matrix diagonal are $M \times M$ in size, so that the overall matrix is quite sparse. Nevertheless, an enormous amount of computational work goes into advancing the solution a single timestep. In this approach, the matrices are $(MN_xN_y \times MN_xN_y)$ in two dimensions and $(MN_xN_yN_z \times MN_xN_yN_z)$ in three dimensions, where N_y and N_z are the number of grid points in the y-direction and z-direction, respectively.

When only convection is involved, implicit methods can be relatively inexpensive. It is then possible to combine the equations to obtain an implicit elliptic equation, which is often much easier to solve than the general case given in Eq. (4–6.6). In these global-implicit algorithms, first-order numerical damping (or diffusion) is required to maintain positivity. Because the equations are linearized, iterations are needed to include nonlinear terms which are partially or fully implicit.

In general, the level of accuracy in an implicit Eulerian convection calculation is equivalent to that obtained using a monotonic algorithm with half the number of grid points in each spatial dimension. Thus the global-implicit approach costs more by about a factor of four in two dimensions, and eight in three dimensions, if monotone transport algorithms cannot be used.

In complex kinetics problems with no spatial variation, the M fluid variables are the species number densities plus the temperature of the homogeneous volume of interest. The Gear method, discussed in Chapter 5, is an example of a global-implicit approach for integrating sets of coupled ordinary differential equations.

4–6.3 Timestep Splitting or Fractional-Step Coupling

The second coupling method is the fractional-step approach, also called timestep splitting, in which the individual processes are solved independently and the changes resulting from the separate partial calculations are coupled (added) together. The processes and the interactions among them may be treated by analytic, asymptotic, implicit, explicit, or other methods. An advantage of this approach is that it can avoid many costly matrix operations and allow the best method to be used for each type of term. The exact way the processes are coupled depends on the individual properties of the different algorithms used.

Timestep splitting is described in detail by Yanenko (1971). The qualitative criterion for its validity is that the values of the physical variables must not change too quickly over a timestep from any of the individual processes. This requirement was also noted earlier for global-implicit algorithms. Consider writing Eq. (4–6.1) in a form suited to timestep splitting,

$$\frac{d}{dt}\rho(\mathbf{x},t) = \mathbf{G}_1 + \mathbf{G}_2 + \mathbf{G}_3 + \cdots + \mathbf{G}_M , \qquad (4-6.7)$$

where $\mathbf{G}$ has been broken into its constituent processes,

$$\mathbf{G} = \mathbf{G}_1 + \mathbf{G}_2 + \mathbf{G}_3 + \cdots + \mathbf{G}_M . \qquad (4-6.8)$$

Each of the functions $\{\mathbf{G}_i\}$ contributes a part of the overall change in ρ during the numerical timestep. Thus

$$\left.\begin{aligned} \Delta\rho_1^{n-1} &= \Delta t\, \mathbf{G}_1^{n-1} \\ \Delta\rho_2^{n-1} &= \Delta t\, \mathbf{G}_2^{n-1} \\ \vdots\quad & \quad \vdots \\ \Delta\rho_M^{n-1} &= \Delta t\, \mathbf{G}_M^{n-1} \end{aligned}\right\} . \qquad (4-6.9)$$

For example, $\mathbf{G}_1^{n-1}$ might be the chemical kinetics terms, $\mathbf{G}_2^{n-1}$ the diffusion terms, $\mathbf{G}_3^{n-1}$ the thermal conduction terms, and so on.

The solution for the new values ρ^n is found by summing all of the partial contributions,

$$\rho^n = \rho^{n-1} + \sum_{m=1}^{M} \Delta\rho_m^{n-1} . \qquad (4-6.10)$$

If each of the processes is simulated individually, Eq. (4–6.10) gives a simple prescription for combining the results. Recognizing at this point that there are some caveats, we note that this approach works quite well. This is the generic approach taken in most of this book. Its exact implementation in several programs is described in Chapter 13.

If the change in ρ contributed by each separate process G_i is small, explicit algorithms are stable for each of the processes individually. However, the timestep needed to satisfy all of the separate explicit stability conditions may be prohibitively small. Because many matrix operations are avoided in timestep splitting, it is often not very expensive to advance the model even though the timestep is short. Usually there is great value in being able to increase the effective timestep used in the fractional-step method.

4–6.4 Trade-offs in the Choice of a Coupling Method

When choosing a method to couple the physical processes, we also determine the overall structure of the simulation model itself. It is important to understand the trade-offs between these two approaches, global-implicit coupling and fractional-step coupling, at the beginning of a simulation project. For each approach, we consider stability, accuracy, and ease of use.

The global-implicit formalism nominally treats almost any set of coupled partial differential equations. This approach maximizes the strain on computer resources and minimizes the strain on the modeler. Stable solutions are guaranteed by added numerical damping which smooths out the profiles. Because of this damping, however, it is often hard to know if the answers have converged correctly. It is relatively easy for solutions to be wrong, yet stable. Convergence of the computed solutions must be tested by increasing the spatial and temporal resolution, which is a particularly expensive approach in multidimensions.

Timestep splitting puts less strain on computational resources, but demands more effort and thought on the part of the modeler. Stability is not guaranteed; it depends on the separate criteria of the individual algorithms and the properties of their coupling. Using timestep splitting, it is often easy to see when the answer is wrong, because then the solutions degrade in catastrophic ways. For example, in a convection calculation, the supposedly positive densities become negative. Convergence is usually tested by increasing the spatial and temporal resolution. When some aspect of the program uses asymptotic methods, however, situations arise for which reducing the timestep makes the answers less accurate. This failing of a model is usually obvious in the timestep-split calculation. For example, when an asymptotic calculation of a chemical kinetics problem fails, mass is no longer conserved.

A major advantage of timestep splitting is that it encourages modular simulation models. Each module can be programmed using the best available technique for that process. Various candidates for these algorithms are considered in greater depth in Chapters 5 through 11. If a new technique becomes available, it can often be incorporated without changing the entire implementation of a timestep-split model. Thus convection, diffusion, equation of state calculations, and chemical kinetics are all in separate packages. Particular care in testing the coupling between the software modules is required.

The pros and cons of using global-implicit methods versus timestep-splitting methods are roughly balanced with respect to accuracy of the solution. If a significant time scale is not resolved, neither solution method can give detailed profiles of phenomena occurring on that scale. Similarly, to compute spatial gradients accurately, these gradients must be resolved with enough grid points in either type of calculation. The potential of easily incorporating monotone convection algorithms is a plus for the fractional-step approach.

The fact that timestep splitting demands more thought and effort is counterbalanced by the fact that the resulting simulations are generally faster than those using global-implicit methods. Often the reduction in computing time is substantial. For example, solving a set of chemical kinetics equations for M species requires inverting $M \times M$ matrices with approximately M^3 operations. Other methods, such as asymptotic integration techniques, scale as M or M^2.

It is useful here to mention several other general approaches, such as the Method of Lines and the finite-element method. The Method of Lines approach focuses on the idea of separating the spatial and temporal parts of the problem regardless of the processes or nonlinearity. The full nonlinear set of equations is reduced to a set of stiff ordinary differential equations in time, which can be solved by one of the standard methods discussed in Chapter 5. This method is more tractable than the global-implicit methods when the equations are highly nonlinear. It does, however, preclude the use of monotone methods for convection (although there is now some research into hybrid algorithms) and it is not easily implemented in more than one dimension. Roughly double the spatial resolution is required in the Method of Lines to resolve shocks and steep gradients as is needed in the best monotone methods.

Finite-element methods (FEM) have yet another focus, though some of their features are common to both the Method of Lines and global-implicit methods. The finite-element approach finds the "best" solution by error minimization techniques. Thus FEMs are extremely appealing even though the computational expense is comparable to, or can exceed, global-implicit methods. One added feature of FEMs is that the locations of the grid points as well as the values of the dependent variables at these points can be made into degrees of freedom for the error minimization process. This added flexibility and power is discussed in Chapter 6 in the description of adaptive gridding.

Throughout this book we emphasize timestep splitting because it allows more flexibility, that is, it allows the use of different algorithms for solving each type of process. Ways of applying timestep splitting to flame and detonation calculations are described in the last chapter of this book.

References

Anderson, D.A., J.C. Tannehill, and R.H. Pletcher, 1984, *Computational Fluid Mechanics and Heat Transfer*, McGraw-Hill, New York.

Brackbill, J.U., and B.I. Cohen, 1985, *Multiple Time Scales*, Academic, New York.

Oran, E.S., and J.P. Boris, 1981, Detailed Modelling of Combustion Systems, *Prog. Ener. Comb. Sci.* 7: 1–72.

Potter, D., 1973, *Computational Physics*, Wiley, New York.

Roache, P.J., 1982, *Computational Fluid Dynamics*, Hermosa, Albuquerque, NM.

Yanenko, N.N., 1971, *The Method of Fractional Steps*, Springer-Verlag, New York.

Chapter 5

ORDINARY DIFFERENTIAL EQUATIONS: CHEMICAL KINETICS AND OTHER LOCAL PHENOMENA

Algorithms for integrating ordinary differential equations (ODEs) were originally derived by scientists interested in solving these equations for their particular applications. Bashforth and Adams (1883), for example, developed a method for their studies of capillary action. One of the first algorithms to cope with the difficulties of integrating stiff ODEs was suggested by Curtiss and Hirschfelder (1952) for chemical kinetics studies. Ten years after Curtiss and Hirschfelder identified the stiffness problem in ODEs, Dahlquist (1963) exposed numerical instability as the cause of the difficulty and provided basic definitions and concepts that are still helpful in classifying and evaluating algorithms. Since then, developing and testing integration methods for solving coupled ODEs has been an active field of research. The recent efforts of applied mathematicians have put numerical solution of ODEs on a sounder theoretical basis and have provided insights into the constraints imposed by stability, convergence, and accuracy requirements.

Our interest here is primarily in solving systems of ODEs describing chemical reactions in complex reactive flows. Coupled, nonlinear, first-order ODEs for such chemically reactive systems were shown in Eq. (2–1.2). The part of the equations describing chemical kinetics was isolated in Eq. (2–2.22), which describes the production and loss of reacting species:

$$\frac{\partial n_i}{\partial t} = Q_i - n_i L_i, \qquad i = 1, \ldots, N_s. \qquad (5-0.1)$$

Here $\{n_i\}$ are the species number densities and $\{Q_i\}$ and $\{n_i L_i\}$ are production and loss terms, respectively. The $\{Q_i\}$ and $\{L_i\}$ are functions of the $\{n_i\}$ and provide the nonlinear coupling among the various species.

Equations of this form also describe many other local processes besides chemical kinetics. ODEs appear when the Method-of-Lines, spectral, and other expansion methods are used on time-dependent partial differential equations. In these cases, spatial derivatives are converted to algebraic relationships leaving ODEs to be integrated in time. ODEs also describe the motions of projectiles and orbiting bodies, population dynamics, heat flow, electrical circuits, local temperature equilibration, external forces, momentum interchange in multiphase flows, the decomposition of radioactive material, and energy level and species conversion processes in atomic, molecular, and nuclear physics. Thus there are many applications of the methods and algorithms described in this chapter.

We want efficient, accurate algorithms for solving Eqs. (5–0.1) that can bes used for coupled sets of ODEs when they are integrated alone and for ODE systems which are part of a reactive flow calculation where many physical processes interact. As described below, these two types of applications often require different approaches.

In a reactive flow model, solution of the chemical kinetics is often the most expensive part of the calculation. In flame and detonation simulations, integrating the chemical kinetics equations may take an order of magnitude longer than solving the convective and diffusive transport terms. The computational cost is directly related to the number of species, the number of reactions among them, and the number of spatial cells in the computational representation. The cost also depends strongly on the particular form of the expressions for $\{Q_i\}$ and $\{L_i\}$. When these involve exponentials or fractional powers, as is usually the case with chemical rates, they become even more expensive to evaluate. Other problems arise when the stepsize required to obtain an accurate solution becomes prohibitively small.

We recommend a number of books and articles on ODEs which are more detailed than the presentation given here. These include introductory texts such as McCalla (1967), Carnahan, Luther and Wilkes (1969), and Ralston (1978), which describe the rudiments of integrating classical ODEs. More advanced texts include the books by Gear (1971) and Lambert (1973), which review methods for solving stiff equations. We also recommend the article by Shampine et al. (1976) which presents a practical evaluation of the methods for solving ODEs which are not stiff, and the review article by Bui et al. (1984) which describes recent advances in the numerical solution of ODEs. In particular, we recommend the recent review article by May and Noye (1984) who give a clear overview aimed at the researcher who needs to solve sets of ODEs. Portions of this chapter and much of the philosophy of our presentation are based on this article.

Because our primary interest in ODEs derives from a need to describe chemical kinetics in reactive flows, we use a notation which differs from those used in standard books and papers on ODEs. When discussing the generic properties of ODEs, we describe the evolution of the quantity $y(t)$ and call the independent variable t, instead of x, to emphasize the variation in time. When specializing to chemical kinetics problems, we call the dependent variables $\{n_i(t)\}$, the number densities of the chemical species at time t.

5–1. DEFINITIONS AND PROPERTIES

5–1.1 The Initial Value Problem

A first-order nonliner ODE can be written in the form

$$\frac{dy}{dt} = f(t,y) , \qquad y(t_o) = y_o , \tag{5 – 1.1}$$

where the independent variable, t, increases from an initial value t_o. The analogous set of coupled ODEs can be written in vector form,

$$\frac{d\mathbf{y}}{dt} = \mathbf{f}(t,\mathbf{y}) , \qquad \mathbf{y}(t_o) = \mathbf{y}_o . \tag{5 – 1.2}$$

These are called initial value problems. Given the initial conditions $\mathbf{y}_o$, we want to integrate this system forward in time to determine the values of $\mathbf{y}(t)$ at later times. Equations of the form

$$\frac{d\mathbf{y}}{dt} = \mathbf{f}(\mathbf{y}) , \qquad \mathbf{y}(t_o) = \mathbf{y}_o , \tag{5 – 1.3}$$

without the functional dependence on t appearing on the right hand side, are called *autonomous* systems of ODEs. Any system governed by Eq. (5–1.2) can be put into the autonomous form by adding one more equation for t to the system. The general solutions of Eq. (5–1.1) or Eq. (5–1.2) are families of curves and the initial conditions single out particular solutions in this family.

One approach to computationally solving sets of ODEs is to use expansions in terms of basis functions for representing the variations of a function. The vector $\mathbf{y}(t)$ is expanded in a series of functions of t, and the coefficients of each term in the series are chosen so that the sum of the terms representing $\mathbf{y}(t)$ satisfies the initial conditions, $\mathbf{y}_o$. If the expansion function coefficients are suitably chosen, evaluating these expansions at subsequent values of t gives an approximation to the solution. A common expansion method uses a Taylor series for determining $\mathbf{y}(t + \Delta t)$ in terms of $\mathbf{y}(t)$. Difficulties arise because the expansions can be extrapolated forward only a finite time before they become hopelessly inaccurate, and the methods do not work well for stiff equations.

In this chapter we concentrate on finite-difference methods for solving ODEs which divide the interval of time t over which we wish to integrate the equations into a discrete set of values $\{t_n\}$, where $n = 0, 1, \ldots, N$. The intervals, or timesteps, between the various values t_n are denoted as $\{h_n\}$, where

$$h_n \equiv t_n - t_{n-1} . \tag{5 – 1.4}$$

The exact solution to Eq. (5–1.1) at the nth time t_n is denoted by $y(t_n)$. The solution obtained though a finite-difference method is denoted y_n. Given y_n, we need to know how to determine the value at the next step, y_{n+1}, as accurately and efficiently as possible. The different timesteps are denoted by subscripts here because in this chapter there are no spatial variations or grids for which subscripts are usually used, and because the subscripts will then not be confused with exponents which are used.

5 1.2 Accuracy, Convergence, and Stability

We can express a scalar ODE algorithm as

$$y_{n+1} = y_n + h_n \mathcal{F}\left(h_n, \{y_i\}, \left\{\left.\frac{dy}{dt}\right|_i\right\}, \ldots\right), \qquad (5-1.5)$$

where $i = 0, 1, \ldots, n+1$ indicates the various discrete steps at which the solution has been calculated. The numerical scheme which advances y_n is embodied in the function $\mathcal{F}$. The most useful numerical schemes depend on at most a few previous values, that is, $i = n,\ n-1,\ n-2,\ \ldots$.

Different kinds of errors can appear during this procedure, several of which are illustrated in Figure 5–1. The upper curve in the figure is the exact solution $y(t)$ as a function of t in the interval t_n to t_{n+1}. The values of $y(t_n)$ and $y(t_{n+1})$ are the correct values for the solutions at t_n and t_{n+1} using y_o, the exact initial conditions specified at t_o. The lower solid curve, $y^*(t)$, is the exact solution of the equations obtained by using y_n, the numerically calculated value at t_n, as the initial condition for subsequent integration. Because y_n can deviate appreciably from the correct solution $y(t_n)$, systematic error will also appear in the exact solution $y^*(t_{n+1})$, as shown in the figure. The dashed curve connecting y_n and y_{n+1} is the numerical solution obtained from Eq. (5–1.6) using an appropriate function $\mathcal{F}$. This numerical solution also passes through the correct initial condition y_o because the numerical solution was initialized to that value.

The quantity E in Figure 5–1 is the *global truncation error* at t_n, defined by

$$E_n \equiv y(t_n) - y_n . \qquad (5-1.6)$$

The quantity E_n is the difference between the exact solution at t_n based on the original initial conditions and the current numerical approximation. At t_{n+1} the global error is E_{n+1}, as we show in the figure.

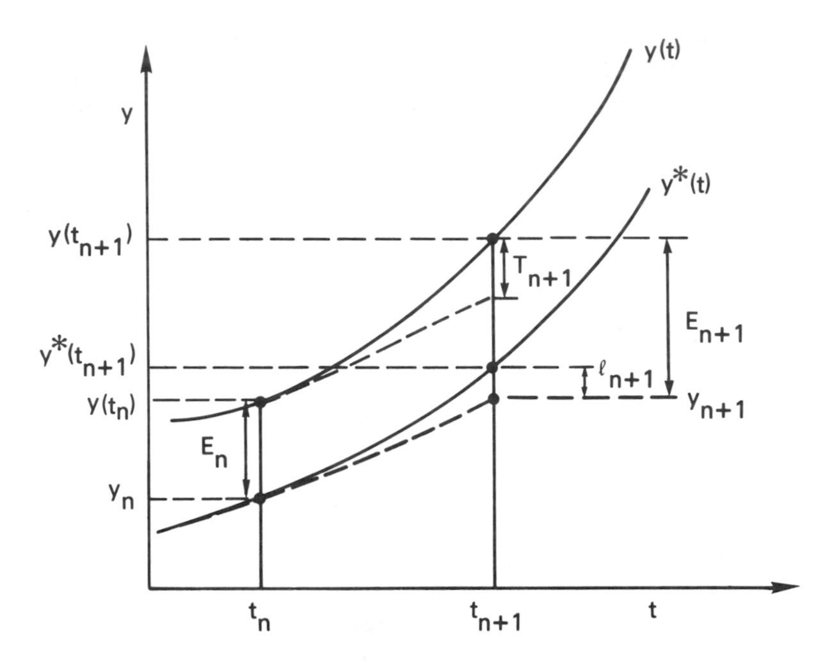

Figure 5–1. Types of errors in a finite-difference timestep. The exact solution, found from integrating from the starting condition at $y(t_o)$, is denoted $y(t)$. The exact solution starting from the numerical value at $y(t_n)$ is denoted $y^*(t)$. The global truncation error is denoted E_n. The local truncation error is denoted T_n. The local error is denoted ℓ_n.

A *local truncation error*, T_{n+1}, is defined by

$$\begin{aligned} T_{n+1} &\equiv y(t_{n+1}) - y(t_n) - h_n \mathcal{F} \\ &= E_{n+1} - E_n \ . \end{aligned} \qquad (5-1.7)$$

This is an estimate of the amount by which the finite-difference approximation, Eq. (5–1.5), deviates from the analytic solution $y(t)$. The global truncation error is the sum of the local truncation errors made at each step.

Finally, there is a *local error*, ℓ_n, which is the error made in one step using a particular algorithm, assuming the values of the dependent variable at the previous step are used as initial conditions,

$$\ell_{n+1} = y^*(t_{n+1}) - y_{n+1} \ . \qquad (5-1.8)$$

The local truncation error T_n is approximately equal to the local error ℓ_n because $y(t) \approx y^*(t)$. Although the exact solution is unavailable in practice, these different errors are important because they guide our thinking about errors and error propagation during the integration. In many situations, the

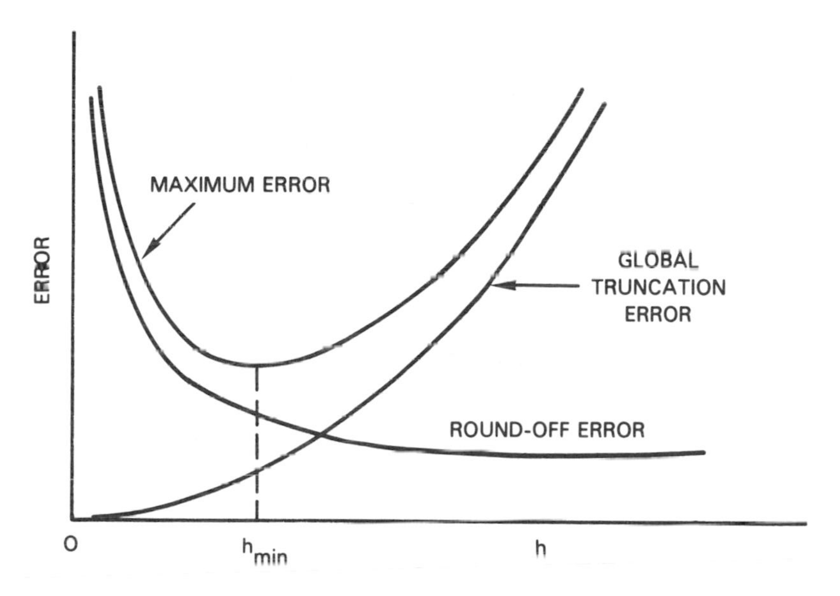

Figure 5–2. Schematic diagram showing error due to roundoff and truncation as a function of timestep.

way errors accumulate over many steps is far more important than the error made at any particular step.

Errors are also introduced by machine-dependent numerical roundoff. The global truncation error decreases as h decreases for a fixed integration interval, but the maximum roundoff error increases. This is illustrated in Figure 5–2. Roundoff error can contribute significantly to the global error when very short timesteps must be used. Thus there is an optimal stepsize, shown as $h_{\min}$ in Figure 5–2, where truncation and roundoff errors are comparable.

The *order of accuracy* of a method is defined by expressing the measure of error in powers of the stepsize h. If the global truncation error behaves as $\mathcal{O}(h^p)$ for $p > 0$, the method is said to be pth order. In general, a pth order method has a local truncation error which scales as $\mathcal{O}(h^{p+1})$. The global truncation error scales as h^{-1} of the local truncation error when the errors accumulate proportional to the number of steps, which is often the case. We generally use higher-order methods because they converge faster and are more accurate. However, there are many problems for which a lower-order method, perhaps with a smaller stepsize, is the best approach to use. A lower-order method whose local truncation error varies in sign and tends to cancel from step to step may be more accurate globally than a high-order method whose truncation error accumulates.

Stability and convergence are both concerned with the behavior of the computed solution in the limit as the stepsize h goes to zero. A method *converges* if

$$y_n \to y(t_n), \qquad t \in [t_o, b]$$
$$\text{as } h \to 0 \qquad \text{and} \qquad y(t_o) \to y_o \, .$$

A method is said to be *unstable* if an error introduced at some stage in the calculation becomes unbounded, that is, the overall global error continually increases instead of decreases. Note that the numerical solution can be unstable and at the same time relatively accurate if the real solution also grows indefinitely.

A method can also be stable, but not converge. For example,

$$y_n = y_{n-1}, \qquad n = 1, 2, \ldots$$

does not converge unless $\mathcal{F} = 0$, but is stable. However, a convergent method is stable. Thus stability does not require convergence, but convergence requires stability.

One reasonable way to control global error is to control the local error at each step. The strategy is usually to choose an accurate, stable method that allows a reasonable stepsize. In general, stepsize control is the mechanism used to keep local errors acceptably small. However, another strategy is to vary the order or even the type of method during the course of the integration depending on the nature of $\mathcal{F}(t, y)$.

During relatively short but important periods in the evolution of many reactive flows, fluctuations in the fluid dynamics, transport, chemistry, or some combination of these, grow rapidly. Usually these periods of unstable growth are short because most physical instabilities disrupt the system which drives them. Nevertheless, understanding a numerical simulation during these periods of growth may require temporarily rethinking the meanings of stability and convergence. Because the relative errors are measured against a constantly growing scale, a weakly unstable algorithm may be accurate enough in a physically unstable environment. Conversely, in a system that is rapidly decaying to an equilibrium, a convergent algorithm may appear unstable because of increasing relative errors, even though the absolute errors are getting smaller.

In Chapter 4 we introduced stability concepts through a linear test problem given in Eq. (4–2.1). If a numerical method is unstable for such a simple problem, it must be judged unreliable at best for more complex problems. Also, most ODEs can be linearized locally by expanding in a

Taylor series out to the first derivative. For small intervals about t_n, the values of $\mathbf{f}(\mathbf{y})$ will be close to $\mathbf{f}(\mathbf{y}_n)$, and thus

$$\mathbf{f}(\mathbf{y}) = \mathbf{f}(\mathbf{y}_n) + (\mathbf{y} - \mathbf{y}_n) \cdot \left.\frac{\partial \mathbf{f}}{\partial \mathbf{y}}\right|_{\mathbf{y}_n} + \text{higher order terms} . \qquad (5-1.9)$$

When this is substituted into Eq. (5–1.2), we obtain

$$\begin{aligned} \frac{d\mathbf{y}}{dt} &\approx \mathbf{y} \cdot \left.\frac{\partial \mathbf{f}}{\partial \mathbf{y}}\right|_{\mathbf{y}_n} + \left[\mathbf{f}(\mathbf{y}_n) - \mathbf{y}_n \cdot \left.\frac{\partial \mathbf{f}}{\partial \mathbf{y}}\right|_{\mathbf{y}_n}\right] \\ &\approx \mathbf{\Lambda} \cdot \mathbf{y} + \mathbf{g}(\mathbf{y}_n) , \end{aligned} \qquad (5-1.10)$$

which is the vector extension of the linear problem treated in Chapter 4. The term in square brackets defines $\mathbf{g}(\mathbf{y}_n)$, an additive constant vector which could be included in the analysis of Chapter 4. This term shifts the zero point of the dependent variables $\{\mathbf{y}\}$. By considering small enough intervals, the integral of the approximate equation, Eq. (5–1.10), is arbitrarily close to the correct solution.

In this linear case, $\mathbf{\Lambda}$ is the $N \times N$ Jacobian matrix $\mathbf{J}$, whose elements are given by

$$J_{ij} \equiv \frac{\partial f_i}{\partial y_j} , \qquad (5-1.11)$$

where N is the number of independent variables in the autonomous system of equations. For the scalar case, or for each of the eigenvectors of the matrix problem given in Eq. (5–1.10), we can write

$$(y_{n+1} - y_\infty) = r(\lambda h)(y_n - y_\infty) , \qquad (5-1.12)$$

where

$$y_\infty \equiv \frac{h g(y_n)}{(1 - r(\lambda h))} , \qquad (5-1.13)$$

and r is a function of λh that depends on the particular method.

When the real part of λ is negative, y_∞ is the value that y approaches as t approaches $+\infty$. When the real part of λ is positive, so that the problem is unstable, y_∞ is the value that y approaches as t approaches $-\infty$. The limiting value y_∞ is undefined when $r(\lambda h)$ is identically unity. The function $r(\lambda h)$ is the numerical approximation to the analytic solution, that is,

$$r(\lambda h) = e^{\lambda h} + \mathcal{O}(h^{p+1}) , \qquad (5-1.14)$$

where p is the order of the method. That is, the numerical method gives the analytic exponential solution plus a local error of order h^{p+1}. The local and truncation errors are bounded and decay if

$$|r(\lambda h)| \leq 1 \; . \tag{5-1.15}$$

Absolute stability is defined as

$$|r(\lambda h)| < 1 \; . \tag{5-1.16}$$

The region in the complex λh plane where Eq. (5–1.16) is satisfied is called the *region of absolute stability.*

5–1.3 Implicit and Explicit Solutions

An algorithm for Eq. (5–1.1) or (5–1.2) is said to be *explicit* if the function $\mathcal{F}$ does not involve the variable y_{n+1}, and *implicit* if it does. Implicit methods are usually more stable and slightly less accurate than stable, explicit methods and are more expensive to implement. They generally require an iteration, and each iteration step can be as costly as the explicit algorithm. When the solution is slowly varying, however, implicit methods allow long stepsizes that would make an explicit algorithm unstable. Thus they are particularly useful when the ODEs are stiff.

5–1.4 Stiff Ordinary Differential Equations

Practically, ODEs are *stiff* when numerical stability rather than accuracy dictates the choice of stepsize. Consider two coupled equations, perhaps in the form of Eq. (5–1.2), which have the solution

$$\begin{aligned} y_1(t) &= +e^{-2000t} + e^{-2t} + 1 \\ y_2(t) &= -e^{-2000t} + e^{-2t} + 1 \; . \end{aligned} \tag{5-1.17}$$

There are two transients here. The fast one, $\exp(-2000t)$, decays by time $t = 0.01$. The slow one, $\exp(-2t)$, decays by $t = 10$, leaving the steady-state solution $y_1 = y_2 = 1$. An accurate numerical integration uses a small stepsize for the period $t < 0.01$ because the solutions are changing rapidly and the algorithm has to resolve that variation. After this fast transient, however, it would be useful to be able to use a larger stepsize.

Unfortunately, stability criteria do not care whether the unstable components are important or not, so many methods require a small stepsize throughout the integration. Even though the numerical solution converges as $h \to 0$, h must be intolerably small. Here $h \sim 5 \times 10^{-4}$ s is required throughout the integration, so that it would take 20,000 steps to reach ten seconds of real time. The system is practically and mathematically stiff. Often h must be so small for stability that roundoff errors become critical.

Stiffness occurs because there are a wide range of time scales in the solution. This leads to another more rigorous definition of stiffness (Lambert, 1980), namely that a set of equations of the form

$$\dot{\mathbf{y}} = \mathbf{J} \cdot \mathbf{y}, \tag{5-1.18}$$

where $\mathbf{J}$ is the Jacobian matrix, is stiff if

$$\begin{aligned} &(i) \quad \Re(\lambda_j) < 0, \qquad \text{for } j = 1, ..., N \\ &(ii) \quad \frac{\max|\Re(\lambda_j)|}{\min|\Re(\lambda_j)|} \gg 1 , \end{aligned} \tag{5-1.19}$$

where $\Re(\lambda)$ indicates the real part of λ. This definition of stiff does not apply when the independent modes of the system are unstable.

Equation (5–1.18) is stiff mathematically if $\mathbf{J}$ has at least one eigenvalue whose real part is negative and large compared to the time scales of variation displayed by the solution. A system is stiff if it contains transients which decay fast compared to the typical scale of integration (Gear, 1971; Curtis, 1978). In practical computation, a system is stiff if the stepsize we would like to take, based on cost or running time, is too large to get an accurate answer. This is a statement about the minimum stepsize we can tolerate because of computing time in a particular application and about the method of solution chosen.

Sets of coupled equations may be stiff, but a single equation can also be stiff (Gear, 1971). Consider the equation

$$\frac{dy}{dt} = \lambda[y - F(t)] + \frac{dF(t)}{dt} , \tag{5-1.20}$$

where $\lambda < 0$, $|\lambda| \gg 1$, and $F(t)$ is a smooth, slowly varying function. The analytic solution is

$$y = [y_o - F(0)]e^{\lambda t} + F(t) . \tag{5-1.21}$$

The λt exponent soon becomes negative enough so that the first term is negligibly small compared to the second. Nevertheless, λ controls the timestep. Compare this solution with Eq. (5–1.10). Here again, the constant term is not zero and is associated with a particular inhomogeneous solution to the equations, but the stability and stiffness are determined by the homogeneous solutions. Given a numerical method for solving Eq. (5–1.20), the local truncation error is determined by h and stability depends on the value of λh, even though the solution, Eq. (5–1.21), does not. This is clearly unacceptable. We consider special methods for solving stiff equations in Section 5–3.

5–2. OVERVIEW OF CLASSICAL METHODS OF SOLUTION

In this section we describe common classical methods for solving ODEs. Whereas these methods are not directly useful for treating the chemical kinetics terms in reactive flow calculations, they provide the background needed for understanding methods for stiff equations. In addition, the algorithms described here are useful in other contexts.

5–2.1 One-Step Methods

In a one-step method, the value of y_{n+1} at t_{n+1} is calculated knowing only h, t_n, and y_n. Information about the solution at only the most recent step is used. In two- or three-dimensional simulations, there are only a few fluid variables at each grid point but there can be many species present. Each species density has to be integrated and stored at each grid point for each timestep. Thus for a large reactive flow simulation, one-step methods are the methods of choice because they require a minimal amount of information to be stored in the computer memory. If additional values of these variables from steps before t_n have to be saved for the integration algorithm, memory requirements would increase drastically. Even when computer memory is not a restriction, the nonlocal and fluid dynamic changes occurring in the simulation mean that extrapolations of species densities from previous timesteps are suspect if they are not updated to account for the fluid dynamic changes. Thus most large reactive flow calculations use one-step methods.

A general one-step method can be written as

$$y_{n+1} = y_n + h\mathcal{F}(t_n, y_n, h), \qquad (5-2.1)$$

from Eq. (5–1.5) above. The derivative approximation $\mathcal{F}$ can be evaluated either from a Taylor series expansion about t_n,

$$y_{n+1} = y_n + h\frac{dy_n}{dt} + \frac{h^2}{2}\frac{d^2y_n}{dt^2} + \cdots , \qquad (5-2.2)$$

or from approximations of the function $\mathcal{F}$ in the integral formula

$$y_{n+1} = y_n + \int_{t_n}^{t_{n+1}} \mathcal{F}(t, y(t))\, dt . \qquad (5-2.3)$$

In this case, approximations to $\mathcal{F}$ typically take the form of polynomials or exponentials.

The simplest one-step method is the explicit, first-order Taylor algorithm or the *Euler method*,

$$y_{n+1} = y_n + hf(t_n, y_n) \ . \qquad (5-2.4)$$

Taylor series algorithms for higher orders can be written analogously from Eq. (5–2.2). Usually the higher derivatives which appear are found from differentiating Eqs. (5–1.1)–(5–1.3) analytically and substituting these expressions for the derivatives in the Taylor series.

Runge-Kutta methods are efficient, easily programmed, one-step algorithms that generally give higher-order accuracy at lower cost than Taylor series methods. Their gains come from evaluating the function $f(t, y)$, at more than one point in the neighborhood of (t_n, y_n) instead of evaluating higher derivatives. The function $\mathcal{F}$ is expressed as a weighted average of first derivatives obtained numerically at points in the region $[t_n, t_{n+1}]$.

The generic form for an R-stage Runge-Kutta algorithm is

$$k_r = f\left(t_n + ha_r,\ y_n + h\sum_{s=1}^{R} b_{rs}k_s\right), \qquad r = 1, 2, \ldots, R\ , \qquad (5-2.5a)$$

where r labels the stage and

$$y_{n+1} = y_n + h\sum_{r=1}^{R} c_r k_r \ . \qquad (5-2.5b)$$

The sequences of quantities $\{a_r\}$, $\{b_{rs}\}$, and $\{c_r\}$ are constants. Also,

$$\sum_{r=1}^{R} c_r \equiv 1\ , \qquad (5-2.6)$$

and there are a number of other constraints on these coefficients. Equations (5–2.5) give explicit Runge-Kutta algorithms when all k values used as arguments to f are calculated at an earlier step, that is, when $b_{rs} = 0$, for all $s \geq r$. Otherwise, Eqs. (5–2.5) give implicit Runge-Kutta algorithms. Many possible algorithms come from Eqs. (5–2.5) which Gear (1971) and May and Noye (1984) discuss in some detail. Because most of the ODEs that describe chemical reactions become stiff and these explicit methods are not applicable to stiff equations, only a few explicit Runge-Kutta algorithms are reproduced here.

Table 5–1 shows four Runge-Kutta algorithms. The first two are second-order accurate and the last two are fourth-order accurate. The first is the

Table 5–1. Runge-Kutta Integration Algorithms

Modified Euler Method $\mathcal{O}(h^2)$

$$k_1 = f(t_n, y_n)$$
$$k_2 = f(t_n + \tfrac{1}{2}h, y_n + \tfrac{1}{2}hk_1)$$
$$y_{n+1} = y_n + hk_2$$

Improved Euler Method $\mathcal{O}(h^2)$

$$k_1 = f(t_n, y_n)$$
$$k_2 = f(t_n + h, y_n + hk_1)$$
$$y_{n+1} = y_n + \tfrac{1}{2}h(k_1 + k_2)$$

Classical Runge-Kutta Method $\mathcal{O}(h^4)$

$$k_1 = f(t_n, y_n)$$
$$k_2 = f(t_n + \tfrac{1}{2}h,\ y_n + \tfrac{1}{2}hk_1)$$
$$k_3 = f(t_n + \tfrac{1}{2}h,\ y_n + \tfrac{1}{2}hk_2)$$
$$k_4 = f(t_n + h,\ y_n + hk_3)$$
$$y_{n+1} = y_n + \tfrac{h}{6}[k_1 + 2k_2 + 2k_3 + k_4]$$

Runge-Kutta-Gill Method $\mathcal{O}(h^4)$

$$k_1 = f(t_n, y_n)$$
$$k_2 = f(t_n + \tfrac{1}{2}h,\ y_n + \tfrac{1}{2}hk_1)$$
$$k_3 = f(t_n, +\tfrac{1}{2}h,\ y_n + (-\tfrac{1}{2} + \tfrac{1}{\sqrt{2}})hk_1 + (1 - \tfrac{1}{\sqrt{2}})hk_2)$$
$$k_4 = f(t_n + h,\ y_n - \tfrac{1}{\sqrt{2}}hk_2 + (1 + \tfrac{1}{\sqrt{2}})hk_3)$$
$$y_{n+1} = y_n + \tfrac{h}{6}\left[k_1 + 2(1 - \tfrac{1}{\sqrt{2}})k_2 + 2(1 + \tfrac{1}{\sqrt{2}})k_3 + k_4\right]$$

modified Euler algorithm, an explicit two-stage ($R = 2$) method with $a_2 = \frac{1}{2}$, $b_{21} = \frac{1}{2}$, $c_2 = 1$, and all of the other constants equal to zero. The modified Euler method is sometimes called the *midpoint rule* or the *modified midpoint rule*. The *improved Euler method* is also an explicit two-stage method, but now $a_2 = 1, b_{21} = 1, c_1 = c_2 = \frac{1}{2}$, and all other constants are zero.

Second-order accuracy is obtained in the modified Euler and improved Euler methods by estimating the derivative f at the center of the interval $[t_n, t_{n+1}]$. In the modified Euler method, Eq. (5–2.4) is used to generate a first-order estimate of the new value of y at the half step, $t_n + \frac{h}{2}$. The average estimate of the derivative is a step-centered quantity. The improved Euler method takes the average of the old derivative and the first-order estimate of the new derivative at the end of the step.

The two-stage modified Euler method is the basis for the second-order time integration in the Lax-Wendroff method and the Flux-Corrected Transport algorithms described in Chapters 8 and 9. It is used for these because it is straightforward to program and requires only two derivative evaluations per timestep. The improved Euler method also has these properties but requires additional storage because k_1 is used twice and hence cannot be overwritten with the k_2 values, as in the modified Euler method. In complicated reactive flow calculations with many computational cells, speed comes from being able to change the stepsize quickly and efficiently, rather than from obtaining high order or a slightly longer step. When the solution changes its nature quickly, the method that can quickly vary the timestep to the correct size with the fewest wasted evaluations of the derivative function is usually the method of choice.

The most commonly used Runge-Kutta methods are the fourth-order explicit methods listed in Table 5–1. The last algorithm in the table is the Runge-Kutta-Gill algorithm (Gill, 1951; see also May and Noye, 1984) which we have found particularly useful. The computational form of the Runge-Kutta-Gill algorithm is given in Table 5–2. The method uses one additional vector, $\mathbf{q}$, that is passed along with the dependent variables $\mathbf{y}$. This is an error control vector which is initially zero, but then changes from step to step to reflect the accumulation of error in the calculation.

In implicit Runge-Kutta methods, at least one of the coefficients b_{rs} is nonzero for $s \geq r$, so that at least one of the k_r's must be found implicitly. These implicit Runge-Kutta methods require iterative procedures but are extremely accurate and stable. They are also applicable to stiff ODEs, as discussed in the next section.

5–2.2 Linear Multistep Methods

A k-step linear multistep method can be written in the general form

$$y_{n+k} = h\beta_k f_{n+k} + \sum_{j=0}^{k-1} (h\beta_j - \alpha_j y_{n+j}) \; . \qquad (5-2.7)$$

We typically need to know the set of past quantities

$$(t_n, y_n), \; (t_{n-1}, y_{n-1}), \ldots, \qquad (5-2.8)$$

at equally spaced intervals in h to compute y_{n+1}. Thus results must be stored for several steps back. Adams methods, the explicit Adams-Bashforth methods, and the implicit Adams-Moulton methods are all linear multistep

Table 5–2. Computational Form of Gill's Method

Stage 1	given (t_n, y_n, q_4, and h)
	$k_1 = f(t_n, y_n)$
	$q_1 = q_4 + \frac{3}{2}(hk_1 - 2q_4) - \frac{1}{2}hk_1$
	$w_1 = y_n + \frac{1}{2}(hk_1 - 2q_4)$
Stage 2	$k_2 = f(t_n + \frac{1}{2}h,\ w_1)$
	$q_2 = q_1 + 3(1 - \frac{1}{\sqrt{2}})(hk_2 - q_1) - (1 - \frac{1}{\sqrt{2}})hk_2$
	$w_2 = w_1 + (1 - \frac{1}{\sqrt{2}})(hk_2 - q_1)$
Stage 3	$k_3 = f(t_n + \frac{1}{2}h,\ w_2)$
	$q_3 = q_2 + 3(1 + \frac{1}{\sqrt{2}})(hk_3 - q_2) - (1 + \frac{1}{\sqrt{2}})hk_3$
	$w_3 = w_2 + (1 + \frac{1}{\sqrt{2}})(hk_3 - q_2)$
Stage 4	$k_4 = f(t_n + h,\ w_3)$
	$q_4 = q_3 + \frac{1}{2}(hk_4 - 2q_3) - \frac{1}{2}hk_4$
	$y_{n+1} = w_3 + \frac{1}{6}(hk_4 - 2q_3)$

methods. Examples of these methods up to $k = 4$ are given in Table 5–3. The $k = 1$ Adams-Bashforth method is the same as the first-order Euler method. The $k = 1$ Adams-Moulton method is called the *trapezoidal method* because it is the trapezoidal quadrature formula. These are *predictor-corrector methods* because they use a lower-order method to predict the answers until enough timesteps have accumulated to carry out the full multistep procedure.

In practical applications, predictor-corrector methods compare favorably to Runge-Kutta methods. For equal computational effort, predictor-corrector methods can usually be made more accurate, but Runge-Kutta methods generally require less storage and are *self-starting*, because they only require data at one time level to begin the integration. Predictor-corrector methods often use a one-step method to accumulate enough values from previous times to proceed. In Runge-Kutta methods, it is easier to vary the stepsize. The Runge-Kutta methods are preferred for low accuracy requirements when the derivative evaluation is not expensive.

Linear multistep methods are generally not used to solve the ODE parts of large reactive flow problems. This is because they require storing values of $\mathbf{y}$ from more than one previous timestep, which would have to be updated

Table 5–3. Adams Methods

Adams-Bashforth (explicit)

$$
\begin{aligned}
y_{n+1} &= y_n + h f_n \\
y_{n+2} &= y_{n+1} + \tfrac{h}{2}(3f_{n+1} - f_n) \\
y_{n+3} &= y_{n+2} + \tfrac{h}{12}(23f_{n+2} - 16f_{n+1} + 5f_n) \\
y_{n+4} &= y_{n+3} + \tfrac{h}{24}(55f_{n+3} - 59f_{n+2} + 37f_{n+1} - 9f_n)
\end{aligned}
$$

Adams-Moulton (implicit)

$$
\begin{aligned}
y_{n+1} &= y_n + \tfrac{h}{2}(f_{n+1} + f_n) \\
y_{n+2} &= y_{n+1} + \tfrac{h}{12}(f_{n+2} + 8f_{n+1} - f_n) \\
y_{n+3} &= y_{n+2} + \tfrac{h}{24}(9f_{n+3} + 19f_{n+2} - 5f_{n+1} + f_n) \\
y_{n+4} &= y_{n+3} + \tfrac{h}{720}(251f_{n+4} + 646f_{n+3} - 264f_{n+2} \\
&\qquad - 106f_{n+1} - 19f_n)
\end{aligned}
$$

by other processes. If timestep splitting is used to combine solutions of the various physical processes, if the chemistry must be subcycled on the fluid dynamic or diffusion timestep, and if the ODEs are not stiff, multistep methods could be used.

5–2.3 Extrapolation Methods

In these methods, the equations are integrated several times by a lower-order method, and then a more accurate, higher-order solution is found by extrapolating the more accurate approximations to the zero timestep limit. The full interval h is divided into N_o intervals of size h_o, N_1 intervals of size h_1, N_2 intervals of size h_2, etc.

The Euler-Romberg method computes the solution to Eq. (5–1.2) by such an iterative procedure. Euler's method is applied repeatedly, but each integration over the same interval is done twice as many times with half the timestep of the previous integration. Then an extrapolation is made to $h = 0$ using the increasingly accurate integrations with smaller and smaller h values. The Euler-Romberg method is self-starting and rivals the best predictor-corrector methods for efficiency and accuracy. The choice of stepsize is fairly arbitrary, because the method successively halves the stepsize until the required accuracy is achieved at each step.

Given the stepsize h_o and (t_n, y_n), the method approximates y_{n+1} by constructing the table of values shown in Table 5–4. The procedure is started by computing an initial estimate for y_{n+1} (denoted Y_o^o in Table 5–4) using

Table 5–4. Euler-Romberg Coefficients

Iteration	Interval	Y_m^k				
	h_o	Y_o^o				
1	$h_o/2$	Y_o^1	Y_1^o			
2	$h_o/2^2$	Y_o^2	Y_1^1	Y_2^o		
3	$h_o/2^3$	Y_o^3	Y_1^2	Y_2^1	Y_3^o	
⋮	⋮	⋮	⋮	⋮	⋮	...

Euler's method and the full step h_o. By halving h_o and integrating two steps to reach h_o, values of Y_o^1 are generated.

The interpolation part of the procedure is based on the linear interpolation formula

$$Y_m^k(h) = \frac{Y_{m-1}^k[h - h_{k+m}] - Y_{m-1}^{k+1}[h - h_k]}{h_k - h_{k+m}} . \tag{5 – 2.9}$$

This expression is linear in h because the Euler method is only first-order accurate. It would be quadratic with even powers of h if the underlying method were second order. The best numerical approximation is found by extrapolating this equation to $h = 0$, giving

$$Y_m^k = \frac{2^m \, Y_{m-1}^{k+1} - Y_{m-1}^k}{2^m - 1} . \tag{5 – 2.10}$$

Equation (5–2.10) is used to fill each row of Table 5–4 after Y_o^k is computed using the Euler method with stepsize $h_o/2^k$. We continue iterating and generating the Y_m^k until the extrapolated solution converges, as determined by the criterion

$$|Y_m^k - Y_m^{k-1}| \leq \epsilon , \tag{5 – 2.11}$$

where ϵ is a predetermined small quantity. A detailed description is given in McCalla (1967). Using a similar procedure with a method that is higher order than Euler's method produces extrapolations that converge much more quickly.

Extrapolation methods are optimal when the evaluation of the derivative function is relatively expensive, the timestep for a given accuracy varies greatly in the course of the calculation, the equations are not stiff, or high accuracy is required. These methods are often based on the analysis by Neville (1934) that was later developed by Gragg (1965). One extension

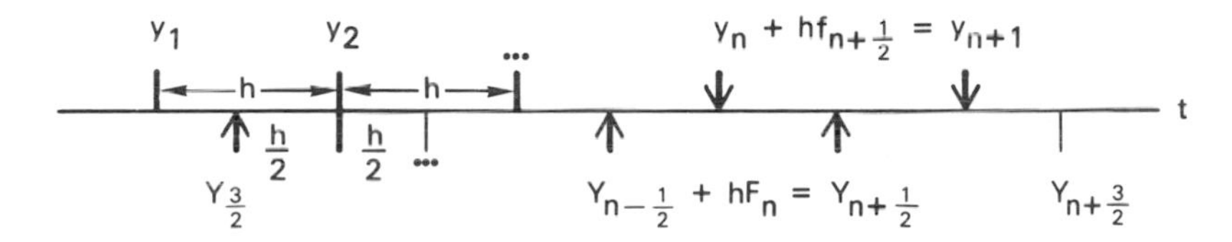

Figure 5–3. Time-line diagram illustrating the leapfrog integration method on a staggered mesh.

of the polynomial extrapolation methods shown above is a *rational extrapolation* method (Bulirsch and Stoer, 1964, 1966) which corresponds to an extrapolation to $h = 0$ using rational functions of polynomials. Classical extrapolation methods do not work well for stiff equations, but recent extensions for stiff equations have been proposed by Deuflhard (1982).

5–2.4 Leapfrog Integration Methods

The leapfrog algorithm is an explicit, second-order integration method. It is also reversible in time, a property which is particularly useful in integrating systems of ODEs which exhibit reversibility. The main uses of the method are for dynamical systems represented as coupled, first-order differential equations rather than for chemical kinetic systems. Because the equations governing oscillations and waves are time-reversible, the leapfrog algorithm mirrors that physical property accurately. Waves propagate without growth or damping. Although phase errors can be present, the amplitude of each of the harmonics of the solution remains unity throughout a leapfrog integration.

The centered time integration of the leapfrog method ensures at least second-order accuracy. Reversibility is achieved by staggering the discretizations of the dynamical variables in time. The process is shown schematically as a time-line diagram in Figure 5–3. Two sets of variables are shown, $\mathbf{y}_n$ and $\mathbf{Y}_{n+\frac{1}{2}}$. The vertical marks in the figure above the time line indicate the discrete times $t_{n-1}, t_n, t_{n+1}, \ldots$ at which the vector $\mathbf{y}_n$ is specified. Another vector, $\mathbf{Y}_{n+\frac{1}{2}}$, is indicated below the time line. The discretization of $\mathbf{Y}$ is staggered in time from that of $\mathbf{y}$ by a half step, $h/2$.

We wish to integrate the set of equations

$$\frac{d\mathbf{y}}{dt} = \mathbf{f}(\mathbf{Y}, t) \tag{5-2.12}$$

$$\frac{d\mathbf{Y}}{dt} = \mathbf{F}(\mathbf{y}, t) \; . \tag{5-2.13}$$

Because the derivative of $\mathbf{y}_n$ depends on the dependent variables $\mathbf{Y}_{n+\frac{1}{2}}$ and vice versa, the centered derivative for each of the variables can always be evaluated explicitly using the most recently updated values for the other set of variables. For Eqs. (5–2.12) and (5–2.13) the second-order leapfrog algorithm is

$$\mathbf{Y}_{n+\frac{1}{2}} = \mathbf{Y}_{n-\frac{1}{2}} + h\mathbf{F}(\mathbf{y}_n, t_n) , \qquad (5-2.14)$$

and

$$\mathbf{y}_{n+1} = \mathbf{y}_n + h\mathbf{f}(\mathbf{Y}_{n+\frac{1}{2}}, t_{n+\frac{1}{2}}) , \qquad (5-2.15)$$

where $t_{n+\frac{1}{2}} = t_n + \frac{h}{2}$. Because the algorithm is reversible, these two equations can be equally well solved for $\mathbf{y}_n$ and then $\mathbf{Y}_{n-\frac{1}{2}}$ by stepping through the algorithm in reverse order. Thus long integrations can be retraced to their initial conditions.

Leapfrog algorithms can be generalized to include an implicit dependence of the variables,

$$\mathbf{Y}_{n+\frac{1}{2}} = \mathbf{Y}_{n-\frac{1}{2}} + h\mathbf{F}(\mathbf{y}_n, \frac{\mathbf{Y}_{n-\frac{1}{2}} + \mathbf{Y}_{n+\frac{1}{2}}}{2}, t_n) , \qquad (5-2.16)$$

$$\mathbf{y}_{n+1} = \mathbf{y}_n + h\mathbf{f}(\mathbf{Y}_{n+\frac{1}{2}}, \frac{\mathbf{y}_n + \mathbf{y}_{n+1}}{2}, t_{n+\frac{1}{2}}) . \qquad (5-2.17)$$

The form written in Eqs. (5–2.16) and (5–2.17) is similar to the modified Euler method in that the derivative functions are evaluated at the center of the interval using the averaged dependent variables. As long as the implicit dependence on $\mathbf{Y}_{n+\frac{1}{2}}$ and $\mathbf{y}_{n+1}$ can be inverted analytically in the equations, the algorithm is reversible. If the implicit equations must be iterated, the intrinsic reversibility of the algorithm is lost unless the iteration has converged perfectly.

An alternate form of the implicit algorithm, based on the improved Euler method, may be easier to invert,

$$\mathbf{Y}_{n+\frac{1}{2}} - \frac{h}{2}\mathbf{F}(\mathbf{y}_n, \mathbf{Y}_{n+\frac{1}{2}}, t_n) = \mathbf{Y}_{n-\frac{1}{2}} + \frac{h}{2}\mathbf{F}(\mathbf{y}_n, \mathbf{Y}_{n-\frac{1}{2}}, t_n) \qquad (5-2.18)$$

and

$$\mathbf{y}_{n+1} - \frac{h}{2}\mathbf{f}(\mathbf{Y}_{n+\frac{1}{2}}, \mathbf{y}_{n+1}, t_{n+\frac{1}{2}}) = \mathbf{y}_n + \frac{h}{2}\mathbf{f}(\mathbf{Y}_{n+\frac{1}{2}}, \mathbf{y}_n, t_{n+\frac{1}{2}}) . \qquad (5-2.19)$$

Reversible generalizations of this method to higher order are also possible. As with all of the classical techniques described so far, leapfrog methods break down for stiff equations. Implicit leapfrog algorithms may eventually be applied to stiff systems, but this has not occurred to date. An important property of an implicit leapfrog algorithm is that it is more stable than an explicit leapfrog algorithm.

One of the major uses of the leapfrog algorithm is in orbit and particle dynamics calculations where the vector positions, $\mathbf{x} \equiv \mathbf{y}$, depend on the velocities, $\mathbf{v} \equiv \mathbf{Y}$, and the velocities depend on a force which is only a function of the instantaneous positions of the particles. This application has been described by Hockney (1965), and an example of leapfrog algorithms in both classical particle dynamics calculations and semiclassical quantum mechanical calculations is given in Page et al. (1985). The implicit generalizations of the leapfrog algorithm are useful in orbit dynamics calculations with a velocity dependence in the expression for the force, such as when there are relativistic corrections and magnetic fields (Buneman, 1967; Boris, 1971).

5–3. SOME APPROACHES TO SOLVING STIFF EQUATIONS

Stiffness in ordinary differential equations is related to a basic computational problem: the presence of a wide range of time scales affecting the system dynamics. Mathematically, a system is stiff when the Jacobian matrix has eigenvalues whose magnitudes differ by a large ratio (see Eq. (5–1.18)). This means that at least two independent homogeneous solutions vary with time at greatly disparate rates. Suppose, for example, that there are four orders of magnitude difference between the largest and the smallest eigenvalues. If the fastest changing mode of behavior can be resolved with ten steps per characteristic time, then 10^5 steps are required to integrate one characteristic time of the slowest mode.

Consider problems whose largest eigenvalues correspond to rapidly relaxing modes, so that their actual contribution to the evolution of the composite solution is negligible. In such problems, the timestep may have to be very small to accommodate the stability conditions of these modes. Although this case is mathematically stiff, an integration technique could use a long timestep if the high-frequency modes were treated stably. When we use such an algorithm, the problems are mathematically stiff but not practically stiff. There is no need to resolve the high-frequency modes because only their averaged effects are important. The stiff problems encountered in physics and reactive flows are usually of this form.

In some cases the large eigenvalues of the Jacobian are positive so that the high-frequency modes grow rather than damp. These physically unstable situations lead to variations on short time scales which must be accurately resolved to determine the composite solution. Such problems are stiff both mathematically and practically. Fortunately, the stiffest modes usually do not have growth rates. When they do, these violently unstable states saturate quickly and the growth becomes stabilized.

As a coupled nonlinear ODE system evolves, its Jacobian matrix also changes. This leads to changing patterns of stiffness in the system as the rapid, unstable transients die away. Once the period of fast growth is over, the stiffest modes in the system show damping solutions in the vicinity of the new or quasi-static equilibrium. An exception occurs when high-frequency oscillations exist in the system for a long time. Then the composite solution changes rapidly nearly everywhere nearly all the time. In these cases the stiff modes are essentially undamped and their eigenvalues have a zero real part.

The problems with stiffness cause us to seek methods that do not restrict the stepsize for stability reasons, and which can treat widely disparate scales in some reasonable manner. This need to solve stiff equations reliably has led to the development of nonclassical methods, which address the obstacles of accuracy versus stability.

Curtiss and Hirschfelder (1952) proposed special multistep formulas that produced acceptable solution to stiff ODEs. Their methods did not have the severe stepsize constraint imposed by stability of the more traditional multistep and Runge-Kutta methods. Dahlquist (1963) proposed the trapezoidal rule and extrapolations as a suitable technique for integrating stiff equations. Survey papers on techniques for solving stiff equations have been written by Cooper (1969), Bjurel et al. (1970), Seinfeld et al. (1970), Gelinas (1972), Enright and Hull (1976), and most recently by May and Noye (1984) and Bui et al. (1984). Radhakrishnan (1984) has recently compared a number of these methods.

5–3.1 Stability and Stiff ODEs

When a set of equations is solved numerically, the stepsize h must be chosen according to two criteria:

1. The unstable modes with $\Re(\lambda_j) > 0$ must be represented accurately regardless of their intrinsic time scales, if they contribute significantly to the composite solution.
2. The values of $h\lambda_j$ must be within a region of absolute numerical stability for all j when $\Re(\lambda_j) < 0$.

If the second condition determines the stepsize h, the system is stiff. Below we introduce several aspects of stability in stiff systems of equations. This is important because relaxed stability conditions can allow for better accuracy, particularly when there are growing modes.

An implicit method such as the backward Euler method,

$$y_{n+1} = y_n + hf(y_{n+1}, t_{n+1}) , \qquad (5-3.1)$$

if applied to a test problem of the form

$$\frac{dy}{dt}(t) = \lambda y(t) , \qquad (5-3.2)$$

may be stable with a single-step error amplification factor of the form $(1 - h\lambda)^{-1}$. If $\Re(\lambda)$ is less than zero, the error amplification is less than one. This is an example of *A-stability*, as defined by Dahlquist (1963). A numerical method is A-stable if, when applied to Eq. (5–3.2) with complex λ with a negative real part, the error due to the numerical approximation tends to zero as the number of timesteps goes to infinity.

This means that all physically decaying modes are also decaying numerically but not necessarily at the correct rate. Though the numerical solution may be orders of magnitude larger than the physical solution, the error goes to zero because the numerical solution decays to zero. The A-stability condition says nothing about the accuracy of the physical modes that grow. For A-stability, the values of $h\lambda$ can fall anywhere in the negative half of the $h\lambda$ plane, as shown in the crosshatched region of Figure 5–4a. In this case, and in cases where all of the $\Re(\lambda_j)$ are less than zero, any value of $h > 0$ is within the region of absolute stability.

Because A-stability is such a a severe requirement, it is hard to find methods which are A-stable. A more relaxed criterion, *A(α)-stability*, was proposed by Widlund (1967). A method is A(α)-stable, where α is in the range $(0, \pi/2)$, if all numerical approximations to $dy/dt = \lambda y$ converge to zero as the number of timesteps goes to infinity, with fixed h, for all $|\arg(-\lambda)| < \alpha$, $|\lambda| \neq 0$. This means that the region of absolute stability is a wedge or, for coupled ODEs, a series of wedges in the negative $h\lambda$ plane. This is illustrated in the crosshatched regions of Figure 5–4b.

Dahlquist (1963) has shown that an A-stable multistep method can be at most second-order accurate. The trapezoidal and backward Euler methods, mentioned above, are examples of A-stable methods. For stiff ODEs, the trapezoidal method has the advantage of being precisely A-stable, but because it is only second order, it requires rather small stepsizes. Gear (1971) has summarized classes of methods which are A-stable. These include R-stage implicit Runge-Kutta methods of order 2R (Ehle, 1969) and methods which are implicit extensions of Taylor series methods. For the lowest order, these reduce to the trapezoidal rule and the implicit midpoint rule, described below.

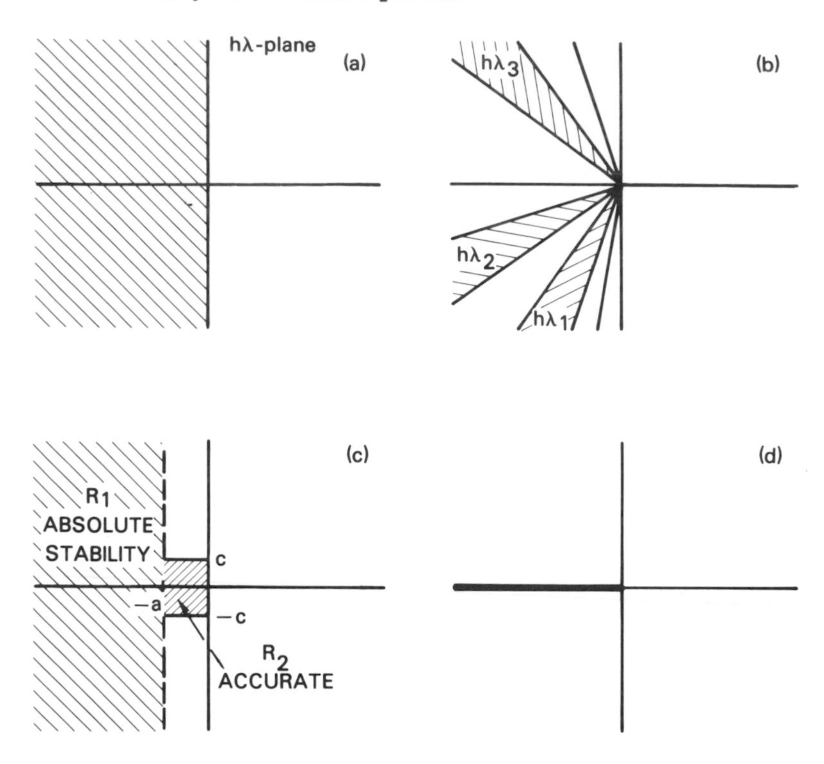

Figure 5–4. Regions of absolute stability in the $h-\lambda$ plane for (a) A-stable, (b) stiffly-stable, (c) A(α)-stable, (d) A_o-stable algorithms.

A method is *stiffly stable* (Gear, 1969) if it is absolutely stable in the region R_1 of the complex $h\lambda$ plane for $\Re(\lambda) \leq -a$, and is accurate in the region R_2 for $-a < \Re(h\lambda) < b, -c < \Im(h)\lambda \leq c$, where $\Im$ indicates the imaginary part and a, b, and c are positive constants. This is shown as the crosshatched region in Figure 5–4c, and is a more restrictive criterion than A(α)-stability. Gear has shown that there are k-step methods of order k which are stiffly stable for $k \leq 6$. These methods are used in the backward differentiation parts of the Gear packages for solving ODEs.

The most relaxed criterion is *A_o-stability* (Cryer, 1973), that applies when the region of absolute stability is the whole negative real axis. Figure 5–4 shows the regions of stability for the four criteria, A-stability, stiff-stability, A(α)-stability, and A_o-stability, given in order of descending restrictiveness. It is easier to find less restrictive methods, though they generally require shorter timesteps.

5–3.2 Implicit Methods for Stiff ODEs

An implicit multistep method for stiff equations differs from methods for nonstiff equations. Consider a linear multistep method described by a set of equations of the form of Eq. (5–2.7) and rewrittten as

$$\mathbf{y}_{n+k} = h\beta_k \, \mathbf{f}(t_{n+k}, \mathbf{y}_{n+k}) + \mathbf{B}_{k-1} \,, \qquad (5-3.3)$$

where $\mathbf{B}_{k-1}$ is known at the kth step of the method,

$$\mathbf{B}_{k-1} = \sum_{j=0}^{k-1} (h\beta_j \, \mathbf{f}(t_{n+j}, \mathbf{y}_{n+j}) - \alpha_j \mathbf{y}_{n+j}) \,. \qquad (5-3.4)$$

When the equations are not stiff, we can use a direct iteration of the form

$$\mathbf{y}_{n+k}^{(m+1)} = h\beta_k \, \mathbf{f}(t_{n+k}, \mathbf{y}_{n+k}^{(m)}) + \mathbf{B}_{k-1} \,, \qquad (5-3.5)$$

where m is the iteration index. Evaluate $\mathbf{B}_{k-1}$, guess the value of $\mathbf{y}_{n+k}^{o}$, and then evaluate successive values of $\mathbf{y}_{n+k}$, that is, $\mathbf{y}_{n+k}^{(1)}$, $\mathbf{y}_{n+k}^{(2)}$, ..., until the iteration converges.

This direct approach does not work for stiff equations, which require $h < 1/|J_{ij}|\beta_k$ for convergence because

$$|J_{ij}| = \left\| \frac{\partial f_i}{\partial y_j} \right\| \geq \max|\lambda_j| \qquad (5-3.6)$$

would be very large, and therefore h would have to be very small.

This stiffness problem has to be handled by using Newton's method to solve Eq. (5–3.3). To solve the equation

$$\mathbf{F}(\mathbf{y}) = \mathbf{0} \,, \qquad (5-3.7)$$

using Newton's method, we have

$$\mathbf{y}^{(m+1)} = \mathbf{y}^{(m)} - \mathbf{F}(\mathbf{y}^{(m)}) \cdot \mathbf{J}^{-1}(\mathbf{y}^{(m)}), \qquad m = 0, 1, \ldots \,. \qquad (5-3.8)$$

Using this gives the matrix equation

$$\mathbf{y}_{n+k}^{(m+1)} = \mathbf{y}_{n+k}^{(m)} - \left[\mathbf{y}_{n+k}^{(m)} - h\beta_k \, \mathbf{f}(t_{n+k}, \mathbf{y}_{n+k}^{(m)})\right] \cdot \left[\mathbf{I} - h\beta_k \frac{\partial \mathbf{f}}{\partial \mathbf{y}}(t_{n+k}, \mathbf{y}_{n+k}^{(m)})\right]^{-1}, \qquad m = 0, 1, 2, \ldots \,, \qquad (5-3.9)$$

where $\mathbf{I}$ is the unit matrix. The iteration requires a good initial estimate of $\mathbf{y}_{n+k}^{(o)}$. Generally the equation is solved in the form where Eq. (5–3.9) is multiplied through by the last term on the left side, because advantage can then be taken of the sparseness of the Jacobian matrix in finding the solutions.

Table 5–5. Coefficients for Backward Differentiation Methods

k	β_k	α_o	α_o	α_o	α_o	α_o	α_o	α_o
1	1	-1	1					
2	$\frac{2}{3}$	$\frac{1}{3}$	$-\frac{4}{3}$	1				
3	$\frac{6}{11}$	$-\frac{2}{11}$	$\frac{9}{11}$	$-\frac{18}{11}$	1			
4	$\frac{12}{25}$	$\frac{3}{25}$	$-\frac{16}{25}$	$\frac{36}{25}$	$-\frac{48}{25}$	1		
5	$\frac{60}{137}$	$-\frac{12}{137}$	$\frac{75}{137}$	$-\frac{200}{137}$	$\frac{300}{137}$	$-\frac{300}{137}$	1	
6	$\frac{600}{147}$	$\frac{10}{147}$	$-\frac{72}{147}$	$\frac{225}{147}$	$-\frac{400}{147}$	$\frac{450}{147}$	$-\frac{360}{147}$	1

5–3.3 Methods for Solving Stiff Equations

We now introduce a number of the more commonly used algorithms for solving stiff ODEs. Others are covered in the review article by May and Noye (1984).

Backward Differentiation Formulas

These methods are the most common methods for solving systems of stiff equations and are described in detail by Gear (1971). They are linear multistep methods written in the form of Eq. (5–2.7) with $\beta_j = 0$,

$$y_{n+k} = h\beta_k f_{n+k} - \sum_{j=0}^{k-1} \alpha_j y_{n+j} \,, \tag{5 - 3.10}$$

where $\alpha_o \neq 0$ and $\beta_k \neq 0$, and the order is equal to the step number, k. For orders one through six these methods are stiffly stable. The first order method is the backward Euler method. This method, as well as the second and third order methods, are absolutely stable in the right half-plane close to the origin. The higher-order methods are not absolutely stable in a region of the left-hand plane near the imaginary axis and may give poor results for a system with an eigenvalue near the imaginary axis. In that case they either damp a solution which should be increasing in value, or they allow a solution to grow when it should decay. Table 5–5 gives values of the parameters for these formulas. Gear used these formulas in the program DIFSUB (Gear, 1971), discussed below in Section 5–4.

Exponential Methods

Implicit methods for stiff equations (see, for example, May and Noye, 1984) can be derived by *curve-fitting* (Lambert, 1973). In this process, a particular form of interpolating function is adopted with free parameters determined by requiring the interpolant to satisfy certain conditions on the approximate solutions and their derivatives. For example, we can choose the two-parameter polynomial function

$$I(t) \;=\; A + Bt \tag{5 – 3.11}$$

as the interpolant in time, and require that $I(t)$ satisfy the constraints

$$I(0) \;=\; y_n, \quad I'(0) \;=\; f_n, \quad I(h) \;=\; y_{n+1} \tag{5 – 3.12}$$

on the interval $[0,h] = [t_n, t_{n+1}]$. This results in the Euler approximation

$$y_{n+1} \;=\; y_n + hf_n \; . \tag{5 – 3.13}$$

Other constraints and other forms of the interpolants result in other previously described explicit and implicit methods.

Keneshea (1967) suggested using exponential functions to approximate the solution in the equilibration regime of stiff ODEs describing chemical kinetics. Liniger and Willoughby (1970), Brandon (1974), Babcock et al. (1979), and Pratt and Radhakrishnan (1984) then extended the curve-fitting approach to consider exponential interpolants. This approach is based on the idea that the exact solutions of stiff linear ODEs behave like decaying exponential functions. Because exponentials are poorly approximated by polynomials when the stepsize is larger than the characteristic decay rate, using exponential approximations should allow considerably longer timesteps. For example, consider the three-parameter exponential interpolant

$$I(t) \;=\; A + Be^{Zt} \tag{5 – 3.14}$$

for which A, B, and Z must be determined. The conditions in Eq. (5–3.12) determine A and B in terms of Z,

$$A \;=\; y_n - \frac{f_n}{A} \qquad\qquad B \;=\; \frac{f_n}{Z} \, , \tag{5 – 3.15}$$

yielding

$$y_{n+1} \;=\; y_n + hf_n \left[\frac{e^{Zh} - 1}{Zh} \right] . \tag{5 – 3.16}$$

There are a number of ways to choose the parameter Z, for example,

$$\begin{aligned} \text{explicit}: \quad Z &= f'_n/f_n \\ \text{explicit}: \quad Z &= \frac{1}{h_{n-1}} \ln\left(\frac{f_n}{f_{n-1}}\right) \\ \text{implicit}: \quad Z &= \frac{1}{h} \ln\left(\frac{f_{n+1}}{f_n}\right) \\ \text{implicit}: \quad Z &= \frac{1}{2}\left[\frac{f'_n}{f_n} + \frac{f'_{n+1}}{f_{n+1}}\right] \end{aligned} \tag{5-3.17}$$

where f' is used to represent d^2y/dt^2.

Another useful approach to solving the equations implicitly is by exponentially fitting a trapezoid rule (Brandon, 1974; Babcock et al., 1979),

$$y_{n+1} = y_n + h\{\theta f_{n+1} + (1-\theta) f_n)\} \, , \tag{5-3.18}$$

where θ is an implicitness parameter written in terms of Z,

$$\theta = \frac{1}{Zh} + \frac{1}{1 - e^{Zh}} \, . \tag{5-3.19}$$

This exponential-fitted trapezoidal rule is A-stable. It is also equivalent to polynomial interpolants of at least order two and as great as six to eight. Tests by Radhakrishnan (1984) have shown that exponential methods can be at least comparable in speed and accuracy to the backwards differentiation methods for stiff ODEs.

Asymptotic Methods

In Chapter 4 we described an asymptotic method applied to a single ODE. There it appeared that such methods had a decided advantage over explicit and even implicit methods because of greater accuracy at large timesteps. This approach to solving stiff ODEs has been developed by Young and Boris (1977) and Young (1979), whose method comes from a second-order asymptotic expansion. To understand this approach, rewrite Eq. (2–2.22) as

$$\frac{dy}{dt} = Q(t) - y(t)\, \tau(t)^{-1} = f(y,t) \, . \tag{5-3.20}$$

The quantity $\tau(t)$, the reciprocal of the loss coefficient $L(t)$, is the characteristic relaxation time describing how quickly the single variable y reaches its equilibrium value. For simplicity we have dropped the subscript i on y, or equivalently, the vector notation indicating a coupled set of equations.

The arguments given below apply equally well to a coupled set representing many reacting species.

In this asymptotic approach, stiff equations, characterized by a sufficiently small value of τ, are solved in the form

$$\frac{y_{n+1} - y_n}{h} = \frac{Q_{n+1} + Q_n}{2} - \left(\frac{y_{n+1} + y_n}{\tau_{n+1} + \tau_n}\right) . \qquad (5-3.21)$$

The numerical problem here is that Q_{n+1} and τ_{n+1} are also implicit functions of y_{n+1}. The formal solution of Eq. (5–3.21) for y_{n+1} is

$$y_{n+1} = \frac{y_n(\tau_{n+1} + \tau_n - h) + \frac{h}{2}(Q_{n+1} + Q_n)(\tau_{n+1} + \tau_n)}{(\tau_{n+1} + \tau_n + h)} . \qquad (5-3.22)$$

We use Eq. (5–3.22) as a corrector formula, and use

$$y_1 = \frac{y_o(2\tau_o - h) + 2hQ_o\tau_o}{2\tau_o + h} \qquad (5-3.23)$$

as the predictor.

This method produces the best results when the solution is slowly varying, the rapidly relaxing models are nearly at their asymptotic state, but the time constants are prohibitively small. This occurs when both Q and y/τ are large but nearly equal. The method is stiffly stable and tends to damp out small oscillations caused by very small time constants. One problem with the asymptotic method is that it does not necessarily conserve y. Because conservation cannot be guaranteed everywhere, it becomes a convenient check on accuracy. Various approaches to assuring conservation have been devised, but are usually not very satisfactory. When exact conservation is important, this method should not be used.

The asymptotic methods are very fast and moderately accurate. Their primary advantages are: (1) they do not require evaluation or use of the Jacobian or any matrix operations; (2) they are self-starting, requiring only initial values from one timestep; (3) they can be used with easily implemented stepsize estimation techniques; and (4) they can be combined with a simple, explicit algorithm for equations which are not stiff in ways that permit parallel processing of the equations. These features make the asymptotic methods convenient for combining with algorithms for convection and diffusion processes. This approach is implemented in the program CHEMEQ, discussed below.

Implicit Extrapolation Methods

Because explicit extrapolation methods do not solve stiff ODEs any better than other explicit methods, implicit extrapolation methods have been developed. The simplest of these is the implicit or "backward Euler" method given by Eq. (5–3.1), written here in vector form,

$$\mathbf{y}_{n+1} = \mathbf{y}_n + h\,\mathbf{f}(\mathbf{y}_{n+1}) \,. \tag{5 - 3.24}$$

Because this formula is implicit, it requires the iterative numerical solution of a system of algebraic equations. For stiff equations, we use a Newton-type iteration of the form

$$\begin{aligned} (\mathbf{I} - h\,\mathbf{J}) \cdot \mathbf{y}_{n+1}^{(m+1)} &= \mathbf{y}_n + h\,\bar{\mathbf{f}}(\mathbf{y}_{n+1}^{(m)}) \qquad m = 0, 1, \ldots \\ \bar{\mathbf{f}}(\mathbf{y}) &= \mathbf{f}(\mathbf{y}) - \mathbf{J} \cdot \mathbf{y} \end{aligned} \tag{5 - 3.25}$$

and $\mathbf{J}$ is evaluated at $\{\mathbf{y}_n\}$. This method is stable on the whole left half-plane, which is desirable, but it is too stable on the right half-plane, the property of superstability. This means that increasing analytical solutions may be approximated by decreasing numerical solutions.

Other basic algorithms for implicit extrapolation methods include the implicit trapezoidal method,

$$\mathbf{y}_{n+1} = \mathbf{y}_n + \frac{h}{2}\Big(\mathbf{f}(\mathbf{y}_n) + \mathbf{f}(\mathbf{y}_{n+1})\Big) \tag{5 - 3.26}$$

or the implicit midpoint rule (Dahlquist, 1963; Dahlquist and Lindberg, 1973; and Lindberg 1973),

$$\mathbf{y}_{n+1} = \mathbf{y}_n + h\,\mathbf{f}\left(\frac{\mathbf{y}_n + \mathbf{y}_{n+1}}{2}\right) . \tag{5 - 3.27}$$

These methods have asymptotic expansion and are both A(α)-stable. However, none of the implicit extrapolations compare favorably with the backward differentiation methods.

Another extrapolation approach using the semi-implicit midpoint rule was introduced by Bader and Deuflhard (1983). They combined this with algorithms for order and stepsize control to provide a reasonable application of extrapolation methods to stiff ODEs. The term "semi-implicit" in this context means that only one Newton iteration is used per timestep. This approach is implemented in the program LARKIN, described below. In general these methods are promising but not yet suitable for solving the ODEs in the context of reactive flows.

5–3.4 Summary: Some Important Issues

There has been a substantial effort recently to develop methods for the numerical solution of stiff ODEs. This effort is driven by a number of applications, but most strongly by the types of chemical kinetics problems which interest us here. Several important issues have emerged from the efforts to find efficient, accurate algorithms.

The first important issue, control of the stepsize, has given rise to many stepsize estimation techniques. Stepsize control is crucial in all methods except those which use constant stepsizes or stepsizes which are fixed fractions of those previously used. Ideally we should take the largest stepsize compatible with the required accuracy. For nonstiff equations, this is not too great a problem. However, for stiff equations, unless implicit or asymptotic methods are used, the stepsize requirement may be so small that the answer is dominated by roundoff error.

Even when the method used to solve a set of stiff equations is A-stable, the answers may have large errors if the stepsize is large because gradients are not resolved. One advantage of using a less stable method (for example, one that is stiffly stable) is that it can be unstable for large stepsize, thus giving a warning that the stepsize is too large for accuracy.

The second issue is the use of the Jacobian matrix. It is expensive to solve a large matrix equation because inversion requires of order M^3 operations, where M is the size of the matrix. A key point in having an efficient algorithm is to save computer time by using the old Jacobian as long as possible. Recalculating the Jacobian is expensive, but an inefficient convergence rate is a signal to reevaluate the Jacobian and to decompose the new matrix.

The third issue concerns how much accuracy is needed. For reactive flow problems, where the chemical kinetics is coupled to fluid dynamic flow, the solutions almost never get into the equilibrium regime where the equations are so stiff that cancellations exceed roundoff accuracy. Inevitable fluctuations in temperature and pressure ensure slight deviations from equilibrium, and this generally allows leeway to use less expensive methods. This aspect of coupling fluid dynamics and chemical kinetics algorithms is discussed in Chapter 13.

5–4. INTEGRATION PACKAGES

Many of the classical methods described in Section 5–2 conserve the dependent variables to the limits of numerical roundoff errors, but these algorithms are generally not adequate for stiff equations. Most of the algorithms discussed in Section 5–3 for integrating stiff equations are implicit, and these can be expensive. The cost of a method generally scales with system size. If a full Jacobian must be inverted, the method scales as M^3, otherwise it scales as M or M^2, where M is the dimension of $\mathbf{y}$.

In general, there are two classes of methods for solving coupled ODEs, some of which are stiff:

1. Methods which treat all of the differential equations, both stiff and nonstiff, identically, and
2. *Hybrid methods* which identify the stiff equations and integrate them with an implicit or asymptotic algorithm. The rest of the equations are treated by a faster but less stable, classical algorithm.

Hybrid methods try to take advantage of the structure of the particular set of ODEs. This is usually at the expense of exact conservation, which is difficult to guarantee when different formulas are used for different subsets of equations. The ODE integration packages described below include methods of both types.

In hybrid methods, two important problems must be addressed. The first problem is to establish a criterion for determining which method to use for which equation. This required determining which equations are stiff at any step. The second problem is determining when, during the integration, the method should be changed. Changing the timestep can shift some equations from being normal to being stiff. We also need to consider how much nonconservation can be tolerated to say in advance how much imbalance in mass or particles can be allowed in an answer. Hybrid methods are fast but because conservation errors may occur after many steps, they often must be independently renormalized.

Here some of the currently used software packages for solving coupled sets of ODEs, some or all of which are stiff, are discussed. For problems involving a modest-sized set of ODEs, any of the methods described below are generally good. For coupling to reactive flow problems, we recommend the hybrid asymptotic approach.

5–4.1 Programs Based on the GEAR Method

The FORTRAN program DIFSUB was written by Gear (1969) (see Gear, 1971) to solve sets of coupled ODEs, some of which are stiff. DIFSUB treats the entire set of equations as either stiff or nonstiff, and therefore uses the same method for all equations in a set. The stiff equations are solved by the variable-order backward differentiation formulas described in Section 5–3. The normal equations are solved by a variable-order Adams predictor-corrector method described in Section 5–2. These routines were rewritten by Hindmarsh (1974) to create the GEAR package, but this version was useful only for regular or fairly small sets of stiff equiations because it used the full $M \times M$ form of the Jacobian. Later variants, including GEARB, GEARS, and GEARBI (Hindmarsh, 1976a,b, 1977), were developed for large, stiff problems having sparse structure in the Jacobian or where good approximations to the Jacobian are sparse.

The latest and most easily obtained version of the GEAR programs have been incorporated in the package ODEPACK, assembled by Hindmarsh (1983). These programs use an implicit Adams-Moulton formula,

$$\mathbf{y}_n = \mathbf{y}_{n-1} + h \sum_{l=0}^{k-1} \beta_l \mathbf{f}(t_{n-l}, \mathbf{y}_{n-l}) , \qquad 1 \leq k \leq 12 \qquad (5-4.1)$$

to advance the regular nonstiff equations, where k is the order of accuracy and the coefficients β_l depend only on k and sum to unity. Equation (5–4.1) is implicit since β_o is greater than zero. In the implicit solution, $\mathbf{y}$ at the advanced time is found by iterating the equation

$$\mathbf{y}_n^{(m+1)} = \mathbf{y}_{n-1} + h\beta_0 f(t_n, \mathbf{y}_n^{(m)}) + h \sum_{l=1}^{k-1} \beta_l \mathbf{f}(t_{n-l}, \mathbf{y}_{n-l}) . \qquad (5-4.2)$$

The initial prediction $\mathbf{y}_n^{(o)}$ is found from an analogous explicit formula. Note that no matrix operations are required.

The stiff equations are solved using

$$\begin{aligned} \mathbf{y}_n &= \sum_{l=1}^{k} \alpha_i \mathbf{y}_{n-l} + h\beta_o \mathbf{f}(\mathbf{y}_n, t_n) \\ &= a_n + h\beta_o f(t_n, \mathbf{y}_n), \\ &\quad \text{with } \beta_o > 0 , \quad 1 \leq k \leq 5 . \end{aligned} \qquad (5-4.3)$$

As in the formulas for regular equations given above, the higher-order fomulas can only be used after there have been several previous timesteps of the same size. Because the equations are stiff, they do not converge for reasonable stepsizes. A modified Newton iteration,

$$\mathbf{P} \cdot \left[\mathbf{y}_n^{(m+1)} - \mathbf{y}_n^{(m)}\right] = \mathbf{y}_n^{(m)} - \mathbf{a}_n - h\beta_o \mathbf{f}(t_n, \mathbf{y}_n^{(m)}) \qquad (5-4.4)$$

is used where $\mathbf{P}$ is a matrix approximating the Jacobian,

$$\mathbf{P} \simeq \mathbf{I} - h\beta_o \, \mathbf{J} \, . \qquad (5-4.5)$$

The prediction is from an analogous explicit formula. This is different from a true Newton iteration. The Jacobian is only evaluated at the initial predicted values $\mathbf{y}_o$ and only on steps where a new value is necessary based on criteria such as convergence failure. Therefore the same value of $\mathbf{P}$ is used over all iterations in any one step and over several steps. In large, stiff systems, we can often take advantage of the sparsity of the $\mathbf{P}$ matrix.

The package ODEPACK contains a number of ODE solvers. The solver LSODE works for stiff or regular equations. In addition to solving the equations with a full Jacobian, it also allows solutions with various approximations to the Jacobian and is more flexible than the original GEAR solvers. LSODA is a variant of LSODE which does automatic switching between stiff and nonstiff methods. This version is well-suited to problems whose nature changes during the integration. LSODI is specifically designed for the implicit form of ODE,

$$\mathbf{A}(t, \mathbf{y}) \cdot \mathbf{f}(\mathbf{y}, t) = \mathbf{g}(t, \mathbf{y}) \, . \qquad (5-4.6)$$

This form is useful for the the method of lines approach to solving partial differential equations, which results in equations of the form of Eq. (5–4.6).

5–4.2 An Asymptotic Integration Method: CHEMEQ and VSAIM

The Selected Asymptotic Integration Method (SAIM) was designed specifically for use in reactive flow simulations where the chemical reactions must be described for each spatial cell. In these cases the storage and time costs are major factors. However, it has also been used extensively for one point problems. The SAIM is implemented in a hybrid integrator, CHEMEQ (Young and Boris, 1977; Young, 1979), which solves regular equations by an explicit Euler method and uses the asymptotic method discussed in Section 5–3 for stiff equations. VSAIM, a version of the same program designed specifically for reactive flow models, vectorizes the solution procedure over a number of computational cells. A chemical rate processor is included which reads chemical equations in their usual symbolic notation and sets up a FORTRAN program to evaluate the derivatives automatically. Software interfaces, such as this chemical rate processor, are valuable additions to all reactive flow modeling efforts.

A crucial part of the program determines criteria for the initial stepsize and identifies the stiff equations. The initial stepsize is

$$\Delta t = \epsilon \min\left\{ \frac{n_i}{Q_i - \mathcal{L}_i n_i}\bigg|_{t=0} \right\} . \qquad (5-4.7)$$

Here ϵ is a scale factor, typically the same as the convergence criterion, and i indexes the initial values of each component of the vector $\mathbf{y}$. If this stepsize is greater than a specified value τ_i, the equation is considered stiff and is integrated according to the asymptotic formula. Equations considered stiff at the beginning are treated as stiff throughout the integration step. A fixed, small number of iterations are done each step, and whether or not convergence is achieved determines how the stepsize will change in the next step. It is best to reduce the stepsize sharply (a factor of two or three) when the equations do not converge, and to increase the stepsize only by 5-10% when convergence is achieved. During the integration of several successive steps, the appropriately modified stepsize from the converged integration cycle is used as the trial stepsize for the next integration cycle rather than using the starting formula, Eq. (5–4.7).

The method is generally efficient, accurate, and stable. It is self-starting, requiring data only from the previous timestep to initiate the current timestep. It does not use the Jacobian, and therefore does not require matrix operations or inversions. It is not inherently conservative. Conservation is controlled by adjusting the convergence criterion or by adjusting the criterion for choosing which equations to treat as stiff. As described in Chapter 4, an asymptotic method may actually be more accurate for large stepsizes. Thus the stepsize must be carefully monitored to insure accuracy, convergence, and adequate conservation. The greatest problem with this method is integrating very stiff systems as they approach equilibrium. In this regime, the production and loss terms cancel almost perfectly. In the CHEMEQ method, this situation can lead to large fluctuations in the estimated derivatives.

5–4.3 An Exponentially Fitted Method: CREK1D

CREK1D (see, for example, Pratt and Radhakrishnan, 1984) is a computer code designed for transient chemical kinetics problems, and combustion kinetics problems in particular. It is based on the exponentially-fitted trapezoidal method described in Section 5–3. It consists of a predictor-corrector algorithm and includes automatic stepsize selections. Given a prescribed error tolerance, it selects either a Newton-Jacobi or a Newton iteration for convergence. It compares favorably with LSODE for lower accuracy requirements, and costs somewhat more when higher accuracy is required.

The decision whether to solve the implicit equations using a Newton-Raphson iteration or Jacobi-Newton iteration (which does not require evaluation of the Jacobian matrix) is automated. The trade-offs involve accuracy, convergence rate, and total computation time. A Newton-Raphson iteration converges quadratically, but the Jacobian and its inverse must be evaluated frequently to ensure this convergence.

In CREK1D, Jacobi-Newton iteration is used during the early part of the integration of chemical equations involving the chemical induction time and the early heat release stages. During this time, when small stepsizes are required for stability, this iteration minimizes the work involved. During the late heat release stages and the equilibration region, the Newton-Raphson iteration is preferred since it has better convergence properties and larger stepsizes must be used. A test is performed to identify each regime based on the differences between the forward and reverse reactions.

5–4.4 Semi-Implicit Extrapolation Methods: METAN1 and METAS1

The polynomial extrapolation method using the semi-implicit midpoint rule described in Section 5–3 has been implemented in two integrators, METAN1, and the sparse matrix version, METAS1. The idea is to apply Richardson extrapolation in the quantity h^2 by repeating the integration over a basic interval, H. A table of polynomials, such as described in Section 5–2, is constructed.

The integrator METAS1 has been incorporated into a user-oriented package, LARKIN, designed to handle large systems of stiff chemical kinetics equations. Given the chemical equations and chemical rates, LARKIN contains algorithms which generate the functions and the Jacobian, and then produces numerical output of the concentrations at prescribed times. Deuflhard et al. (1981) have shown that LARKIN, and thus METAS1, performs comparably to the implicit multistep backward differentiation used by Gear.

5-4.5 The Kregel Method

The Kregel method (or the K-method) and the software package developed to implement it are described by Coffee et al. (1980). The basic method is a third-order implicit linear multistep predictor-corrector algorithm. As such, it encounters the standard difficulty of treating the Jacobian matrices. Specifically, we need to solve

$$\mathbf{y}_n^p - \mathbf{y}_n^c = \left(h\beta_o \mathbf{J}_n - \mathbf{I}\right) \cdot \mathbf{d}_n . \qquad (5-4.8)$$

The major innovation of the Kregel integrator is the method of solving this equation. Generally the matrix must be evaluated at each step. However, the integrator attempts to reduce the size of the matrix that must be solved by finding suitable approximations to the values of some of the difference vectors, $\mathbf{d}_n$. Then for each value of the $\mathbf{d}$ that can be determined, a row and column can be eliminated from the matrix.

First, consider the approximation to $\mathbf{d}_n$

$$d_{n,k} = \frac{\left(y_{n,k}^p - y_{n,k}^c\right)}{\left(h\beta_0 J_n^{kk} - 1\right)} . \qquad (5-4.9)$$

This is a reasonable approximation when the stepsize is very small, such as at the beginning of the integration, so that

$$|h\beta_o \sum_{j\neq k} J_n^{k,j} d_{n,j}| \ll |d_{n,k}| , \qquad (5-4.10)$$

and off-diagonal elements can then be neglected. Even when h is very large, the equation for $\mathbf{y}_n$ could be weakly coupled to the other equations, i.e.,

$$|J_n^{kk} d_{n,k}| \gg |\sum_{j\neq k} J_n^{k,j} d_{n,j}| . \qquad (5-4.11)$$

Finally, at any value of h, the predictor and corrector values may be close, so that

$$|y_{n,k}^p - y_{n,k}^c| \simeq 0 \simeq d_{n,k} . \qquad (5-4.12)$$

The reduced system is solved by standard matrix decomposition methods so that the problem then rests in determining which of the sets of equations can be solved accurately enough by Eq. (5–4.9). This is done by monitoring error terms, as described by Coffee et al. (1980).

5–5. SENSITIVITY ANALYSIS

The sensitivity of the behavior of a system of ordinary differential equations to uncertainties in its input parameters, such as chemical rates or initial conditions, is particularly important. For example, the chemical rates are often semiquantitative formulas extracted from experiments or fundamental calculations. They can be strong functions of both temperature and pressure. Often their values are not well known, as illustrated in Figure 2–1. Understanding the uncertainties in these parameters and how these uncertainties affect the reaction mechanism is important in determining the validity of the reaction model.

The complexity and size of many proposed chemical reaction mechanisms make it expensive and difficult to assess the effects of the uncertainties in the reaction rates on the predicted concentrations and temperatures. *How important are the uncertainties in chemical reaction rates or initial conditions to the predicted answer? Where should we focus attention in order to improve the accuracy of or reduce the size of the reaction mechanism? How sensitive are the solutions to the uncertainties in the parameters of the equations?*

Two classes of methods have evolved for evaluating the effects of uncertainties. The *local* or *deterministic* methods produce information on how the uncertainty in one parameter, for example, k_i, effects the calculated results of one of the dependent variables, for example, y_j. The *global* or *stochastic* methods consider the effects of simultaneously varying parameters over a range of values. The sensitivity measure that results from this global approach is an average effect of the uncertainties.

A number of reviews have been written about sensitivity methods. We recommend those by Tilden et al. (1981) and Rabitz et al. (1983). Below we describe some of the principles of these methods, leaving details to be found in the references cited.

5–5.1 Local Sensitivity Methods

Rewrite Eq. (5–1.1) as

$$\frac{d\mathbf{y}}{dt} = \mathbf{f}(\mathbf{y},\mathbf{k}) \,, \qquad (5-5.1)$$

$$\mathbf{y}(0) = \mathbf{y}_o \,,$$

where $\mathbf{y}(t)$ is the solution vector of N species concentrations, and $\mathbf{k}$ is a vector of M input parameter such as chemical rate constants and activation energies. This form of the equations emphasizes the functional dependence of $\mathbf{y}(t)$ on the parameters $\mathbf{k}$. There is usually an estimated value, $\hat{k}_j$, and uncertainties associated with each value k_j.

If only small deviations of k_j are considered about $\hat{k}_j$, a truncated Taylor series expansion can be used to estimate $y_i(t, \mathbf{k}) = y_i(t, \hat{\mathbf{k}} + \Delta\mathbf{k})$,

$$y_i(t, \mathbf{k}) = y_i(t, \hat{\mathbf{k}}) + \sum_{j-1}^{m} \frac{\partial y_i(t, \hat{\mathbf{k}})}{\partial k_j} \Delta k_j \,. \qquad (5-5.2)$$

The first-order sensitivity coefficient matrix is defined as

$$\beta_{ij} \equiv \left. \frac{\partial y_i}{\partial k_j} \right|_{\hat{\mathbf{k}}} \,. \qquad (5-5.3)$$

When the variation of y_i with k_j is being considered, all other rates k_l, $l \neq j$, are held fixed. Local sensitivity analysis focuses on the calculation and interpretation of these sensitivity coefficients.

Figure 5–5, a schematic diagram of a solution surface $y_i(t, \mathbf{k})$, shows the region of uncertainty in two parameters, k_1 and k_2 (Gelinas and Vajk, 1978). The estimated parameter values are $\hat{k}_1$ and $\hat{k}_2$. The assumed upper and lower limits of variation produce a region of uncertainty in the $k_1 - k_2$ plane, which result in a range of uncertainty in y_i. The sensitivity coefficients, β_{ij}, evaluated at $\hat{k}_1$ and $\hat{k}_2$, represent the slopes of the surface in the two coordinate directions at point Q.

There are several approaches to calculating first-order sensitivity coefficients. The simplest is a "brute-force" finite-difference approach. This involves varying each parameter, one at a time, to compute the deviation of y_i arising from a small change in k_j at the beginning of and throughout the integration. Then

$$\Delta y_i(t, \hat{\mathbf{k}}) = y_i(t, \hat{k}_1, \ldots, \hat{k}_j + \Delta k_j, \ldots, \hat{k}_m) - y_i(t, \hat{k}_1, \ldots, \hat{k}_j, \ldots, \hat{k}_m) \,. \qquad (5-5.4)$$

The y_i are obtained by integrating Eqs. (5–5.1). This way of determining $\{\Delta y_i(t, \mathbf{k})\}$ depends on the particular problem chosen for integration, and thus cannot lead to any generalizations about sensitivity. This approach can be used for arbitrary variations in Δk_j and is not limited to small variations in the uncertainties. It leads to the approximate first-order sensitivity coefficients,

$$\beta_{ij}^{(o)} \equiv \frac{\Delta y_i(t, \hat{\mathbf{k}})}{\Delta k_j} \,. \qquad (5-5.5)$$

This is the most direct approach to finding the effect of varying a parameter, and is certainly the simplest and most often used method.

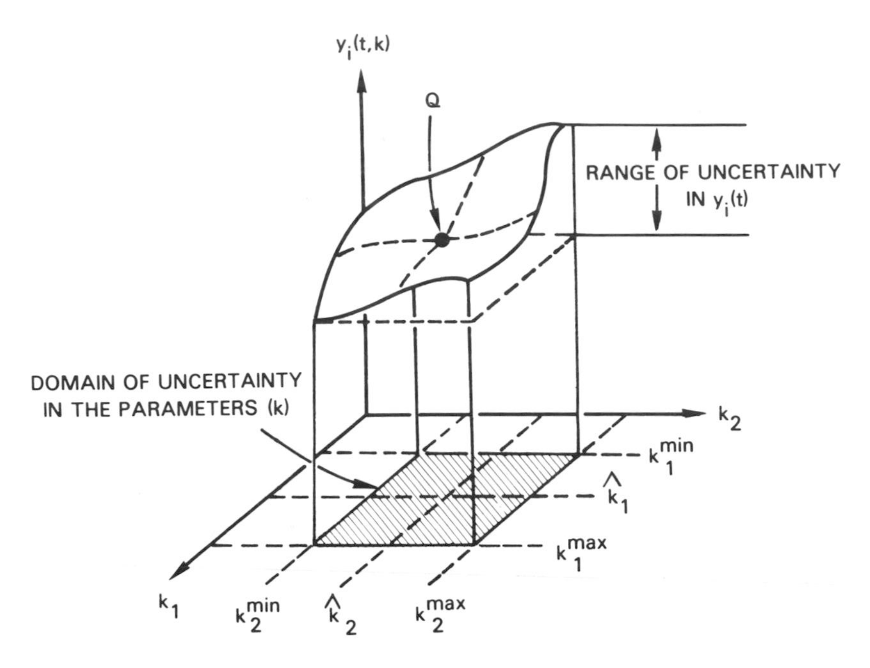

Figure 5–5. Schematic of the solution surface, $y_i(t,\mathbf{k})$, showing the region of uncertainty in two parameters, k_1 and k_2 (Gelinas and Vajk, 1978).

Alternative local approaches involve "direct solution," where the sensitivity coefficients themselves are considered as dynamic variables (see, for example, Dickinson and Gelinas, 1976). Differential equations for first-order sensitivity coefficient vectors can be derived using

$$\begin{aligned} \frac{d\boldsymbol{\beta}_j}{dt} &= \mathbf{J} \cdot \boldsymbol{\beta}_j(t) + \mathbf{b}_j(t) \\ \boldsymbol{\beta}_j(0) &= 0\,. \end{aligned} \tag{5 – 5.6}$$

Equations (5–5.6) are a set of M, N-dimensional vector ODEs where $\mathbf{J}$ is the $N \times N$ Jacobian,

$$J_{il} \equiv \frac{\partial f_i}{\partial y_l}\,. \tag{5 – 5.7}$$

The quantities $\{\boldsymbol{\beta}_j\}$ and $\{\mathbf{b}_j\}$ are N-dimensional vectors, such that $\boldsymbol{\beta}_j = (\beta_{1j}, \ldots, \beta_{Nj})$ and $\mathbf{b}_j = (\partial f_i/\partial k_j, \ldots, \partial f_N/\partial k_j)$. We could solve Eqs. (5–5.6) directly in conjunction with Eqs. (5–5.1). This amounts to solving $2N$ equations M times. Otherwise, we can solve Eqs. (5–5.1) to produce $\mathbf{y}(t,(k))$ and then interpolate values of $\mathbf{y}(t,(k))$ to use in $\mathbf{J}$. These approaches are discussed in Tilden et al. (1981). Solving these equations provides systematic information on the importance of small variations about $\mathbf{k}$ for a particular problem, but the results are not necessarily valid for large uncertainties or for different problems.

Another local method uses a Green's function. This has been shown (Kramer et al., 1984a,b) to be more efficient than direct solutions of Eqs. (5–5.6) when there are more parameters whose sensitivity is in question than there are dependent variables. The solution to Eqs. (5–5.6) can be written as

$$\boldsymbol{\beta}_j(t) = \boldsymbol{\psi}(t,0)\boldsymbol{\beta}_j(0) + \int_o^t \boldsymbol{\psi}(t,\tau)\mathbf{b}_j(\tau)d\tau \,, \qquad (5-5.8)$$

where $\boldsymbol{\psi}$ is determined by the equation

$$\begin{aligned} \frac{d\boldsymbol{\psi}(t,\tau)}{d\tau} + \boldsymbol{\psi}(t,\tau)\mathbf{J}(\tau) &= 0 \\ \boldsymbol{\psi}(t,t) &= (I) \,. \end{aligned} \qquad (5-5.9)$$

Efficient computational methods involving recursion relations in time have been developed for using this method (Rabitz et al., 1983; Kramer et al., 1984a,b).

5–5.2 Global Sensitivity Methods

Consider Figure 5–5 again. For small displacements about the estimated parameter values, the tangent plane at Q differs by only a small amount from the actual solution surface. The sensitivity coefficients at Q do not contain information on the behavior of the surface away from Q, nor do they indicate the full range of variation of y_i in the region of uncertainty of the parameters.

As we have seen, the most straightforward approach to studying the sensitivity of Eqs. (5–5.1) is repeatedly integrating the ODEs using different values for the parameters. This allows us to construct a solution surface in parameter space with the parameters varied systematically (see Box et al., 1978). However, this approach becomes cumbersome and expensive as the number of species and parameters increase. When the expense of this brute-force method becomes prohibitive, alternative global methods can be used.

Global methods calculate averaged sensitivities which quantify the effects of simultaneous, large variations in $\mathbf{k}$. These sensitivities are a qualitatively different measure from the local sensitivity coefficients discussed above. They are based on the assumption that the uncertainty in k_j can be expressed in terms of a probability density function, p_{y_i} for the dependent variable y_i. These can be used to evaluate statistical properties of the system, such as the expected value,

$$< y_i(t) > = \int_{k_M^{min}}^{k_M^{max}} \cdots \int_{k_1^{min}}^{k_1^{max}} p_{y_i}(t,\mathbf{k})p_1(k_1)p_2(k_2)\cdots dk_1 dk_1 \cdots dk_M \tag{5-5.10}$$

where $\{p_i(k_i)\}$ are the probability distribution functions of the $\{k_i\}$. The sensitivity analysis then involves computing the functions $\mathbf{p}$. Suppose we have some knowledge of the probability distributions of two parameters, p_{k_1} and p_{k_2}. Then the probability distribution of y_i can be calculated. Whether or not the probability distributions for k_1 and k_2 are given, the solution surface for y_i can be determined by systematically selecting test points in the $k_1 - k_2$ plane and solving the system to determine $y(t; k_1, k_2)$. This is the approach shown in Figure 5–6 (Gelinas and Vajk, 1978). A global sensitivity analysis requires sampling over the range of uncertainty of the parameters.

Costanza and Seinfeld (1981) developed the *stochastic approach* based on the probability distribution function. First rewrite Eqs. (5–5.1) as

$$\frac{d\mathbf{x}}{dt} = \mathbf{F}(\mathbf{x},t) \quad \text{and} \quad \mathbf{x}(0) = \mathbf{x}_o\,, \tag{5-5.11}$$

where $\mathbf{x}$ is a vector whose $N + M$ components are the N species concentrations $\mathbf{y}$ and the M parameters $\mathbf{k}$. Then $F_i(\mathbf{x},t) = f_i(\mathbf{y},t)$ for $i = 1, \ldots, N$ and $F_{i+N} = 0$ for $i = 1, \ldots, M$. If we assume that the uncertainties in $\mathbf{k}$, including those in initial conditions, can be represented by a probability distribution, then the initial conditions $\mathbf{x}_o$ are random variables with the probability distribution $p_o(\mathbf{x}_o)$ for $i = N+1, \ldots, N+N$.

Let $p(\mathbf{x},t)$ be the probability distribution of $\mathbf{x}(t)$. Then

$$< g(\mathbf{x}(t)) > = \int d\mathbf{x}\, p(\mathbf{x},t) g(\mathbf{x})\,, \tag{5-5.12}$$

Costanza and Seinfeld (1981) show that $p(\mathbf{x},t)$ satisfies

$$\frac{\partial p(\mathbf{y},t)}{\partial t} + \nabla_{\mathbf{y}} \cdot (p(\mathbf{y},t)\mathbf{f}(\mathbf{y},t)) = 0 \tag{5-5.13}$$

$$p(\mathbf{y},0) = p_o(\mathbf{y})\,.$$

Thus, given $p_o(\mathbf{y}) = p_o(\mathbf{x}_o, \mathbf{k})$ and the set of ODEs, Eqs. (5–5.13) can be solved for $p(\mathbf{y},t)$. Once we know p($\mathbf{x}$,t), this surface in $\mathbf{x}$-space can be studied for fixed t. However this is difficult for large $N + M$, and so it is

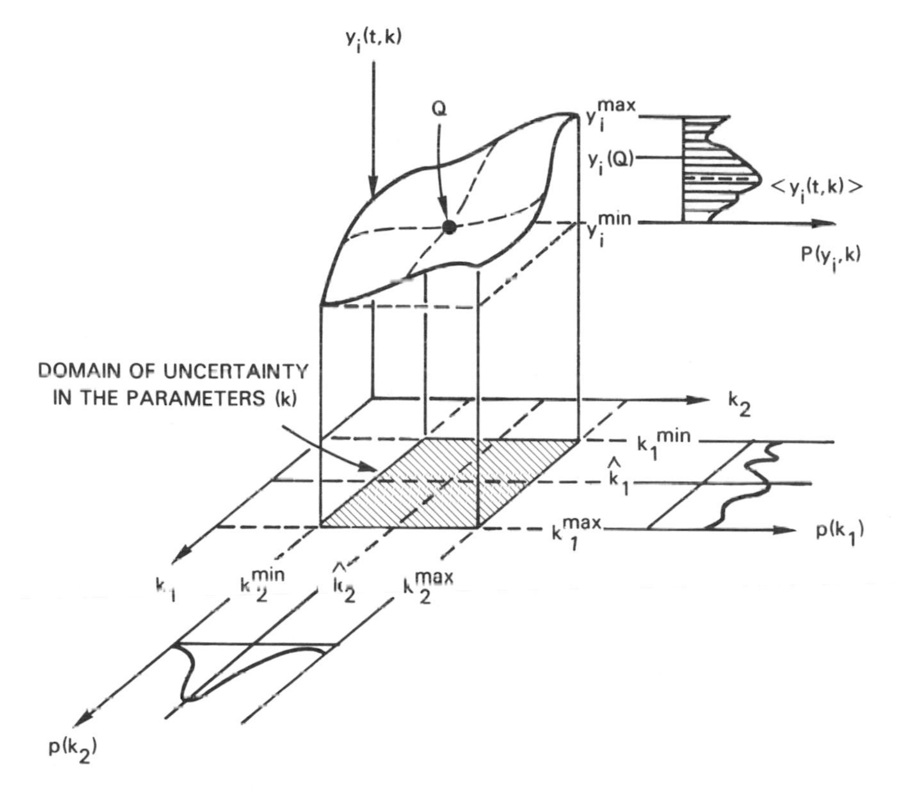

Figure 5–6. Schematic of the solution surface, $y_i(t,\mathbf{k})$, showing the probability distribution function approach to finding sensitivities (Gelinas and Vajk, 1978).

useful to study reduced probability distributions obtained by integrating p over a subset of the components of $\mathbf{x}$.

Sometimes there is not enough information about the initial state or the parameters to define p_o, but there are the estimated values of the parameters and an estimate of their range of uncertainty. Then each parameter can be sampled over its range to assess the range of variation of $\mathbf{y}(\mathbf{t})$. This leads to the pattern search procedures: Monte Carlo methods (see, for example, Stolarski et al., 1978); pattern methods (see, for example, Sobol, 1979); and Fourier methods (see, for example, Cukier et al., 1973; McRae et al., 1982). These methods choose the specific sampling point differently. In general, Fourier methods require considerably fewer sampling points than Monte Carlo methods, but there are more questions about the interpretation of Fourier sensitivity analyses.

5–5.3 General Thoughts on Sensitivity Analysis

Sensitivity methods can provide relationships among the variables and nominal constants in the system. If a set of equations is presented to someone who has little feeling for the important parameters controlling that system, sensitivity analysis is one way of helping to understand the system. When the system being studied is fairly well understood, directly varying selected parameters and recalculating might be the best, and is certainly the simplest, way to proceed.

Even though the results of a sensitivity analysis might be the optimal information to have, it is expensive. Direct methods, local methods, and global methods are all expensive. In the case of large systems of stiff equations, varying just one parameter can be inordinately expensive. In linear methods, important longer-range information could be missed while in global methods, it might also be difficult to interpret the results.

Although knowing the probability distribution functions completely allows the calculation of everything about the system, this is extremely expensive to implement in practice. Therefore most practical methods try to extract more limited pieces of information, obtaining computational efficiency by trading off other things. It is important to realize exactly what the benefits of each sensitivity analysis methodology are before choosing one.

Perhaps the most popular methods are the brute-force method and the Green's function approach. The brute-force methods are often combined with clever ways of extracting information. For example, the method described by Miller and Frenklach (1983) and Frenklach (1984) combines chemical kinetics calculations with experimental design optimization methods. Points in parameter space are used to construct a series of nearby simulations. The results are used to fit an interpolating hypersurface in parameter space, and the surface provides the desired information. This method has been applied to several systems, and seems promising. The Green's function method is also easily accessible to researchers, and might be the most economical method for large systems where local analysis is adequate.

5–6. CHEMICAL KINETICS CALCULATIONS

Equations (5–0.1) describe the time evolution of a number of chemical species in a homogeneous, premixed volume. These equations may be solved with physically reasonable constraints such as constant temperature, constant pressure, or constant volume. From the set of solutions $\{n_j(t)\}$, a number of quantities can be derived which may be compared directly with available experimental data. These derived quantities include chemical induction times, energy release rates, emission intensities from excited states of reacting species, and bulk properties such as overall activation energy and total energy release.

In the previous sections of this chapter we dealt with generic sets of ordinary differential equations and said relatively little specific to chemical kinetics. Here we consider some of the problems that occur when modeling chemical kinetics in reactive flows.

5–6.1 Chemical Kinetics in Reactive Flows

Solving complicated rate equations coupled to a fluid dynamic calculation is very different from solving a set of chemical rate equations alone. First, in a reactive flow calculation, the chemical rate equations must be integrated for many timesteps interspersed with computations for the nonlocal diffusion and convective terms in the equations. The integration formulas can only be applied for relatively short periods of time between the times when the fluid dynamic variables are updated. Thus, there is a benefit in using single-step, self-starting algorithms with low computational cost for restarting an integration.

Second, the large number of computational cells, perhaps tens or hundreds of thousands in a multidimensional domain, usually implies that we cannot afford to store auxiliary information about all the species in each cell between timesteps. Because it requires much computer time to start and restart a high-order method, fast flexible, low-order methods are preferred.

Third, fluid dynamic errors are rarely less than a few percent, so it is probably not necessary to compute the chemistry to very high accuracy. Reaction rates are at best poorly known. Again, low-order methods are the logical choice over complicated high-order methods.

Fourth, reactive flow programs are generally changed frequently. Thus a general integration technique is usually preferable to one that is highly specialized for a specific set of rates.

Finally, multidimensional reactive flow calculations are usually expensive even using the fastest methods, so that it is best to avoid evaluating transcendental functions which often arise in expressions for chemical reaction rates. Whenever analytic solutions are approximated, a polynomial, rational-function approximation, or a table lookup routine should be considered instead.

5–6.2 Solving the Temperature Equation

The evolution equation for the temperature has to be solved in conjunction with Eqs. (5–0.1) when chemical energy is being released. This equation does not appear explicitly in the conservation equations, but can be derived from the energy equation by relating the internal energy to the temperature

$$\epsilon = \frac{f}{2}kT , \qquad (5-6.1)$$

where f is the number of degrees of freedom. Every chemical reaction occuring in a local thermodynamic equilibrium releases or consumes a known amount of energy. Therefore the time rate of change of the temperature is closely tied to the rate equations for the individual species as well as fluid dynamic effects.

Alternatively, because of these constraints, we can use the expression

$$\epsilon = H(T, \{n_j\}) - P , \qquad (5-6.2)$$

where H is the enthalpy of the system. If we know the value of the enthalpy, we can find an expression for $\partial T/\partial t$. The expression for the temperature derivative involves all of the $\{\partial n_j/\partial t\}$ derivatives as well as expressions of the form $\{\partial h_j/\partial t\}$ which may be expressed as powers of T. Solving for temperature by this approach is usually expensive computationally.

The correct expression for the enthalpy also involves complicated sums over excited states of the molecules. When each of the species is in local thermodynamic equilibrium, the individual $h_j(T)$ can be evaluated as a function of temperature and fit to a polynomial expansion. This has been done for the JANAF tables (Stull and Prophet, 1971) and in the work of Gordon and McBride (1976).

By using tabulated values of the enthalpies $\{h_i(T)\}$ and the heats of formation $\{h_{oi}(T)\}$ (see Chapter 2), the tedious temperature integration can often be avoided completely. During the chemical reaction portions of each timestep, let us assume that the total internal energy of the system does not change but may be redistributed among chemical states. Then Eq. (5–6.3) can be solved iteratively for a new temperature which is consistent with the new number densities calculated. A simple Newton-Raphson iteration, as described in Carnahan et al. (1969), is usually good enough to solve Eq. (5–6.3) for the temperature. When incorporated in simulations, the iteration converges in one or two iterations per timestep when relatively little change is allowed in the system variables per timestep.

5–6.3 Applications to Chemical Kinetics

With improving knowledge of elementary chemical reaction rates, it is reasonable to construct detailed chemical mechanisms describing complex intermediate reaction paths in global reaction systems. For example, consider the global system of methane and oxygen:

$$CH_4 + 2O_2 \rightarrow CO_2 + H_2O \ .$$

This expresses the global process of oxidation of methane to produce carbon dioxide and water. However, this process really takes place as a sequence of elementary reactions describing the removal of a hydrogen from the CH_4 to produce methyl radical, the reaction of methyl radical with molecular oxygen, and finally the production of CO_2 and H_2O.

Chemical reaction mechanisms consisting of elementary reaction rates with increasing levels of complexity have been constructed through the years. For methane, the recent mechanisms of Westbrook and Dryer (1984) and Frenklach and Bornside (1984) consist of as many as 150 forward reactions. Recently proposed chemical reaction mechanisms for soot formation consist of as many as 500 reactions.

Several stages of development are necessary to construct such a chemical reaction mechanism. First, a set of elementary reaction rates has to be formulated for the global mechanism. These mechanisms are constructed by considering possible reaction pathways, or sets of reaction pathways, and by determining the best sets of reaction rate coefficients that are available. This is usually done selectively by a knowledgeable chemist with the insight to know which pathways are likely to be important and which are probably not. It can also be done by writing down every possible pathway and then using a combination of experience, numerical analysis, and intuition to eliminate some of them.

Once the governing reaction pathways are proposed, values of the corresponding chemical rate constants must be assembled. As discussed in Chapter 2, these are generally taken from experiments or from *ab initio* calculations. Often the heats of formation of the materials and their heat capacities as a function of temperature are needed, in addition to the chemical rates. The proposed mechanisms must also be subjected to analyses that estimate, for a given temperature and pressure, which reactions may be eliminated from the overall scheme or which must be measured more carefully.

The systems of ordinary differential equations representing these mechanisms are then integrated and the numerical results compared to existing experimental data. In the best cases, such data exist over a wide range of temperature and pressure. There are uncertainties in the choice of reactions in the mechanism, the choice of reaction rate coefficients, and in the experimental data used to calibrate the reaction mechanism. Because of these uncertainties, physical intuition about the system and the numerical simulations are required to study the consequences of assumptions about the rates and to compare pathways. Sensitivity analyses are useful in determining when more accurate data are needed. All functions for reaction rates and enthalpies must be represented numerically by continuous expressions, and these expressions cannot be extended beyond their range of validity. It can take several years to develop and calibrate the reaction mechanism in detail.

In spite of these problems, a number of complete reaction mechanisms have been developed. For example, there are compilations and studies of hydrogen combustion by Burks and Oran (1981), and a compilation and tests by Frenklach and Bornside (1984) for methane kinetics. There is a discussion and study of the rates needed for modeling hydrocarbon combustion chemistry by Westbrook and Dryer (1984), and a survey of rates coefficients in the $C/H/O$ system by Warnatz (1984). Collections of individual rates for the hydrogen-oxygen system have been compiled and analyzed by Baulch (for example, Baulch et al., 1972, 1981). In addition, there are large collections of tabulated rate constants compiled at the National Bureau of Standards, for example, by Hampson (1980) and Westley (1981).

References

Babcock, P.D., L.F. Stutzman, and D.M. Brandon, Jr., 1979, Improvements in a Single-Step Integration Algorithm, *Simulation* 33: 1–10.

Bader, G., and P. Deuflhard, 1983, A Semi-Implicit Mid-Point Rule for Stiff Systems of Ordinary Differential Equations, *Numer. Math.* 41: 373–398.

Bashforth, F., and J.C. Adams, 1883, *An Attempt to Test the Theories of Capillary Action by Comparing the Theoretical and Measured Forms of Drops of Fluid. With an Explanation of the Method of Integration Employed in Constructing the Tables which Give the Theoretical Forms of Such Drops*, University Press, Cambridge (England).

Baulch, D.L., D.D. Drysdale, D.G. Horne, and A.C. Lloyd, 1972, *Evaluated*

Kinetic Data for High Temperature Reactions, Butterworth, London.

Baulch, D.L., J. Duxbury, S.J. Grant, and D.C. Montague, 1981, *Evaluated Kinetic Data for High Temperature Reactions*, *J. Phys. Chem. Ref. Data* 10, Supplement No. 1.

Bjurel, G., G. Dahlquist, B. Lindberg, S. Linde, and L. Oden, 1970, *Survey of Stiff Ordinary Differential Equations*, Report NA 70.11, Department of Information Processing, Royal Institute of Technology, Stockholm.

Boris, J.P., 1971, Relativistic Plasma Simulation — Optimization of a Hybrid Code, in J.P. Boris and R.A. Shanny, eds., *Numerical Simulation of Plasmas*, Naval Research Laboratory, Washington, DC, 3–67.

Box, G.E.P, W.G. Hunter, and J.S. Hunter, 1978, *Statistics for Experimenters*, Wiley, New York.

Brandon, D.M., Jr., 1974, A New Single-Step Implicit Integration Algorithm with A-stability and Improved Accuracy, *Simulation* 23: 17–29.

Bui, T.D., A.K. Oppenheim, and D.T. Pratt, 1984, Recent Advances in Methods for Numerical Solution of ODE Initial Value Problems, *J. Comput. Appl. Math.* 11: 283–296.

Bulirsch, R., and J. Stoer, 1964, Fehlerabschätzungen und Extrapolation mit rationalen Funktionen bei Verfahren vom Richardson-Typus, *Numer. Math.* 6: 413–427.

Bulirsch, R., and J. Stoer, 1966, Numerical Treatment of Ordinary Differential Equations by Extrapolation Methods, *Numer. Math.* 8: 1–13.

Buneman, O., 1967, Time-Reversible Difference Procedures, *J. Comput. Phys.* 1: 517–535.

Burks, T.L., and E.S. Oran, 1981, *A Computational Study of the Chemical Kinetics of Hydrogen Combustion*, NRL Memorandum Report 4446, Naval Research Laboratory, Washington, DC [AD A094384]

Carnahan, B., H.A. Luther, and J.O. Wilkes, 1969, *Applied Numerical Methods*, Wiley, New York.

Coffee, T.P., J.M. Heimerl, and M.D. Kregel, 1980, *A Numerical Method to Integrate Stiff Systems of Ordinary Differential Equations*, Technical Report ARBRL-TR-02206, U.S. Army Ballistic Research Laboratory, Aberdeen, MD. [AD A080988]

Cooper, G.J., 1969, The Numerical Solution of Stiff Differential Equations, *FEBS. Lett.* 2: S22–S29 (Supplement).

Costanza, V., and J.H. Seinfeld, 1981, Stochastic Sensitivity Analysis in Chemical Kinetics, *J. Chem. Phys.* 74: 3852–3858.

Cryer, C.W., 1973, A New Class of Highly-Stable Methods: A_o-Stable Methods, *BIT* 13: 153–159.

Cukier, R.I., C.M. Fortuin, K.E. Shuler, A.G. Petschek, and J.H. Schaibly,

1973, Study of the Sensitivity of Coupled Reaction Systems to Uncertainties in Rate Coefficients. I. Theory, *J. Chem. Phys.* 59: 3873–3878.

Curtis, A.R., 1978, Solution of Large, Stiff Initial Value Problems — The State of the Art, in D. Jacobs, ed., *Numerical Software — Needs and Availability*, Academic, New York, 257–278.

Curtiss, C.F., and J.O. Hirschfelder, 1952, Integration of Stiff Equations, *Proc. Nat. Acad. Sci.* 38: 235–243.

Dahlquist, G. G., 1963, A Special Stability Problem for Linear Multistep Methods, *BIT* 3: 27–43.

Dahlquist, G.G., and B. Lindberg, 1973, *On Some Implicit One-Step Methods for Stiff Differential Equations*, Technical Report TRITA-NA-7302, Royal Institute of Technology, Stockholm.

Deuflhard, P., G. Bader, and U. Nowak, 1981, Larkin — A Software Package for the Numerical Simulation of Large Systems Arising in Chemical Reaction Kinetics, in K.H. Ebert, P. Deuflhard, and W. Jäger, eds., *Modelling of Chemical Reaction Systems*, Springer-Verlag, New York.

Deuflhard, P., 1982, Recent Progress in Extrapolation Methods for Ordinary Differential Equations, invited talk at the SIAM 30th Anniversary Meeting, Stanford, CA, July, 1982.

Dickinson, R.P., and R.J. Gelinas, 1976, Sensitivity Analysis of Ordinary Differential Equation Systems, *J. Comput. Phys.* 21: 123–143.

Ehle, B.L., 1969, *On Pade Approximations to the Exponential Function and A-Stable Methods for the Numerical Solution of Initial Value Problems*, Research Report No. CSRR 2010, Department of Applied Analysis and Computer Science, University of Waterloo.

Enright, W.H., and T.E. Hull, 1976, Comparing Numerical Methods for the Solution of Stiff Systems of ODEs Arising in Chemistry, in L. Lapidus and W.E. Schiesser, eds., *Numerical Methods for Differential Systems*, Academic, New York, 45–66,

Frenklach, M., and D.E. Bornside, 1984, Shock-Initiated Ignition in Methane-Propane Mixtures, *Combust. Flame* 56: 1–27.

Frenklach, M., 1984, Modeling, in W.C. Gardner, ed., *Combustion Chemistry*, Springer-Verlag, New York, 423–453.

Gear, C.W., 1969, The Automatic Integration of Stiff Ordinary Differential Equations, in A.J.H. Morrell, ed., *Information Processing 68*, Proceedings of the IFIP Congress, 1968, North-Holland, Amsterdam (Netherlands), 187–193.

Gear, C.W., 1971, *Numerical Initial Value Problems in Ordinary Differential Equations*, Prentice-Hall, Englewood Cliffs, NJ.

Gelinas, R.J., 1972, Stiff Systems of Kinetics Equations — A Practitioner's

View, *J. Comput. Phys.* 9: 222–236.

Gelinas, R.J., and J.P. Vajk, 1978, *Systematic Sensitivity Analysis of Air Quality Simulation Models*, Science Applications Inc., Pleasanton, CA. [PB80-112162]

Gill, S., 1951, A Process for the Step-by-Step Integration of Differential Equations in an Automatic Digital Computing Machine, *Proc. Cambridge Philos. Soc.* 47: 96—108.

Gordon, S., and B.J. McBride, 1976, *Computer Program for Calculation of Complex Chemical Equilibrium Compositions, Rocket Performance, Incident and Reflected Shocks, and Chapman-Jouguet Detonations*, NASA SP-273, National Aeronautics and Space Administration, Washington, DC. [N78-177243]

Gragg, W.B., 1965, On Extrapolation Algorithms for Ordinary Initial Value Problems, *SIAM J. Numer. Anal.* 2: 384–403.

Hampson, R.F., 1980, *Chemical Kinetic and Photochemical Data Sheets for Atmospheric Reactions*, Report No. FAA/EE-80-17, Department of Transportation, Washington, DC. [AD H091631]

Hindmarsh, A.C., 1983, ODEPACK, A Systematized Collection of ODE Solvers, in R.S. Stepleman, ed., *Numerical Methods for Scientific Computation*, North-Holland, New York, 55–64. Also see Lawrence Livermore National Laboratory Report, UCRL-88007, 1982.

Hindmarsh, A.C., 1974, *GEAR: Ordinary Differential Equation System Solver*, LLNL Report UCID-30001, Rev. 3, Lawrence Livermore National Laboratories, Livermore, CA.

Hindmarsh, A.C., 1976a, *Preliminary Documentation of GEARB: Solution of ODE Systems with Block-Iterative Treatment of the Jacobian*, LLNL Report UCID-30149, Lawrence Livermore National Laboratories, Livermore, CA. [UCID-30130]

Hindmarsh, A.C., 1976b, *Preliminary Documentation of GEARBI: Solution of Implicit Systems of Ordinary Differential Equations with Banded Jacobian*, Lawrence Livermore National Laboratories, Livermore, CA.

Hindmarsh, A.C., 1977, *GEARB: Solution of Ordinary Differential Equations Having Banded Jacobian*, LLNL Report UCID-30059, Rev. 2, Lawrence Livermore National Laboratories, Livermore, CA.

Hockney, R.W., 1965, A Fast Direct Solution of Poisson's Equation Using Fourier Analysis, *J. Ass. Comput. Mach.* 12: 95–113.

Keneshea, T.J., 1967, *A Technique for Solving the General Reaction-Rate Equations in the Atmosphere*, AFCRL-67-0221, Air Force Cambridge Research Laboratories, Hanscom Field, MA. [AD 654010]

Kramer, M.A., H. Rabitz, J.M. Calo, and R.J. Kee, 1984a, Sensitivity Analysis in Chemical Kinetics: Recent Developments and Computational Comparisons, *Int. J. Chem. Kin.* 16: 559–578.

Kramer, M.A., R.J. Kee, and H. Rabitz, 1984b, *CHEMSEN: A Computer Code for Sensitivity Analysis of Elementary Chemical-Reaction Models*, Sandia Report SAND-82-8230, Sandia National Laboratories, Albuquerque, NM.

Lambert, J.D., 1973, *Computational Methods in Ordinary Differential Equations*, Wiley, New York.

Lambert, J.D., 1980, Stiffness, in I. Gladwell and D.K. Sayers, eds., *Computational Techniques for Ordinary Differential Equations*, Academic, New York, 19–46.

Lindberg, B., 1973, *IMPEX2 — A Procedure for Solution of Systems of Stiff Differential Equations*, Technical Report TRITA-NA-7303, Royal Institute of Technology, Stockholm.

Liniger, W., and R.A. Willoughby, 1970, Efficient Integration Methods for Stiff Systems of Ordinary Differential Equations, *SIAM J. Num. Anal.* 7: 47–66.

McCalla, T.R., 1967, *Introduction to Numerical Methods and FORTRAN Programming*, Wiley, New York.

McRae, G.J., J.W. Tilden, and J.H. Seinfeld, 1982, Global Sensitivity Analysis – A Computational Implementation of the Fourier Amplitude Sensitivity Test (FAST), *Comp. and Chem. Eng.* 6: 15–25.

May, R., and J. Noye, 1984, The Numerical Solution of Ordinary Differential Equations: Initial Value Problems, in J. Noye, ed., *Computational Techniques for Differential Equations*, North-Holland, New York, 1–94.

Miller, D., and M. Frenklach, 1983, Sensitivity Analysis and Parameter Estimation in Dynamic Modeling of Chemical Kinetics, *Int. J. Chem. Kin.* 15: 677–696.

Neville, E.H., 1934, Iterative Interpolation, *J. Ind. Math. Soc.* 20: 87–120.

Page, M., E. Oran, D. Miller, R. Wyatt, H. Rabitz, and B. Waite, 1985, A Comparison of Quantum, Classical and Semiclassical Descriptions of a Model, Collinear, Inelastic Collision of Two Diatomic Molecules, *J. Chem. Phys.* 83: 5635–5646.

Pratt, D.T., and K. Radhakrishnan, 1984, *CREKID: A Computer Code for Transient, Gas-Phase Combustion Kinetics*, NASA Technical Memorandum 83806, National Aeronautics and Space Administration, Washington, DC. [N85-10068/3/XAB]

Rabitz, H., M. Kramer, and D. Dacol, 1983, Sensitivity Analysis in Chemical Kinetics, *Ann. Rev. Phys. Chem.* 34: 419–461.

Radhakrishnan, K., 1984, A Comparison of the Efficiency of Numerical

Methods for Integrating Chemical Kinetic Rate Equations, Proceedings of the JANNAF Propulsion Meeting, New Orleans; also NASA Technical Memorandum 83590, National Aeronautics and Space Administration, Washington, DC. [N85-21162/1/XAB]

Ralston, A., 1978, *A First Course in Numerical Analysis*, McGraw-Hill, New York.

Seinfeld, J.H., L. Lapidus, and M. Hwang, 1970, Review of Numerical Integration Techniques for Stiff Ordinary Differential Equations, *I & EC Fund.* 9: 266–275.

Shampine, L.F., H.A. Watts, and S.M. Davenport, 1976, Solving Nonstiff Ordinary Differential Equations — The State of the Art, *SIAM Rev.* 18: 376–411.

Sobol, I.M., 1979, On the Systematic Search in a Hypercube, *SIAM J. Numer. Anal.* 16: 790–793.

Stolarski, R.S., D.M. Butler, and R.D. Rundel, 1978, Uncertainty Propagation in a Stratospheric Model, 2. Monte Carlo Analysis of Imprecisions due to Reaction Rates, *J. Geophys. Res.* 83: 3074–3078.

Stull, D.R., and H. Prophet, 1971, *JANAF Thermochemical Tables*, 2nd ed., National Standard Reference Data Series, U.S. National Bureau of Standards, No. 37, Gaithersburg, MD. Government Printing Office, Washington, DC.

Tilden, J.W., V. Costanza, G.J. McRae, and J.H. Seinfeld, 1981, Sensitivity Analysis of Chemically Reacting Systems, in K.H. Ebert, P. Deuflhard, and W. Jäger, eds., *Modelling of Chemical Reaction Systems*, Springer-Verlag, New York, 69–91.

Warnatz, J., 1984, Rate Coefficients in the C/H/O System, in W.C. Gardner, ed., *Combustion Chemistry*, Springer-Verlag, New York, 173–197.

Westbrook, C.K., and F.L. Dryer, 1984, Chemical Kinetic Modeling of Hydrocarbon Combustion, *Prog. Energy Combust. Sci.* 10: 1–57.

Westley, F., 1981, *Tables of Experimental Rate Constants for Chemical Reactions Occurring in Combustion (1971–1977)*, Report NBSIR-81-2254, National Bureau of Standards, Washington, DC. [PB81-205429]

Widlund, O.B., 1967, A Note on Unconditionally Stable Linear Multistep Methods, *BIT* 7: 65–70.

Young, T.R., and J.P. Boris, 1977, A Numerical Technique for Solving Stiff Ordinary Differential Equations Associated with the Chemical Kinetics of Reactive-Flow Problems, *J. Phys. Chem.* 81: 2424–2427.

Young, T.R., 1979, *CHEMEQ — Subroutine for Solving Stiff Ordinary Differential Equations*, NRL Memorandum Report 4091, Naval Research Laboratory, Wash., DC. [AD A083545]

Chapter 6

REPRESENTATIONS, RESOLUTION, AND GRIDS

Choosing a computational representation is just as important as choosing a mathematical model to describe the system, or as choosing the algorithms to implement the model. In this chapter we return to the problems of representing a continuous physical variable by a discrete set of numerical quantities. We introduce Eulerian and Lagrangian representations and discuss the differences between them. The choice of numerical algorithms and gridding methods are both constrained by this choice of representation. We introduce the ideas of variable and adaptive gridding, and show how localized improvements in resolution can substantially increase accuracy. This chapter could usefully be reread at the conclusion of Chapter 10.

6–1. REPRESENTATIONS OF CONVECTIVE FLOWS

Chapter 2 introduced some of the physical complexities that arise in convection problems and Chapter 4 presented several straightforward finite-difference algorithms for representing convection. Now we reconsider the difficult numerical problem of representing and solving the equations for convection.

6–1.1 Why Convection is Difficult to Simulate

Calculating the convection of a quantity ρ from one location to another usually demands more of numerical techniques than calculating the effects of local processes, diffusive effects, or wave and oscillatory phenomena. This is true even when the variables are all smoothly varying and well resolved in some computational discretization.

One of the problems is that convection does not naturally smooth out fluctuations and gradients and thus lacks the numerical smoothing quality of diffusion and relaxation found in local processes. In fact, it can cause the opposite effect. Shear, for example, leads to stretching in the fluid that can steepen gradients and lead to appreciably more fine-scale structure as time advances. Convection does not damp the short-wavelength components of flow structures, but can even increase their amplitudes. Thus the short wavelengths as well as long wavelengths must be treated relatively accurately.

Convection is subject to constraints of causality, positivity, conservation, and time reversibility. Therefore these qualitative properties should be built into numerical convection algorithms. Both the phases and the amplitudes of the modes must be integrated accurately. When the flow is incompressible, and thus the sound waves are very fast compared to velocities of interest in the flow, distant fluid elements become strongly coupled. Correlating these motions over large distances introduces new numerical problems.

The demands on convection algorithms are accentuated by the *secular*, or growing, nature of phase errors. These errors become apparent when a structure composed of a number of different wavelength components is convected across a discrete computational domain. As shown in Chapter 4, errors in the amplitude of a given mode are bounded and tend to decrease as long as an algorithm and the physical situation represented are stable. Phase errors, on the other hand, can increase indefinitely because they are not limited by stability conditions, only by accuracy conditions. The phasing between different harmonics comprising the profile can be important when we want to resolve structures with small wavelengths. For example, assume that the longest and shortest significant wavelength components in a structure have phase velocities that differ by only 3%. Then a flow structure incorrectly doubles in thickness by the time it is convected only about $\sim 1/.03 = 30$ times its original size. Furthermore, the profile is significantly distorted and nonphysical undershoots and overshoots may appear.

Chapter 4 showed a computed profile departing steadily from the actual profile as the short and long scales in the computed approximation become uncorrelated, and also showed that diffusion can reduce growing phase errors by damping out the short wavelengths. A subset of the numerical literature is devoted to algorithms for *convective-diffusion problems*, so that they can take advantage of the presence of physical diffusion to cover up the errors of the numerical convection algorithm. Unfortunately, the amount of diffusion required to mask secular phase errors in numerical convection algorithms is most often orders of magnitude greater than physical diffusion. In practice, we wish to retain as much small-scale information about the flow as possible. One of the major issues in computational fluid dynamics is how to minimize this extra smoothing.

6–1.2 The Continuity Equation

The basic equation in convective flow is the continuity equation. We often treat the continuity equation in *conservation form*,

$$\frac{\partial \rho}{\partial t} + \nabla \cdot \rho \mathbf{v} = 0 . \qquad (6-1.1)$$

However, there are other ways to write the continuity equation. The *convection form* is

$$\frac{\partial \rho}{\partial t} + \mathbf{v} \cdot \nabla \rho = \frac{d\rho}{dt} = -\rho \nabla \cdot \mathbf{v} , \qquad (6-1.2)$$

where $\mathbf{v} \cdot \nabla \rho$ represents convection and $-\rho \nabla \cdot \mathbf{v}$ represents compression. The full derivative, $d\rho/dt$, as used here is the Lagrangian derivative, the rate of change of the density ρ in the frame of reference moving with the local fluid velocity. The *integral form* of the continuity equation is

$$\frac{\partial}{\partial t} \int_{\text{volume}} \rho dV = - \int_{\text{surface}} \rho \mathbf{v} \cdot d\mathbf{A} , \qquad (6-1.3)$$

where dV is a volume element and $d\mathbf{A}$ is an area element. Equation (6–1.3) states that the amount of material $\int \rho dV$ in a volume can change only through flows of material $\rho \mathbf{v} \cdot d\mathbf{A}$ through the boundary of that volume.

These continuity equations are a statement about the behavior of the continuous, infinite-dimensional vector ρ of which each component is a point in the space $\mathbf{x}$ through which ρ passes. Numerical approximations to ρ are projections of this vector onto a finite-dimensional subspace. When we approximate Eq. (6–1.1) numerically, we devise a means of predicting, on the basis of a finite approximation at some time t, the equivalent finite approximation to ρ at time $t' > t$. The inherent problems of representing a continuum function by a finite number of degrees of freedom have been introduced in Chapter 4 and are discussed further below in Section 6–2. The art in inventing numerical methods involves defining algorithms which closely track the evolution of the continuum function in the time and space ranges of interest.

The continuity equation, written as Eqs. (6–1.1) or (6–1.2), has several qualitative physical properties which we also want numerical algorithms to have. One of these properties is causality, which means that the material currently at a given point arrives there only by previously leaving another place and passing through all points in between. Another property is conservation, which means that the total quantity of ρ in a closed system does not change. In most situations the conservation properties of physics have to be mirrored in the numerical method, or else the solutions show nonphysical instabilities or unacceptable secular errors. A third property is positivity, which states that if the quantity ρ was originally positive, it will never become negative due to convection alone. Although positivity is frequently as important as conservation, not all methods for simulating continuity equations enforce positivity. Many of the common and annoying problems which arise in the solution of coupled systems of convective equations occur because the quantity convected becomes negative incorrectly.

As a basis for evaluating representations and the algorithms implemented in those representations, we list some general properties which the continuity equation algorithms should have. These algorithms should:

1. Be stable for all cases of interest.
2. Mirror conservation properties of the physics.
3. Ensure the positivity property when appropriate.
4. Be reasonably accurate.
5. Be computationally efficient.
6. Be broadly applicable, not problem dependent.

The algorithms described in the following chapters are discussed in terms of these criteria.

6–1.3 Lagrangian vs Eulerian Representations

The convection of the variable ρ is governed by the appropriate continuity equation in conservation form, Eq. (6–1.1), or in Lagrangian form, Eq. (6–1.2). Note that when ρ represents an *intrinsic* rather than a *conserved* quantity, the right hand side of Eq. (6–1.2) is zero. For example, an intrinsic quantity could be a ratio of hydrogen atoms to oxygen atoms, or a progress variable marking the evolution of an interface. The *Eulerian derivative*, $\partial\rho/\partial t$, represents the time rate of change of ρ at a fixed location in the coordinate system. The *Lagrangian derivative*, $d\rho/dt$, is the time rate of change of ρ in an element of fluid moving with the fluid. Whether to use an Eulerian or a Lagrangian frame of reference is an important choice.

In a discrete *Eulerian representation*, the spatial grid points are kept fixed and the fluid is convected through the cells. Convoluted interfaces in the flow cannot be localized more precisely than a single cell without introducing additional degrees of freedom in the representation. The most persistent problem arising in Eulerian representations is numerical diffusion which has the effect of moving material across cells faster than any physical process. At other times numerical diffusion appears as premature mixing throughout a cell when physically the mixing should have just begun at one side of the computational cell.

In a *Lagrangian representation*, the grid points and computational cells move with the fluid so the continuity equation reduces to an ordinary differential equation for ρ in the moving frame of reference. If ρ is a mass density composed of many species, each species density, ρ_i, is individually conserved as it moves with the fluid. Even when there are sound waves present, entropy is conserved along the fluid path lines when nonisentropic heating and cooling can be neglected. From the point of view of numerical simulation, there are enormous simplifications and potential advantages to the Lagrangian approach. The equations for each individual element are simpler and interface locations can be tracked automatically. These advantages are counterbalanced by corresponding disadvantages. The geometry becomes more complex because the fluid elements can wrap around each other. The trade-off involves choosing between accuracy with geometric complexity and computational simplicity with numerical diffusion.

There is no answer to the question of which representation, Eulerian or Lagrangian, is better. Once one is chosen for a calculation, different types of algorithms and gridding procedures follow. Both representations are discussed extensively in the remaining chapters of this book.

6–2. REPRESENTING A CONTINUOUS VARIABLE

6–2.1 Grid-Based Representations

In *grid-based* representations, the mathematical expansion imposes a grid on the physical space and the continuous fluid variable $\rho(\mathbf{x}, t)$ is expanded as a linear combination of independent basis functions $\{\rho_j(\mathbf{x}, t)\}$,

$$\rho(\mathbf{x}, t) = \sum_{j=1}^{\infty} \rho_j(t) f_j(\mathbf{x}) . \qquad (6-2.1)$$

The coefficients $\rho_j(t)$ are functions of time only and the basis functions $f_j(\mathbf{x})$ vary with the spatial variable $\mathbf{x}$.

To have highly accurate representations of even well behaved, continuous variables requires that the number of expansion functions approach infinity. To make the representation computable, the number of discrete degrees of freedom, the number of coefficients in the set $\{\rho_j\}$, must be finite. This discretization is often accomplished by associating one of the expansion functions with each cell of a computational grid. For this association to be meaningful, the expansion functions $\{f_j(\mathbf{x})\}$ have to be reasonably localized in space, each about one of the cells. As soon as this association is made in the representation, and regardless of the formal order of accuracy of subsequent algorithms, a nonphysical quantity, the cell size, is impressed on the numerical solution. If more than one degree of freedom is associated with each cell, so that spatial variations within a cell can be resolved to some extent, the cell-size effects can be reduced but not removed. There are whole families of expansions which are progressively more delocalized in the sense that they use more values from cells which are further away. Totally delocalized or global expansions, called *spectral methods*, are valuable in situations where it is useful to couple the entire computational domain. Intermediate representations, called *spectral elements*, have a number of degrees of freedom associated with variably shaped patches in the computational domain.

The Eulerian approach uses a linear expansion to express the spatial variations of continuous variables throughout the domain of interest. The expansions are generally related to a fixed *Eulerian grid* and the fluid dynamic variables are viewed as being transported or moved from cell to cell by fluxes of these variables.

The Lagrangian approach tracks the motion of a number of nodes which move with the flow of the fluid. This simplifies the equations by separating the problem into a geometrical part, tracking the relative configuration of the nodes, and a physical part, calculating how the physical properties of the fluid at each node change in time. The moving nodes are clustered such that local gradients and their properties can be approximated accurately. Connecting these nodes together in some sensible order related to their locations gives another kind of grid, a *Lagrangian grid*. In Lagrangian representations, the flow itself determines the grid structure. Lagrangian representations are generally treated as local expansions.

Finite differences are the most commonly used procedures for constructing solution algorithms in either Eulerian or Lagrangian representations. As described in Chapter 4, the values of the physical variables characteristic of small elements of the fluid are defined on a time- and space-discretized grid. These values are computed at the end of the timestep from their values at the start of the timestep, perhaps from a few previous timesteps. At each cell, time is usually updated to the same value so the concept of a time level for the entire spatial grid is meaningful. Usually the numerical treatment of time and space derivatives is quite different.

Consider a typical Eulerian representation consisting of a set of $J+1$ locations $\{x_{j-\frac{1}{2}}\}$ that describe a one-dimensional spatial domain bounded by $x_{\rm left} \le x_{j-\frac{1}{2}} \le x_{\rm right}$. These locations are ordered to form a grid, also called a mesh or a lattice, such that

$$x_{j-\frac{1}{2}} < x_{j+\frac{1}{2}} , \quad j = 1,2,...,J . \qquad (6-2.2)$$

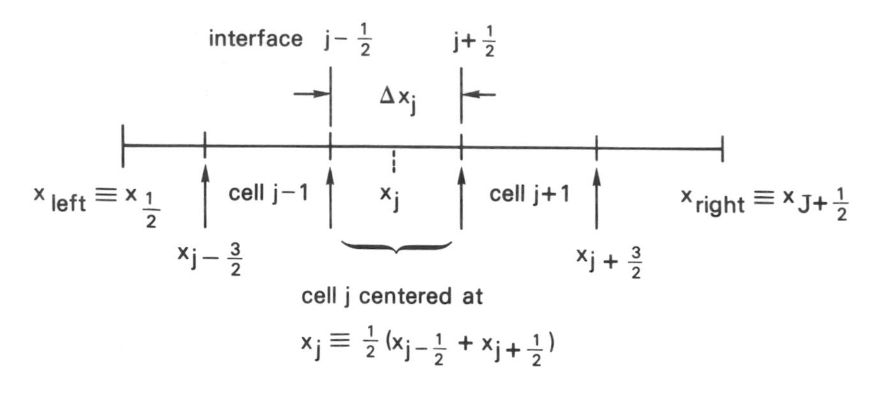

Figure 6–1. The computational domain of a typical one-dimensional Eulerian grid.

The typical Eulerian grid, shown in Figure 6–1, divides the computational domain into J cells centered at $\{x_j\}$, where

$$x_j \equiv \frac{1}{2}(x_{j-\frac{1}{2}} + x_{j+\frac{1}{2}}) \, . \tag{6 – 2.3}$$

The cell sizes, $\{\Delta x_j\}$, are given by

$$\Delta x_j \equiv (x_{j+\frac{1}{2}} - x_{j-\frac{1}{2}}) \, . \tag{6 – 2.4}$$

As can be seen in the figure, the original set of locations $\{x_{j+\frac{1}{2}}\}$ are the positions of the interfaces bounding the discrete computational cells.

We consider three simple, discrete representations on this grid. The first specifies a real, finite-precision value ρ_j for each cell. We assume that this value is representative of the values of the variable throughout the cell. In this *piecewise constant* representation, the sequence of values $\{\rho_j\}$ can be viewed as giving the average value of $\rho(x)$ in each cell. Such a piecewise constant numerical approximation to the continuous variable $\rho(x)$ has discontinuities at each cell interface. Conservation of total mass M in the whole system is expressed as the sum

$$M = \sum_{j=1}^{J} \int_{x_{j-\frac{1}{2}}}^{x_{j+\frac{1}{2}}} \rho(x) dx = \sum_{j=1}^{J} \rho_j \, \Delta x_j \, . \tag{6 – 2.5}$$

This representation of $\rho(x)$ is shown schematically in Figure 6–2a. One typical expansion function for this approximation is the localized *top hat* distribution that is one cell wide and of unit height. This function is shown

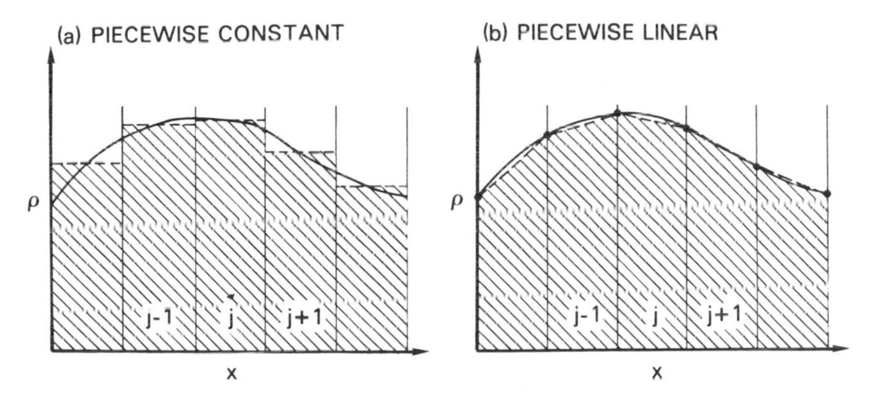

Figure 6–2. Two representations of a smooth function.

in Figure 6–3a. The numerical approximation, under the dashed lines in Figure 6–2a, can be written as a linear combination of local basis functions of this type. This approximation to the continuous variable is not smooth.

A *piecewise linear* approximation assumes that the set of discrete values represent values of $\rho(x)$ at the cell interfaces. The values throughout each cell are then found by interpolating linearly between the two bounding interface values,

$$\rho(x) = \frac{[\rho_{j-\frac{1}{2}}(x_{j+\frac{1}{2}} - x) + \rho_{j+\frac{1}{2}}(x - x_{j-\frac{1}{2}})]}{\Delta x_j} \tag{6 – 2.6}$$

for $x_{j-\frac{1}{2}} \leq x \leq x_{j+\frac{1}{2}}$. Conservation of total mass is now

$$\begin{aligned} M &= \sum_{j=1}^{J} \int_{x_{j-\frac{1}{2}}}^{x_{j+\frac{1}{2}}} \tilde{\rho}(x)dx \\ &= \rho_{\frac{1}{2}} \frac{\Delta x_1}{2} + \sum_{j=1}^{J-1} \rho_{j+\frac{1}{2}} \frac{(\Delta x_j + \Delta x_{j+1})}{2} + \rho_{J+\frac{1}{2}} \frac{\Delta x_J}{2} \ . \end{aligned} \tag{6 – 2.7}$$

These two representations have the same number of degrees of freedom and essentially the same mass integration formulas. However, the piecewise linear representation of a curve, shown in Figure 6–2b, is apparently more accurate than the piecewise constant representation. The reason for this is related to the width and hence relative smoothness of the expansion functions used. The expansion function for the piecewise linear representation is the

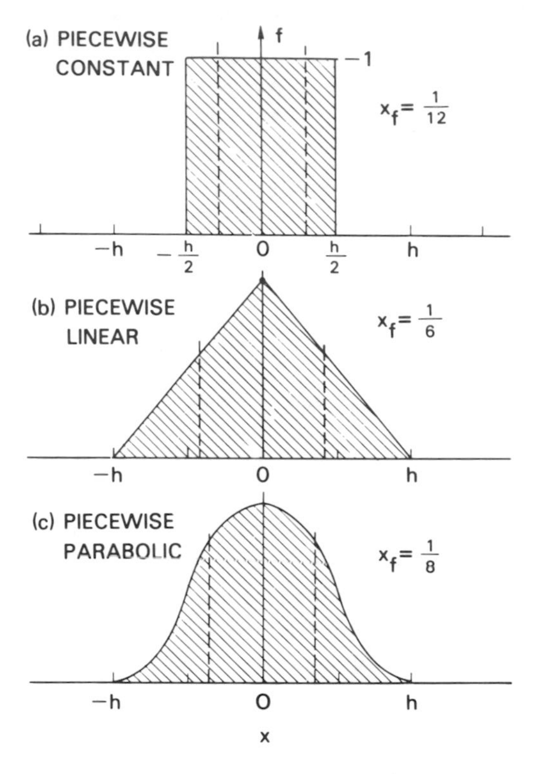

Figure 6–3. Expansion functions for three representations. The vertical dashed lines show the locations of x_f, the characteristic width of the three local expansion functions.

"teepee," "tent," or "roof" function of unit height at one interface that drops linearly to zero at the edges of the two adjacent cells. This is shown in Figure 6–3b.

Figure 6–3c shows one of the expansion functions in a piecewise parabolic representation. Though this would seem to be a higher-order approximation than the piecewise constant and piecewise linear representations, the spatial extent is actually somewhat less than the extent of the teepee functions. If the quantity x_f^2, the square of the extent of the basis function, is used to measure the effective width,

$$x_f^2 \equiv \frac{\int_{-\infty}^{\infty} x^2 f(x) dx}{\int_{-\infty}^{\infty} f(x) dx} , \tag{6-2.8}$$

the three local expansion functions shown in Figure 6–3 have

$$x_f^2 = \begin{cases} \frac{1}{12} & \text{piecewise constant} \\ \frac{1}{6} & \text{piecewise linear} \\ \frac{1}{8} & \text{piecewise parabolic .} \end{cases} \tag{6 – 2.9}$$

The piecewise linear basis functions are actually the widest. The piecewise parabolic basis functions are not as wide because linear terms have been left out of the expansion functions to keep them localized. This excess localization for a higher-order method reduces its mathematical accuracy and makes it a relatively unattractive expansion to use. When the linear terms and the quadratic terms are included, information from more than one or two adjacent cells is needed to determine the coefficients of the expansion basis functions. Such higher-order, less localized representations are considered in greater detail in Chapter 9 for solving continuity equations using finite-element and spectral methods.

6–2.2 Expansion and Spectral Representations

In practice, $\rho(\mathbf{x}, t)$ is represented as as a finite, linear sum of expansion functions, $\{f_j(\mathbf{x})\}$, as shown in Eq. (6–2.1). The time-dependent coefficients $\{\rho_j(t)\}$ are determined initially from the starting density profile, and are then advanced in time by a numerical algorithm. When the basis functions are local, as in the three cases above, each of the J time-dependent coefficients can be associated with a local value or with an average value of ρ over a cell-sized region. The local value of ρ now multiplies this basis function in the expansion.

It is also possible to use representations such as the one in Eq. (6–2.1) with expansions based on even less localized eigenfunctions than the piecewise constant and piecewise linear functions. Finite-element methods, discussed in Chapter 9, use slightly less localized expansions which couple all of the cells together implicitly. Information from over the whole grid contributes to the evaluation of derivatives and to the coefficient of each of the expansion functions. As higher and higher derivatives of the solution are matched at the element interfaces, the width and the smoothness of the equilvalent basis functions increase. As the basis functions extend farther and farther, it makes less sense to think of the coefficients as the values of ρ at the grid points or discrete node locations, and more sense to think of them as defining extended degrees of freedom.

Piecewise cubic splines, for example, use a different cubic polynomial in each of the grid intervals. The four coefficients in the cubic expansion are chosen so that the composite spline function, its first derivative, and its second derivative are continuous at the boundaries of the cells. To ensure this, the problem must be solved implicitly, again coupling information from the whole grid. Even in an explicit finite-difference formula, obtaining third-order accuracy typically requires information from at least four grid points.

The assertion that higher-order accuracy, obtained by using less localized expansion functions, is better than lower-order accuracy can be carried to its logical extreme. The analysis presented in Chapter 4 showed the usefulness of Fourier expansions for representing and analyzing continuous functions. A Fourier expansion is completely nonlocal. It uniformly weights information from all over the spatial grid to construct the Fourier coefficients. The expansion coefficients are the amplitudes of Fourier harmonics and the basis functions are trigonometric functions. In Fourier representations, it is conventional to think of a discrete grid in *k-space* or *wavenumber space*, rather than *x-space* or *configuration space*.

Because the Fourier harmonic is an average containing information from all locations with the same size weights, it is the antithesis of a local value of a physical variable. For studies of large-scale collective phenomena or other phenomena where information can be assumed to propagate everywhere at essentially infinite speed, Fourier representations are excellent methods to use (see for example, Gottlieb and Orszag (1977) and other references in Chapter 9). However, acoustic waves, flame fronts, or detonations involve phenomena which propagate at finite speeds. Using spatially extended expansion functions in such cases allows the numerical solution in a region to react to an approaching discontinuity before the discontinuity physically arrives. This artifact of the finite width of the expansion functions is not physical and often produces unacceptable results. For example, ripples can incorrectly appear at the upstream boundary in a finite-element or spectral solution of a gas dynamic shock problem because these methods implicitly couple the solution together throughout the grid. The physical delay time for the speed of information propagation in the medium is then incorrect.

The top hat basis function in the piecewise constant representation is the closest that a discrete Eulerian representation can come to a delta-function, although it is still one cell wide. Thus it is reasonable to expect that discontinuous behavior would be reproduced more faithfully by methods and representations with minimal instantaneous transfer of numerical information over large distances in a single timestep. We also expect that smooth, potential flow or incompressible flows would be well treated by spectral methods because the fluid behavior is globally coupled through the relatively fast acoustic waves.

6–2.3 Lagrangian Representations

A different approach to the problem of representing a continuous function is to identify finite regions of elements of fluid and to treat them as separate entities. In this Lagrangian approach, the physical properties of a moving fluid element are tracked. Conservation laws are then formulated in terms of interacting fluid elements. Lagrangian approaches span a broad range of methods which include tracking moving fluid elements and macroparticles. At the end of a timestep in a Lagrangian calculation, we must update not only the physical variables, but also the locations, $\{\mathbf{x}_l\}$, of the fluid elements or particles, according to

$$\frac{d}{dt}\mathbf{x}_l(t) = \mathbf{v}_l(t) . \qquad (6-2.10)$$

These locations are now degrees of freedom in the solution because they change every timestep. For a given spatial resolution, a Lagrangian representation has extra degrees of freedom relative to Eulerian representations because the locations of the nodes, as well as the dependent variables, are free to vary.

Now consider the problem of the motion of a small cloud which moves through the surrounding medium without diffusion. Suppose that we represent the cloud in a two-dimensional simulation by a number of cells, each containing a mass m_l at a location $\mathbf{x}_l$. A shear flow has been imposed across the center of the cloud, so that as the flow evolves, the adjacent cells distort and move. However, no fluid crosses Lagrangian cell boundaries because these boundaries are assumed to be moving with the flow. After a time, the Lagrangian nodes which were initially close may no longer be close, and some nodes that were initially distant, approach each other. When general rotational motions or shears are important in flows, a Lagrangian grid tends to become tangled and thus needs continual restructuring.

To restructure such a grid requires either a remapping procedure which moves the cell interfaces relative to the fluid, or restructuring the grid connectivity to allow each Lagrangian node to change its neighbors as the flow evolves. The first process is numerically diffusive. The second process is not necessarily diffusive, and has been implemented by Fritts and Boris (1979) in a dynamically restructuring, triangular grid. These methods are discussed further in Section 6–4 of this chapter, which describes adaptive gridding, and also in Chapter 10 where extensive references are given.

We have so far discussed the Lagrangian representations in terms of small but finite volumes of the fluid which move, rotate, stretch, and interact with their neighbors. There are, however, other ways to interpret Lagrangian representations. One interpretation is in terms of a collection of *macroparticles*. The term macroparticle is sometimes used to identify finite elements or volumes of the medium when two or more of them are allowed to coexist in a volume of space. Macroparticles move at the local fluid velocity, and their superposition determines the local value of ρ. By releasing a large number of numerical macroparticles into the simulation volume, the value of ρ in a region can be approximated by counting all the particles nearby. For example, if ρ is the fluid density, the density in a region can be calculated by counting all of the particles nearby and taking their respective masses into account. Note that there are not nearly as many macroparticles used to represent the fluid as there are real particles in the fluid, and thus their size and mass has to be much bigger. How these macroparticles move, the forces they are subject to, and the manner in which continuum fluid properties are derived from the properties and locations of all the macroparticles allow many different algorithms with various good and bad attributes.

There are also Monte Carlo (see, for example, Bird, 1976, 1983) and cellular automata (see, for example, Frisch et al., 1986) variants of Lagrangian representations to solve complicated fluid problems. These variants are particle- and bit-oriented Lagrangian models that resemble the way fluid elements are treated in the Lagrangian methods discussed above but differ in the way the macroparticles are treated. As soon as a statistical aspect of the basic representation enters, however, the usefulness of a method for complex reactive flows diminishes. Thus we will not consider these particle approaches further here and return to the finite-difference formulations with spatial grids and discrete timesteps. Macroparticle approaches are discussed further in Chapter 10.

6–2.4 The Method of Lines

The Method of Lines combines a computational representation with an algorithmic approach giving a rather general procedure for solving partial differential equations. We discuss the approach here, but could have just as appropriately discussed it in Chapter 9 on coupled continuity equations, or in a discussion of general coupling techniques. In the Method of Lines, the temporal and spatial representations of a system of partial differential equations are separated and treated independently. Leaving the time variable nominally continuous is sometimes called the *semianalytic method of lines*, a technique first introduced by Liskovets (1965) then Hicks and Wei

(1967), and applied by Bledjian (1973) to the structure of laminar flames. The method has been reviewed by Carver (1976) and Hyman (1979). A number of the terms used here to describe the Method of Lines are defined in Chapters 8 and 9. We therefore suggest rereading this section after Chapter 9.

To develop a Method of Lines model, the system of conservation equations and its associated boundary conditions are first discretized in space. This usually takes the form of finite-difference formulas of higher than second order. Then a suitable algorithm is chosen to integrate the resulting system of coupled ordinary differential equations in time, so that existing techniques and packages for ODEs can be used.

The spatial discretization and temporal integration must both be chosen with consideration of the the form of numerical dissipation which has to be incorporated. As with a number of the earlier methods we discuss in Chapter 8, it is necessary to include unphysical diffusion terms (artificial dissipation) to prevent unacceptable overshoots and undershoots at shocks (see, for example, Hyman, 1979). This appears to be something of an art form, and a considerable amount of effort must go into choosing when and how to switch artificial dissipation terms off and on. There does not appear to be any way to include the nonlinear flux- and slope-limiting procedures in positivity preserving monotone methods. Therefore more spatial resolution is required for a given accuracy than for the monotone convection algorithms described in Chapters 8 and 9. Nevertheless, good results are obtained in many one-dimensional problems. The method is more tractable than global implicit finite-difference representations when the equations are highly nonlinear, but multidimensional situations seem to present comparable or worse difficulties.

The Method of Lines has been modified by Galant (1981), who uses implicit multistep methods to integrate the resulting stiff ODEs in time because they do not involve evaluation and inversion of the Jacobian matrix. His results concentrate on the ozone-oxygen flame system using Gear integration schemes (see Chapter 5). Hyman (1979) also recommends an iterated implicit multistep method to handle the potentially stiff time integrations. This approach appears to deal adequately with the moderate stiffness arising from steep spatial gradients and shocks as long as a modified Courant-Friedrichs-Lewy condition is satisfied and shocks move less than a cell per timestep (Hyman, 1979). It is not clear how well the Method of Lines formalism overcomes the much more severe explicit timestep restrictions accompanying stiff chemistry or really stiff fluid dynamics, such as integrating very fast sound waves stably in nearly incompressible flows.

6–3. RESOLUTION AND ACCURACY ON A GRID

Quantitative measures of the resolution and accuracy of an algorithm are usually based on its applications to idealized problems. These measures are a useful guide, but have less meaning when the problems are complicated by irregular gridding, complex geometries, and complex flow patterns. Because of this, we are also concerned with more subtle and qualitative properties of the algorithms, such as their ability to ensure global physical properties such as conservation, positivity, and causality.

6–3.1 Errors Introduced by Grid Representations

In Table 4–4 we listed four types of errors which occur in finite-difference simulations: local errors, amplitude errors, phase errors, and Gibbs errors. The last three are nonlocal errors that arise because of the choice of representation, discretization, or algorithm to represent the derivative. It is useful to look again at these types of errors in the light of additional information presented in this chapter.

Discretizing the spatial variations usually generates the largest errors in reactive flows. Thus local errors, which include roundoff, truncation, and lack of conservation in solutions of local phenomena, are not directly relevant to our current discussion of errors induced by gridding and spatial resolution. We considered several aspects of local errors in detail in Chapter 5.

Local errors can, however, be important indirectly through their coupling to spatial variations. For example, local errors could be important when the results of solving chemical equations contribute strongly to the overall energy flow in a system. Another way local errors can be important is through random local roundoff errors in density which appear as random waves in the pressure and eventually in the velocity. If the material is nearly incompressible, these waves resemble the continuous generation of fairly large sound waves.

Often derivatives are approximated as differences between neighboring cell values. When these values are nearly equal, roundoff errors are enhanced because the differences of the physical values nearly cancel. We would like to ensure that roundoff errors do not become larger than the truncation errors in the numerical algorithms. To ensure this, roundoff errors should be several orders of magnitude smaller than the accuracy required of the various algorithms. These requirements translate into a need for about seven decimal digits of accuracy in the real numbers calculated. Practically, this means that 32-bit, floating-point numbers are generally adequate to suppress roundoff effects, although care must be taken to estimate the magnitude of the errors in each specific problem.

Fourier transforms are useful ways to analyze algorithms for nonlocal derivatives on uniformly spaced grids, as shown in Chapter 4. For each Fourier harmonic, errors appear in the phase and amplitude of the complex amplification factor, $A(k)$, due to both truncation errors in the approximation to derivatives and to the finite grid. Amplitude errors are important in the diffusion, convection, and wave equations. Phase errors occur in solutions to convection and in wave equations, but not in local integrations and diffusion equations.

As shown in Chapter 4, amplitude errors introduce too much smoothing, called numerical diffusion, or too much steepening, called numerical instability. Because numerical instability can be so disasterous in a solution, it is important to ensure that the algorithms used are always stable. In practice, this usually means that some numerical diffusion is included.

Additional resolution can reduce numerical diffusion, but it increases the computational cost. In two dimensions, doubling the spatial resolution in each direction costs a factor of eight in the execution time of the program because the timestep must also be reduced proportionately. Thus a first-order algorithm, with amplitude errors scaling as $\epsilon = v\Delta t/\Delta x$, becomes only twice as accurate for an eightfold increase in computational cost. The scaling for three-dimensional problems is even worse. Because of this unfavorable scaling, we recommend algorithms which keep amplitude errors second order or smaller in both the cell size and the timestep.

When physical diffusion processes are large, they can mask numerical diffusion. If the effective numerical diffusion coefficient is known, the presence of the physical diffusion can even be put to good use. For example, if physical diffusion is large enough, lower-order convection algorithms may be adequate. For this reason, a combined *convective-diffusion equation* is often considered as a single entity in the computational fluid dynamics literature. Remember, however, that an algorithm constructed using this crutch will most likely perform poorly in situations where the physical diffusion is relatively small.

Phase errors are a nonphysical form of dispersion in which different wavelength harmonics are numerically advanced at different speeds. Many types of physical waves are dispersive, but ideal convection and sound waves, whose properties we wish to reproduce, are not dispersive. Phase errors can adversely affect causality, positivity, and accuracy. They are of particular concern in algorithms for convection, because they continually accumulate and may not reach a limiting level.

In general, it is difficult to represent and advance sharp gradients or small-scale structure with nonlocal expansions. The basic problem is that the shortest wavelengths are responsible for the steepest portions of the profile, and are also subject to the greatest phase errors. To use a nonlocal expansion requires using many basis functions and ensuring that their contributions are all properly phased so that they add up to a steep transition with smooth regions on either side. If the phases of any of the superimposed harmonics are not maintained properly, the solution deteriorates quickly.

Chapter 9 discusses a number of methods for solving the continuity equation. Comparing a number of methods shows that fourth-order accuracy in convective phases seems to give appreciable improvement over second-order accuracy. When the phase errors are fourth order, as a practical matter, they no longer produce a steadily eroding convective profile.

The final type of error we consider is the Gibbs phenomenon or Gibbs error that occurs in the solution of wave and convection equations. These errors look very much like dispersion errors but have a different source. They are not due to a particular algorithm, but to the discrete nature of the representation itself. Thus they cannot be eliminated by improving the phase and amplitude properties of the numerical algorithm.

The Gibbs errors originate in the very short-wavelength harmonics, shorter than two computational cells, that are not included in the representation. Thus the structure of the continuous solutions within each cell cannot be resolved. These unresolved spatial variations are an inherent error associated with representing a continuous function on a discrete grid. In numerical simulations, Gibbs errors appear as undershoots and overshoots which are larger near steep gradients and discontinuities and become apparent as soon as a sharp profile moves a fraction of a cell from its initial location.

Gibbs errors set a bound on the accuracy possible in discrete approximations. Such errors can be reduced only by improving the resolution in an Eulerian representation. By moving the nodes in a Lagrangian manner, the Gibbs errors are suppressed at the cost of additional degrees of freedom and more expensive algorithms.

When representing a discontinuity numerically, a balance should be established between the number of degrees of freedom which are used to represent the discontinuity and the accuracy with which each degree of freedom is integrated. We have found that best overall results are obtained by methods of intermediate order, with fairly local expansion functions, and with fourth- or higher-order phase errors. Residual second-order amplitude errors seem to be acceptable in many circumstances, but phase errors accumulate in time and so must be kept small.

6–3.2 Resolution and Accuracy

The piecewise constant and piecewise linear representations of Figure 6–2 are both useful and have the same grid. In each case the continuous function is expanded as a superposition of relatively well localized expansion functions. The number of degrees of freedom are almost the same and the resolution is virtually identical. Nevertheless, the piecewise linear discretization seems to be a better approximation because the approximating function is continuous everywhere, even though its derivatives are not. By using the same resolution and accuracy, but by interpreting the representation differently, a more accurate approximation is possible.

In general, higher-order methods are better than lower-order methods. However, there is a point of diminishing returns at which the extra work of computing the higher-order solution and of keeping the method stable does not really bring accompanying increases in accuracy. The higher-order methods can only yield better answers for scale lengths which are longer than a few cells. Interpolation, no matter how high the order, cannot reproduce unresolved variations occurring within a computational cell. This is a fundamental limitation set by the resolution of the representation. It is independent of the order of accuracy, the type of expansion, or the algorithms used. The Gibbs errors discussed above arise from this fundamental computational limitation.

The representation of time and space variations may be optimized in different ways depending on the particular problem. Whether point values, volume-averaged values, or Fourier coefficients are given, the goal is a good representation of a particular function with a minimum number of degrees of freedom. When the real function is smooth with the exception of a few sharp steps, the optimal representation is very different from the case when it has an oscillatory structure spread over many spatial scales.

6–3.3 Introduction to Multidimensional Grids

Determining a suitable grid in two or three dimensions is often a difficult problem. Some of the difficulty arises because the computational domain can be geometrically complex, and some because the flow structure itself is usually more complex in multidimensions than in one dimension.

In recent years a number of books and conferences have focused on gridding techniques because more can be gained by intelligently gridding a complex problem than by extending the integration algorithms to higher order. The logistics of multidimensional grids becomes an important factor in large simulation models. Computer memory can be at a premium because of the large number of cells required for good resolution. High resolution is

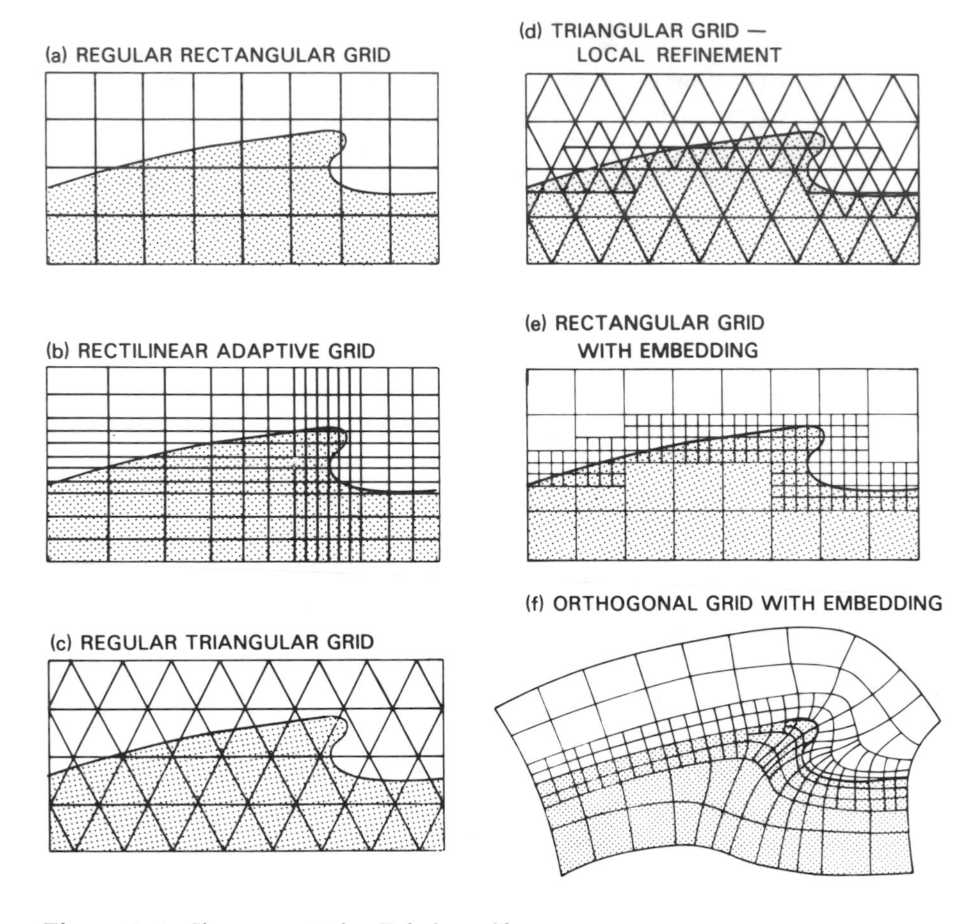

Figure 6–4. Six representative Eulerian grids.

usually not needed everywhere, and lower-order formulas are generally more robust and easier to program and use.

Figures 6–4 and 6–5 introduce a few of the more useful types of multidimensional Eulerian and Lagrangian gridding techniques. The flow pattern that the grids are trying to represent consists of two regions separated by a convoluted interface. The interface could be, for example, the interface between oil and water, the surface of the ocean, an advancing thin flame front, or a steep region around a cloud boundary. Although the physical flow pattern is the same in each of the twelve cases, the two-dimensional grids are progressively more complicated.

The simplest grids in multidimensions are the uniformly spaced Eulerian grids shown in Figures 6–4a and 6–4c. In these figures, the computational

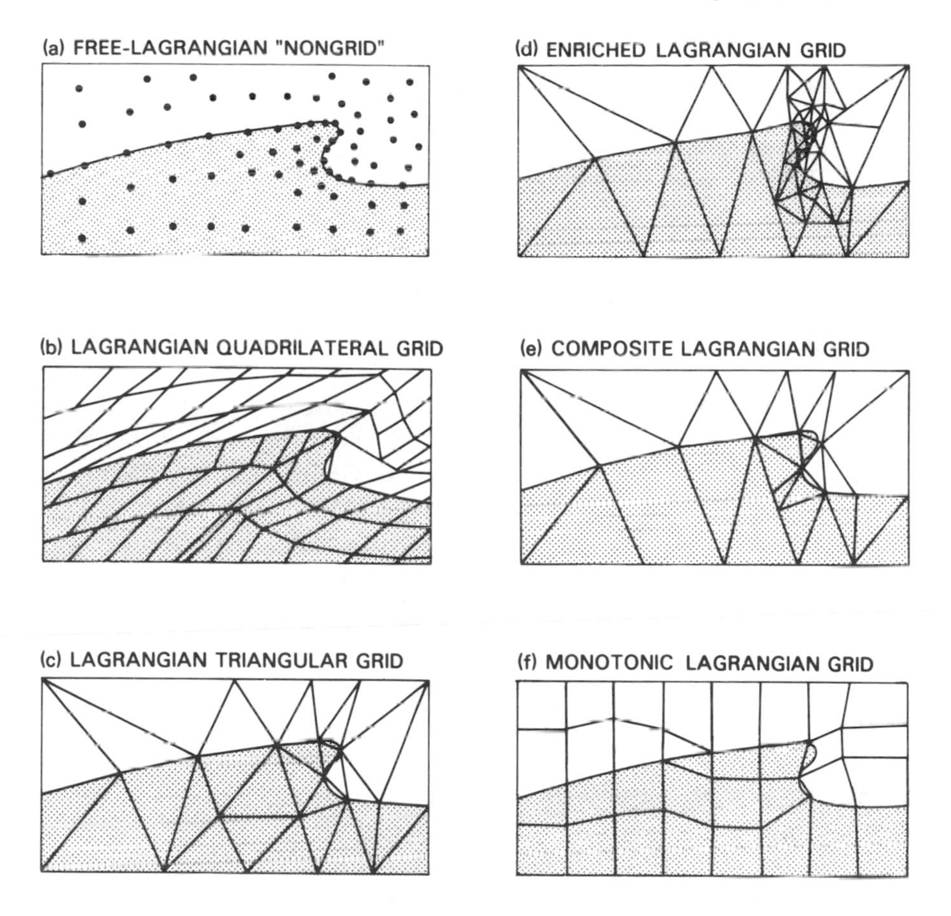

Figure 6–5. Six representative Lagrangian grids.

cells are rectangles and triangles, respectively. The spacing may vary in each of the grid directions, as shown in Figure 6–4b. Variably spaced, rectilinear lines in multidimensions still define a locally orthogonal grid, for which standard solution methods can be used. The same methods can, in fact, be used for distorted grids which are locally orthogonal, as shown in Figure 6–4f. In such an orthogonal grid, a region of higher resolution is shown on either side of the interface, aligned conveniently with the coarser orthogonal grid. This type of variable or *adaptive gridding* can be very effective. It is now of major importance, for example, in aerodynamic calculations where the geometry is fixed. Refinement along a surface is also a useful method for treating boundary layers which conform to smoothly varying surface.

Figures 6–4d, e, and f show improved resolution in the region around the discontinuity. By confining the refinement to local regions, the effective resolution can be increased while the number of cells does not become exorbitant. In Figures 6–4d and e, the boundaries of the fine cells are not aligned with the interface, as they are in Figure 6–4f. In most cases, the general orthogonal grid, with extra cells along the interface, is more accurate but also more computationally expensive. Therefore if the grid only has to be set up once, it is particularly useful, but if it has to be recomputed at each timestep, it can be a computational burden.

The grids shown in Figures 6–4b,d, and e are relatively simple to change in the course of a calculation. However, Figure 6–4b, in which the grid lines are rectilinear and orthogonal everywhere, is perhaps the easiest grid to set up and modify. Fine zones can be clustered where they are needed and the grid logic and finite-difference formulas are simpler than in the other variably-gridded Eulerian cases because the grid lines are straight. The price for this accuracy is carrying fine zones across the whole system to the boundaries in each spatial dimension, if they are needed anywhere.

Lagrangian cells change as the flow progresses. Thus as fluid elements become tangled in rotational or shear flows, the geometry can rapidly become very distorted. The only really effective procedure to handle such distortions is to reconnect the moving nodes into a new grid, thus altering the *grid connectivity*. This is difficult to do in a reliable, automatic way. Nevertheless, there are advantages to doing this, and there is a sizable effort to develop reconnecting, adaptive Lagrangian grids. Early progress includes the Particle and Force (PAF) algorithm (Harlow, 1962). In the 1970s these early concepts were extended to grids of triangles (Crowley, 1971), MHD algorithms over a triangular mesh (Brackbill, 1976), and adaptive triangular meshes (Fritts and Boris, 1979). During this same period, methods using Voronoi meshes were developed (Peskin, 1977) in which the computational cell is formed by connecting the centroids of triangles. These and related methods are discussed in more detail in Chapter 11.

General connectivity grids change with the flow patterns during the course of a calculation. In the last few years there has been major progress for Eulerian and Lagrangian calculations. The most recent efforts for Lagrangian grids are summarized in the conference proceedings edited by Crowley et al. (1985). Applications now include finite-difference and finite-element calculations of classical fluid dynamic instabilities, tokamak and high-temperature plasma physics modeling, heat conduction, water-wave interactions with structures, impact deformations, the oscillation and breakup of droplets, and hydrodynamic problems for both compressible and incompressible fluids. In Chapter 13 we describe a calculation of complex shock structures, in extremely complicated geometries, using the Eulerian technique with general connectivity grids.

The ideal *free Lagrangian* grid, which is really a "nongrid," is shown in Figure 6–5a. This grid is free in the sense that the representation does not imply any particular grid connecting the nodes. Vortex dynamics and boundary integral formulations fall in this general category. The price for removing the grid is the need to compute all of the node interactions by pairs instead of only interactions with nearby nodes.

Lagrangian methods now use quadrilateral, triangular and mixed meshes in two dimensions, tetrahedral meshes in three dimensions, Voronoi meshes in both two and three dimensions, and methods which are mesh-free. In addition, the cell interfaces can no longer be orthogonal. Thus many of the finite-difference operators we would like to represent become more difficult and more expensive.

Figures 6–5b and 6–5c illustrate the standard Lagrangian grids composed of quadrilaterals and triangles, respectively. In the quadrilateral grid, the accuracy-reducing distortions which usually accompany rotational flow are shown where the interface is becoming tangled in the center of the figure. The advantage of the triangular grid is that the nodes can be reconnected to produce a better grid whenever the triangles become too elongated or compressed, and the grid tangling and "bowties" which occur with quadrilateral Lagrangian grids are absent.

Figure 6–5d shows a Lagrangian triangular-grid representation that has been refined locally by dividing cells. The price for this finely focused, high resolution is the need to work with algorithms which are not easily vectorized for use on parallel processing or multiprocessing computers. Their relative computational inefficiency is generally compensated for by the very small number of zones needed for excellent resolution.

The *Monotonic Lagrangian Grid* shown in Figure 6–5f combines some of the better features of Lagrangian and Eulerian representations (Boris, 1986). It has the regular indexing and global structure of the Eulerian grids and the interface and node-tracking capabilities of a Lagrangian grid. This potentially useful concept is discussed further in Chapter 10.

6–4. TECHNIQUES FOR ADAPTIVE GRIDDING

Adaptive gridding methods try to concentrate the computational cells in regions where there is structure in the flow and leave the grid coarse in regions where there is little structure. Because of the tremendous potential adaptive gridding has for reducing computational costs while maintaining the same level of accuracy, it is a forefront area in computational physics. There are not many general references. We recommend the books by Thompson et al. (1985) and Crowley et al. (1985), the review article by Eiseman (1985), and the recent articles by Löhner et al. (1985, 1987). Because this entire area is in a stage of rapid development, the reader should be wary of blindly adopting any but the most straightforward technique without evaluating the associated difficulties.

6–4.1 Rezoning to Improve Resolution

It is often necessary to resolve discontinuities or localized steep gradients in the computational domain, yet we cannot afford the cost of using enough closely spaced zones to resolve those regions as they move throughout the entire computational domain. Furthermore, such fine resolution is of no benefit in most of the domain. In these cases, we use *adaptive gridding* algorithms, in which small computational cells are placed only in the regions with large gradients.

Refined localized gridding techniques for boundary layers in steady state calculations have been used for years. An early example is the work of Mitchell and Thomson (1958). For many transient fluid problems with a fixed but complicated boundary geometry, adaptive gridding is used to obtain accurate answers with a modest number of grid points. These problems arise, for example, in interior flows in automobile, jet, and rocket engines, and also in external aerodynamic flows. Adaptive gridding is also important in models of bodies of water where the shore of a river, bay, or ocean must be simulated with prescribed accuracy.

In the simplest application of adaptive gridding, variably spaced grids are established initially and then held fixed through the course of the calculation. They are adapted to the domain boundaries and to average properties of the flow. When the regions in the flow requiring the most accuracy are localized and move relative to the boundaries, we must adopt algorithms that change the grid in the course of the calculation to keep resolution localized where it is needed. This is the case, for example, in models of propagating shocks or flame fronts and of dense drops or bubbles moving through a liquid. In any particular problem, it is usually not difficult to know approximately where the finely gridded regions must be placed. There are a number of ad hoc criteria for this that are usually straightforward to program.

Consider an unsteady flame or detonation computation that requires an adaptive gridding procedure. The criterion for clustering the closely spaced cells is usually related to a local temperature or species concentration gradient. The changes in the variables, their derivatives, and their curvatures may be estimated to determine whether more or less resolution is needed, and the cell spacing can be adjusted to minimize the overall error. Examples of such calculations include, for example, the detonation calculations by Oran et al. (1982a,b) and Kailasanath et al. (1983), and the flame calculations by Margolis (1978), Dwyer et al. (1979), and Kailasanath et al. (1982a,b).

Currently more general formalisms are being developed for placing the adaptive grid nodes according to a variational, finite-element formulation. For example, a variational principle may be established to minimize an error defined for the systems of equations being solved. Gelinas et al. (1981) have taken this approach with finite elements, and their results are encouraging. They use an error measure for the overall accuracy of the solution and solve one-dimensional problems with exceptional resolution and accuracy. Their two-dimensional extensions, which use grids with triangular cells, require considerably more development before they can be used easily and inexpensively. In general, finite-element adaptive gridding methods that use a variational principle to determine the location of the grid points are complicated and expensive on arbitrarily reconnecting multidimensional grids. The auxiliary calculations that must be performed whenever the grid is changed become more expensive than integrating the variables. Winkler et al. (1984) and Mihalas et al. (1984) have, however, made significant progress by using a very finely resolved, one-dimensional method for radiation hydrodynamics. They move existing zones so that the variations are distributed as evenly as possible over the cells. Also, Löhner et al. (1985, 1987) have developed general methods for gridding two- and three-dimensional transient gas dynamics methods with propagating shocks, again using a finite-element approach. The location of their grid refinements is determined by important but ad hoc properties and structure in the flow. Their strategy is to insert cells as needed ahead of advancing structured regions in the flow, and to remove cells in regions where the spatial gradients are diminishing.

6–4.2 Eulerian Adaptive Gridding

In Eulerian calculations, there are simple, effective adapting gridding algorithms that use a continuous *sliding rezone* in which both the cell interfaces and the fluid move simultaneously. The basic idea is to increase resolution where it is needed by sliding in zones from nearby regions where high resolution is not needed.

Figure 6–6 illustrates a sliding rezone in which a region with a steep gradient, such as a shock or flame, moves through a region gridded with twelve cells and thirteen interfaces. Clustering the cells around the gradients maintains finer resolution with fewer cells. In the sequence shown, the gradient is always surrounded by cells that are finer than average by a factor of two. Thus equal spacing throughout the entire domain would require 24 cells rather than twelve, and thus would double the cost of the calculation. Often fine zones are used which are one tenth to one twentieth of the size of the coarse zones, saving over an order of magnitude in computer time.

The figure also shows a transition region between the coarse and fine zones. The size of the cells in the grid should change gradually. Rapid changes in cell sizes usually reduce accuracy in the approximation of the derivatives, and hence reduce the accuracy of solving nonlocal processes. Another problem with large cell variations is that propagating waves and fluctuations can reflect off these variations numerically, and this can generate other spurious effects. Accuracy is improved appreciably by gradually varying the cell sizes; the more gradual the variation, the greater the accuracy. Generally we recommend at most a 10–20% change in size between adjacent cells.

As an example of the placement of the finely gridded regions in a time-dependant calculation, we consider the gridding in a reactive shock model (Oran et al., 1978) that uses a number of sliding closely spaced regions. For example, consider the situation in which a shock overtakes first a contact discontinuity and then a leading shock, as shown in Figure 6–7. Three regions, one surrounding the contact discontinuity and one surrounding each shock, are gridded with fine, evenly spaced cells for short distances on each side. Coarser cells are used in regions of uniform or near uniform flow. The figure shows the positions of the cell interfaces, $\{x_i\}$, as a function of the cell indices. The system is bounded by the interfaces x_L on the left and x_R on the right. There are seven regions, the three closely gridded ones and the four surrounding these that consist of moderate-sized cells. The moderate-sized cells must change size as the faster shock approaches and passes the contact discontinuity and then passes the leading shock. Two coarse regions are placed between the finely gridded regions and the boundaries. The fine

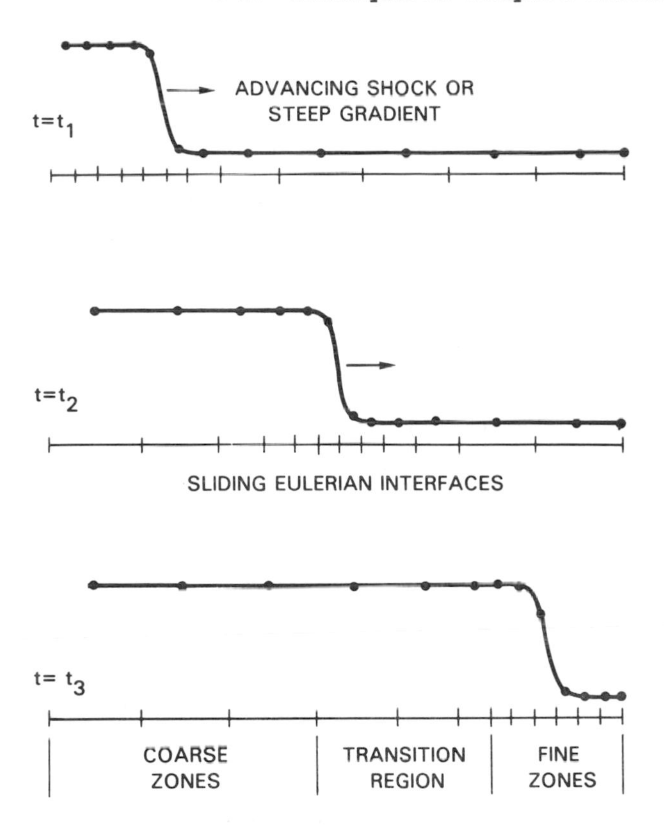

Figure 6–6. Schematic of an Eulerian sliding rezone for adaptively gridding an advancing steep gradient in a continuous fluid profile.

cells are one third the size of the moderate cells, which are one third the size of the coarse cells.

In principle there can be any number of finely spaced regions, each with its own preferred cell size. These regions can grow, shrink, split, and merge as required by the evolving flow. The entire grid is completely specified by the sequence of boundary index pairs $\{x_r, I_r\}$, where x is a location, I is the corresponding boundary index, and $r = 1, ..., 8$ labels the region boundaries. The actual cell spacings, $\{\Delta x_{r+\frac{1}{2}}\}$, are defined by

$$\Delta x_{r+\frac{1}{2}} \equiv \frac{x_{r+1} - x_r}{I_{r+1} - I_r} , \qquad (6-4.1)$$

where $I_1 = 1$ and $I_8 = N_x + 1$ fix the limits of the physical grid. The boundary indices $\{I_r\}$ are real numbers, so that the boundary between two

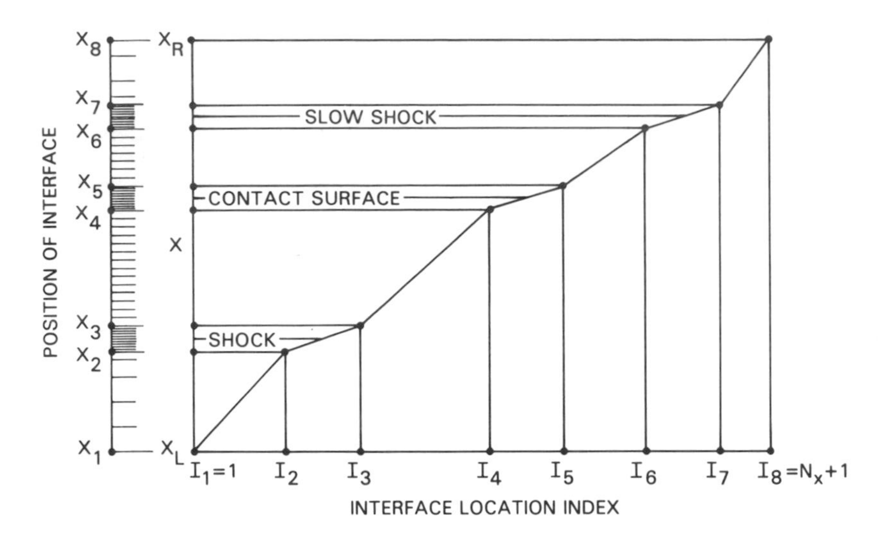

Figure 6–7. Index mapping for a multiregion sliding Eulerian rezone.

regions of different resolution can occur anywhere in a cell as well as at cell interfaces. If the boundary index pairs are varied continuously as the flow evolves, the closely gridded regions can be moved in a fully adaptive, controlled manner.

As shown in Figure 6–7, the cell sizes in the closely spaced regions must transition smoothly into the coarser regions. This is done by taking the piecewise uniformly spaced grid and numerically smoothing the locations according to

$$x_i^{(m+1)} = \frac{1}{4}\left(x_{i+1}^{(m)} + x_{i-1}^{(m)}\right) + \frac{1}{2}x_i^{(m)} , \qquad i = 2, 3, ..., N_x , \qquad (6-4.2)$$

where the superscript m indicates an iteration number. The smoothing averages the locations but should not be allowed to move the edges of the grid at x_L and x_R. Usually m is large enough to stretch cells at the edges of the finely gridded regions, but not large enough to erode all of the fine cells.

If a finely gridded region is initially ten cells wide, only three or four iterations of the smoothing algorithm can be used before the small cells at the center of the closely spaced region are stretched by the larger cells outside. Three or four iterations is only good enough, therefore, for about a factor of two change in cell size. If the closely spaced region is extended to twenty cells, eight or nine iterations can be used and still leave a few uniform cells in the center. This is generally enough for a factor ten change in cell size.

Sometimes the rate at which the cell sizes vary is limited by the physical situation itself. A shock propagating from a region of coarse resolution to one of fine resolution shows nonphysical oscillations if the cell size changes faster than the shock can physically steepen. Strong shocks steepen quickly, but weak shocks must travel a great distance to steepen appreciably. As a shock propagates across the coarsely gridded region, it has a thickness of about one cell. As it enters the finely gridded region, this becomes a ramp several cells across. This ramp steepens up to a shock while shedding oscillations (Boris and Book, 1973). These improper oscillations can be suppressed by changing the cell size more slowly than the shock steepens naturally.

We have given a prescription for smoothly adjusting the mesh, but have not yet defined the boundaries of the grid regions. As a shock moves along the system, the finely gridded region is programmed to surround the shock. The shock front is located by determining where the local acceleration,

$$a_{i+\frac{1}{2}} = \frac{1}{\overline{\rho}_{i+\frac{1}{2}}} \frac{P_{i+1} - P_i}{x_{i+1} - x_i} , \qquad (6-4.3)$$

is maximum, where P is the pressure, x is the position coordinate, and $\overline{\rho}$ is an average mass density. Similarly, the contact surface can be tracked by looking for the maximum density gradient in a region where the pressure gradient is nearly zero.

Enough closely spaced cells must be kept around each physical discontinuity to maintain uniform spacing in the vicinity of the gradient. This increases the calculation accuracy and eliminates local instabilities due to the interaction of the cell positions with steepening gradients. This multiregion adaptive gridding procedure can give comparable results to error-minimization approaches with comparable minimum cell sizes.

The Eulerian sliding rezone and other adaptive regridding procedures can have difficulties at material interfaces. Resolution is increased where needed by sliding in extra zones from nearby regions where they are not required. However, as a closely spaced region comes close to an interface between a liquid and a solid, cells from the solid cannot slide into the liquid without involving complicated logic to keep the solid and liquid from interpenetrating diffusively . To partially circumvent this limitation at interfaces, DeVore et al. (1984) have treated a few of the cell interfaces as Lagrangian. This defines the material interfaces, and the remaining zones in each of the shells of distinct material move as an Eulerian, sliding rezone to resolve each of the regions.

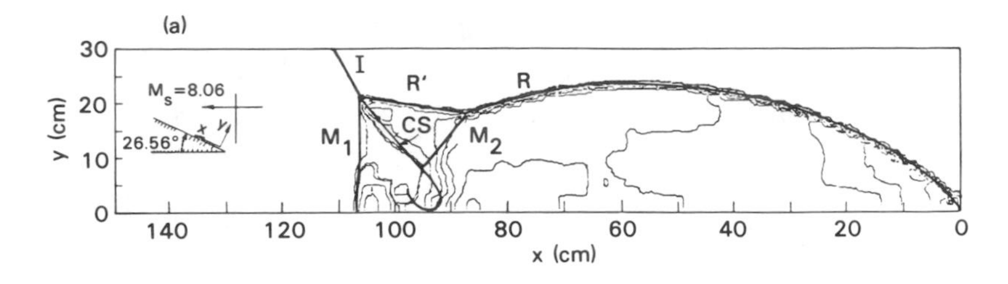

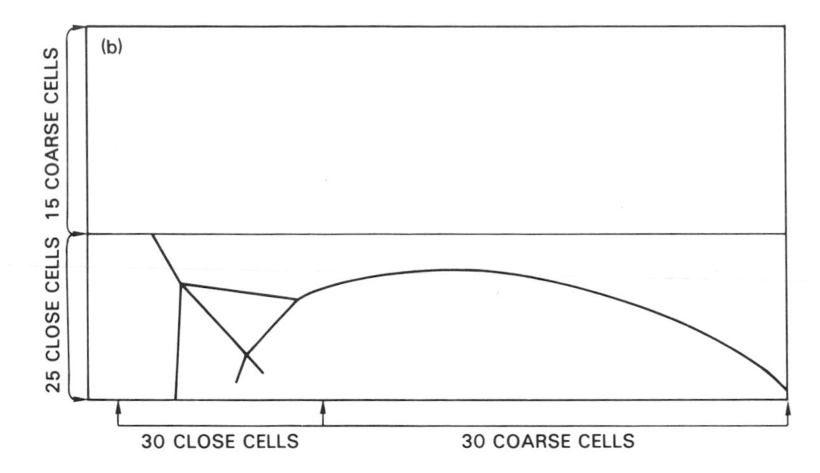

Figure 6–8. Density contours (a) and schematic of the computational grid (b) for a calculation of the reflection of a shock from a smooth surface (Book et al., 1981).

The one-dimensional, multiregion, adaptive gridding procedure described above can be extended to multidimensions when orthogonal, rectilinear grids are used. This approach, shown in Figure 6–4b, adaptively changes the grid in one direction without affecting the location of the grid lines in the other direction. Thus the spacing in, for example, x and y could be changed independently. Figure 6–8, another example of a calculation using such a grid, was used to study reflection of stong shocks off of smooth surfaces. The gridding in the y-direction is fixed, but the region of fine cells in the x-direction is programmed to move with the advancing Mach stem (Book et al., 1981). In addition to maintaining high order, another major advantage of using rectilinear grids is that the algorithms can be highly optimized. For these reasons, rectilinear grids are widely used despite some shortcomings.

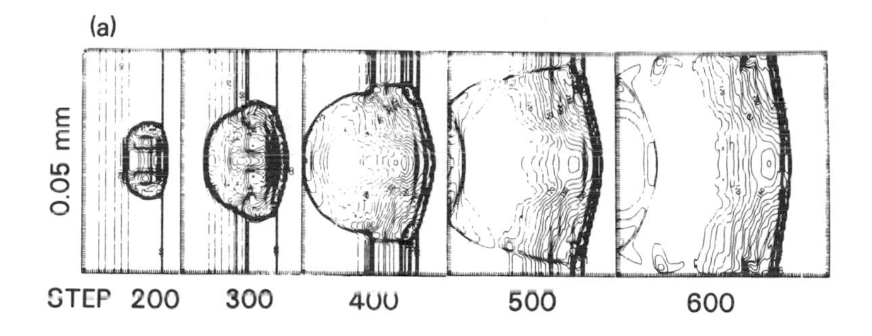

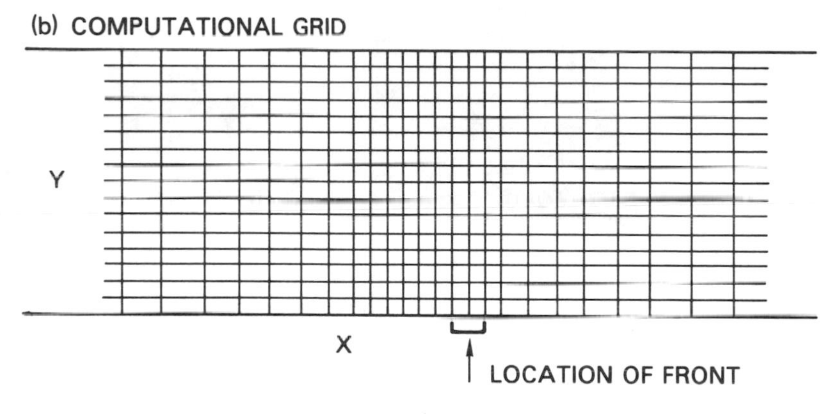

Figure 6–9. (a) Pressure contours showing the development and propagation (left to right) of a two-dimensional detonation (Guirguis et al., 1986). (b) Possible adaptive grid for this type of calculation, showing approximately every 8th grid line (Kailasanath et al., 1985).

There is one major disadvantage of the approach just described. The fine cells extend all the way across the system, whether or not they are needed across that whole area. This means that resolving complicated structures in one part of the grid requires keeping closely spaced cells where there may be little structure in the flow. However, there are many problems where having these added cells is not prohibitive. For example, the grid used to resolve the structure of a propagating detonation in a tube is shown in Figure 6–9 (Kailasanath et al., 1985; Guirguis et al., 1986). In this case the flow structure extends across the entire vertical height of the tube and moves horizontally with the detonation.

One interesting and useful application of adaptive gridding is to smooth higher-order, grid-induced irregularities which can appear, for example, as

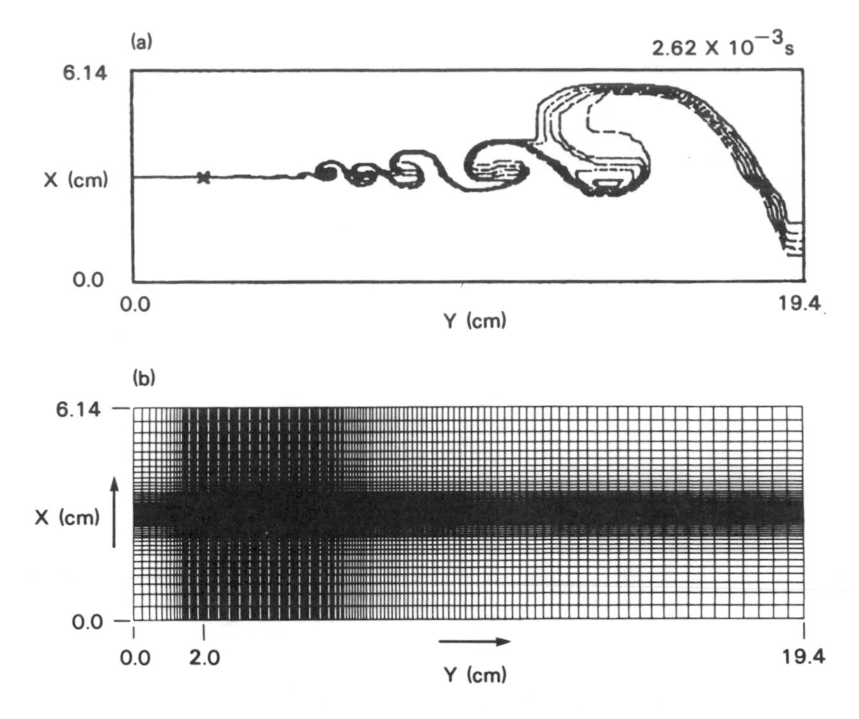

Figure 6–10. Calculations of the Kelvin-Helmholtz instability in a splitter-plate flow. The faster fluid enters from the lower half of the left boundary. The slower fluid enters from the upper half of the left boundary. (a) Contours of mixedness, defined here as the ratio of number density of faster to slower fluid. (b) Grid used in the calculation (Grinstein et al., 1986).

small ripples near oblique shocks. At every other timestep, jiggling alternate grid lines back and forth about 15–25% of a cell width can appreciably reduce errors arising from the discreteness of the grid. We might say that the fluid becomes "confused" about where the grid is, and cannot respond as easily to the grid in a nonphysical manner.

Figure 6–10 shows another example of a two-dimensional stretched grid used for studies of the Kelvin-Helmholtz instability (Grinstein et al., 1986). In this case the grid is stationary. Closely spaced cells are clustered at the shear layer, and are stretched as the distance from the shear layer increases. The grid is held fixed because the entire structure along the shear layer must be well resolved at the same time. Structures originate near the shear layer origin at the splitter plate, and grow as they move downstream.

6–4.3 Lagrangian Adaptive Gridding

It is tempting to use a sliding rezone in an Eulerian calculation to produce a Lagrangian calculation. There are, however, problems that appear in Eulerian calculations when the cell interfaces move exactly at the fluid velocity, as in a Lagrangian calculation. We can understand this by noting that in an Eulerian calculation with a sliding rezone moving at the fluid velocity, there is still some flux that leaks through the cell interfaces. The mass in each cell does not stay exactly fixed, even though the fluid and interface velocities are nominally the same. The result is that fluid can slip into adjacent cells, and a slowly growing instability appears in which alternate cells slowly expand and shrink. Craxton and McCrory (1979) have proposed a method for avoiding this problem by using a mass integral technique to determine the interface motion.

The procedure for adaptive gridding is more complicated in the Lagrangian case than in the Eulerian case because the Lagrangian framework itself provides some natural concentration of grid points in regions where the fluid is compressing. There are important situations, however, such as a flame front, where fine resolution is required but the fluid is expanding. A practical problem exists: how to arbitrarily concentrate grid points in a Lagrangian calculation and still maintain the desirable property of no numerical diffusion. The regridding procedures used in Eulerian calculations are numerically diffusive as the grid usually moves through the fluid.

In order to avoid introducing diffusion in a Lagrangian model through the regridding procedure, we use a method based on *splitting* and *merging* existing cells. These two discontinuous Lagrangian rezone operations are illustrated schematically for one-dimensional grids in Figure 6–11. When these operations are used in regions with no steep gradients in the fluid variables, we can add or subtract one or more cell interfaces without causing diffusion through any of the already existing cell interfaces. The loss or gain of information is strictly localized to the place and time where the interface configuration is changed and is fully consistent with the changes in information content desired in the representation.

A single cell can be split by adding another interface anywhere between two existing interfaces, and setting the values of the variables in the two new cells equal to those in the old cell. This does not, however, improve the resolution of steep gradients because no intermediate, averaged values are generated. This also leaves terraces of two adjacent cells with equal values. A more satisfying composite regrid operation composed of two splittings and a merging is shown on the left of Figure 6–11. Fractions of adjacent cells are split off and then merged together to form a smaller cell of intermediate

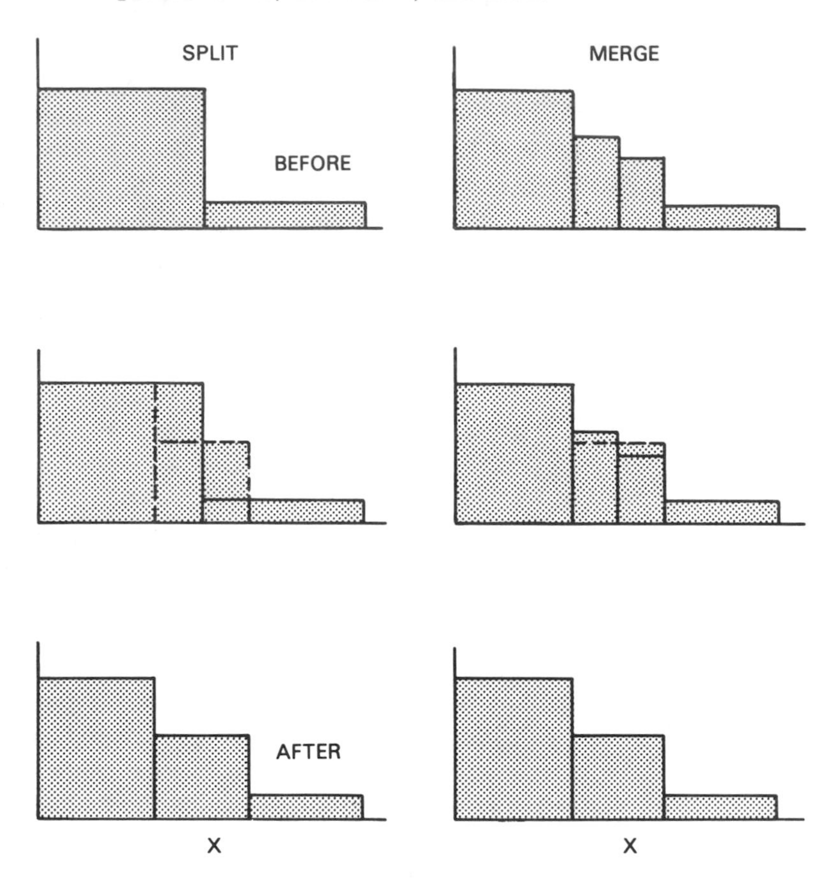

Figure 6–11. Schematic showing a discontinuous Lagrangian rezoning procedure.

value. Merging diminishes the resolution and speeds up the calculation by reducing the number of cells and by allowing a larger timestep. It also has much the same effect as numerical diffusion.

Splitting and merging results in discontinuous changes in the number of cells and interfaces in the calculation. We can program criteria so that the splitting and merging is done automatically in a way that adapts to the resolution needed in the evolving flow. However, the bookkeeping for this method is complicated, since the number of grid points changes and the location in computer memory of data referring to a given physical point in space also changes as cells are subdivided or removed. In one dimension, however, this added complexity is not severe. This method has been used, for example, in studies of ignition and extinction properties of flames (Oran and Boris, 1981; Kailasanath et al., 1982a,b).

In flame calculations, the grid must be refined *before* a gradient enters the cell or else numerical diffusion is added by the interpolation process. The conditions chosen for splitting cells ahead of an advancing gradient, and merging them after the gradient passes, determine the accuracy and cost of the overall algorithm. If the splitting is not performed far enough ahead of the approaching gradients, some of the region needing fine resolution moves into a region where the cells are too large. If the splitting is done too far ahead of the gradient over too large a region, a large amount of computing is wasted. When the merging occurs too soon after the transient has passed, sharp gradients diffuse faster than physical transport would allow. When the merging occurs too late, the advancing steep gradient region leaves more and more fine cells behind and the calculation slowly grinds to a halt. We recommend allowing only one level of splitting or merging per timestep at a given location.

The problem in Lagrangian adaptive gridding in multidimensions is that rotational components of the flow quickly distort the Lagrangian grid to the point where the calculation becomes meaningless. Few problems using multidimensional Lagrangian grids can be treated satisfactorily without some form of adaptive gridding. The common approach is to apply an interpolative regridding procedure every few cycles. When this regridding is done too often, the composite effect may be as diffusive as starting with an Eulerian algorithm. By intermittently regridding the calculation, numerical diffusion may be reduced below that present in an Eulerian calculation. However, when short wavelength shear flows are excited, even this potential improvement is of limited use.

The distortion in a Lagrangian grid can be handled by using a finite-difference grid of triangles in two dimensions and tetrahedrons in three dimensions. This type of gridding, illustrated in Figure 6–5c and 6–5d and described briefly above, has been used in many finite-element representations, and was adapted to finite differences by Crowley (1971) and Fritts and Boris (1979). There are two reasons for improvements in performance. First, a variable number of triangles at each vertex permits a smoother representation of complicated shapes. Second, because the number of lines meeting at a vertex is not fixed, it can be varied during a calculation by automatic grid reconnection procedures. This can prevent severe local grid distortions. In addition, there are algorithms which can be used to split or merge cells selectively to change the resolution in multidimensions as well as one dimension.

Using a Lagrangian triangular grid is the most physically appealing approach and least numerically diffusive technique employed so far. Its drawbacks center around the complexity of the program and the fact that the arithmetic operations depend on linked lists, random access, and scalar processing. We also note that as the dimensionality of the problem increases, the logic of splitting and merging, and in general of a reconnecting grid, becomes much more complicated. The interpolations and transformations needed to conserve all the conserved quantities are not obvious because the grid is moving. We describe this method in more detail in Chapter 10.

6–4.4 Embedded Grid and Multigrid Techniques

So far we have assumed that the representation consists of a single grid with different resolution in different regions but all of the cells are advanced with the same timestep. There are several ways of relaxing this timestep restriction.

First, consider some of the more complex grids shown in Figures 6–4 and 6–5, in which localized high-resolution grids are embedded around gradients or interfaces. One general approach would be to integrate the finely spaced regions with smaller timesteps, using as boundary conditions the results at the borders bounding the less resolved regions. When the less resolved regions are advanced a timestep, the more resolved regions are integrated a number of timesteps to catch up. Work in this field of embedded grids is being pursued, for example, by Berger (1985) and Thompson et al. (1985). The major computational problems center around matching these different grids at their interfaces. Also, conservation is usually difficult to ensure throughout a grid with a changing embedded grid. The major issues are determining when to advance which regions of the grid.

Another idea that is being explored is *intermittent embedding* of an adaptive grid. This approach depends strongly on there being two distinct space scales. At intermittent times, a finely gridded region is embedded in the larger grid and a calculation is done for several timesteps using the fine zones. For example, the calculation for a thin transition region between two phases could be embedded and integrated with the large-scale calculation only long enough to relax to a new equilibrium. The parameters of the embedded calculation, such as the energy release rate and flame temperature, could then be used in the less resolved calculation as source terms for a suitable phenomenology. These source terms could be held fixed for a number of macroscopic steps until the embedded calculation needs to be updated.

An embedded grid could actually contain a physically different calculation from the one performed on the larger grid. For example, suppose that only a few large cells are used to represent a region so that phenomena described by short time and space scales are not adequately resolved. Then the embedded calculation can resolve the smaller scales, and the appropriate averaged quantities can be formed to include in coarser calculations.

Because the region which requires resolution is only calculated during a small fraction of the time, a method of connecting and disconnecting this portion of the grid to the coarse grid must be developed. This problem is analogous to the problems of matching two different spatial resolutions. Furthermore, when the fine grid is disconnected, a method for representing the discontinuity in the coarsely resolved calculation is required. This requires control over jump and interface conditions that are at least as complex as the shock jump conditions. These requirements are also problem dependent because, for example, a shock will have different jump and conservation conditions from a flame front.

The intermittent embedded-grid approach is attractive for systems where the separation of space and time scales is large. In such cases it might not be possible to resolve all the scales in any practical computation, yet they all are important to the overall behavior of the system. Such a model, however, would not be valid in the regime in which the small-scale embedded structures interact directly rather than asymptotically with the coarse scales.

Another approach based on relaxing the timestep constraint mentioned above is to let the timestep as well as the space step vary through the entire calculation on the whole grid. Normally, when the timestep is chosen for the calculation, it does not vary as a function of position. The concept of a time level is well defined. However, the timestep must then be small everywhere when it is small at some location on the grid. The major computational gain from integrating each cell with a different timestep is to reduce the total number of point-steps in the calculation. Thus in a flame calculation, small space and time cells could be clustered only along the advancing flame front. This more equal treatment of time and space comes naturally out of some types of finite-element methods. It could, in theory, be applied to finite-difference methods.

As an example, consider using an evolving triangular grid in $x-t$ space. Then the variable connectivity of the grid could be used to allow for initiation and termination of cell interfaces as well as the greatly disparate timesteps in adjacent cells. Such an approach could incorporate many of the ideas and algorithms developed for dealing with relativistic fluids and plasma, and undoubtedly will be tried in due course. It is important to note that the software for such a calculation is going to be even more complex than it is for the adaptive Lagrangian triangular grids.

References

Berger, M.J., and A. Jameson, 1985, An Adaptive Multigrid Method for the Euler Equations, in Soubbaramayer and J.P. Boujot, eds., *Ninth International Conference on Numerical Methods in Fluid Dynamics*, Springer-Verlag, New York, 92–97.

Bird, G.A., 1976, *Molecular Gas Dynamics*, Clarendon Press, Oxford (England).

Bird, G.A., 1983, Definition of Mean Free Path for Real Gases, *Phys. Fluids* 26: 3222–3223.

Bledjian, L., 1973, Computation of Time Dependent Laminar Flame Structure, *Combust. Flame* 20: 5–17.

Book, D., J. Boris, A. Kuhl, E. Oran, M. Picone, and S. Zalesak, 1981, Simulation of Complex Shock Reflections from Wedges in Inert and Reactive Gaseous Mixtures, in W.C. Reynolds and R.W. MacCormack, eds., *Seventh International Conference on Numerical Methods in Fluid Dyanmics*, Springer-Verlag, New York, 84–90.

Boris, J.P., 1986, A Vectorized 'Nearest-Neighbors' Algorithm of Order *N* Using A Monotonic Logical Grid, *J. Comp. Phys.* 66: 1–20.

Boris, J.P., and D.L. Book, 1973, Flux-Corrected Transport I: SHASTA — A Fluid Transport Algorithm that Works, *J. Comp. Phys.* 11: 38–69.

Brackbill, J.U., 1976, Numerical Magnetohydrodynamics for High-Beta Plasmas, *Methods in Comput. Phys.*, 16: 1–41.

Carver, M.B., 1976, The Choice of Algorithms in Automated Method of Lines Solution of Partial Differential Equations, in L. Lapidus and W. Schiesser, eds., *Numerical Methods in Differential Equations*, Academic Press, New York, 243–265.

Craxton, R.S., and R.L. McCrory, 1979, A Simple Rezoning Technique for Use with the Flux-Corrected Transport Algorithm, *J. Comp. Phys.* 33: 432–440.

Crowley, W.P., 1971, FLAG: A Free-Lagrange Method for Numerically Simulating Hydrodynamic Flows in Two Dimensions, in M. Holt, ed., *Second International Conference on Numerical Methods in Fluid Dynamics*, Springer-Verlag, New York, 37–43.

Crowley, W.P., M.J. Fritts, and H. Trease (Eds.), 1985, *The Free Lagrange Method*, *Lecture Notes in Physics*, Vol. 238, Springer-Verlag, New York.

DeVore, C.R., J.H. Gardner, J.P. Boris, and D. Mosher, 1984, Hydrodynamic Simulations of Light Ion Beam-Matter Interactions: Ablative Acceleration of Thin Foils, *Laser Particle and Beams* 2: 227–243.

Dwyer, H.A., R.J. Kee, and B.R. Sanders, 1979, Calculation of a Complex Combustion Wave with Adaptive Grids, *Prog. Astro. Aero.* 76: 172 183.

Eiseman, P.R., 1985, Grid Generation for Fluid Mechanics Computations, *Ann. Rev. Fluid Mech.* 17: 487–522.

Frisch, U., B. Hasslacher, and Y. Pomeau, 1986, A Lattice Gas Automaton for the Navier-Stokes Equation, *Phys. Rev. Lett.*, 56: 1505–1508.

Fritts, M.J., and J.P. Boris, 1979, The Lagrangian Solution of Transient Problems in Hydrodynamics Using a Triangular Mesh, *J. Comp. Phys.* 31: 173–215.

Galant, S., 1981, An Improved Method of Lines to Compute Time Dependent Laminar Flame Structure, *Eighteenth Symposium (International) on Combustion*, pp. 1451–1459, The Combustion Institute, Pittsburgh, PA.

Gelinas, R.J., S.K. Doss, and K. Miller, 1981, The Moving Finite Element Method: Applications to General Partial Differential Equations with Multiple Large Gradients, *J. Comp. Phys.* 40: 202–249.

Gottlieb, D., and S.A. Orszag, 1977, *Numerical Analysis of Spectral Methods: Theory and Applications*, SIAM, Philadelphia.

Grinstein, F.F., E.S. Oran, and J.P. Boris, 1986, Numerical Simulations of Asymmetric Mixing in Planar Shear Flows, *J. Fluid Mech.* 165: 201–220.

Guirguis, R., E.S. Oran, and K. Kailasanath, 1986, Numerical Simulations of the Cellular Structure of Detonations in Liquid Nitromethane — Regularity of the Cell Structure, *Combust. Flame* 65: 339–366.

Harlow, F.H., 1962, *Theory of the Correspondence Between Fluid Dynamics and Particle-and-Force Models*, Report LA-2806, Los Alamos National Laboratory, Los Alamos, NM.

Hicks, J.S., and J.Wei, 1967, Numerical Solution of Parabolic Partial Differential Equations with Two-Point Boundary Conditions by the Method of Lines, *J. Assoc. Comp. Mech.* 14: 549–562.

Hyman, J.M., 1979, *A Method of Lines Approach to the Numerical Solution of Conservation Laws*, LANL Report LA-UR 79-837, Los Alamos National Laboratory, Los Alamos, NM.

Kailasanath, K., E.S. Oran, J.P. Boris, and T.R. Young, 1982a, Time-Dependent Simulation of Flames in Hydrogen-Oxygen-Nitrogen Mixtures, in N. Peters and J. Warnatz, eds., *Numerical Methods in Laminar Flame Propagation*, Friedr. Wieweg, Wiesbaden, West Germany, 152–166.

Kailasanath, K., E.S. Oran, and J.P. Boris, 1982b, A Theoretical Study of the Ignition of Premixed Gases, *Combust. Flame* 47: 173–190.

Kailasanath, K., and E.S. Oran, 1983, Ignition of Flamelets behind Incident Shock Waves and the Transition to Detonation, *Combust. Sci. Tech.* 34: 345–362.

Kailasanath, K., E.S. Oran, J.P. Boris, and T.R. Young, 1985, Determination of Detonation Cell Size and the Role of Transverse Waves in Two-Dimensional Detonations, *Combust. Flame* 61: 199–209.

Liskovets, O.A., 1965, The Method of Lines, *Differentsial 'nye Uravneniya* 1: 1662–1678; English translation in *Differ. Equations* 1: 1308-1323.

Löhner, R., K. Morgan, and O.C. Zienkiewicz, 1985, An Adaptive Finite Element Procedure for Compressible High Speed Flows, *Comp. Methods App. Mech. Eng.*, 51: 441–465.

Löhner, R., K. Morgan, J. Peraire, and M. Vahdati, 1987, Finite Element Flux-Corrected Transport (FEM-FCT) for the Euler and Navier-Stokes Equations, *Int. J. Num. Meth. Fluids,* to appear.

Löhner, R., K. Morgan, M. Vahdati, J.P. Boris, and D.L. Book, 1986, FEM-FCT: Combining Unstructured Grids with High Resolution, *J. Comp. Phys.*, to appear.

Margolis, S.B., 1978, Time Dependent Solution of a Premixed Laminar Flame, *J. Comp. Phys.* 27: 410–427.

Mihalas, D., K-H.A.Winkler, and M.L. Norman, 1984, Adaptive-Mesh Radiation Hydrodynamics — II. The Radiation and Fluid Equations in Relativistic Flows, *J. Quant. Spec. Rad. Trans.* 31: 479–490.

Mitchell, A.R., and J.Y. Thomson, 1958, Finite Difference Methods of Solution of the Von Mises Boundary Layer Equation with Special Reference to Conditions Near a Singularity, *Z. angew. Math. Phys.* 9:26–37.

Oran, E.S., T.R. Young, and J.P. Boris, 1978, Application of Time-Dependent Numerical Methods to the Description of Reactive Shocks, *Seventeenth Symposium (International) on Combustion*, The Combustion Institute, Pittsburgh, PA, 43–53.

Oran, E.S., and J.P. Boris, 1981, Theoretical and Computational Approach to Modelling Flame Ignition, *Prog. Astro. Aero.* 76: 154–171.

Oran, E.S., T.R. Young, J.P. Boris, and A. Cohen, 1982a, Weak and Strong Ignition, I. Numerical Simulations of Shock Tube Experiments, *Combust. Flame* 48: 135–148.

Oran, E.S., T.R. Young, J.P. Boris, J.M. Picone, and D.H. Edwards, 1982b, A Study of Detonation Structure: The Formation of Unreacted Pockets, *Nineteenth Symposium (International) on Combustion*, The Combustion Institute, Pittsburgh, PA, 573–582.

Peskin, C.S., 1977, Numerical Analysis of Blood Flow in the Heart, *J. Comp. Phys.* 25: 220–252.

Thompson, J.F., Z.U.A. Warsi, and C.W. Mastin, 1985, *Numerical Grid Generation — Foundations and Applications*, Elsevier Science Publishing Co., New York.

Winkler, K-H.A., M.L. Norman, and D. Mihalas, 1984, Adaptive-Mesh Radiation Hydrodynamics — I. The Radiation Transport Equation in a Completely Adaptive Coordinate System, *J. Quant. Spec. Rad. Trans.* 31: 473–478.

Chapter 7

DIFFUSIVE TRANSPORT PROCESSES

The term *diffusive transport* encompasses molecular diffusion, viscosity, thermal conduction, thermal diffusion, and radiation transport. These are the parts of Eqs. (2–1.1) – (2–1.18) represented by the expressions

$$\frac{\partial n_i}{\partial t} = -\nabla \cdot n_i \mathbf{v}_{di}, \qquad (7-0.1)$$

$$\frac{\partial \rho \mathbf{v}}{\partial t} = -\nabla \cdot \delta \mathbf{P}, \qquad (7-0.2)$$

and

$$\frac{\partial E}{\partial t} = -\nabla \cdot (\delta \mathbf{q} + \mathbf{q}_r) \, . \qquad (7-0.3)$$

Each equation is a divergence of a flux which, though not explicitly displayed, is related to a local gradient. The convection terms are omitted here to focus on the diffusive transport processes.

In Eq. (7–0.1), $\mathbf{v}_{di}$ is the diffusion velocity of species i, defined as

$$\mathbf{v}_{di} = \mathbf{v}_i - \mathbf{v} \qquad (7-0.4)$$

where $\mathbf{v}$ is the fluid velocity averaged over all species and $\{\mathbf{v}_i\}$ are the velocities of the fluid species in the same frame of reference as $\mathbf{v}$.

The diffusion velocities $\{\mathbf{v}_{di}\}$ are found by inverting the matrix equation

$$\mathbf{G}_i = \sum_k \frac{n_i n_k}{N^2 D_{ik}} (\mathbf{v}_{dk} - \mathbf{v}_{di}) \, , \qquad (7-0.5)$$

where the source terms $\{\mathbf{G}_i\}$ are defined as

$$\mathbf{G}_i \equiv \nabla \left(\frac{n_i}{N} \right) - \left(\frac{\rho_i}{\rho} - \frac{n_i}{N} \right) \frac{\nabla P}{P} - K_i^T \frac{\nabla T}{T} \, . \qquad (7-0.6)$$

Equation (7–0.1) describes diffusion, although it superficially resembles convection because the fluxes are linear in the spatial gradients through Eqs. (7–0.5) and (7–0.6). The diffusion velocities are defined relative to the net momentum of the multispecies mixture, and thus are constrained to carry no net momentum,

$$\sum_{i=1}^{N_s} \rho_i \mathbf{v}_{di} = 0 \, . \qquad (7-0.7)$$

This constraint is the additional relation needed to remove the singularity arising in the inversion of Eq. (7–0.5).

In Eq. (7–0.2), the divergence is taken of the pressure tensor, $\mathbf{P}$, minus the local diagonal contribution, $P(N,T)\mathbf{I}$,

$$\delta\mathbf{P} \;=\; \left(\frac{2}{3}\mu_m - \kappa\right)(\nabla\cdot\mathbf{v})\mathbf{I} - \mu_m[(\nabla\mathbf{v}) + (\nabla\mathbf{v})^T]. \qquad (7-0.8)$$

These are the diffusive viscous terms, in which μ_m and κ multiply a gradient of the velocity.

Equation (7–0.3) represents diffusion of heat due to gradients in temperature and pressure. Subtracting the zeroth-order convective energy flux gives

$$\delta\mathbf{q}(N,T) \;=\; -\lambda_m\nabla T + \sum_i n_i h_i \mathbf{v}_{di} + P\sum_i K_i^T \mathbf{v}_{di}\,. \qquad (7-0.9)$$

The quantity $\mathbf{q}_r$ is the heat flux due to radiation.

This chapter describes methods for solving these equations and the difficulties associated with these methods. It also presents formulas for the diffusion coefficients appropriate for combustion. Radiation transport is treated in Chapter 12, again emphasizing the diffusion approximation.

7–1. THE DIFFUSION EQUATION

7–1.1 The Physical Origin of Diffusive Effects

Diffusion is a macroscopic effect arising from the microscopic thermal motions of particles. The particles bounce back and forth between collisions in a random walk about the average fluid velocity, $\mathbf{v}$. Electromagnetic radiation, in the form of photons, can also be repeatedly absorbed and reemitted as it interacts with the molecules. Thus radiation also random walks as it goes through matter. These random motions about an average flow spread gradients, mix materials at sharp interfaces, create motion in the fluid, and transport mass, momentum, and energy. Thus diffusion processes are important macroscopically, despite the fact that no net mass flow is involved.

The generic representation of diffusion considered here is the second-order partial differential equation

$$C(\rho,\mathbf{x},t)\frac{\partial\rho}{\partial t} \;=\; \nabla\cdot D(\rho,\mathbf{x},t)\nabla\rho + S(\mathbf{x},t) \qquad (7-1.1)$$

where ρ indicates a generic conserved variable. The quantities C and D may be constant or may depend on the system variables. When the energy equation is rewritten to become a temperature equation, ρ is a temperature and the function C is a specific heat. When the ρ represents the momentum, C is the density weighting that relates the velocity to the momentum. For species diffusion equations, ρ represents the density of the individual species, a vector of length N_s at each spatial location. In this case the diffusion coefficient is a tensor, $\mathbf{D}$, and Eq. (7–1.1) becomes

$$\frac{\partial \rho}{\partial t} = \nabla \cdot \mathbf{D}(\rho, \mathbf{x}, t) \nabla \rho \, . \qquad (7-1.2)$$

In general, $\mathbf{D}$ is a function of ρ, and separately of position, and time.

The diffusion equation is a simple macroscopic model of the complicated manybody dynamics that works so well that many different microscopic phenomena are represented by the same macroscopic equation. These phenomena include particle diffusion in gases and liquids, momentum diffusion in gases and liquids, thermal energy diffusion in gases, liquids, and solids, radiation transport in optically thick media, turbulent transport on large spatial scales, the repeated averaging of observations, and numerical smoothing from cell uncertainty.

In each case, the diffusion equation arises from an average of many interacting degrees of freedom. This average collapses complicated physics occurring on small space and fast time scales into a simpler macroscopic model. However, the macroscopic representation is imperfect because it has lost information. The separation between the microscopic and macroscopic scales can never be complete, and higher-order effects from the microscopic physics always extend into the macroscopically averaged fluid equations. Such effects are neglected in the diffusion approximation, but they can have important effects.

It is no surprise that the mathematical formulas that represent diffusion as a continuum process break down in some regimes. This can be insidious because the mathematical solution appears to evolve in a reasonable way but the answer is wrong. The solutions calculated using the diffusion equation as a model of the physics do not reproduce the real behavior of the physical system.

It is not surprising that the mathematical expression for molecular diffusion breaks down on scales comparable to the mean free path between particle collisions. These errors become small, however, when gradients extend over many mean free paths. It is more surprising that the diffusion equation also breaks down at large distances and finite times.

To help understand the intrinsic errors in the diffusion equation model, we examine a δ-function test problem whose analytic solution encompasses all wavelengths and time scales, and therefore can be evaluated in various limits.

7–1.2 The δ-Function Test Problem

A finite amount of a conserved quantity whose density $\rho(\mathbf{x},t)$ is deposited at $\mathbf{x} = \mathbf{x}_o$ at time $t = t_o$. It spreads diffusively in time, and the value of ρ at $\mathbf{x} = \mathbf{x}_o$ decreases. The total integral of ρ over all space does not change with time. We assume that there are no sources and that D is constant in time and space. The solution for $\rho(\mathbf{x},t)$ at later times becomes Gaussian, and the characteristic length scale $k^{-1}(t)$ and amplitude $\rho(t)$ vary with time.

By substituting a solution of the form

$$\rho(x,t) = \rho(t)e^{-k^2(t)x^2} \tag{7-1.3}$$

into Eq. (7–1.2), we obtain two equations that must be satisfied,

$$k(t)\frac{d}{dt}k(t) = -2k^4(t)D\ , \tag{7-1.4}$$

and

$$\frac{d}{dt}\rho(t) = -2k^2(t)D\rho(t)\ . \tag{7-1.5}$$

The solution of these two equations gives a closed-form result,

$$\rho(x,t) = \rho(t_1)\left(\frac{t}{t_1}\right)^{-\frac{1}{2}} \exp\left(-\frac{x^2}{4D(t-t_o)}\right)\ , \tag{7-1.6}$$

where the characteristic wavenumber, $k(t)$, evolves according to

$$k^2(t) = \frac{1}{4D(t-t_o)}\ . \tag{7-1.7}$$

The time $t_1 > t_o$ is the starting time when the density at $x = 0$ is $\rho(t_1)$.

The analytic solution for ρ in d dimensions ($d = 1, 2$, or 3) is

$$\rho(\mathbf{x},t) = \rho_o\left(\frac{k}{\sqrt{\pi}}\right)^d e^{-k^2(t)\ (\mathbf{x}-\mathbf{x}_o)^2}\ . \tag{7-1.8}$$

We call this a δ-function solution because the limit of Eq. (7–1.8), as t approaches t_o, is one definition of the singular Kronecker δ-function,

$$\delta(\mathbf{x}-\mathbf{x}_o) = \lim_{k\to\infty}\left(\frac{k}{\sqrt{\pi}}\right)^d e^{-k^2(\mathbf{x}-\mathbf{x}_o)^2}. \tag{7-1.9}$$

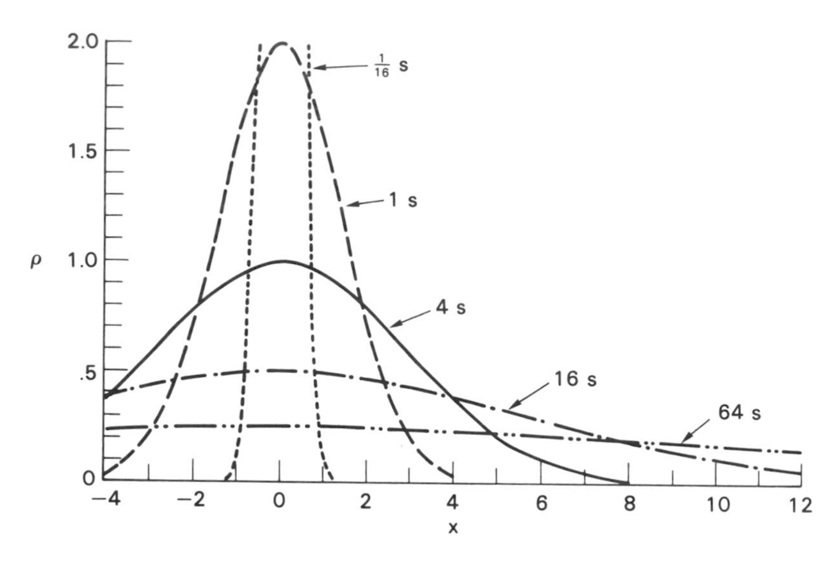

Figure 7–1. A decaying Gaussian solution with D=1 cm/s². The integrated area under each curve has the same value.

For all times $t > t_o$, this solution is Gaussian in shape and is also conservative because the area under the curve $\rho(\mathbf{x}, t)$ is constant in time. A decaying Gaussian is shown in Figure 7–1.

When $t = t_o$, the solution is singular and $k^2(t)$ from Eq. (7–1.7) is infinite. This corresponds to all of ρ, in this case

$$\rho_o = \int_{-\infty}^{\infty} d\mathbf{x}\ \rho(\mathbf{x}, t_o) \ , \qquad (7-1.10)$$

deposited initially at $\mathbf{x}_o$. The wavenumber k is well defined for all times greater than t_o. The distribution of ρ is still very localized, essentially a δ-function when t is slightly larger than t_o, but the profiles are finite.

Even for more complicated and dispersed initial conditions, the solutions approach a Gaussian asymptotically. Equation (7–1.8) has a wide range of applications for a constant diffusion coefficient. When a structured initial distribution of $\rho(\mathbf{x}, t_o)$ is localized in a region of space, it can be represented as a superposition of δ-functions. Because the diffusion equation is linear, each δ-function component individually follows Eq. (7–1.3) or (7–1.8) at subsequent times. The composite solution also approaches a Gaussian when $1/k(t)$ becomes larger than L_o, characterizing the size of the region of initial sources. The composite solution is

$$\rho(\mathbf{x},t) = \int_{-L_o}^{L_o} d\mathbf{x}_o \, \rho(\mathbf{x}_o,0) \left(\frac{k(t)}{\sqrt{\pi}}\right)^d e^{-k^2(t)\,(\mathbf{x}-\mathbf{x}_o)^2} \qquad (7-1.11)$$

The integral limits signify that the initial source distribution is localized within a distance L_o of the origin and involves only a finite amount of ρ, given by

$$\rho_o = \int_{-L_o}^{L} d\mathbf{x}_o \, \rho(\mathbf{x}_o,0) \, . \qquad (7-1.12)$$

After enough time elapses, $k(t)L_o$ becomes less than unity and each δ function element of the overall solution spreads until the initial source region is overlapped by all the δ-function components. The equation for $\rho(\mathbf{x},t)$, Eq. (7–1.11), approaches the Gaussian solution for ρ_o initially placed at the center of mass of the source distribution $\rho(\mathbf{x}_o,0)$. Because $k(t)$ decreases as time increases, for any distance $|\mathbf{x}|$ there is always a finite time t_x after which $k|\mathbf{x}| < \frac{1}{2}$,

$$t_x = t_o + \frac{L_o|\mathbf{x}|}{2D} \, . \qquad (7-1.13)$$

For times less than t_x, diffusion has not spread the distributions sufficiently to eliminate the influence of the initial conditions. After t_x has elapsed, however, there is an expanding shell about the source region. The profile inside this shell, $\rho(\mathbf{x},t)$, is essentially indistinguishable from a composite one-Gaussian source in which the equivalent amount of ρ_o is deposited at the center. Outside this shell the solution is exponentially small, but large differences are possible between the lumped-source solution and the multi-Gaussian solution with a nonlocalized initial density profile.

The important point is that initially localized diffusion problems approach profiles where deviations from the appropriate Gaussian decrease rapidly at any location. These deviations are relatively large only in the exponentially small tail of the solution, where the diffusion equation breaks down anyway. Figure 7–2 shows the evolution of such a superposition graphically. Three δ-functions are initially located at $x = -1$, 0, and 1 cm with twice the amount of ρ in the central Gaussian as in the other two. The exact solution and the single composite Gaussian solution are shown at later times to show the merging of the separate Gaussians. The lower panel of the figure shows the difference between the lumped-paramter one-Gaussian solution and the three-Gaussian solution.

There are some important numerical implications of these tendencies of a localized diffusing system to approach a Gaussian. If errors in the solution tend to decrease in time as the calculation proceeds, the solution

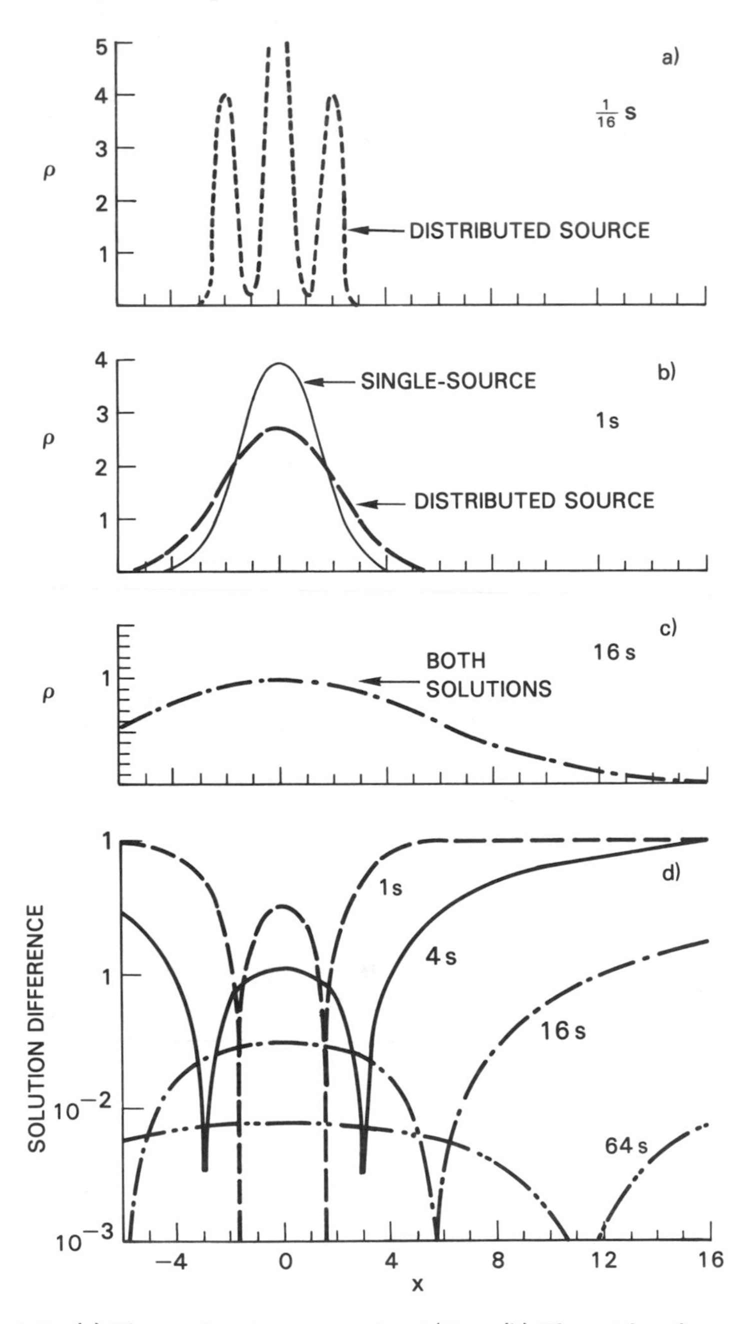

Figure 7–2. (a) Three δ-function source after 1/16 s. (b) Three δ-function source and equivalent composite δ-function source after 1 s. (c) The two solutions after 16 s are identical on the scale of the figure. (d) The difference between the two solutions at several times.

becomes more accurate rather than less accurate. The short-wavelength components of the solution diffuse the fastest, and their amplitudes quickly approach zero if there are no sources of short-wavelength structures. This property is useful because numerical simulations using finite spatial grids usually have the worst errors at short wavelengths. If physical diffusion is large enough, it actually reduces the continuing short-wavelength errors generated by numerical representations with finite resolution.

7–1.3 Short- and Long-Wavelength Breakdowns

In Chapter 4 the diffusion flux was defined as

$$\mathbf{f}_D \equiv -D \, \nabla \rho, \tag{7 – 1.14}$$

proportional to the gradient of the dependent fluid variable $\rho(\mathbf{x}, t)$. On the scale of particle mean free paths, this gradient is relatively meaningless and thus the flux, given by Eq. (7–1.14), gives too smooth a solution. The diffusion equation is derived by taking macroscopic averages over many individual particles, and so it is not surprising that the representation fails at short scale lengths. Thus it is often attractive to look for higher-order continuum corrections to the diffusion equation at these short scales. Unless these corrections contain a high-frequency source component, however, they cannot represent the qualitative behavior of the random walk process at short scales.

Another way that the diffusion equation fails is far from fluctuating sources at relatively large times after the sources have been deposited. Consider ρ from a single source diffusing through space. Any amount of ρ appearing at a distant location $\mathbf{x}$ arrives there by traveling for a finite time through all intermediate locations. An average diffusive transport velocity can be calculated from the δ-function Gaussian solution,

$$\mathbf{v}_D = \frac{\mathbf{f}_D}{\rho} = \frac{(\mathbf{x} - \mathbf{x}_o)}{2(t - t_o)}, \tag{7 – 1.15}$$

where $\mathbf{v}_D$ is the diffusion velocity.

Thus at long wavelengths it appears that the propagation velocity does not depend on the diffusion constant! This is the macroscopic consequence of the fact that the thermal velocity, which relates to the diffusive propagation speed, does not depend on the mean free path. Hence the diffusion velocity is constant. This result is curious because Eq. (7–1.15) indicates that for any finite time, the velocity of diffusive transport becomes arbitrarily large for $|\mathbf{x} - \mathbf{x}_o|$ approaching ∞. The exponentially small tails which a linear diffusion equation generates at large distances actually move far too fast for the random walk of the particles. This is indicated by the far-field limit of Figure 7–2d.

It is tempting to try to identify this superfast diffusive motion with the very fast particles in the tails of the thermal distribution, but this explanation is incorrect. Some of the superfast particles would have to exceed the speed of light when $\mathbf{x}$ is far enough from $\mathbf{x}_o$. The distance between air molecules at standard temperature and pressure is about 30 Å and their mean free path is about 1000 Å. There are about 5×10^9 collisions per particle per second. In one collision time, 2×10^{-10} s, the continuum Gaussian solution has particles moving at the speed of light at a distance of 12 cm from the source location. Fortunately they are slowing down; they had to travel at an average speed of 6×10^{10} cm/s to get there! This nonphysical behavior of the analytic solution occurs because the tails of the Gaussian become much smaller than can be represented by discrete particles spread out in any smooth way.

7–1.4 Numerical Diffusion and Instability

Numerical diffusion refers to errors which arise both from truncation in the algorithms used and the discrete nature of the representation. In a computation, numerical diffusion behaves like physical diffusion at scale lengths of a few cells and longer, but its origin is not physical. Thus results which are quantitatively incorrect can appear reasonable. The term *numerical diffusion* contributes additional confusion. By naming the unwanted phenomenon of excess numerical damping numerical diffusion, we invest it with a modicum of legitimacy that it does not deserve.

There is an unavoidable uncertainty that exists about where in a computational cell the fluid actually is, and this is itself a major source of numerical diffusion. The average density in a cell of finite volume does not contain information about how the density is distributed within the cell. This uncertainty is particularly important in solving the continuity equation and is discussed further in Chapter 8.

Numerical instability often appears in a computed solution as if a diffusion equation were being run backward, or as if one of the diffusion coefficients were inadvertently made negative. Too large a timestep or too large a gradient in a spatial derivative can change the sign of the damping term in the numerical amplification factor for the algorithm. These errors grow quickly from the initial conditions. A smoothly varying function begins to show small fluctuations which rapidly become spikes. Because the short wavelengths grow fastest, the location of the biggest spike is usually a good indicator of where the stability condition is first violated. At subsequent times, due to finite resolution and nonlinear interactions in the overall model, waves of spikes propagate out from the initial site.

7–2. INTEGRATION OF THE DIFFUSION EQUATION

7–2.1 A Finite-Difference Formula in One Dimension

In Section 4–3 we introduced a finite-difference approximation to the diffusion operator and solved the problem of the diffusive smoothing of a periodic function. Explicit, semi-implicit, and fully implicit algorithms were discussed, and the results were compared to the corresponding analytic solution. We assumed constant timesteps, a uniformly spaced one-dimensional grid, and a constant diffusion coefficient. This section considers one-dimensional diffusion and treats spatial variations in the cell sizes and diffusion coefficients.

Consider the diffusion of $\rho(x,t)$ in a rigid duct whose cross-sectional area, $A(x)$, varies with x, as shown schematically in Figure 7–3. A conservative finite-difference form of the diffusion equation with a variably spaced grid and variable diffusion coefficients is given by

$$\begin{aligned} \Lambda_i \frac{(\rho_i^n - \rho_i^{n-1})}{\Delta t} &= A_{i+\frac{1}{2}} D_{i+1/2} \frac{(\overline{\rho}_{i+1} - \overline{\rho}_i)}{(x_{i+1} - x_i)} \\ &\quad - A_{i-\frac{1}{2}} D_{i-\frac{1}{2}} \frac{(\overline{\rho}_i - \overline{\rho}_{i-1})}{(x_i - x_{i-1})} + \Lambda_i S_i \,, \qquad (7-2.1) \\ &\quad i = 1, 2, \ldots, N. \end{aligned}$$

Here $\{\Lambda_i\}$ are the volumes of cell i, $\{A_{i+\frac{1}{2}}\}$ are the areas of the cell interfaces, and $\{D_{i+\frac{1}{2}}\}$ are the diffusion coefficients evaluated at the cell interfaces. The $\{S_i\}$ are source or sink terms representing the effects of chemical reactions. The bars over the values of ρ on the right side indicate that an average in time can be taken, giving centered, semi-implicit, or fully implicit schemes to improve the order of accuracy of the time integration. In Eq. (7–2.1), cell-interface locations are denoted $\{x_{i+\frac{1}{2}}\}$, with $i = 0, 1, ..., N$. The left boundary of the computational region is $x_{\frac{1}{2}}$ and the right boundary is $x_{N+\frac{1}{2}}$.

This approximation to the diffusion equation conserves the total amount of ρ. Diffusive flux is defined at the cell interfaces, so the exact amount of ρ leaving one cell is automatically added to the next cell. Conservation can

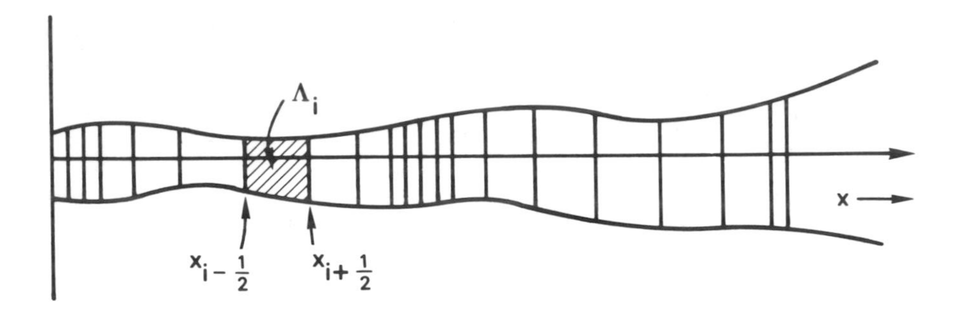

Figure 7–3. Schematic of a rigid duct with variable cross-sectional area.

be shown by summing Eq. (7–2.1) over all cells to obtain

$$\sum_{i=1}^{N} \Lambda_i \rho_i^n = \sum_{i=1}^{N} \Lambda_i \rho_i^{n-1} + \sum_{i=1}^{N} S_i \qquad (7-2.2)$$
$$+ \Delta t \big[A_{N+\frac{1}{2}} D_{N+\frac{1}{2}} (\nabla \overline{\rho})_R - A_{\frac{1}{2}} D_{\frac{1}{2}} (\nabla \overline{\rho})_L \big] \; .$$

The amount of ρ in the computational domain at the new time is equal to (the amount at the old time) + (any that comes in from the boundaries) − (any that goes out) + (the sum of all the source terms).

7–2.2 Implicit Solutions using Tridiagonal Matrix Inversions

When $\overline{\rho}$ in Eq. (7–2.1) is replaced by ρ^{n-1}, the finite-difference formula is explicit and solving for ρ^n is straightforward. Values of ρ at the new time are a function of three values at the previous time. The effects of a local change propagate at one cell per timestep.

When the expression for $\overline{\rho}$ includes some of the ρ^n in addition to ρ^{n-1}, the algorithm is at least partially implicit. For example, if

$$\overline{\rho}_i = \theta \rho_i^n + (1-\theta) \rho_i^{n-1} \qquad (7-2.3)$$

is substituted into Eq. (7–2.1), the equation can be be rearranged into matrix form,

$$\mathbf{M} \cdot \boldsymbol{\rho}^n = \mathbf{s}^o \; . \qquad (7-2.4)$$

Here $\mathbf{M}$ is a tridiagonal matrix of coefficients and $\mathbf{s}^o$ is a vector containing the inhomogeneous and source terms,

$$\begin{aligned}
M_{i,i-1} &= -\theta B_{i-1} \\
M_{i,i} &= \frac{\Lambda_i}{\Delta t} + \theta(B_{i+1} + B_{i-1}) \\
M_{i,i+1} &= -\theta B_{i+1} \\
s_i^o &= (1-\theta)B_{i-1}\rho_{i-1}^{n-1} + \left[\frac{\Lambda_i}{\Delta t} - (1-\theta)(B_{i+1}+B_{i-1})\right]\rho_i^{n-1} \\
&+ (1-\theta)B_{i+1}\rho_{i+1}^{n-1} + S_i \, ,
\end{aligned} \qquad (7-2.5)$$

where

$$\begin{aligned}
B_{i-1} &= \frac{A_{i-\frac{1}{2}} D_{i-\frac{1}{2}}}{x_i - x_{i-1}} \\
B_{i+1} &= \frac{A_{i+\frac{1}{2}} D_{i+\frac{1}{2}}}{x_{i+1} - x_i} \, .
\end{aligned} \qquad (7-2.6)$$

The $\boldsymbol{\rho}^n$ is a vector of all of the values of $\{\rho_i\}$ at the new time. Note that the coefficients representing the specific heats, $C(\rho, \mathbf{x}, t)$ in Eq. (7–1.1), may be included in the Λ_i if they are constant during a timestep.

7–2.3 Large Variations in Δx and D

The advantages and problems associated with using variably spaced zones is discussed in Chapter 6. The motivation for considering such methods is that they improve resolution where there are significant changes in the flow field and they reduce computational effort and cost when only slow variations are present.

There are, however, numerical problems with variably spaced grids. For example, using a variably spaced grid, as Eq. (7–2.1) allows, does not automatically guarantee accuracy when adjacent cell sizes vary significantly. The order of accuracy of an algorithm is often reduced, and the truncation error increases.

There are practical ways to use variable spacing or rapidly varying diffusion coefficients and yet maintain enough accuracy. One approach is to keep the rate of change of the cell sizes small, so the truncation errors are small. The acceptable amount of variation from cell to cell varies with the particular physical problem and with the numerical algorithm. Sometimes it is possible to space the cells to ensure that spatial derivatives are of arbitrary order. This is often not practical, however, especially when high resolution is required to track regions in the flow with large gradients. The practical rule for variable spacing is to allow maximum changes of 10-15% from cell to cell.

Figure 7–4 shows several adjacent cells in which the density, indicated as point values at cell centers, and the diffusion coefficients, indicated as piecewise constant within each cell, vary significantly from cell to cell. Adjacent cell sizes can also be very different. Such situations usually occur at internal interfaces. An example is the interface between a relatively cool liquid droplet and warm background air where the thermal conductivity varies discontinuously across the interface. Even in single-phase media such situations occur in the vicinity of steep gradients where cell sizes and nonlinear diffusion coefficients vary rapidly.

As shown in the figure, the numerical approximation to such systems has very different states on opposite sides of an interface. Choosing average diffusion coefficients at the interface, the $\{D_{i+\frac{1}{2}}\}$, should take these discontinuities into account. The trick to choosing this average is to define a fictitious value of ρ at the interface, called $\rho^*_{i+\frac{1}{2}}$ in the figure. This value guarantees that the flux F_-, which reaches the interface from the left, exactly matches the flux F_+, which leaves the interface on the right. This flux-matching condition ensures that the averaged value of ρ at the interface is physically meaningful. The condition that $F_+ = F_-$ determines $\rho^*_{i+\frac{1}{2}}$,

$$\begin{aligned} F_+ &\equiv A_{i+\frac{1}{2}} \left(\frac{\overline{\rho}_{i+1} - \rho^*_{i+\frac{1}{2}}}{x_{i+1} - x_{i+\frac{1}{2}}} \right) D_{i+1} = \\ F_- &\equiv A_{i+\frac{1}{2}} \left(\frac{\rho^*_{i+\frac{1}{2}} - \overline{\rho}_i}{x_{i+\frac{1}{2}} - x_i} \right) D_i \,. \end{aligned} \qquad (7-2.7)$$

The factors of $A_{i+\frac{1}{2}}$ cancel and Eq. (7–2.7) can be solved for $\rho^*_{i+\frac{1}{2}}$ giving

$$\rho^*_{i+\frac{1}{2}} = \frac{\overline{\rho}_{i+1} D_{i+1}(x_{i+\frac{1}{2}} - x_i) + \overline{\rho}_i D_i (x_{i+1} - x_{i+\frac{1}{2}})}{D_{i+1}(x_{i+\frac{1}{2}} - x_i) + D_i(x_{i+1} - x_{i+\frac{1}{2}})} \,. \qquad (7-2.8)$$

Note that the cell sizes and diffusion coefficients can vary extensively and $\rho^*_{i+\frac{1}{2}}$ still is between $\overline{\rho}_{i+1}$ and $\overline{\rho}_i$, a reasonable expectation of a weighted average.

Using this provisional value $\rho^*_{i+\frac{1}{2}}$, an average diffusion coefficient at the interface can be defined that incorporates the information in $\{\rho^*_{i+\frac{1}{2}}\}$. Inserting Eq. (7–2.8) into Eq. (7–2.7) and setting this equal to the flux $F_{i+\frac{1}{2}}$

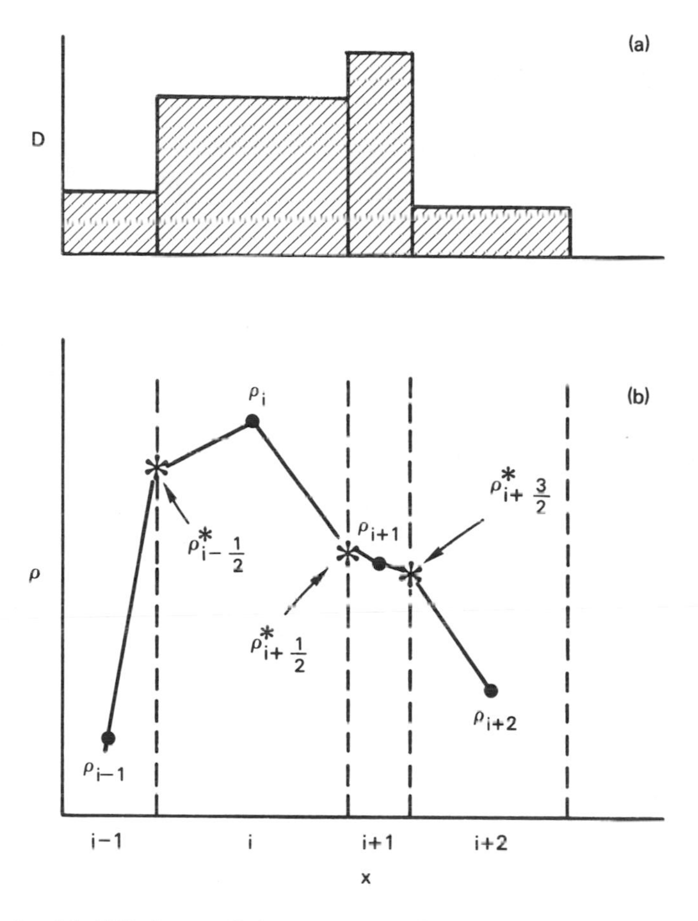

Figure 7–4. (a) Diffusion coefficient variations from cell to cell. (b) Values of the variable ρ at the cell centers (indicated by $\ldots, i-1, i, i+1, \ldots$) and cell interfaces (indicated by $\ldots, i-\frac{1}{2}, i+\frac{1}{2}, \ldots$).

gives

$$
\begin{aligned}
F_{i+\frac{1}{2}} &= A_{i+\frac{1}{2}} D_{i+\frac{1}{2}} \left(\frac{\overline{\rho}_{i+1} - \overline{\rho}_i}{x_{i+1} - x_i} \right) = F_+ \\
&= \frac{A_{i+\frac{1}{2}} D_i D_{i+1} (\overline{\rho}_{i+1} - \overline{\rho}_i)}{D_{i+1}(x_{i+\frac{1}{2}} - x_i) + D_i(x_{i+1} - x_{i+\frac{1}{2}})} \, .
\end{aligned}
\qquad (7-2.9)
$$

The definition of the interface-averaged diffusion coefficients, $\{D_{i+\frac{1}{2}}\}$, given by Eq. (7–2.9), is

$$
D_{i+\frac{1}{2}} = \frac{D_i D_{i+1} (x_{i+1} - x_i)}{D_{i+1}(x_{i+\frac{1}{2}} - x_i) + D_i(x_i - x_{i+\frac{1}{2}})} \, . \qquad (7-2.10)
$$

Using these values of $\{D_{i+\frac{1}{2}}\}$ ensures the fewest inconsistencies in the fluxes at all of the interfaces when using Eq. (7–2.1). Note that this formulation avoids using the provisional values $\{\rho^*_{i+\frac{1}{2}}\}$, so that tridiagonal matrix solvers can still be used to solve implicit forms of Eq. (7–2.1). This flux matching trick is useful in may ways, including nonlinear diffusion and convection. This idea is used a number of times in this book.

7–3. NONLINEAR EFFECTS AND STRONG DIFFUSION

7–3.1 Nonlinear Diffusion Effects

Nonlinearities are impediments to solving the diffusion equation quickly and accurately. Such a nonlinearity exists if, for example, the diffusion coefficient $D(\rho, \mathbf{x}, t)$ in Eq. (7–1.1) depends on the current values of ρ. An important example of a nonlinear diffusion equation is derived from Eq. (7–0.3) considered as an equation for temperature, T. The thermal conduction coefficient $\lambda(T)$ usually increases with temperature at least as fast as $T^{\frac{1}{2}}$. Therefore nonlinear thermal conduction often appears in the form

$$\begin{aligned}\frac{\partial T}{\partial t} &= \frac{\partial}{\partial x}\lambda(T)\frac{\partial T}{\partial x} = \frac{\partial}{\partial x}\,\lambda_{\frac{1}{2}}T^{\frac{1}{2}}\,\frac{\partial T}{\partial x} \\ &= \frac{2\lambda_{\frac{1}{2}}}{3}\,\frac{\partial}{\partial x}\left(\frac{\partial T^{\frac{3}{2}}}{\partial x}\right) .\end{aligned} \qquad (7-3.1)$$

Here it has been possible to bring the explicit temperature dependence of the diffusion coefficient inside the gradient operator. (In cgs units, the coefficient $\lambda_{\frac{1}{2}}$ has units of $\mathrm{K}^{-\frac{1}{2}}\mathrm{cm}^2\mathrm{s}^{-1}$).

The heat flux is stronger than a linear function of the temperature. A thermal conduction coefficient with the temperature raised to the $\frac{1}{2}$ or $\frac{3}{4}$ power is typical in gas-phase combustion and is important in flames. This is a weak nonlinearity when the power law exponent is less than unity, but still the conduction becomes more important as the temperature increases. The electron thermal conductivity in a highly ionized plasma scales as $T^{\frac{5}{2}}$, and radiation transport scales as about T^4. In general, if $\lambda(T)$ is a monotonic, increasing function of T, the temperature dependence can be brought inside the gradient operator. This helps in solving Eq. (7–3.1) conservatively and implicitly.

Equation (7–3.1) has an interesting and instructive analytical solution for a particular propagating, fixed profile. We assume the temperature profile can be expressed as $T(x - Vt)$, where V is a constant velocity. The time derivative of T in Eq. (7–3.1) is then

$$\frac{\partial T}{\partial t} = -V\frac{\partial T}{\partial x} . \qquad (7-3.2)$$

This leaves a second-order ordinary differential equation in x to be solved for the profile,

$$\frac{dT}{dx} = -\frac{2}{3}\frac{\lambda_{\frac{1}{2}}}{V}\frac{d^2T^{\frac{3}{2}}}{dx^2} . \qquad (7-3.3)$$

The first integral of this equation introduces a constant of integration. By equating this constant with the temperature T_∞ far upstream of the advancing nonlinear heat wave, we find

$$T(x) - T_\infty = -\frac{2}{3}\frac{\lambda_{\frac{1}{2}}}{V}\frac{dT^{\frac{3}{2}}}{dx} . \qquad (7-3.4)$$

Equation (7–3.4) can be integrated analytically to obtain

$$x - x_1 = \int_{x_1}^{x} dx = -\frac{2}{3}\frac{\lambda_{\frac{1}{2}}}{V}\int_{T_1}^{T(x)} \frac{d\,T^{\frac{3}{2}}}{(T - T_\infty)} , \qquad (7-3.5)$$

where T_1 is the temperature at x_1. The integral on the right side of Eq. (7–3.5) yields

$$x - x_1 = \frac{\lambda_{\frac{1}{2}}}{V}\left[-2(\sqrt{T} - \sqrt{T_1}) + \sqrt{T_\infty}\ \ln\left|\frac{\sqrt{T} + \sqrt{T_\infty}}{\sqrt{T_1} + \sqrt{T_\infty}}\right|\left|\frac{\sqrt{T_o} - \sqrt{T_\infty}}{\sqrt{T} - \sqrt{T_\infty}}\right|\right] \qquad (7-3.6)$$

Figure 7–5 shows the solution to Eq. (7–3.1) given by Eq. (7–3.6) with $V = 5$ cm/s at an initial time $t_o = 0$ and several subsequent times. This solution is a profile with fixed shape that propagates to the right. The effect of the nonlinearity in the diffusion coefficient is large when the medium into which the heat is being conducted is cold, that is, when T_∞ is small. The gradient of temperature is larger at high temperature and smaller at low temperature. This keeps the heat flux relatively constant except where the material is heating up most rapidly. This solution assumes no motion of the fluid in response to the changing temperature so the problem is unrealistic for a gas. It could, however, represent the propagation of a thermal wave in a solid over a limited temperature range, excluding phase changes.

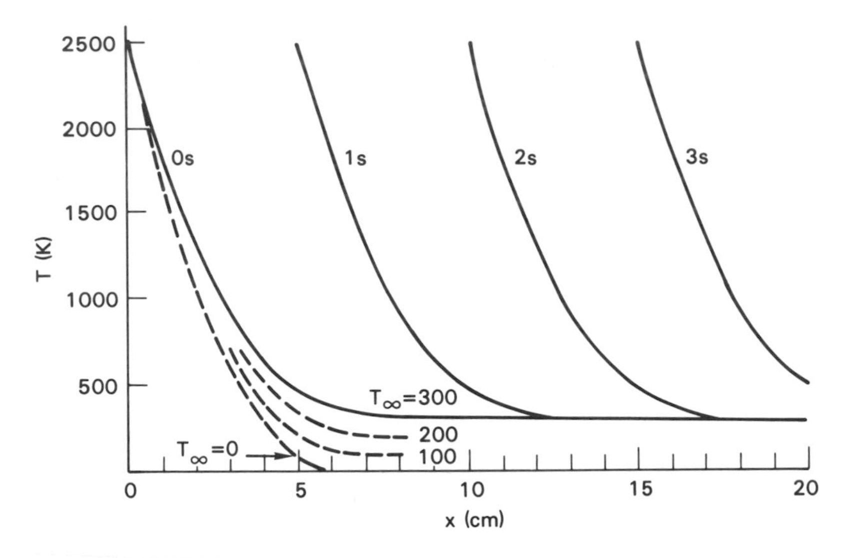

Figure 7–5. Temperature as a function of position for the solution of the thermal conduction front given in Eq. (7–3.6), with V=5 cm/s. The solid lines are the solution for T_∞=300 K, given at several times. The dashed lines represent the solution at t=0 for T_∞=0, 100, and 200 K.

The integral equation analogous to Eq. (7–3.4) can be integrated analytically for a wide range of nonlinear thermal conductivities with the general power law form,

$$\lambda(T) \equiv \lambda_\alpha T^\alpha . \qquad (7-3.7)$$

In the limiting case where the heat is diffusing into a cold medium with $T_\infty = 0$, the solution is

$$T(x) = \left[\frac{\alpha V}{\lambda_\alpha}(x_1 - x) + T_1^\alpha\right]^{1/\alpha} . \qquad (7-3.8)$$

This solution reaches

$$\begin{array}{llll} T_1 & \text{at} & x_1 & \text{and passes through} \\ T = 0 & \text{at} & x_o = x_1 + \dfrac{\lambda_\alpha T_1^\alpha}{\alpha V} , & \end{array} \qquad (7-3.9)$$

a finite distance away. In the vicinity of x_o the temperature approaches zero as $(x - x_o)^{1/\alpha}$. When α is less than unity, that is, $\alpha = \frac{1}{2}$ in Eq. (7–3.6), the leading edge of the temperature in this limiting solution goes smoothly to zero at finite distance with zero slope. The solution is not singular and has infinite slope only as T approaches infinity.

When α is greater than unity, the slope of the front becomes infinite at $x = x_o$ and the thermal conduction wave advances like a shock. The temperature behind the front must have a very steep gradient because the relatively cold medium ahead of the advancing wave has a relatively low conductivity. Such highly nonlinear diffusion equations arise in some models of turbulent transport and turbulent mixing, where they represent zero net convection in regions of the fluid where the vorticity is zero.

The finite temperature of the initial medium keeps the gradient from becoming infinite and causing the diffusion equation to break down. When $T_\infty > 0$, the singular derivative at x_o is replaced by a very thin transition region in which the temperature profile goes smoothly to zero with an exponentially small foot extending to infinity. This behavior applies to the strongly nonlinear case. When $\alpha = 2$ the solution equivalent to Eq. (7–3.6) is

$$x - x_1 = -\frac{\lambda_2}{V}\left[\frac{T^2 - T_1^2}{2} + T_\infty(T - T_1) + T_\infty^2 \ln\left(\frac{T - T_\infty}{T_1 - T_\infty}\right)\right] . \quad (7-3.10)$$

Figure 7–6 shows Eq. (7–3.10) for $V = 5$ cm/s with $x_1 = 0$, $T_1 = 2500$ K, and λ_2 chosen so that the limiting solution goes to zero at 6 cm. These simple nonlinear conduction solutions are closely related to the temperature profiles found in flames and ablation layers.

This analytic solution provides an ideal test of a diffusion algorithm because the correct temperature profile and propagation velocity are known. If the numerical algorithm is conservative, it will propagate the approximate numerical profile at the correct speed. The shape of the profile near T_∞ is generally most sensitive to the algorithms and the grid used.

When the thermal conductivity is highly nonlinear, for example, when $\alpha > 1$, and $T_\infty = 0$, the gradient becomes large to keep the heat flow roughly constant. The rate at which the shock-like thermal conduction wave advances is determined by the rate at which heat is added at the boundary, in this simplified case by the value of λ_α. As shown in Figures 7–5 and 7–6, the solution as T approaches T_∞ has much greater curvature when the nonlinearity is strong ($\alpha > 1$) than when it is weak ($\alpha < 1$). Thus the case $\alpha > 1$ is a more stringent test of nonlinear thermal conduction algorithms. Adequate spatial resolution at the inflection point of the solutions with finite T_∞ is particularly important in obtaining accurate numerical solutions. This region is shaded in Figure 7–6.

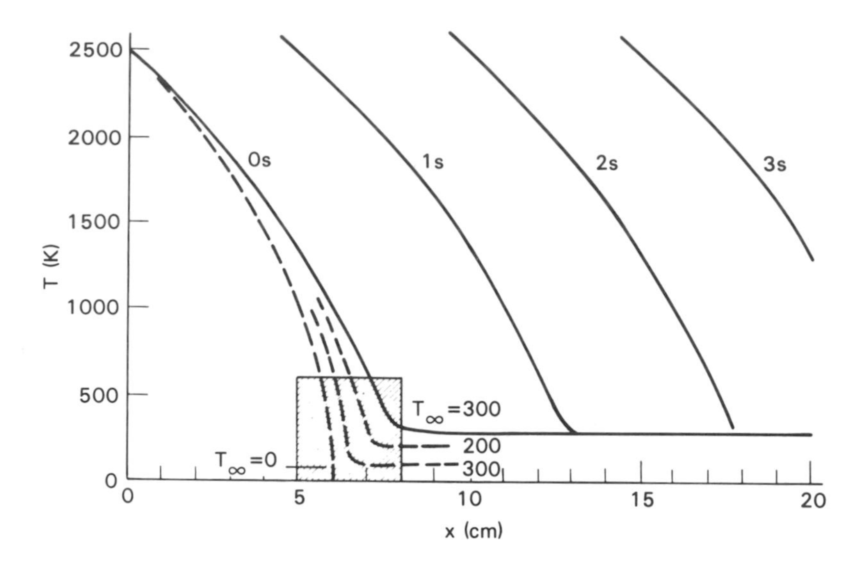

Figure 7–6. Temperature as a function of position for the solution of the thermal conduction front given in Eq. (7–3.10), with $V=5$ cm/s and $\alpha=2$. The solid lines are the solution for $T_\infty=300$ K, given at several times. The dashed lines represent the solution at $t=0$ for $T_\infty=0$, 100, and 200 K. The shaded area indicates that region requiring good numerical resolution when the equation is solved numerically.

7–3.2 Numerical Techniques for Nonlinear Diffusion

Discretizing the partial differential equation describing diffusion can be complicated by the functional form of the diffusion coefficients. However, if the timestep is small enough, the changes in cell values $\{\rho_j\}$ are small and the nonlinear terms in the diffusion coefficients do not change fast enough to present major difficulty. Then the nonlinear terms can be evaluated using known quantities at the old time t^{n-1} and errors are small. One or two iterations to reevaluate the nonlinear coefficients using progressively better estimates of the new variable values usually improves the accuracy appreciably. However, when the timestep is large, such a simple iteration does not necessarily converge. When it does, it may converge very slowly. At even longer timesteps, this prescription is unstable and the computed solution can change appreciably in a single timestep.

When the diffusion problem is highly nonlinear, it is often necessary to use implicit algorithms. Using the general power law thermal conductivity as an example, the fully implicit three-point formula for the temperature at the new time is

$$\frac{T_i^n - T_i^{n-1}}{\Delta t} = \frac{\lambda_\alpha}{\alpha+1}\left[\frac{T_{i+1}^{\alpha+1} - T_i^{\alpha+1}}{\Delta x^2} - \frac{T_i^{\alpha+1} - T_{i-1}^{\alpha+1}}{\Delta x^2}\right]^n , \qquad (7-3.11)$$

where the entire term in brackets on the right side is evaluated at the new time. For this equation, the straightforward use of a tridiagonal solver to find the new temperatures $\{T_i^n\}$ is complicated by the nonlinear terms on the right side.

A quadratically convergent iteration can be used to solve Eq. (7–3.11). A sequence of solutions $T_i^{(m)}$, where m indicates the iteration number, results from replacing the current best estimates $\{T_i^{(m)}\}$ of the desired new values, $\{T_i^n\}$, with

$$T_i^{(m+1)} = T_i^{(m)} + \Delta T_i^{(m)} , \qquad i = 1, \ldots, N_c . \qquad (7-3.12)$$

Equation (7–3.11), with Eq. (7–3.12) substituted for the desired T_i^n values, yields

$$\frac{\Delta T_i^{(m)}}{\Delta t} - \frac{\lambda_\alpha}{\Delta x^2}\left[(T_{i+1}^{(m)})^\alpha \Delta T_{i+1}^{(m)} - 2(T_i^{(m)})^\alpha \Delta T_i^{(m)} + (T_{i-1}^{(m)})^\alpha \Delta T_{i-1}^{(m)}\right]$$
$$= \frac{T_i^o - T_i^{(m)}}{\Delta t} + \frac{\lambda_\alpha}{\Delta x^2(\alpha+1)}\left[(T_{i+1}^{(m)})^{\alpha+1} - 2(T_i^{(m)})^{\alpha+1} + (T_{i-1}^{(m)})^{\alpha+1}\right] . \qquad (7-3.13)$$

With this linearization, the new corrections $\Delta T_i^{(m)}$ can be determined using a tridiagonal solver. In this form each iteration is conservative, and so the overall solution is conservative.

Equation (7–2.10) is a formula for determining the appropriate average values of diffusion coefficients at cell interfaces, given their values at cell centers. Nonlinear diffusion becomes a difficult problem when the coefficients vary greatly from cell to cell, and so it is natural to apply this flux-matching approach for nonlinear diffusion. When the diffusion coefficient is integrable, as in Eq. (7–3.7), or, more generally, when the flux can be written

$$D(\rho)\nabla\rho = \nabla G(\rho) , \qquad (7-3.14)$$

using flux-matching reduces to a linear interpolation of $G(\rho)$ at the cell interface. This occurs because the effective diffusion coefficient is constant once $D(\rho)$ is brought inside the gradient.

Usually the form of real nonlinear diffusion coefficients is not as simple as that assumed in Eq.(7–3.7). Suppose the thermal conduction coefficient varies as $\lambda(\rho, T)$. If the T dependence can be brought inside the gradient, flux-matching can be applied to the variation of λ with ρ that stays outside the gradient. Even when all of the variation of λ can be brought inside the gradient, as in Eq. (7–3.14), flux-matching yields the average determining the cell-interface fluxes which should be used when the cell sizes vary.

7–3.3 Asymptotic Methods for Fast Diffusion

When the diffusion coefficient is large everywhere or convection is relatively slow, the source and sink terms dominate and Eq. (7–1.1) becomes

$$\nabla \cdot D(\rho, x, t) \nabla \rho = -S \; , \qquad (7-3.15)$$

an elliptic equation. Essentially the same structure for the solution matrix results if the time derivative is kept. In the limit of large timestep, a finite-difference approximation for the time derivative is not expensive computationally and increases the diagonal dominance of the resulting sparse matrix equation.

When D is large and S is finite, the gradient of ρ must become small quickly so that the right and left sides of Eq. (7–3.15) are equal. In the limit of fast diffusion when the sources are negligible, ρ becomes nearly constant. When the source term is important, ρ has small variations around the constant value. The limit of fast thermal conduction is an isothermal system in which the temperature can be determined from global energy conservation.

When the source terms are present but change slowly compared to the rate at which diffusion can move ρ through the system, an elliptic or parabolic equation must be solved. In one dimension the resulting tridiagonal matrix is straightforward to solve. In multidimensions, slow changes in the solution can be used to simplify the solution procedure. The variation of the diffusion coefficient, if it depends on the slowly changing solution, may not need to be updated during a timestep. The nonlinearities may be evaluated at the beginning of the cycle, thus reducing a potentially nonlinear elliptic equation to a linear problem with known but varying coefficients.

7–3.4 Flux-Limiters

A flux-limiting formula is an alternate approach to solving the parabolic or elliptic equation discussed above. The idea of flux-limiting is to reduce the flux artificially to a value which is not large enough to evacuate any cells in one timestep. This is equivalent to locally decreasing the timestep below the explicit stability limit. Because the diffusion coefficient is presumably large where flux-limiters are applied, the diffusion is still fast and gradients in the solution are still small, although not as small as they would be if the full diffusive fluxes were allowed.

Veiwed in this way, flux-limiters are a useful trick allowing a larger explicit timestep than numerical stability conditions would otherwise permit. If very fine resolution is required somewhere on the grid for reasons other than resolving the diffusion terms, flux-limiters help to avoid the diffusive stability condition that would require exceptionally small timesteps. Flux-limiting formulas also arise naturally from the physics as well as the numerics.

In Section 7–1.3 we showed how the diffusion velocity approaches infinity sufficiently far from the center of a spreading Gaussian. If a quantity, heat or material, is diffusing, the largest average diffusion speed possible is approximately the thermal velocity of the particle distribution at the point of interest. Thus the flux can be at most $v_{th}\rho_i$ for one of the chemical species, or $Nk_BTv_{th}/(\gamma - 1)$ for the energy, where v_{th} is the local thermal velocity. Actually, the limiting flux is smaller because not all of the particles can be random walking in the same direction simultaneously. Practically, the limiting diffusion velocity is at most a few percent of the thermal velocity.

Flux-limiting in diffusion is particularly important in nearly collisionless plasmas with steep temperature gradients, and on the scale of the mean free path in gas dynamic shocks. In the plasma case, the hot particles have a much longer mean free path than the cold particles and thus can carry mass and heat a long way into a cold medium before they thermalize. Because the local gradients are steep, however, the classical continuum treatment predicts a flux that is far too large. In shocks, the hot particles behind the shock carry mass, momentum, and energy forward into the unshocked medium. If the shock is treated as a discontinuous profile, the diffusive flux may again be too large without flux-limiting.

7–4. MULTIDIMENSIONAL ALGORITHMS

7–4.1 Implicit, Explicit, and Centered Algorithms

Thermal conduction illustrates well the generalizations needed to solve multidimensional diffusion problems. We deal with a system in which the fluid velocity is zero, or assume that the thermal conduction process is being integrated separately from other physical processes. Then the diffusion of heat is described by

$$\frac{\partial}{\partial t} C(T, \mathbf{x}, t) T(\mathbf{x}, t) \;=\; \nabla \cdot \lambda(T, x, t) \nabla T + S(T, \mathbf{x}, t) \; , \qquad (7-4.1)$$

where $C(T, \mathbf{x}, t)$ is the specific heat. The term $S(T, \mathbf{x}, t)$ again represents temperature or energy sources and sinks, and $\lambda(T, x, t)$ is the coefficient of thermal conduction. When the functions C, λ, or S depend on the temperature, as indicated in Eq. (7–4.1), the problem is nonlinear but generalization of the techniques in Section 7–3 can extend the algorithms given there.

When λ and C are constant and S is zero, the one-dimensional solution given in Chapter 4 generalizes easily to two dimensions. Let the temperature profile be a constant everywhere, T_{avg}, with an added sinusoidal fluctuation. We expect the solution at later times to have the form

$$T(x, y, t) \;=\; T_{\text{avg}} + \Delta T_o \; e^{ik_x x} e^{ik_y y} e^{-\Gamma t}, \qquad (7-4.2)$$

where ΔT_o is the amplitude of the temperature fluctuation at $t = 0$. Substituting Eq. (7–4.2) into Eq. (7–4.1) gives

$$\Gamma \;=\; \frac{\lambda}{C}(k_x^2 + k_y^2) \; , \qquad (7-4.3)$$

showing that the decay rate of the various modes increases as the square of the wavenumber.

Although there are many finite-difference approximations to Eq. (7–4.1), we adopt a two-dimensional five-point difference template, shown in Figure 7–7a. Then

$$\begin{aligned} C_{ij} T_{ij}^n &+ \frac{\theta_x \Delta t}{\Delta x_{ij}} F_x^n + \frac{\theta_y \Delta t}{\Delta y_{ij}} F_y^n \\ &= C_{ij} T_{ij}^{n-1} + \frac{(1-\theta_x)\Delta t}{\Delta y_{ij}} F_x^{n-1} + \frac{(1-\theta_y)\Delta t}{\Delta y_{ij}} F_y^{n-1} \; , \end{aligned} \qquad (7-4.4a)$$

where

$$F_x \;=\; \lambda^x_{i+\frac{1}{2},j} \left(\frac{T_{i+1,j} - T_{ij}}{\Delta x_{i+\frac{1}{2},j}} \right) - \lambda^x_{i-\frac{1}{2},j} \left(\frac{T_{ij} - T_{i-1,j}}{\Delta x_{i-\frac{1}{2},j}} \right) \qquad (7-4.4b)$$

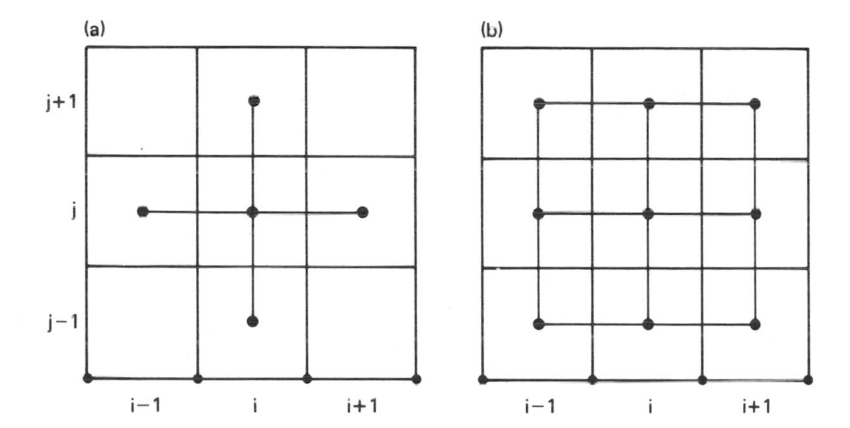

Figure 7–7. Local two-dimensional finite-difference templates. (a) Five-point template. (b) Nine-point template.

and

$$F_y = \lambda^y_{i,j+\frac{1}{2}} \left(\frac{T_{i,j+1} - T_{ij}}{\Delta y_{i,j+\frac{1}{2}}} \right) - \lambda^y_{i,j-\frac{1}{2}} \left(\frac{T_{ij} - T_{i,j-1}}{\Delta y_{i,j-\frac{1}{2}}} \right) . \qquad (7-4.4c)$$

Figure 7–7b shows a nine-point template that can also be used. The extra freedom that exists in nine-point difference formulas can be used to increase the accuracy when cell sizes and diffusion coefficients vary in space.

Equations (7–4.4) ensure conservation of energy when the diffusion coefficient varies across the grid. The grid can be unevenly spaced in both x and y, but the grid lines are assumed to define an orthogonal coordinate system. The cell-centered and interface-centered cell sizes are defined as before,

$$\begin{aligned} \Delta x_{i+\frac{1}{2},j} &\equiv (x_{i+1,j} - x_{i,j}) \\ \Delta x_{ij} &\equiv (x_{i+\frac{1}{2},j} - x_{i-\frac{1}{2},j}) \end{aligned} \qquad (7-4.5a)$$

and

$$\begin{aligned} \Delta y_{i,j+\frac{1}{2}} &\equiv (y_{i,j+1} - y_{ij}) \\ \Delta y_{ij} &\equiv (y_{i,j+\frac{1}{2}} - y_{i,j-\frac{1}{2}}) . \end{aligned} \qquad (7-4.5b)$$

The two directions may have different implicitness parameters, θ_x and θ_y, and these can even vary from cell to cell. These equations are a straightforward generalization of Eq. (7–2.1) and fully second-order accurate when

$\theta_x = \theta_y = \frac{1}{2}$ and the cell spacing is uniform in both directions. A three-dimensional finite-difference equation can also be written as a simple extension of Eqs. (7–4.4).

We now follow basically the same procedure as in Chapter 4: assume that T initially has a sinusoidal spatial distribution consistent with the theoretical solution presented above. Also, we assume that λ is constant in space and time and that the computational cells, Δx and Δy, are uniformly spaced. This produces a two-dimensional analog of Eq. (4–3.9) which relates the new temperature perturbation to the old one,

$$\Delta T^n = \Delta T^{n-1} \frac{\{1 - \frac{\lambda \Delta t}{2C}[(1-\theta_x)K_x^2 + (1-\theta_y)K_y^2]\}}{\{1 + \frac{\lambda \Delta t}{2C}[\theta_x K_x^2 + \theta_y K_y^2]\}} . \qquad (7-4.6)$$

Here we define

$$K_x^2 = \frac{2(1 - \cos k_x \Delta x)}{(\Delta x)^2} \qquad (7-4.7a)$$

$$K_y^2 = \frac{2(1 - \cos k_y \Delta y)}{(\Delta y)^2} \qquad (7-4.7b)$$

as the finite-difference approximations to the continuum wavenumbers k_x and k_y. At long wavelengths, where the spatial variations are well resolved, K is essentially the same as k. Equation (7–4.6) is the two-dimensional analog to the one-dimensional amplification factor in Eq. (4–3.10). The explicit ($\theta = 0$), centered ($\theta = \frac{1}{2}$), and implicit algorithms ($\theta = 1$) behave as they did in the one-dimensional case.

Based on the behavior of the one-dimensional methods and on the two-dimensional numerical amplification factor given by $\Delta T^n / \Delta T^{n-1}$ in Eq. (7–4.6), the fully implicit algorithm appears to perform the best. In particular, it behaves in a physically stable way at all wavelengths. To solve the multi-dimensional fast diffusion problem, it is natural to start off by assuming that the best approach is to generalize the fully implicit algorithm. This is easy to do in principle and the dispersion relation is known, as shown above. Unfortunately, actually solving the implicit system of coupled difference equations in two dimensions is more difficult computationally. However, if the implicit system can be solved, the semi-implicit algorithms can also be solved.

Because the finite-difference template couples together values of the temperature implicitly in more than one direction, the matrix that must be inverted is no longer tridiagonal. In two dimensions the direct solution of Eqs. (7–4.4) involves inverting an $N_x N_y \times N_x N_y$ sparse array and requires about $N_x^2 \times N_y^2$ operations. This is often too expensive for a reactive flow calculation. These large sparse matrices are expensive to invert efficiently, especially in the general case of variable diffusion coefficients and cell sizes. It is necessary either to accept an approximate inverse or an iterative algorithm to find the new temperatures satisfying Eqs. (7–4.4). We refer the reader to Section 11–3 for a more complete discussion of methods for solving the parabolic and elliptic problems.

7–4.2 ADI and Split-Direction Methods

Tridiagonal matrices may be solved quickly and readily by available methods (Section 11–3). The obvious question is whether a tridiagonal solver can be used to solve the multidimensional problem by alternately applying it to the different spatial directions. This approach leads to the *Alternating Direction Implicit* or ADI algorithms. Here we briefly outline the application of ADI to the diffusion problem of interest here.

Two steps are required to apply ADI methods to the idealized thermal conduction problem described above. The first step solves the x-direction derivatives implicitly using the y finite-difference approximations evaluated explicitly at the old time. This step involves solving N_y independent tridiagonal matrix systems, each of length N_x. The next "alternating" step evaluates the x-direction finite-difference approximations explicitly, and solves the y-direction derivatives implicitly. This step involves solving N_x independent tridiagonal solutions. Note that there is a bias in the way the different directions are treated. One semi-implicit way to remove some of the bias is to use the old time values (t^{n-1}), and then average the two intermediate results.

A useful exercise is to write out the equations described for these two approaches, based on Eqs. (7–4.4), and then to derive the dispersion relations for both algorithms. These should then be compared with each other and with the fully implicit dispersion relation which results from solving the sparse matrix system directly.

7–5. SOLVING FOR SPECIES DIFFUSION VELOCITIES

The costs of a reactive flow calculation are compounded when there are many reacting species. If the cost scaled merely as the number of species at each computational cell, N_s, the problem would be tractable. Whenever a matrix of size $N_s \times N_s$ must be inverted, however, the operation count and hence the computational cost scales as $(N_s)^3$ for each cell.

Calculating the species diffusion velocities, $\{\mathbf{v}_{di}\}$, requires solving Eq. (7–0.5) and Eq. (7–0.6), subject to the constraint in Eq. (7–0.7). We are again faced with the problem of the high computational cost of an expensive but straightforward matrix inversion.

7–5.1 The Fickian Diffusion Approximation

The Fickian diffusion approximation is commonly made to reduce this cost, so understanding its assumptions and limitations is worthwhile. Consider an idealized one-dimensional problem in which temperature and pressure are constant, there are no external forces, no chemical reactions, and no fluid velocity, but the species densities can vary with position. The species continuity equation reduces to

$$\frac{\partial n_i}{\partial t} + \frac{\partial}{\partial x}(n_i v_{di}) = 0 , \qquad (7-5.1)$$

and a number-averaged velocity, w, can be defined as

$$w \equiv \frac{1}{N} \sum_{i=1}^{N_s} n_i v_{di} , \qquad (7-5.2)$$

where N is the total number density, $\sum_{i=1}^{N_s} n_i \equiv N$. Then

$$\frac{\partial w}{\partial x} = 0 , \qquad (7-5.3)$$

which can be found by summing Eq. (7–5.1) over all i and noting that $\partial N/\partial t = 0$ is consistent with the assumptions given above. Using Eq. (7–5.3), we see that

$$\frac{\partial (n_i w)}{\partial x} = w \frac{\partial n_i}{\partial x} . \qquad (7-5.4)$$

We can rewrite Eq. (7–5.1) as

$$\frac{\partial n_i}{\partial t} = -\frac{\partial n_i v_{di}}{\partial x} + \frac{\partial w n_i}{\partial x} - w \frac{\partial n_i}{\partial x} . \qquad (7-5.5)$$

Using the definition of w from Eq. (7–5.2), this becomes

$$\frac{\partial n_i}{\partial t} = -\frac{\partial}{\partial x}\left[n_i \sum_{j=1}^{N_s} \left(\delta_{ij} - \frac{n_j}{N} \right) v_{di} \right] - w(t) \frac{\partial n_i}{\partial x} . \qquad (7-5.6)$$

Given the approximations stated at the beginning of this section, the diffusion velocity can be related to the binary diffusion coefficient D_{ij} through the expression for the mass flux (Hirschfelder et al., 1954),

$$j_i = n_i m_i v_{di} = \frac{N^2}{\rho} \sum_{j=1}^{N_s} m_i m_j D_{ij} \frac{\partial}{\partial x} \left(\frac{n_i}{N} \right) . \qquad (7-5.7)$$

The final equation combining Eqs. (7–5.6) and (7–5.7) is

$$\frac{\partial n_i}{\partial t} = -\frac{\partial}{\partial x} \sum_{k=1}^{N_s} \sum_{j=1}^{N_s} \left(\delta_{ij} - \frac{n_j}{N}\right) \frac{n_i N}{n_j \rho} m_k D_{jk} \frac{\partial n_k}{\partial x} - w(t) \frac{\partial n_i}{\partial x} . \quad (7-5.8)$$

Equation (7–5.8) is the *the multicomponent generalization of Fick's second law of diffusion*, and is discussed in more detail in Hirschfelder et al. (1954).

In the limit of only two species, n_1 and n_2, Eq. (7–5.8) reduces to

$$\frac{\partial n_1}{\partial t} = -\frac{\partial}{\partial x} D_{12} \frac{\partial n_1}{\partial x} - w(t) \frac{\partial n_1}{\partial x} \quad (7-5.9a)$$

and

$$\frac{\partial n_2}{\partial t} = -\frac{\partial}{\partial x} D_{12} \frac{\partial n_2}{\partial x} - w(t) \frac{\partial n_2}{\partial x} . \quad (7-5.9b)$$

Summing these gives

$$\frac{\partial (n_1 + n_2)}{\partial t} = \frac{\partial N}{\partial t} = 0 , \quad (7-5.10)$$

consistent with the assumptions of constant temperature and pressure made above.

An expression of the form

$$\frac{\partial C}{\partial t} = \frac{\partial}{\partial x} D \frac{\partial C}{\partial x} , \quad (7-5.11)$$

where C is a concentration, is commonly called *Fick's law* or *Fickian diffusion*. Note that the first-order derivative term found in Eqs. (7–5.9) is absent and the assumptions of constant temperature, pressure, etc., made above may not actually be satisfied in practical applications.

7–5.2 An Efficient Iterative Algorithm

Because the straightforward inversion of Eqs. (7–0.5) – (7–0.7) to find the diffusion velocities requires on the order of N_s^3 arithmetic operations per cell, a more efficient method is needed when there are more than four or five species present. The most efficient algorithm we have found (Jones and Boris, 1981) requires of order N_s^2 operations and is based on a special initial guess. To understand the procedure, we first rewrite Eqs. (7–0.5) and (7–0.6) as

$$G_i = \sum_{k=1}^{N_s} W_{ik} (v_{dk} - v_{di}) \quad (7-5.12)$$

and

$$G_i = \nabla \left(\frac{n_i}{N}\right) - X_i \frac{\nabla P}{P} - Y_i \frac{\nabla T}{T} . \qquad (7-5.13)$$

Here the $\{W_{ik}\}$ and $\{Y_i\}$ are now functions of the transport coefficients and the vector notation has been dropped as each direction can be treated independently.

When these equations are summed over all species, we find that

$$\sum_{k=1}^{N_s} G_k = 0 . \qquad (7-5.14)$$

Because the values of G_i sum to zero, the matrix $\mathbf{W}$ is singular as defined in Eq. (7–5.12) and (7–0.5), and the N_s different diffusion velocities cannot all be independent. The extra equation needed is the contraint Eq. (7–0.7) which is written here as

$$\sum_{i=1}^{N_s} \rho_i v_{di} = 0 . \qquad (7-5.15)$$

If a particular set of solutions $\{v_{di}^p\}$ is known for Eq. (7–5.12), any constant velocity may be added giving another equally correct solution $\{v_{di}^p + \delta v_d\}$ for $i = 1, ..., N_s$. The as yet undetermined constant, δv_d, may be used to enforce the constraint Eq. (7–5.15).

Let the diffusion coefficient D_{is} describe diffusion of species i through the background provided by the sum of all other species. Then an excellent starting approximation for the ith diffusion velocity, v_{di}^o, is

$$v_{di}^o \equiv -\frac{\rho - \rho_i}{\rho} \frac{N^2 D_{is}}{(N - n_i) n_i} G_i . \qquad (7-5.16)$$

Higher-order terms arise as corrections to this starting approximate diffusion velocity,

$$v_{di} \equiv v_{di}^o + \delta v_{di} . \qquad (7-5.17)$$

The equations for $\{\delta v_{di}\}$ are found by substituting Eq. (7–5.17) into Eq. (7–5.12), giving

$$\delta G_i \equiv \sum_{k=1}^{M} A_{ik} G_k = \sum_{k=1}^{M} W_{ik} (\delta v_{dk} - \delta v_{di}). \qquad (7-5.18)$$

The $\{D_{is}\}$ in Eq. (7–5.16) are defined as

$$\frac{D_{is}}{(N - n_i)} \sum_{k \neq i} \frac{n_k}{D_{ik}} \equiv 1 , \qquad (7-5.19)$$

and the matrix elements of **A** are given by

$$A_{ik} \equiv \frac{\rho_i}{\rho}\delta_{ik} + \frac{n_i}{D_{ik}}\frac{(\rho - \rho_k)}{\rho}\frac{D_{ks}}{(N - n_k)}(1 - \delta_{ik}). \qquad (7-5.20)$$

Equation (7–5.18) defines a linear system of equations which can be solved for the $\{\delta v_{di}\}$ and this is basically the same equation we started with, Eq. (7–5.12)! The right side of Eq. (7–5.18) vanishes when summed over i. It is easy to see that the choice of A_{ik} here is not unique. Each row of **A** can have an arbitrary constant added according to

$$\tilde{A}_{ik} \equiv A_{ik} - C_i \qquad (7-5.21)$$

without changing Eq. (7–5.18), which is consistent with Eq. (7–5.15). Such constants leave $\{\delta v_{di}\}$ unchanged. The general form of the complete solution is, therefore,

$$v_{di} = \frac{-(\rho - \rho_i)}{\rho}\frac{N^2 D_{is}}{(N - n_i)n_i} \times [\delta_{ik} + A_{ik} + A_{il}A_{lk} + \cdots]G_k . \qquad (7-5.22)$$

The matrix in square brackets is the formal expansion of $[1 - \mathbf{A}]^{-1}$.

In the numerical evaluation of Eq. (7–5.22), none of the indicated matrix multiplications have to be performed. Since $\{G_i\}$ is known, multiply from the right first. Each additional power of A is obtained by multiplying a vector, rather than a matrix, by A, giving a computational cost of order N_s^2. In practice, it is convenient to take $C_i = 0$ in Eq. (7–5.21), and to truncate the expansion in Eq. (7–5.22) at the A^2 term. At least the first correction A_{ik} needs to be included to get the correct sign for all the diffusion fluxes. The quadratic term sometimes adds significant extra accuracy and further iteration is generally unnecessary. The errors remaining are at most a few percent.

This algorithm converges quickly because of the initial approximation embodied in Eqs. (7–5.16) and (7–5.19). The factor $(\rho - \rho_i)/\rho$ in Eq. (7–5.22) is crucial. When there are only two species, so that $N = n_1 + n_2$, this factor becomes ρ_2/ρ as required to give the exact two species result. Note that terminating the expansion of Eq. (7–5.22) at the δ_{ik} term does not give Fickian diffusion. The effective diffusion coefficient differs from Fickian by the factor $(\rho - \rho_i)/\rho$. Fickian diffusion may not even give the correct sign for the diffusive flux.

7–6. EVALUATING DIFFUSIVE TRANSPORT COEFFICIENTS

The purpose of this section is to provide useful formulas and references for evaluating transport coefficients in neutral gases. Much of the fundamental work in this area was contributed by Chapman and Cowling (1952) and Hirschfelder, Curtiss, and Bird (1954). The material presented below is based on, extends, or applies this work. There are more recent summaries of material relevant to this section by Picone and Oran (1980), Kailasanath et al. (1982), and Dixon-Lewis (1984).

We first present and discuss standard formulas for evaluating the coefficients. Then as an example, we describe the evaluation of the diffusive transport coefficients needed to calculate the properties of a hydrogen flame. These coefficients are used in the flame calculations presented in Section 13–3.

7–6.1 Thermal Conductivity

For multicomponent reactive flows, we need the value of the coefficient of thermal conductivity appropriate for a mixture of neutral gases, λ_m, and λ_i, the coefficient for a pure gas species, i.

For the thermal conductivity of the mixture, an extremely useful equation has been given by Mason and Saxena (1958),

$$\lambda_m = \sum_i \lambda_i \Big[1 + \frac{1.065}{2\sqrt{2}n_i} \sum_{k\neq i} n_k \phi_{ik}\Big]^{-1} , \qquad (7-6.1)$$

where ϕ_{ik} is given by

$$\phi_{ik} = \frac{\Big[1 + (\lambda_i^0/\lambda_k^0)^{1/2}(m_i/m_k)^{1/4}\Big]^2}{[1 + m_i/m_k]^{1/2}} . \qquad (7-6.2)$$

If species i is a monotonic gas, λ_i^0 is the thermal conductivity of species i. If i is a polyatomic gas, λ_i^0 is the thermal conductivity with the internal degrees of freedom considered "frozen." The $\{m_i\}$ are the atomic masses of the different species.

The thermal conductivity of a pure polyatomic gas, λ_i, can be estimated from the λ_i^0 by using the *Euken factor*, E_i (Hirschfelder, 1957b),

$$\lambda_i = E_i \lambda_i^0 , \qquad (7-6.3)$$

where

$$E_i = 0.115 + 0.354 \, \frac{c_{pi}}{k_B} , \qquad (7-6.4)$$

and c_{pi} is the specific heat at constant pressure. The $\{\lambda_i^0\}$ may be evaluated from

$$\lambda_i^0 = \frac{8.322 \times 10^3}{\sigma_i^2 \Omega_{ii}^{(2,2)*}} \left(\frac{T}{m_i}\right)^{1/2} . \qquad (7-6.5)$$

Here σ_i is the *collision diameter*. The quantity $\Omega_{ii}^{(2,2)*}$ is a *collision integral* normalized to its rigid sphere value. It is a function of T_i^*, the reduced temperature

$$T_i^* = \frac{k_B T}{\epsilon_i} , \qquad (7-6.6)$$

where ϵ_i / k_B is the *potential parameter*, and k_B is Boltzmann's constant. The ϵ_i and the σ_i are constants in the intermolecular potential function describing the interaction between two molecules of type i. In this formulation, the collision integral has been evaluated using a Lennard-Jones potential between molecules, which is discussed at the end of this section. In general, tables of the $\Omega_{ii}^{(2,2)*}$ for a Lennard-Jones 6–12 potential can be found in Hirschfelder, Curtiss, and Bird (1954), and for more general Lennard-Jones potentials in Klein et al. (1974).

The expression given for $\{\lambda_i^0\}$ is excellent for nonpolar molecules and probably adequate for polar molecules. The origin of the uncertainty lies in the evaluation of the Euken factor, which is an average over microscopic states. The derivation of E_i assumes that all states have the same diffusion coefficients and does not allow for distortions in the electron density distribution function due to rotational transitions. These may be significant even at room temperature for polar molecules (Hirschfelder, 1957a,b).

7–6.2 Ordinary (or Molecular or Binary) Diffusion

The quantity D_{ik} represents the coefficient of ordinary diffusion of a pair of species (i, k) when there are only two species present. Although not strictly equal to the diffusion coefficient of (i, k) in a mixture with many species present, we assume that they are equal to the accuracy needed. For further discussion of the differences between binary and multicomponent diffusion coefficients, see Hirschfelder, Curtiss, and Bird (1954).

For some binary mixtures of dilute gases, Mason and Marrero (1972) give a semiempirical expression which describes the variation of D_{ik} over a range of temperatures,

$$D_{ik} = \frac{A_{ik} T^{B_{ik}}}{N} . \qquad (7-6.7)$$

When the coefficients A_{ik} and B_{ik} are not available, the D_{ik} may be estimated from

$$D_{ik} = \frac{2.628 \times 10^{-3}}{P\sigma_{ik}^2 \Omega_{ik}^{(1,1)*}} \left[\frac{T^3(m_i + m_k)}{2m_i m_k} \right]^{1/2} . \tag{7-6.8}$$

The $\Omega_{ik}^{(1,1)*}$ is a collision integral normalized to its rigid sphere value, a function of the reduced temperature

$$T_{ik}^* = \frac{k_B T}{\epsilon_{ik}} . \tag{7-6.9}$$

Again collision diameters and potential parameters are needed, this time between two different types of molecules, i and k. In most cases, the values of σ_{ik} and ϵ_{ik} are not available and are usually estimated from

$$\sigma_{ik} = \frac{1}{2}\left(\sigma_i + \sigma_k\right) \tag{7-6.10}$$

and

$$\epsilon_{ik} = \left(\epsilon_i \epsilon_k\right)^{1/2} . \tag{7-6.11}$$

The collision integral used here again assumes a Lennard-Jones potential, and tables are given in Hirschfelder, Curtiss, and Bird (1954) and Klein et al. (1974).

7–6.3 Thermal Diffusion

Thermal diffusion is a second-order effect which is only important when there are large differences between the atomic masses of the constituent species. The thermal diffusion ratio K_i^T is a measure of the relative importance of thermal to ordinary diffusion. It can be written as (Chapman and Cowling, 1952)

$$K_i^T = \frac{1}{5k_B N^2} \sum_k \frac{\left(6C_{ik}^* - 5\right)}{D_{ik}} \left[\frac{n_i m_i a_k - n_k m_k a_i}{m_i + m_k} \right] . \tag{7-6.12}$$

The quantity a_i is the contribution of the ith species to the thermal conductivity of the mixture (assuming that the internal degrees of freedom are frozen):

$$a_i = \lambda_i^0 \left[1 + \frac{1.065}{2\sqrt{2}n_i} \sum_{k \neq i} n_k \phi_{ik} \right]^{-1} . \tag{7-6.13}$$

In Eq. (7–6.13), ϕ_{ik} is given in Eq. (7–6.2) and C^*_{ik} is a ratio of collision integrals,

$$C^*_{ik} = \frac{\Omega^{(1,2)*}_{ik}}{\Omega^{(1,1)*}_{ik}} . \qquad (7-6.14)$$

The expression for K^T_i has been derived by assuming that the gases in the mixture are monatomic or that the internal degrees of freedom are frozen. However, Chapman and Cowling (1952) conjectured that the internal degrees of freedom of the gas molecules have a smaller effect on the thermal diffusion ratio than on the thermal conductivity.

7–6.4 Viscosity

Two viscosity coefficients, the bulk viscosity κ and the shear viscosity μ, appear in Eq. (7–0.8). The bulk viscosity is generally small, and we ignore it throughout this text. The shear viscosity coefficient is more important. In gases and gas mixtures, the shear viscosity coefficient is also relatively small. Its most important effect is to produce the boundary layer near the interface of two materials. In many simulations, the diffusive effects of viscosity are swamped by numerical diffusion. However, to calculate boundary layers from first principles, it is necessary to resolve the effects of physical viscosity near the interface or boundary.

The most convenient form for the viscosity of a gas mixture has been given by Wilke (1950),

$$\mu_m = \sum_i \mu_i \left[1 + \frac{\sqrt{2}}{4n_i} \sum_{k \neq i} n_k \phi'_{ik}\right]^{-1} . \qquad (7-6.15)$$

Here the summations are over species and μ_i is the viscosity of the pure component i. To first order,

$$\phi'_{ik} = \frac{\left[1 + \left(\mu_i/\mu_k\right)^{1/2} \left(m_k/m_i\right)^{1/4}\right]^2}{\left[1 + \left(m_i/m_k\right)\right]^{1/2}} . \qquad (7-6.16)$$

The individual μ_i may be determined approximately from (Hirschfelder, Curtiss, and Bird, 1954, Chapter 8)

$$\mu_i = \frac{2.67 \times 10^{-5} \sqrt{m_i T}}{\sigma_i^2 \Omega^{(2,2)*}} , \qquad (7-6.17)$$

(where in cgs units μ_i has units of g/cm-s). If $\Omega^{(2,2)*} \approx 1$, Eq. (7–6.16) can be reduced to

$$\phi'_{ik} = \frac{[1 + \sigma_k/\sigma_i]^2}{[1 + (m_i/m_k)]^{1/2}} \,. \qquad (7-6.18)$$

Equation (7–6.15) differs from the form given in Hirschfelder, Curtiss, and Bird in that pure component viscosity coefficients are required instead of binary diffusion coefficients.

7–6.5 Diffusive Transport for Hydrogen Combustion

The mechanism of hydrogen combustion involves the eight reactive species, $H_2, O_2, O, H, OH, HO_2, H_2O_2, H_2O$, and often a diluent such as N_2 or Ar. Here we give the diffusive transport coefficients for the species involved in hydogen-oxygen combustion to illustrate typical values of these quantities. The coefficients given below and used in the numerical simulations of laminar flames are described in Chapter 13.

To calculate the thermal conductivity, the binary diffusion coefficients, and the thermal diffusion ratios, we need the collision integrals $\Omega^{(1,1)*}$, $\Omega^{(1,2)*}$, $\Omega^{(2,2)*}$. These depend on the form of intermolecular potential function and are tabulated for Lennard-Jones 6–12 potentials which are generally used. Thus

$$\phi(r) = 4\epsilon\left[\left(\frac{\sigma}{r}\right)^{12} - \left(\frac{\sigma}{r}\right)^{6}\right] , \qquad (7-6.19)$$

where r is the distance between two molecules, ϵ is the depth of the potential well, the maximum energy of attraction, and σ is the collision diameter for low energy collisions. This potential is adequate for nonpolar molecules, and can approximate polar molecules over limited ranges in r. The collision integrals for this potential function have been tabulated (for example, in Hirschfelder, Curtiss, and Bird (1954) and Klein et al. (1974)).

There is some uncertainty in the values of σ and ϵ. Table 7–1 gives those used for the species involved in hydrogen combustion. Except for σ_{H_2O} and ϵ_{H_2O}, the values have been taken from Svehla (1962). Since H_2O is a polar molecule, the Lennard-Jones potential is a poor approximation, and a Stockmayer-type potential is more appropriate. However, we have assumed that the Lennard-Jones potential is valid for H_2O, but have used the values

Table 7–1. Lennard-Jones 6–12 Parameters

Species	ϵ/k_B	σ (10^{-8})
H	37	3.5
O	106.7	3.05
H_2	59.7	2.827
OH	79.8	3.147
H_2O	260.0	2.8
O_2	106.7	3.467
HO_2	106.7	3.467
H_2O_2	289.3	4.196
N_2	71.4	3.798

given by Dixon-Lewis (1979) for σ_{H_2O} and ϵ_{H_2O} which are approximations to Lennard-Jones potentials. We also used the Dixon-Lewis value of σ_H.

The ordinary diffusion coefficients have been calculated using Eq. (7–6.7), where values of A_{ik} and B_{ik} have been taken from Mason and Marrero (1972). However, no data are available for many pairs of species in the system. For these pairs, the diffusion coefficients can be estimated by noting from Eq. (7–6.8) that if two pairs of species, (i,k) and (l,m), have similar values for the collision integral $\Omega^{(1,1)*}$, then their diffusion coefficients are related by

$$\frac{D_{ik}}{D_{lm}} = \frac{\sigma_{lm}^2}{\sigma_{ik}^2}\left[\frac{m_{lm}}{m_{ik}}\right]^{\frac{1}{2}} , \qquad (7-6.20)$$

where

$$m_{ik} = \frac{m_i m_k}{m_i + m_k} , \qquad (7-6.21)$$

and σ_{ik} has been given in Eq. (7–6.10). However, this procedure only modifies the temperature-independent term, A_{ik}. When $\Omega^{(1,1)*}$ is very different for the two pairs of species, the temperature dependence, and therefore the B_{ik}, also must be modified. The values for A_{ik} and B_{ik} for all the pairs of species for hydrogen combustion are given in Tables 7–2a and 7–2b. Using Eq. (7–6.7) instead of Eq. (7–6.8) avoids using tables to evaluate the collision integral for each pair of species at each temperature.

For the thermal diffusion ratio, Eq. (7–6.12), the ratio of collision integrals C_{ik}^* has been assumed to be unity. Although the collision integrals $\Omega^{(1,1)*}$ and $\Omega^{(1,2)*}$ are different from unity and vary with temperature, their ratio is close to unity and hardly varies with temperature. The collision integrals and their ratio for the temperature range 300 K to 3000 K for the $H - N_2$ pair is given in Table 7–3.

Table 7–2a. Parameters† A_{ik} for Coefficient $D_{ik}=A_{ik}T^{B_{ik}}N^{-1}$

Species	O	H_2	OH	H_2O	O_2	HO_2	H_2O_2	N_2
H	6.30	8.29	6.30	6.70	6.70	6.70	4.43	6.10
O		3.61	1.22	2.73	0.97	0.97	1.57	0.97
H_2			3.49	6.41	3.06	3.06	4.02	2.84
OH				2.73	1.16	.969	1.57	.969
H_2O					2.04	2.04	1.57	1.89
O_2						0.87	1.14	0.83
HO_2							1.14	0.82
H_2O_2								1.14

†All numbers must be multiplied by 10^{17}.

Table 7–2b. Parameters†† B_{ik} for Coefficient $D_{ik}=A_{ik}T^{B_{ik}}N^{-1}$

Species	O	H_2	OH	H_2O	O_2	HO_2	H_2O_2	N_2
H	7.28	7.28	7.28	7.28	7.32	7.32	7.28	7.32
O		7.32	7.74	6.32	7.74	7.74	6.32	7.74
H_2			7.32	6.32	7.32	7.32	6.32	7.38
OH				6.32	7.24	7.74	6.32	7.74
H_2O					6.32	6.32	6.32	6.32
O_2						7.24	6.32	7.24
HO_2							6.32	7.24
H_2O_2								6.32

††All numbers must be multiplied by 10^{-1}.

Table 7–3. Collision Integral Parameters for $H-N_2$

T (K)	T^*	$\Omega_{ik}^{(1,2)*}$	$\Omega_{ik}^{(1,2)*}$	C_{ik}^*
300	5.837	0.76458	0.81740	0.9354
500	9.728	0.70387	0.74570	0.9439
1000	19.456	0.63215	0.66696	0.9478
2000	38.912	0.56789	0.59886	0.9483
3000	58.368	0.53309	0.56233	0.9480

References

Chapman, S., and T.G. Cowling, 1952, *The Mathematical Theory of Nonuniform Gases*, Cambridge University Press, Cambridge, England.

Dixon-Lewis, G., 1979, Mechanism of Inhibition of Hydrogen-Air by Hydrogen Bromide and Its Relevance to the General Problem of Flame Inhibition, *Combust. Flame* 36: 1–14.

Dixon-Lewis, G., 1984, Computer Modeling of Combustion Reactions in Flowing Systems with Transport, in W.C. Gardner, Jr., ed., *Combustion Chemistry*, Springer-Verlag, New York, 21–126.

Hirschfelder, J.O., 1957a, Heat Conductivity in Polyatomic Electronically Excited or Chemically Reacting Mixtures, *6th Symposium (International) on Combustion*, The Combustion Institute, Pittsburgh, PA, 351–366.

Hirschfelder, J.O., 1957b, Heat Transfer in Chemically Reacting Mixtures, I., *J. Chem. Phys.* 26: 274–282.

Hirschfelder, J.O., C.F. Curtiss, and R.B. Bird (1954) *Molecular Theory of Gases and Liquids*, John Wiley and Sons, New York.

Jones, W.W., and J.P. Boris, 1981, An Algorithm for Multispecies Diffusion Fluxes, *Comp. Chem.* 5: 139–146.

Kailasanath, K., E.S. Oran, and J.P. Boris, 1982, *A One-Dimensional Time-Dependent Model for Flame Initiation, Propagation and Quenching*, NRL Memorandum Report No. 4910, Naval Research Laboratory, Washington, DC.

Klein, M., H.J.M. Hanley, F.J. Smith, and P. Holland, 1974, *Tables of Collision Integrals and Second Virial Coefficients for the (m,6,8) Intermolecular Potential Function*, National Standard Reference Data Series, National Bureau of Standards, No. 47, Gaithersberg, MD.

Mason, E.A., and S.C. Saxena, 1958, Approximate Formula for the Thermal Conductivity of Gas Mixtures, *Phys. Fluids* 1: 361–369.

Mason, E.A., and T.R. Marrero, 1972, Gaseous Diffusion Coefficients, *J. Phys. Chem. Reference Data* 1: 3–118.

Picone, J.M., and E.S. Oran, 1980, *Approximate Equations for Transport Coefficients of Multicomponent Mixtures of Neutral Gases*, NRL Memorandum Report No. 4384, Naval Research Laboratory, Washington, DC.

Svehla, R.A., 1962, *Estimated Viscosities and Thermal Conductivities of Gases at High Temperatures*, Technical Report No. R-132, NASA, Washington, DC.

Wilke, C.R., 1950, A Viscosity Equation for Gas Mixtures, *J. Chem. Phys.* 18: 517–519.

Chapter 8

THE CONTINUITY EQUATION: EULERIAN METHODS

Solving a system of coupled, multidimensional continuity equations,

$$\frac{\partial \rho}{\partial t} + \nabla \cdot \mathbf{f}(\rho) = 0 , \qquad (8-0.1)$$

is a fundamental part of simulating reactive flows. Such a system of equations is often called a *system of conservation laws.* Chapter 4 described and analyzed simple methods for solving a one-dimensional form of this equation,

$$\frac{\partial \rho}{\partial t} + \frac{\partial}{\partial x}(\rho v) = 0 , \qquad (8-0.2)$$

where the function $\mathbf{f}$ is a scalar,

$$f = \rho v . \qquad (8-0.3)$$

Figure 4–6 shows results of a typical test problem, a moving discontinuity in $\rho(x,t)$, for five finite-difference methods. The first three methods, the simplest explicit, centered, and fully implicit methods, are reasonable first approaches to solving the problem but give very poor results. There are large amplitude and phase errors which rapidly degrade the solutions until they become unusable. The fourth method is the first-order donor-cell method whose obvious fault is that it is extremely diffusive. The last algorithm analyzed, the Lax-Wendroff algorithm, is less diffusive than the donor-cell algorithm, but this also exposes the phase errors which introduce unphysical oscillations.

In this chapter we concentrate on Eulerian methods for solving the continuity equation. These methods are generally stable if the Courant-Friedrichs-Lewy condition (Courant et al., 1928) is used to limit the timestep Δt according to

$$\Delta t \leq \frac{\Delta x}{|\mathbf{v}|} . \qquad (8-0.4)$$

Here $|\mathbf{v}|$ is the largest characteristic velocity in the system, assuming constant Δx. Although there are many implicit methods, it is not clear that they are suitable for solving hyperbolic fluid equations because information propagates across the grid numerically much faster than it can propagate physically. Sometimes there is a difficult matrix equation to be solved at each timestep. Accordingly, explicit Eulerian methods are most commonly used and the easiest to program. This chapter primarily considers a scalar continuity equation, in one dimension or multidimensions, with the generic form

$$\frac{\partial \rho}{\partial t} + \nabla \cdot \mathbf{f} = 0 \, . \qquad (8-0.5)$$

We are concerned with how well various algorithms for continuity equations convect a discontinuity. Much of the interest in this problem arises from solving problems with shocks or steep species gradients using sets of coupled continuity equations. Shocks and the solution of coupled continuity equations are discussed in Chapter 9.

8–1. LAX-WENDROFF METHODS

We begin by describing the class of conservative finite-difference methods called Lax-Wendroff methods, originally studied by Lax and Wendroff (1960, 1964). These consist of a one-step Lax-Wendroff method and a series of two-step methods. One of these two-step methods, the MacCormack method (MacCormack, 1969, 1971), is widely used in aerodynamic calculations. These methods are generally suitable when the convected functions are fairly smooth. Problems arise when steep gradients require adding diffusion to the solutions to avoid large oscillations and numerical instability. Good general references are Richtmyer and Morton (1967), Roache (1982), and Potter (1973).

8–1.1 The Basic One-Step Method

In Chapter 4 we used a simple Lax-Wendroff method to solve the moving step or "square wave" problem. The class of Lax-Wendroff methods is based on a Taylor-Series expansion of the variable ρ in time,

$$\rho(\mathbf{x}, t^{n+1}) = \rho(\mathbf{x}, t^n) + \Delta t \frac{\partial}{\partial t} \rho(\mathbf{x}, t^n) + \frac{1}{2} \Delta t^2 \frac{\partial^2}{\partial t^2} \rho(\mathbf{x}, t^n) + \mathcal{O}(\Delta t^3) \, . \quad (8-1.1)$$

We now change notation to write the expansion for the scalar density ρ as

$$\rho^{n+1}(\mathbf{x}) = \rho^n(\mathbf{x}) + \Delta t \frac{\partial}{\partial t} \rho^n(\mathbf{x}) + \frac{1}{2} \Delta t^2 \frac{\partial^2}{\partial t^2} \rho^n(\mathbf{x}) + \mathcal{O}(\Delta t^3) \, , \quad (8-1.2)$$

where $\rho^{n+1}(\mathbf{x})$ represents $\rho(\mathbf{x}, t^{n+1})$. The next step is to express $\partial \rho / \partial t$ and $\partial^2 \rho / \partial t^2$ in terms of spatial gradients using Eq. (8–0.5).

In one dimension,

$$\frac{\partial \rho}{\partial t} = -\frac{\partial}{\partial x} f(\rho) . \tag{8-1.3}$$

Then

$$\frac{\partial^2 \rho}{\partial t^2} = \frac{\partial}{\partial t}\left(-\frac{\partial f}{\partial x}\right) = -\frac{\partial}{\partial x}\frac{\partial f}{\partial t} = -\frac{\partial}{\partial x}\left(-A\frac{\partial f}{\partial x}\right) = \frac{\partial}{\partial x}A\frac{\partial f}{\partial x} , \tag{8-1.4}$$

where

$$\frac{\partial f}{\partial t} = -A\frac{\partial f}{\partial x} , \tag{8-1.5}$$

using

$$A \equiv \frac{\partial f}{\partial \rho} . \tag{8-1.6}$$

For example, in the density continuity equation,

$$\begin{aligned} f &= \rho v , \\ A &= v , \\ \frac{\partial \rho}{\partial t} &= -\frac{(\partial \rho v)}{\partial x} , \end{aligned} \tag{8-1.7}$$

$$\frac{\partial^2 \rho}{\partial t^2} = \frac{\partial}{\partial x}\left(v\frac{\partial \rho v}{\partial x}\right) .$$

In multidimensions, this generalizes to

$$\frac{\partial \rho}{\partial t} = -\nabla \cdot \mathbf{f} \tag{8-1.8}$$

and

$$\frac{\partial^2 \rho}{\partial t^2} = \nabla \cdot (\mathbf{A} \cdot (\nabla \mathbf{f})) . \tag{8-1.9}$$

In one dimension, the Lax-Wendroff difference approximation to Eq. (8–1.2) becomes

$$\rho^{n+1} \cong \rho^n - \Delta t\frac{\partial f}{\partial x} + \frac{1}{2}\Delta t^2\frac{\partial}{\partial x}A\frac{\partial f}{\partial x} , \tag{8-1.10}$$

with

$$\frac{\partial f}{\partial x} \cong \frac{f_{i+1} - f_{i-1}}{2\Delta x} \tag{8-1.11}$$

and

$$\frac{\partial}{\partial x}A\frac{\partial f}{\partial x} \cong \frac{1}{\Delta x}\left(A_{i+\frac{1}{2}}\frac{(f_{i+1} - f_i)}{\Delta x} - A_{i-\frac{1}{2}}\frac{(f_i - f_{i-1})}{\Delta x}\right) \tag{8-1.12}$$

where

$$A_{i\pm\frac{1}{2}} \equiv \frac{1}{2}(A_i + A_{i\pm 1}) . \tag{8-1.13}$$

8–1.2 Two-Step Richtmyer Methods

As can be seen by considering the one-dimensional expansions, differencing the two-dimensional equations will be even more complicated using the one-step methods. These are discussed by Emery (1968). There is an approach to the Lax-Wendroff method, however, which is easy to program in one dimension and extends to two and three dimensions.

Richtmyer (1963) showed that the Lax-Wendroff methods can be written as two-step procedures in which the first step is a Lax method and the second step is a leapfrog method. The one-dimensional two-step Lax-Wendroff method is

$$\begin{aligned} \rho_i^{n+1} &= \frac{1}{2}\left[\rho_{i+1} + \rho_{i-1}\right]^n - \Delta t\left[\frac{f_{i+1} - f_{i-1}}{2\Delta x}\right]^n, \qquad \text{for odd } i \\ \rho_i^{n+2} &= \rho_i^n - 2\Delta t\left[\frac{f_{i+1} - f_{i-1}}{2\Delta x}\right]^{n+1}, \qquad \text{for even } i, \end{aligned} \tag{8 – 1.14}$$

where the superscript n, as above, indicates that the quantities in brackets are evaluated at the discrete time level t^n. The first step is a provisional step. The $\{f_{i\pm1}^{n+1}\}$ in the second step are evaluated using the $\{\rho_{i\pm1}^{n+1}\}$ in the first step. Substituting the first step into the second step recovers the one-step Lax-Wendroff method for the linear continuity equation with constant velocity v. The more common way of using Eq. (8–1.14) is

$$\begin{aligned} \rho_{i+\frac{1}{2}}^{n+\frac{1}{2}} &= \frac{1}{2}\left[\rho_{i+1} + \rho_i\right]^n - \frac{\Delta t}{2}\left[\frac{f_{i+1} - f_i}{\Delta x}\right]^n \\ \rho_i^{n+1} &= \rho_i^n - \Delta t\left[\frac{f_{i+\frac{1}{2}} - f_{i-\frac{1}{2}}}{\Delta x}\right]^{n+\frac{1}{2}}. \end{aligned} \tag{8 – 1.15}$$

In this form, the half-step values $\{\rho_{i+\frac{1}{2}}^{n+\frac{1}{2}}\}$ are evaluated on a staggered grid and then used to evaluate centered derivatives on the original grid to determine $\{\rho_i^{n+1}\}$.

This two-step method generalizes to multidimensions rather easily. In two-dimensions

$$\begin{aligned}
\rho_{i+\frac{1}{2},j}^{n+\frac{1}{2}} &= \frac{1}{2}\left[\rho_{i,j}+\rho_{i+1,j}\right]^n \\
&\quad - \frac{\Delta t}{2}\left[\frac{(f_x)_{i+1,j}-(f_x)_{i,j}}{\Delta x}+\frac{(f_y)_{i+\frac{1}{2},j+1}-(f_y)_{i+\frac{1}{2},j-1}}{2\Delta y}\right]^n \\
\rho_{i,j+\frac{1}{2}}^{n+\frac{1}{2}} &= \frac{1}{2}\left[\rho_{i,j}+\rho_{i,j+1}\right]^n \\
&\quad - \frac{\Delta t}{2}\left[\frac{(f_x)_{i+1,j+\frac{1}{2}}-(f_x)_{i-1,j+\frac{1}{2}}}{2\Delta x}+\frac{(f_y)_{i,j+1}-(f_y)_{i,j}}{\Delta y}\right]^n \\
\rho_{i,j}^{n+1} &= \rho_{i,j}^n \\
&\quad - \Delta t\left[\frac{(f_x)_{i+\frac{1}{2},j}-(f_x)_{i-\frac{1}{2},j}}{\Delta x}+\frac{(f_y)_{i,j+\frac{1}{2}}-(f_y)_{i,j-\frac{1}{2}}}{\Delta y}\right]^{n+\frac{1}{2}} .
\end{aligned} \tag{8 – 1.16}$$

The half-step values in the first two equations may be defined, for example, as

$$\begin{aligned}
f_{i\pm\frac{1}{2}}^{n+\frac{1}{2}} &= f\left(\rho_{i\pm\frac{1}{2}}^{n+\frac{1}{2}}\right) \\
&\equiv f\left(\frac{1}{2}\left(\rho_i^{n+\frac{1}{2}}+\rho_{i\pm1}^{n+\frac{1}{2}}\right)\right) ,
\end{aligned} \tag{8 – 1.17}$$

or as

$$f_{i+\frac{1}{2}}^{n+\frac{1}{2}} = \frac{1}{2}\left[f\left(\rho_i^{n+\frac{1}{2}}\right)+f\left(\rho_{i+1}^{n+\frac{1}{2}}\right)\right] , \tag{8 – 1.18}$$

where f with no subscripts or superscripts indicates "a function of" the quantity in parentheses. Emery (1968) showed that this method is about four times faster than the one-step Lax-Wendroff in two dimensions. A number of modifications of the basic two-step method are described and referenced in Roache (1982). In particular, Gottlieb and Turkel (1976) extended the method to fourth-order in space, and in a different approach, Zalesak (1984) extended it to arbitrary accuracy.

8–1.3 MacCormack Method

The basic, one-dimensional form of the two-step MacCormack method (1969) is

$$\begin{aligned} \rho_i^* &= \rho_i^n - \Delta t \left[\frac{f_{i+1}^n - f_i^n}{\Delta x} \right] \\ \rho_i^{n+1} &= \frac{1}{2} \left\{ \rho_i^n + \rho_i^* - \Delta t \left[\frac{f_i^* - f_{i-1}^*}{\Delta x} \right] \right\} . \end{aligned} \tag{8 – 1.19}$$

The idea here is to use a forward differencing method in the downwind direction as a predictor, to obtain provisional values ρ^*, and a backward differencing method for the value at the next timestep. This forward-backward order gives a different result from the backward-forward order, and gives somewhat better results than the two-step Lax-Wendroff algorithm for convecting discontinuities.

In the spirit of the Richtmyer methods, the MacCormack method may be used to find provisional values at the half timestep, and then final values at the whole timestep. However, it can also be used over two whole timesteps, where the provisional value is taken as the value at one timestep, and the final value is taken as the value at the next timestep. In the case of a constant velocity, linear problem, this formula reduces to the two-step Lax-Wendroff method given in Eq. (8–1.15). Therefore the Lax-Wendroff calculation in Figure 4–6 shows how the MacCormack method performs on a simple convection test problem.

There are many ways to extend this method to two dimensions. For example, the forward and backward steps can be applied differently in the x and y directions. The process could also be applied cyclically over two or four complete timesteps. Details are given in MacCormack (1971) and a derivation is given in Kutler and Lomax (1971).

8–1.4 Artificial Diffusion

Second-order methods, such as the Lax-Wendroff or MacCormack methods, are only slightly diffusive but are quite dispersive and susceptible to nonlinear instabilities. In Chapter 4, we saw that unphysical oscillations can appear in regions of high gradients. These numerical ripples often cause a positive definite quantity to become negative. In some problems this may be acceptable, but in others, especially problems in which a number of physical processes are coupled, this produces unacceptable solutions.

The traditional approach is to introduce extra diffusion into the algorithm which damps the spurious oscillations in regions of large gradients.

This is, at best, a compromise because numerical diffusion also spreads shocks and discontinuities over more computational cells as overshoots and undershoots are damped. If the oscillations are totally damped, the solution shows the same numerical diffusion as the linear first-order methods. On the other hand, if the oscillations are not totally damped, little has been gained over standard high-order methods.

Artificial viscosity methods were first introduced by von Neumann and Richtmyer (1950). They dealt with a problem in coupled continuity equations representing shock propagation. This approach introduces an added viscosity into the momentum and energy equations and attempts to adjust this viscosity so that short wavelengths are damped but long wavelengths are not affected much. We do not present the artificial viscosity methods because the later development of monotone methods makes them unnecessary. Besides the original von Neumann and Richtmyer (1950) reference, other good references to the artificial viscosity method are Richtmyer and Morton (1967) and Potter (1973). We particularly recommend Potter for a good introduction to the subject.

8–2. POSITIVITY AND ACCURACY

Accurate resolution of steep gradients is important in most of the problems we need to solve. It is particularly important in reactive flows, where the gradients at detonation fronts, flame fronts, and material gradients at interfaces in multiphase flows must be accurately represented. Flame speeds depend on steep species gradients, as do the local energy release profiles. Unless Eulerian calculations have very fine meshes, the continual passage of fluid through an Eulerian grid ensures a substantial amount of numerical diffusion, which arises as a consequence of the requirements that the profiles being convected remain stable while guaranteeing positivity.

Figure 8–1 shows how numerical diffusion enters the first-order *upwind* or *donor-cell* algorithm introduced in Chapter 4. A discontinuity, at $x = 0$ at time $t = 0$, moves at a constant velocity from left to right. The velocity, v, the timestep, Δt, and computational cell size, Δx, are chosen such that $v\Delta t/\Delta x = 1/3$. This means that the correct discontinuity travels one third of a cell per timestep.

The solution obtained using the donor-cell algorithm, given by the solid line, is found from the finite-difference formula

$$\rho_i^{n+1} = \rho_i^n - \frac{v\Delta t}{\Delta x}(\rho_i^n - \rho_{i-1}^n) . \qquad (8-2.1)$$

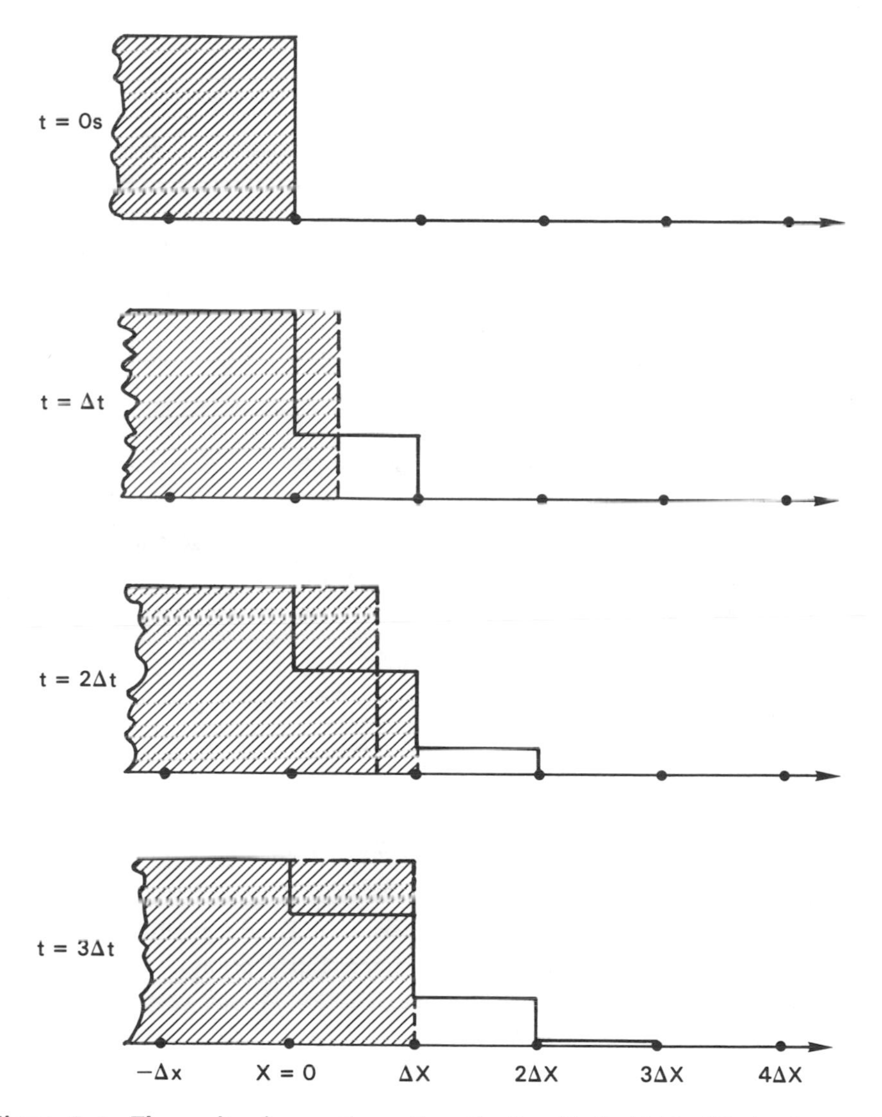

Figure 8–1. The results of convecting a discontinuity with the highly diffusive donor-cell algorithm. The heavy dashed line is the exact profile, which coincides with the numerical solution at t=0 s. The heavy solid line is the numerical solution.

If the $\{\rho_i\}$ are positive at some time $t = n\ \Delta t$ and

$$\left| \frac{v \Delta t}{\Delta x} \right| \leq 1 \qquad (8-2.2)$$

in each cell, the new density values $\{\rho_i^{n+1}\}$ at time $t = (n+1)\Delta t$ are also positive. The price for guaranteed positivity is a severe, unphysical spreading of the discontinuity which should be located at $x = vt$.

In the example shown in Figure 8–1, the initial discontinuity appears to crumble away. This process looks like physical diffusion, but it arises here from a purely numerical source. The numerical diffusion occurs because material that has just entered a cell, and should still be near the left boundary, becomes smeared over the whole cell when the transported fluid elements are interpolated back onto the Eulerian grid. Higher-order approximations to the convective derivatives are required to reduce this diffusion.

Now consider a three-point explicit finite-difference formula for advancing $\{\rho_i^n\}$ one timestep to $\{\rho_i^{n+1}\}$,

$$\rho_i^{n+1} = a_i \rho_{i-1}^n + b_i \rho_i^n + c_i \rho_{i+1}^n \, . \qquad (8-2.3)$$

This general form includes the donor-cell algorithm and other common higher-order algorithms. As before, Δx and Δt are constants. Equation (8–2.3) can be rewritten in a form that guarantees conservation,

$$\begin{aligned} \rho_i^{n+1} &= \rho_i^n - \frac{1}{2}\left[\epsilon_{i+\frac{1}{2}}(\rho_{i+1}^n + \rho_i^n) - \epsilon_{i-\frac{1}{2}}(\rho_i^n + \rho_{i-1}^n)\right] \\ &\quad + \left[\nu_{i+\frac{1}{2}}(\rho_{i+1}^n - \rho_i^n) - \nu_{i-\frac{1}{2}}(\rho_i^n - \rho_{i-1}^n)\right] , \end{aligned} \qquad (8-2.4)$$

where

$$\epsilon_{i+\frac{1}{2}} \equiv v_{i+\frac{1}{2}} \frac{\Delta t}{\Delta x} , \qquad (8-2.5)$$

and the $\{\nu_{i+\frac{1}{2}}\}$ are nondimensional numerical diffusion coefficients which appear as a consequence of considering adjacent grid points. Conservation of ρ in Eq. (8–2.4) also constrains the coefficients a_i, b_i, and c_i in Eq. (8–2.3) by the condition

$$a_{i+1} + b_i + c_{i-1} = 1 \, . \qquad (8-2.6)$$

Positivity of $\{\rho_i^{n+1}\}$ for all possible positive profiles $\{\rho_i^n\}$ requires that $\{a_i\}$, $\{b_i\}$, and $\{c_i\}$ be positive for all i.

Matching corresponding terms in Eqs. (8–2.4) and (8–2.3) gives

$$\begin{aligned} a_i &\equiv \nu_{i-\frac{1}{2}} + \frac{1}{2}\epsilon_{i-\frac{1}{2}} , \\ b_i &\equiv 1 - \frac{1}{2}\epsilon_{i+\frac{1}{2}} + \frac{1}{2}\epsilon_{i-\frac{1}{2}} - \nu_{i+\frac{1}{2}} - \nu_{i-\frac{1}{2}} , \\ c_i &\equiv \nu_{i+\frac{1}{2}} - \frac{1}{2}\epsilon_{i+\frac{1}{2}} \, . \end{aligned} \qquad (8-2.7)$$

If the $\{\nu_{i+\frac{1}{2}}\}$ are positive and large enough, they ensure that the $\{\rho_i^{n+1}\}$ are positive. The positivity conditions derived from Eqs. (8–2.7) are

$$
\begin{aligned}
|\epsilon_{i+\frac{1}{2}}| &\leq \frac{1}{2} , \\
\frac{1}{2} \geq \nu_{i+\frac{1}{2}} &\geq \frac{1}{2} |\epsilon_{i+\frac{1}{2}}| ,
\end{aligned}
\tag{8 – 2.8}
$$

for all i. Thus the condition in Eq. (8–2.8) for *positivity* leads directly to numerical diffusion in addition to the desired convection,

$$
\begin{aligned}
\rho_i^{n+1} &= \rho_i^n + \nu_{i+\frac{1}{2}}(\rho_{i+1}^n - \rho_i^n) - \nu_{i-\frac{1}{2}}(\rho_i^n - \rho_{i-1}^n) \\
&\quad + \text{convection} ,
\end{aligned}
\tag{8 – 2.9}
$$

where Eq. (8–2.8) holds. This first-order numerical diffusion rapidly smears a sharp discontinuity. If algorithms are used with $\nu_{i+\frac{1}{2}} < \frac{1}{2}|\epsilon_{i+\frac{1}{2}}|$, positivity is not necessarily destroyed but can no longer be guaranteed. In practice, the positivity conditions are almost always violated by strong shocks and discontinuities unless the inequalities stated in Eq. (8–2.8) hold. Nevertheless, the numerical diffusion implied by Eq. (8–2.8) is unacceptable. The diffusion coefficient $\{\nu_{i+\frac{1}{2}}\}$, however, cannot be zero, because the explicit three-point formula, Eq. (8–2.4), is subject to a numerical stability problem. Finite-difference methods which are higher than first order, such as the Lax-Wendroff methods, reduce the numerical diffusion but sacrifice assured positivity. This is a dilemma for which some remedy is needed.

To examine the problem of stability and positivity, we consider a stability analysis of the type used in Chapter 4. Consider convecting test functions of the form

$$
\rho_i^n = \rho_o^n e^{\mathrm{i}i\beta} ,
\tag{8 – 2.10}
$$

where

$$
\beta \equiv k\,\Delta x = \frac{2\pi\,\Delta x}{\lambda} ,
\tag{8 – 2.11}
$$

and i indicates $\sqrt{-1}$. Substituting this solution into Eq. (8–2.4) gives

$$
\rho_o^{n+1} = \rho_o^n \left[1 - 2\nu(1 - \cos\beta) - \mathrm{i}\epsilon \sin\beta\right] ,
\tag{8 – 2.12}
$$

where we assume that

$$
\begin{aligned}
\{\nu_{i+\frac{1}{2}}\} &= \nu \\
\{\epsilon_{i+\frac{1}{2}}\} &= \epsilon .
\end{aligned}
\tag{8 – 2.13}
$$

The exact theoretical solution to this linear problem is

$$
\rho_o^{n+1}|_{\text{exact}} = \rho_o^n e^{-\mathrm{i}kv\Delta t}
\tag{8 – 2.14}
$$

Therefore the difference between the exact solution and Eq. (8–2.12) is the numerical error generated at each timestep.

The amplification factor was defined as

$$A \equiv \frac{\rho_o^{n+1}}{\rho_o^n} , \qquad (8-2.15)$$

and an algorithm is always linearly stable if

$$|A|^2 \leq 1 . \qquad (8-2.16)$$

From Eq. (8–2.12),

$$|A|^2 = 1 - (4\nu - 2\epsilon^2)(1 - \cos\beta) + (4\nu^2 - \epsilon^2)(1 - \cos\beta)^2 , \qquad (8-2.17)$$

which must be less than unity for all permissible values of β between 0 and π. In general, $\nu > \frac{1}{2}\epsilon^2$ ensures stability of the linear convection algorithm for any Fourier harmonic of the disturbance, provided that Δt is chosen so that $|\epsilon| \leq 1$. This stability condition is a factor of two less stringent than the positivity conditions $|\epsilon| \leq \frac{1}{2}$. When $\nu > \frac{1}{2}$, there are combinations of ϵ and β where $|A|^2 > 1$, for example $\epsilon = 0$ with $\beta = \pi$. Thus the range of acceptable diffusion coefficients is quite closely prescribed,

$$\frac{1}{2} \geq \nu \geq \frac{1}{2}|\epsilon| \geq \frac{1}{2}\epsilon^2 . \qquad (8-2.18)$$

Even the minimal numerial diffusion required for linear stability, $\nu = \frac{1}{2}\epsilon^2$, may be substantial when compared to the physically correct diffusion effects such as thermal conduction, molecular diffusion, or viscosity. Figure 8–2 shows the first few timesteps from the same test problem as in Figure 8–1, but using $\nu = \frac{1}{2}\epsilon^2$ rather than $\nu = \frac{1}{2}\epsilon$ required for positivity. The spreading of the profile is only one third as much as in the previous case where positivity was assured linearly, but a numerical precursor still reaches two cells beyond the correct discontinuity location. Furthermore, the overshoot between $x = -\Delta x$ and $x = 0$ in Figure 8–2 is a consequence of underdamping the solution. The loss of *monotonicity* indicated by the overshoot can be as bad as violating positivity. A new, nonphysical maximum in ρ has been introduced into the solution. When the convection algorithm is stable but not positive, the numerical diffusion is not large enough to mask either numerical dispersion or the Gibbs phenomenon so the solution is no longer necessarily monotone. New ripples, that is, new maxima or minima, are introduced numerically.

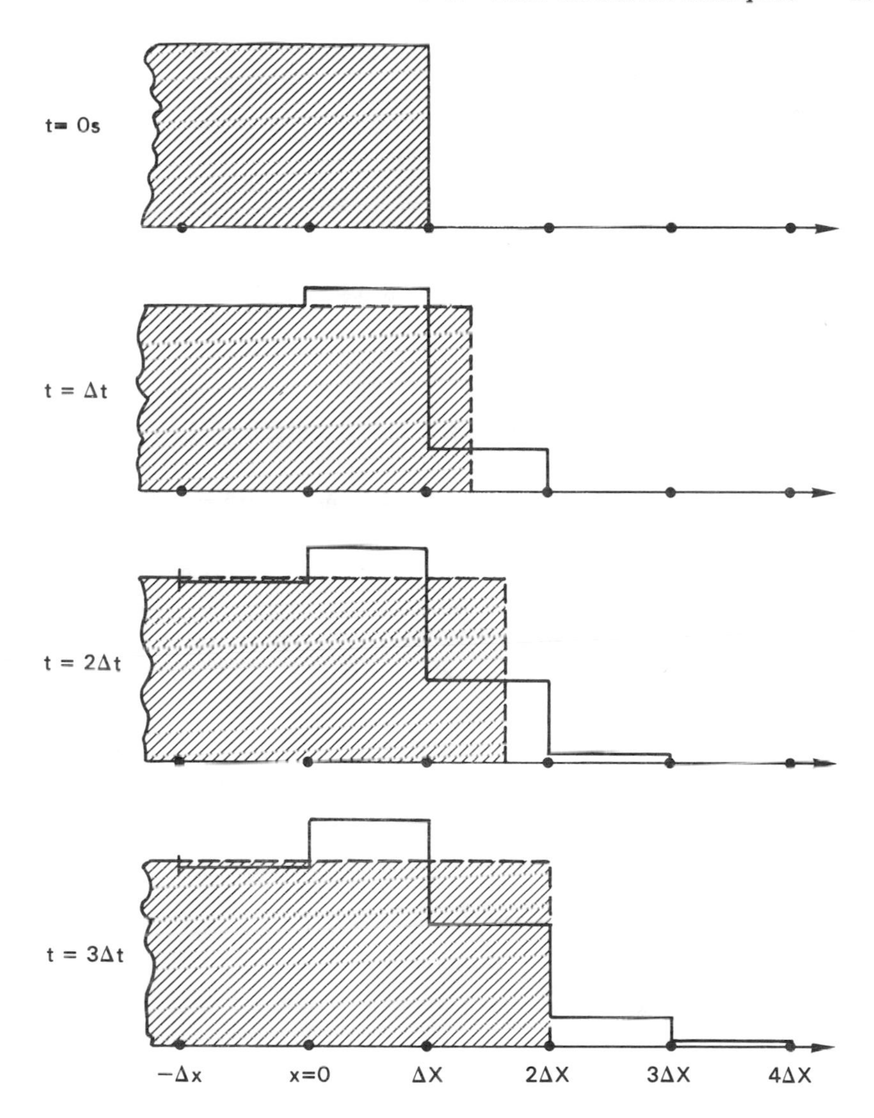

Figure 8–2. Results of convecting a discontinuity with an algorithm with enough diffusion to maintain stability, but not enough to hide the effects of dispersion. Note the growing nonphysical overshoot and the diffusive numerical precursor at times after t=0 s.

8–3. FLUX-CORRECTED TRANSPORT

From the discussion above, the requirements of positivity and accuracy seem to be mutually exclusive. *Nonlinear monotone methods* were invented to circumvent this dilemma. These methods use the stabilizing $\nu = \frac{1}{2}\epsilon^2$ diffusion where monotonicity is not threatened, and increase ν to values approaching

$\nu = \frac{1}{2}|\epsilon|$ when required to assure that the solution remains monotone. Different criteria are imposed in the same timestep at different locations on the computational grid according to the local profiles of the physical solution. *The dependence of the local smoothing coefficients ν on the solution profile makes the overall algorithm nonlinear.*

To prevent negative values of ρ which could arise from dispersion or Gibbs errors, a minimum amount of numerical diffusion must be added to assure positivity and stability at each timestep. We write this minimal diffusion as

$$\nu \approx \frac{|\epsilon|}{2}\left(c + |\epsilon|\right) \tag{8-3.1}$$

where c is a *clipping factor*, $0 \leq c \leq 1 - |\epsilon|$, controlling how much extra diffusion, over the stability limit, must be added to ensure positivity. In the vicinity of steep discontinuities, $c \approx 1 - |\epsilon|$, and in smooth regions away from local maxima and minima, $c \approx 0$.

Monotone algorithms are a reliable, robust way to calculate convection. The first specifically monotone positivity-preserving techniques was the Flux-Corrected Transport (FCT) algorithm developed by Boris and Book (see, for example, Boris, 1971; Boris and Book, 1973, 1976). This was followed closely by the work of van Leer (see, for example, van Leer, 1973, 1979), and Harten (1974, 1983). There have been a series of other monotone methods, several of which are described below. General references to these methods are given in Woodward and Colella (1984a), Baer (1986), and Rood (1987).

8–3.1 The Basic Idea of Flux-Corrected Transport

We now rewrite the explicit three-point approximation to the continuity equation given in Eq. (8–2.3) to determine provisional values, $\tilde{\rho}$, and previous timestep or "old" values, ρ^o,

$$\tilde{\rho}_i = a_i\rho^o_{i-1} + b_i\rho^o_i + c_i\rho^o_{i+1} \,. \tag{8-3.2}$$

Again, Eq. (8–2.6) must be satisfied for conservation and $\{a_i\}$, $\{b_i\}$, and $\{c_i\}$ must all be greater than or equal to zero to assure positivity.

Equation (8–3.2), in conservative form, again becomes

$$\begin{aligned}\tilde{\rho}_i &= \rho^o_i - \frac{1}{2}\left[\epsilon_{i+\frac{1}{2}}(\rho^o_{i+1} + \rho^o_i) - \epsilon_{i-\frac{1}{2}}(\rho^o_i + \rho^o_{i-1})\right] \\ &\quad + \left[\nu_{i+\frac{1}{2}}(\rho^o_{i+1} - \rho^o_i) - \nu_{i-\frac{1}{2}}(\rho^o_i - \rho^o_{i-1})\right] \\ &= \rho^o_i - \frac{1}{\Delta x}\left[f_{i-\frac{1}{2}} - f_{i+\frac{1}{2}}\right] \,.\end{aligned} \tag{8-3.3}$$

The values of variables at interface $i + \frac{1}{2}$ are averages of values at cells $i + 1$ and i, and the values at $i - \frac{1}{2}$ are averages of values at cells i and $i - 1$. At every cell i, the $\tilde{\rho}_i$ differs from ρ_i^o as a result of the inflow and outflow of the fluxes of ρ, denoted $f_{i\pm\frac{1}{2}}$. The fluxes are successively added and subtracted along the array of densities ρ_i^o so that the conservation of ρ condition is satisfied by construction. Summing all the provisional densities gives the sum of the old densities. The expressions involving $\epsilon_{i\pm\frac{1}{2}}$ are called the *convective fluxes.*

By comparing Eqs. (8–3.3) and (8–3.2), we obtain the conditions relating the a, b, and c's to the ϵ's and ν's, essentially as in Eqs. (8–2.7). In Eqs. (8–3.3), the $\{\nu_{i+\frac{1}{2}}\}$ are dimensionless diffusion coefficients included to ensure positivity of the provisional values $\{\tilde{\rho}_i\}$. The positivity condition for the provisional $\{\tilde{\rho}_i\}$ is given in Eq. (8–2.8).

However, after Eq. (8–3.2) is imposed, two of the three coefficients in Eqs. (8–3.3) are still to be determined. One of these sets of coefficients must ensure an accurate representation of the mass flux terms. Thus

$$\epsilon_{i+\frac{1}{2}} = v_{i+\frac{1}{2}} \frac{\Delta t}{\Delta x} , \qquad (8-3.4)$$

where, $\{v_{i+\frac{1}{2}}\}$ is the fluid velocity approximated at the cell interfaces. The other set of coefficients, $\{\nu_{i+\frac{1}{2}}\}$, are chosen to maintain positivity and stability.

The provisional values $\tilde{\rho}_i$ must be strongly diffused to ensure positivity. If $\nu_{i+\frac{1}{2}} = \frac{1}{2}|\epsilon_{i+\frac{1}{2}}|$ in Eq. (8–2.8), we have the diffusive donor-cell algorithm. A correction to remove this strong diffusion uses an additional *antidiffusion* stage,

$$\rho_i^n = \tilde{\rho}_i - \mu_{i+\frac{1}{2}}(\tilde{\rho}_{i+1} - \tilde{\rho}_i) + \mu_{i-\frac{1}{2}}(\tilde{\rho}_i - \tilde{\rho}_{i-1}) , \qquad (8-3.5)$$

in the algorithm to get the new values of $\{\rho_i^n\}$. Here $\{\mu_{i+\frac{1}{2}}\}$ are positive *antidiffusion coefficients.* Antidiffusion reduces the strong diffusion implied by Eq. (8–2.8), but also reintroduces the possibility of negative values or nonphysical overshoots in the new profile. If the values of $\{\mu_{i+\frac{1}{2}}\}$ are too large, the new solution $\{\rho_i^n\}$ will be unstable.

To obtain a positivity-preserving algorithm, we modify the antidiffusive fluxes in Eq. (8–3.5) by a process that we call *flux correction.* The antidiffusive fluxes,

$$f_{i+\frac{1}{2}}^{ad} \equiv \mu_{i+\frac{1}{2}} \, (\tilde{\rho}_{i+1} - \tilde{\rho}_i) , \qquad (8-3.6)$$

appearing in Eq. (8–3.5) are *limited* as described below to ensure positivity and stability.

The biggest linear choice of the antidiffusion coefficients $\{\mu_{i+\frac{1}{2}}\}$ that still guarantees positivity is

$$\mu_{i+\frac{1}{2}} \approx \nu_{i+\frac{1}{2}} - \frac{1}{2} |\epsilon_{i+\frac{1}{2}}| \, . \qquad (8-3.7)$$

However, this is not large enough. To reduce the residual diffusion $(\nu - \mu)$ even further, the flux correction must be nonlinear, depending on the actual values of the density profile $\{\tilde{\rho}_i\}$.

The idea behind the nonlinear flux-correction formula is as follows: Suppose the density $\tilde{\rho}_i$ at grid point i reaches zero while its neighbors are positive. Then the second derivative is locally positive and any antidiffusion would force the minimum density value $\tilde{\rho}_i = 0$ to be negative. Because this cannot be allowed physically, the antidiffusive fluxes should be limited so minima in the profile are made no deeper by the antidiffusive stage of Eq. (8–3.5). Because the continuity equation is linear, we could equally well solve for $\{-\rho_i^n\}$. Hence, we also must require that antidiffusion not make the maxima in the profile any larger. These two conditions form the basis for FCT and are the core of other monotone methods. *The antidiffusion stage should not generate new maxima or minima in the solution, nor accentuate already existing extrema.*

This qualitative idea for nonlinear filtering can be quantified. The new values $\{\rho_i^n\}$ are given by

$$\rho_i^n = \tilde{\rho}_i - f_{i+\frac{1}{2}}^c + f_{i-\frac{1}{2}}^c \, , \qquad (8-3.8)$$

where the corrected fluxes $\{f_{i+\frac{1}{2}}^c\}$ satisfy

$$f_{i+\frac{1}{2}}^c \equiv S \cdot \max \left\{ 0, \, \min \left[S \cdot (\tilde{\rho}_{i+2} - \tilde{\rho}_{i+1}), \, |f_{i+\frac{1}{2}}^{ad}|, \, S \cdot (\tilde{\rho}_i - \tilde{\rho}_{i-1}) \right] \right\} . \qquad (8-3.9)$$

Here $|S| = 1$ and sign $S \equiv$ sign $(\tilde{\rho}_{i+1} - \tilde{\rho}_i)$.

To see what this flux-correction formula does, assume that $(\tilde{\rho}_{i+1} - \tilde{\rho}_i)$ is greater than zero. Then Eq. (8–3.9) gives either

$$\begin{aligned} f_{i+\frac{1}{2}}^c &= \min \left[(\tilde{\rho}_{i+2} - \tilde{\rho}_{i+1}), \, \mu_{i+\frac{1}{2}} (\tilde{\rho}_{i+1} - \tilde{\rho}_i), \, (\tilde{\rho}_i - \tilde{\rho}_{i-1}) \right] \\ &\text{or} \\ f_{i+\frac{1}{2}}^c &= 0 \, , \end{aligned} \qquad (8-3.10)$$

whichever is larger. The "raw" antidiffusive flux, $f_{i+\frac{1}{2}}^{ad}$ given in Eq. (8–3.6), always tends to decrease ρ_i^n and to increase ρ_{i+1}^n. The flux-limiting formula merely ensures that the corrected flux cannot push ρ_i^n below ρ_{i-1}^n, which would produce a new minimum, or push ρ_{i+1}^n above ρ_{i+2}^n, which would produce a new maximum. Equation (8–3.9) is constructed to take care of all cases of sign and slope.

The formulation of an FCT algorithm therefore consists of the following four sequential stages:

1. Compute the transported and diffused values $\tilde{\rho}_i$ from Eqs. (8–3.3), where the $\nu_{i+\frac{1}{2}} > \frac{1}{2}|\epsilon_{i+\frac{1}{2}}|$ to satisfy monotonicity. Add in any additional source terms, for example, $-\nabla P$.
2. Compute the raw antidiffusive fluxes from Eq. (8–3.6).
3. Correct or limit these fluxes using Eq. (8–3.9) to assure monotonicity.
4. Perform the indicated antidiffusive correction through Eq. (8–3.8).

Stages 3 and 4 are the new components introduced by FCT. There are many modifications of this prescription that accentuate various properties of the solution. Some of these are summarized in Boris and Book (1976) and more recently by Zalesak (1979, 1981).

8–3.2 The Subroutine LCPFCT

We now discuss the program LCPFCT (Boris et al., 1987), implemented as a Fortran subroutine for solving the continuity equation. This is an updated version of the program ETBFCT (Boris, 1976) and is available on request. It gives an explicit solution of the general one-dimensional continuity equation,

$$\frac{\partial \rho}{\partial t} = -\frac{1}{r^{\alpha-1}}\frac{\partial}{\partial r}(r^{\alpha-1}\rho v) - \frac{1}{r^{\alpha-1}}\frac{\partial}{\partial r}(r^{\alpha-1}D_1) + C_2\frac{\partial D_2}{\partial r} + D_3, \quad (8-3.11)$$

that allows the possibility of a variable and moving grid. Additional source terms are included by means of the terms D_1, D_2, and D_3. Finally, variable, one-dimensional geometries are permitted: $\alpha = 1$ is Cartesian or planar geometry, $\alpha = 2$ is cylindrical geometry, and $\alpha = 3$ is spherical geometry.

Because more complexity is demanded of the algorithm, its description is necessarily more complicated than the convection algorithms given in Chapter 4. It is surprising that only two or three times more computing is necessary.

Figure 8–3 shows a one-dimensional geometry in which the fluid is constrained to move along a tube. The variable r measures length along the tube. The velocity v^f is the fluid velocity along r. The points at the interfaces between cells are the finite-difference grid points. The interface positions at the beginning of a numerical timestep are denoted by $\{r_{i+\frac{1}{2}}^o\}$,

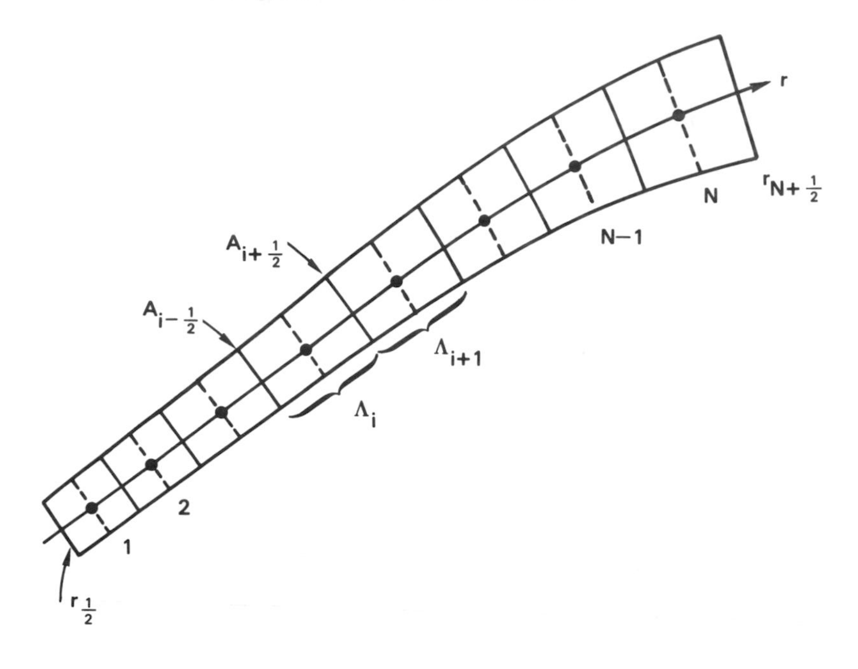

Figure 8–3. Geometry for the LCPFCT grid.

where $i = 0, 1, \ldots, N$. At the end of a timestep Δt, the interfaces are at $\{r^n_{i+\frac{1}{2}}\}$, where

$$r^n_{i+\frac{1}{2}} = r^o_{i+\frac{1}{2}} + v^g_{i+\frac{1}{2}} \, \Delta t \, , \tag{8 – 3.12}$$

where the quantities $\{v^g_{i+\frac{1}{2}}\}$ are the velocities of the cell interfaces. Figure 8–3 also indicates the basic cell volumes $\{\Lambda_i\}$, and the interface areas $\{A_{i+\frac{1}{2}}\}$. The interface areas are assumed to be perpendicular to the tube and hence to the velocities $\{v^f_{i+\frac{1}{2}}\}$. The change in the total amount of a convected quantity in a cell is the algebraic sum of the fluxes of that quantity into and out of the cell through the interfaces. Both the cell volumes $\{\Lambda_i\}$ and the interface areas $\{A_{i+\frac{1}{2}}\}$ that bound them have to be calculated consistently using new and old grid positions.

The positions of the cell centers are denoted $\{r^{o,n}_i\}$ and are related to the cell interface locations by

$$r^{o,n}_i = \frac{1}{2} \, [r^{o,n}_{i+\frac{1}{2}} + r^{o,n}_{i-\frac{1}{2}}] \, , \quad i = 1, 2, \ldots, N \, . \tag{8 – 3.13}$$

The superscripts o or n indicate the old and new grid at the beginning and the end of the timestep. The cell centers could also be computed as some weighted average of the interface locations and are generally only used for calculating diffusion terms added to Eq. (8–3.11). The boundary interface positions, $r^{o,n}_{\frac{1}{2}}$ and $r^{o,n}_{N+\frac{1}{2}}$, have to be specified by the user. For example, they might be the location of bounding walls. Then by programming $r_{\frac{1}{2}}$ as a function of time and forcing the adjacent grid points to move correspondingly, we can simulate the effect of a piston or flexible container.

To calculate convection, we need the flux of fluid through the interface as it moves from $r^o_{i+\frac{1}{2}}$ to $r^n_{i+\frac{1}{2}}$ during a timestep. The velocities of the fluid are assumed known at the cell centers and the velocity of the fluid at the interfaces is given by

$$v^f_{i+\frac{1}{2}} = \frac{1}{2}(v^f_{i+1} + v^f_i) , \qquad i = 1, 2, ..., N-1 . \qquad (8-3.14)$$

Again, other weighted averages are possible but this choice works well for all three geometries.

Because the fluxes out of one cell into the next are needed on the interfaces, we define

$$\Delta v_{i+\frac{1}{2}} = v^f_{i+\frac{1}{2}} - v^g_{i+\frac{1}{2}}, \qquad i = 1, 2, ..., N-1 . \qquad (8-3.15)$$

The boundary interface values $\Delta v_{\frac{1}{2}}$ and $\Delta v_{N+\frac{1}{2}}$ are calculated using the locations, $r^{o,n}_{\frac{1}{2}}$ and $r^{o,n}_{N+\frac{1}{2}}$, and the two endpoint velocities $v^f_{\frac{1}{2}}$ and $v^f_{N+\frac{1}{2}}$. These velocities must also be specified as part of the problem definition because they require information from beyond the computational domain. Then

$$\Delta v_{\frac{1}{2}} = v^f_{\frac{1}{2}} - \frac{r^n_{\frac{1}{2}} - r^o_{\frac{1}{2}}}{\Delta t} ,$$

$$\Delta v_{N+\frac{1}{2}} = v^f_{N+\frac{1}{2}} - \frac{r^n_{N+\frac{1}{2}} - r^o_{N+\frac{1}{2}}}{\Delta t} . \qquad (8-3.16)$$

To determine the flux, we also need the density at the cell interfaces. This is taken as

$$\rho^o_{i+\frac{1}{2}} = \frac{1}{2}\left[\rho^o_{i+1} + \rho^o_i\right], \qquad i = 1, 2, ..., N-1 . \qquad (8-3.17)$$

A more complicated weighted average could also be used here. As with the velocities, we need the values on the interfaces $i = \frac{1}{2}$ and $i = N + \frac{1}{2}$. The formulas

$$\begin{aligned} \rho_o &= L_B \rho_1 + \rho_L , \\ \rho_{N+1} &= R_B \rho_n + \rho_R , \end{aligned} \qquad (8-3.18)$$

are used to calculate densities at the guard cells 0 and $N+1$ beyond the computational domain. The quantities L_B and R_B are multiplicative factors used to specify the left and right boundary conditions, and ρ_L and ρ_B are corresponding additive constants. These formulas are reevaluated several times using the updated ρ values. Thus $\rho_1 - \rho_o$ and $\rho_{N+1} - \rho_N$ are always defined at the first and last interfaces, and Eq. (8–3.17) gives

$$\begin{aligned} \rho^o_{\frac{1}{2}} &= \left(\frac{1}{2} + \frac{1}{2} L_B\right) \rho^o_1 + \frac{1}{2}\rho_L \,, \\ \rho^o_{N+\frac{1}{2}} &= \left(\frac{1}{2} + \frac{1}{2} R_B\right) \rho^o_N + \frac{1}{2}\rho_R \,. \end{aligned} \tag{8 – 3.19}$$

Using these definitions, the convective transport part of the continuity equation is written as

$$\begin{aligned} \Lambda^o_i \rho^*_i &= \Lambda^o_i \rho^o_i - \Delta t \ \rho^o_{i+\frac{1}{2}} \ A_{i+\frac{1}{2}} \ \Delta v_{i+\frac{1}{2}} + \Delta t \ \rho^o_{i-\frac{1}{2}} \ A_{i-\frac{1}{2}} \ \Delta v_{i-\frac{1}{2}} \,, \\ i &= 1,\, 2,\, ...,\, N \,. \end{aligned} \tag{8 – 3.20}$$

The left side, $\Lambda^o_i \rho^*_i$, has not yet undergone the compression or expansion that changes Λ^o_i to Λ^n_i . The source terms have not yet been incorporated and the diffusion and antidiffusion portions of flux correction still have to be included.

The source terms in Eq. (8–3.11) are added into Eq. (8–3.20),

$$\begin{aligned} \Lambda^o_i \rho^T_i &= \Lambda^o_i \rho^*_i + \frac{1}{2}\Delta t \ A_{i+\frac{1}{2}}(D_{1,i+1} + D_{1,i}) - \frac{1}{2}\Delta t \ A_{i-\frac{1}{2}}(D_{1,i} + D_{1,i-1}) \\ &+ \frac{1}{4}\Delta t \ C_{2,i}(A_{i+\frac{1}{2}} + A_{i-\frac{1}{2}}) \ (D_{2,i+1} - D_{2,i-1}) \\ &+ \Delta t \ \Lambda^o_i D_{3,i} \,, \qquad i = 2,\, ...,\, N-1 \,. \end{aligned} \tag{8 – 3.21}$$

The end values, $i = 1$ and $i = N$, are treated using D_R and D_L, the problem-dependent right and left boundary values of D, in place of the interface average at the boundaries. Other source terms can be added easily to the formalism, but the three source terms in Eq. (8–3.11) are adequate to treat many important fluid dynamics problems.

The diffusion stage of this FCT algorithm also includes the compression,

$$\begin{aligned} \Lambda^n_i \tilde{\rho}_i &= \Lambda^o_i \rho^T_i + \nu_{i+\frac{1}{2}} \Lambda_{i+\frac{1}{2}} (\rho^o_{i+1} - \rho^o_i) \\ &- \nu_{i-\frac{1}{2}} \Lambda_{i-\frac{1}{2}} (\rho^o_i - \rho^o_{i-1}) \,, \quad i = 1,\, 2,\, ...,\, N \,. \end{aligned} \tag{8 – 3.22}$$

The quantities $\{\tilde{\rho}_i\}$ are the transported-diffused quantities defined in stage 4 of the previous section. The diffusion coefficients $\{\nu_{i+\frac{1}{2}}\}$ can be chosen to reduce phase errors from second to fourth order. The interface volumes $\{\Lambda_{i+\frac{1}{2}}\}$ multiply the $\{\nu_{i+\frac{1}{2}}\}$ and in Eq. (8–3.22) are defined as

$$\Lambda_{i+\frac{1}{2}} = \frac{1}{2}(\Lambda_{i+1}^n + \Lambda_i^n) , \quad i = 1, 2, ..., N-1 . \tag{8-3.23}$$

The boundary interface volumes are chosen as

$$\begin{aligned} \Lambda_{\frac{1}{2}} &= \Lambda_1^n , \\ \Lambda_{N+\frac{1}{2}} &= \Lambda_N^n . \end{aligned} \tag{8-3.24}$$

The convection, additional source terms, compression, and diffusion have been broken into the two stages shown in Eqs. (8–3.20), (8–3.21), and (8–3.22) because we need to compute the antidiffusive fluxes using $\{\rho_i^T\}$. If we compute the antidiffusive flux using $\{\tilde{\rho}_i\}$, that is, after the diffusion has been added, the algorithm has residual diffusion both when the grid is Lagrangian, $v^f = v^g$, and in the special case when both the grid and the fluid are stationary. Therefore the transported but not diffused values, $\{\rho_i^T\}$ are used to calculate the raw, uncorrected antidiffusive fluxes,

$$f_{i+\frac{1}{2}}^{ad} = \mu_{i+\frac{1}{2}}\Lambda_{i+\frac{1}{2}}[\rho_{i+1}^T - \rho_i^T] , \quad i = 0, 1, ..., N . \tag{8-3.25}$$

The antidiffusion is designed so that when the grid is Lagrangian and $\{\Delta v_{i+\frac{1}{2}}\}$ vanishes in Eqs. (8–3.20),

$$\Lambda_i^n \rho_i^n = \Lambda_i^o \rho_i^o . \tag{8-3.26}$$

Substituting Eq. (8–3.20) and (8–3.21) into Eq. (8–3.22) in the Lagrangian case with no sources gives

$$\Lambda_i^n \tilde{\rho}_i = \Lambda_i^o \rho_i^o + \nu_{i+\frac{1}{2}}\Lambda_{i+\frac{1}{2}}(\rho_{i+1}^o - \rho_i^o) - \nu_{i-\frac{1}{2}}\Lambda_{i-\frac{1}{2}}(\rho_i^o - \rho_{i-1}^o) , \tag{8-3.27}$$

because $\rho_i^T = \rho_i^o$. The antidiffusion procedure, applied to Eq. (8–3.27), gives

$$\begin{aligned} \Lambda_i^n \rho_i^n = &\Lambda_i^o \rho_i^o + (\nu_{i+\frac{1}{2}} - \mu_{i+\frac{1}{2}})\Lambda_{i+\frac{1}{2}}(\rho_{i+1}^o - \rho_i^o) \\ &- (\nu_{i-\frac{1}{2}} - \mu_{i-\frac{1}{2}})\Lambda_{i-\frac{1}{2}}(\rho_i^o - \rho_{i-1}^o) . \end{aligned} \tag{8-3.28}$$

When the grid is Lagrangian, the desired result of Eq. (8–3.26) can be achieved as long as

$$\nu_{i+\frac{1}{2}} = \mu_{i+\frac{1}{2}} . \tag{8-3.29}$$

Boris and Book (1976) explain that the choices

$$\begin{aligned} \nu_{i+\frac{1}{2}} &\equiv \frac{1}{6} + \frac{1}{3}\epsilon_{i+\frac{1}{2}}^2 , \\ \mu_{i+\frac{1}{2}} &\equiv \frac{1}{6} - \frac{1}{6}\epsilon_{i+\frac{1}{2}}^2 , \end{aligned} \tag{8-3.30}$$

reduce the relative phase errors in convection on a locally uniform grid to fourth order. By defining

$$\epsilon_{i+\frac{1}{2}} \equiv A_{i+\frac{1}{2}} \Delta v_{i+\frac{1}{2}} \frac{\Delta t}{2} \left[\frac{1}{\Lambda_i^n} + \frac{1}{\Lambda_{i+1}^n} \right] , \quad i = 0, 1, ..., N , \quad (8-3.31)$$

the diffusion and antidiffusion coefficients are equal in the Lagrangian case and Eqs. (8–3.30) are satisfied for the portion of the fluid motion that convects material through the moving interfaces.

As in Eq. (8–3.9) above, the quantites $\{S_{i+\frac{1}{2}}\}$ can be defined with the sign of $[\tilde{\rho}_{i+1} - \tilde{\rho}_i]$ and magnitude unity. Using $\{f^{ad}_{i+\frac{1}{2}}\}$ from Eq. (8–3.25) as the raw antidiffusive fluxes and $\{\tilde{\rho}_i\}$ from Eq. (8–3.22), the corrected antidiffusive flux is

$$\begin{aligned} f^c_{i+\frac{1}{2}} = S_{i+\frac{1}{2}} \max \Big\{ 0, \min \big[|f^{ad}_{i+\frac{1}{2}}|, \; & S_{i+\frac{1}{2}} \Lambda^n_{i+1} (\tilde{\rho}_{i+2} - \tilde{\rho}_{i+1}), \\ & S_{i+\frac{1}{2}} \Lambda^n_i (\tilde{\rho}_i - \tilde{\rho}_{i-1}) \big] \Big\} , \quad i = 1, 2, ..., N-1 . \end{aligned} \quad (8-3.32)$$

For correcting the boundary fluxes $f^c_{\frac{1}{2}}$ and $f^c_{N+\frac{1}{2}}$, the min[..., ..., ...] term in Eq. (8–3.32) contains only two terms. The correction coming from a difference reaching beyond the boundary is simply dropped from the calculation.

The result, $\{\rho^n_i\}$, is then computed as in Eq. (8–3.8), where the corrected fluxes $\{f^c_{i+\frac{1}{2}}\}$ replace $\{f^{ad}_{i+\frac{1}{2}}\}$. The final density at the new time is

$$\rho^n_i = \tilde{\rho}_i - \frac{1}{\Lambda^n_i} \left[f^c_{i+\frac{1}{2}} - f^c_{i-\frac{1}{2}} \right] . \quad (8-3.33)$$

A few of the geometric variables used above have yet to be defined. The obvious choice of volume elements, at the beginning and end of the timesteps, in Cartesian, cylindrical, and spherical geometries are

$$\Lambda^{o,n}_i = \begin{cases} [r^{o,n}_{i+\frac{1}{2}} - r^{o,n}_{i-\frac{1}{2}}], & \text{Cartesian} \\ \pi[(r^{o,n}_{i+\frac{1}{2}})^2 - (r^{o,n}_{i-\frac{1}{2}})^2] & \text{cylindrical} \\ \frac{4}{3}\pi[(r^{o,n}_{i+\frac{1}{2}})^3 - (r^{o,n}_{i-\frac{1}{2}})^3] & \text{spherical.} \end{cases} \quad (8-3.34)$$

The corresponding interface areas are

$$A_{i+\frac{1}{2}} = \begin{cases} 1 & \text{Cartesian} \\ \pi[r^o_{i+\frac{1}{2}} + r^n_{i+\frac{1}{2}}] & \text{cylindrical} \\ \frac{4}{3}\pi[(r^o_{i+\frac{1}{2}})^2 + r^o_{i+\frac{1}{2}} r^n_{i+\frac{1}{2}} + (r^n_{i+\frac{1}{2}})^2] & \text{spherical.} \end{cases} \quad (8-3.35)$$

The interface areas are time and space centered. Though other centered choices are also possible, these particular definitions ensure that a constant density ρ remains constant and unchanged when the fluid is at rest but the grid is rezoned arbitrarily. If L_B and R_B are unity, the rezone can even move fluid into and out of the system, and the density will still be constant.

Boris et al. (1987) describe how to use this algorithm, show results of representative calculations, and discuss its use in periodic geometry and general curvilinear coordinates. This algorithm vectorizes easily for pipeline and parallel processing, requiring of order 30 floating-point operations per continuity equation per timestep per spatial direction. Chapter 9 shows how the algorithm can be extended to multidimensions using splitting methods and how it can be used for coupled sets of continuity equations.

8–4. ZALESAK'S GENERALIZATION

Zalesak (1979) developed a generalized, alternative approach to FCT that has a number of particularly useful features. One important result of this approach is a prescription for how previously written programs that use nonmonotone methods can be modified to ensure monotonicity. Because the approach does not have a bias in direction, the algorithm can be made fully multidimensional so that the problems arising from direction-splitting in LCPFCT are remedied. The Zalesak generalization of FCT has some features similar to the hybridization scheme of Harten and Zwas (1972).

Zalesak noted that the FCT algorithm could be viewed as six sequential operations at each point i:

1. Compute $f^L_{i+\frac{1}{2}}$, the transportive flux given by some low-order method guaranteed to give monotonic results. The diffusive contribution has to be at least as large as $\frac{1}{2}|\epsilon|$.
2. Compute $f^H_{i+\frac{1}{2}}$, the transportive flux given by some high-order method. This flux is mathematically more accurate, but can lead to physically unacceptable ripples in the solution.
3. Compute the updated low-order, *transported and diffused* solution,

$$\rho_i^{td} = \rho_i^o - \frac{\Delta t}{\Delta x_i}\left[f^L_{i+\frac{1}{2}} - f^L_{i-\frac{1}{2}}\right] .$$

4. Define the *antidiffusive flux* which becomes the amount of the monotone transportive flux that we would like to limit before correcting the transported and diffused densities $\{\rho_d^{td}\}$ of Step 3,

$$f^{ad}_{i+\frac{1}{2}} = f^H_{i+\frac{1}{2}} - f^L_{i+\frac{1}{2}} .$$

5. Limit the antidiffusive fluxes $\{f^{ad}_{i+\frac{1}{2}}\}$ so that ρ^n, as computed in Step 6, is free of the overshoots and undershoots which also do not appear in $\{\rho^{td}_i\}$,

$$f^c_{i+\frac{1}{2}} = C_{i+\frac{1}{2}} f^{ad}_{i+\frac{1}{2}}, \quad 0 \le C_{i+\frac{1}{2}} \le 1 \,.$$

6. Apply the limited antidiffusive fluxes to get the new values ρ^n_i,

$$\rho^n_i = \rho^{td}_i - \frac{\Delta t}{\Delta x_i}\left[f^c_{i+\frac{1}{2}} - f^c_{i-\frac{1}{2}}\right] .$$

The critical step is Step 5. Without Step 5, that is, when $f^c_{i+\frac{1}{2}} = f^{ad}_{i+\frac{1}{2}}$, the $\{\rho^n_i\}$ reduces to the time-advanced higher-order method without the required monotonicity correction. This prescription allows a variety of methods, such as the MacCormack method discussed in Section 8–1, to be flux corrected. The FCT algorithm described in Section 8–3 is computationally efficient because the high-order and low-order algorithms differ only by the diffusion added to the lower-order algorithm.

This formulation generalizes easily to multidimensions. Consider the two-dimensional scalar continuity equation

$$\frac{\partial \rho}{\partial t} + \nabla \cdot \mathbf{f} = 0 \,, \qquad (8-4.1)$$

where ρ and $\mathbf{f}$ are functions of x, y, and t, and $\mathbf{f} = (f_x, f_y)$. In finite-difference form, we have

$$\rho^n_{i,j} = \rho^o_{ij} - \frac{\Delta t}{A_{i,j}}\left[(f_x)_{i+\frac{1}{2},j} - (f_x)_{i-\frac{1}{2},j} + (f_y)_{i,j+\frac{1}{2}} - (f_y)_{i,j-\frac{1}{2}}\right] \,, \quad (8-4.2)$$

with ρ and $\mathbf{f}$ defined at locations $\{x_i,\, y_j\}$, at time levels t^o, and t^n, and A_{ij} a two-dimensional area element centered on grid point (i,j). Now there are two sets of transportive fluxes, f_x and f_y. The algorithm proceeds as before:

1. Compute $(f_x)^L_{i+\frac{1}{2},j}$ and $(f_y)^L_{i,j+\frac{1}{2}}$, the transportive fluxes by a low-order monotonic algorithm.
2. Compute $(f_x)^H_{i+\frac{1}{2},j}$ and $(f_y)^H_{i,j+\frac{1}{2}}$, the transportive fluxes by a high-order method.
3. Compute the previously updated low-order, transported and diffused solution,

$$\rho^{td}_{i,j} = \rho^n_{i,j} - \frac{\Delta t}{A_{i,j}}\left[(f_x)^L_{i+\frac{1}{2},j} - (f_x)^L_{i-\frac{1}{2},j} + (f_y)^L_{i,j+\frac{1}{2}} - (f_y)^L_{i,j-\frac{1}{2}}\right] .$$

4. Define the vector components of the raw *antidiffusive fluxes*,

$$\begin{aligned} f^{ad}_{i+\frac{1}{2},j} &= (f_x)^H_{i+\frac{1}{2},j} - (f_x)^L_{i+\frac{1}{2},j} , \\ f^{ad}_{i,j+\frac{1}{2}} &= (f_y)^H_{i,j+\frac{1}{2}} - (f_y)^L_{i,j+\frac{1}{2}} . \end{aligned} \qquad (8-4.3)$$

5. Limit the antidiffusive fluxes so that there are no overshoots or undershoots in $\{\rho^n_{i,j}\}$ of Step 6 below that do not appear in $\{\rho^{td}_{i,j}\}$ of Step 3,

$$\begin{aligned} f^c_{i+\frac{1}{2},j} &= C_{i+\frac{1}{2},j} f^{ad}_{i+\frac{1}{2},j}, \quad 0 \le C_{i+\frac{1}{2},j} \le 1 , \\ f^c_{i,j+\frac{1}{2}} &= C_{i,j+\frac{1}{2}} f^{ad}_{i,j+\frac{1}{2}}, \quad 0 \le C_{i,j+\frac{1}{2}} \le 1 , \end{aligned} \qquad (8-4.4)$$

where the C's are calculated as described below.

6. Apply the limited antidiffusive fluxes to get the new values $\rho^n_{i,j}$,

$$\rho^n_{i,j} = \rho^{td}_{i,j} - \frac{\Delta t}{A_{i,j}} \left[f^c_{i+\frac{1}{2},j} - f^c_{i-\frac{1}{2},j} + f^c_{i,j+\frac{1}{2}} - f^c_{i,j-\frac{1}{2}} \right] . \qquad (8-4.5)$$

Again, except for Step 5, this algorithm is straightforward.

We want to limit the antidiffusive fluxes $f^{ad}_{i+\frac{1}{2},j}$ and $f^{ad}_{i,j+\frac{1}{2}}$ by choosing the cell-interface flux-correcting factors, $\{C_{i+\frac{1}{2},j}\}$ and $\{C_{i,j+\frac{1}{2}}\}$, such that the combination of four fluxes acting together, $f^c_{i+\frac{1}{2},j}$, $f^c_{i,j+\frac{1}{2}}$, $f^c_{i-\frac{1}{2},j}$, $f^c_{i,j-\frac{1}{2}}$, through Eq. (8–4.5) does not allow $\{\rho^n_{i,j}\}$ to exceed some maximum value $\{\rho^{\max}_{i,j}\}$, or to fall below some minimum value $\{\rho^{\min}_{i,j}\}$.

To accomplish this numerically, we define six quantities:

$$\begin{aligned} P^+_{i,j} &= \text{sum of antidiffusive fluxes into grid point } (i,j) \\ &= \max(0, f^{ad}_{i-\frac{1}{2},j}) - \min(0, f^{ad}_{i+\frac{1}{2},j}) \\ &\quad + \max(0, f^{ad}_{i,j-\frac{1}{2}}) - \min(0, f^{ad}_{i,j+\frac{1}{2}}) \end{aligned} \qquad (8-4.6)$$

$$Q^+_{i,j} = (\rho^{\max}_{i,j} - \rho^{td}_{i,j}) A_{i,j} \qquad (8-4.7)$$

$$R^+_{i,j} = \begin{cases} \min(1, Q^+_{i,j}/P^+_{i,j}), & P^+_{i,j} > 0 , \\ 0, & P^+_{i,j} = 0 . \end{cases} \qquad (8-4.8)$$

Because $\rho^{\max}_{ij} \ge \rho^{td}_{ij}$ is required, all three of the above quantities are positive. The quantity $R^+_{i,j}$ represents the least upper bound on the fraction which must multiply all antidiffusive fluxes into location (i,j) to guarantee no nonphysical overshoot at location (i,j).

We also define the corresponding quantities,

$$\begin{aligned} P^-_{i,j} &= \text{sum of antidiffusive fluxes out of } (i,j) \\ &= \max(0, f^{ad}_{i+\frac{1}{2},j}) - \min(0, f^{ad}_{i-\frac{1}{2},j}) \\ &\quad + \max(0, f^{ad}_{i,j+\frac{1}{2}}) - \min(0, f^{ad}_{i,j-\frac{1}{2}}) \end{aligned} \qquad (8-4.9)$$

$$Q^-_{i,j} = (\rho^{td}_{i,j} - \rho^{\min}_{i,j}) A_{i,j} \qquad (8-4.10)$$

$$R^-_{i,j} = \begin{cases} \min(1,\ Q^-_{i,j}/P^-_{i,j}), & P^-_{i,j} > 0\,, \\ 0, & P^-_{i,j} = 0\,. \end{cases} \qquad (8-4.11)$$

Again, requiring that $\rho^{\min}_{ij} \leq \rho^{td}_{ij}$, R^-_{ij} represents the least upper bound on the fraction which must multiply all antidiffusive fluxes away from location (i,j) to guarantee that there is no nonphysical undershoot at grid point (i,j).

Finally, we see that all antidiffusive fluxes are directed out of one grid point and into an adjacent one. The flux-limiting process therefore applies to undershoots for the antidiffusive fluxes out of a grid point, and to the overshoots for the antidiffusive fluxes into a grid point. A guarantee that there are no overshoots or undershoots requires taking a minimum,

$$C_{i+\frac{1}{2},j} = \begin{cases} \min(R^+_{i+1,j},\ R^-_{i,j})\,, & f^{ad}_{i+\frac{1}{2},j} \geq 0 \\ \min(R^+_{i,j},\ R^-_{i+1,j})\,, & f^{ad}_{i+\frac{1}{2},j} < 0 \end{cases} \qquad (8-4.12)$$

$$C_{i,j+\frac{1}{2}} = \begin{cases} \min(R^+_{i,j+1},\ R^-_{i,j})\,, & f^{ad}_{i,j+\frac{1}{2}} \geq 0 \\ \min(R^+_{i,j},\ R^-_{i,j+1})\,, & f^{ad}_{i,j+\frac{1}{2}} < 0\,. \end{cases} \qquad (8-4.13)$$

Zalesak also suggests setting

$$f^{ad}_{i+\frac{1}{2},j} = 0 \qquad (8-4.14)$$

if simultaneously

$$f^{ad}_{i+\frac{1}{2},j}(\rho^{td}_{i+1,j} - \rho^{td}_{i,j}) < 0\,, \qquad (8-4.15)$$

and either

$$f^{ad}_{i+\frac{1}{2},j}(\rho^{td}_{i+2,j} - \rho^{td}_{i+1,j}) < 0\,, \qquad (8-4.16)$$

or

$$f^{ad}_{i+\frac{1}{2},j}(\rho^{td}_{i,j} - \rho^{td}_{i-1,j}) < 0\,. \qquad (8-4.17)$$

Similarly, set

$$f^{ad}_{i,j+\frac{1}{2}} = 0 \qquad (8-4.18)$$

if simultaneously

$$f^{ad}_{i,j+\frac{1}{2}}(\rho^{td}_{i,j+1} - \rho^{td}_{i,j}) < 0 , \qquad (8-4.19)$$

and either

$$f^{ad}_{i,j+\frac{1}{2}}(\rho^{td}_{i,j+2} - \rho^{td}_{i,j+1}) < 0 , \qquad (8-4.20)$$

or

$$f^{ad}_{i,j+\frac{1}{2}}(\rho^{td}_{i,j} - \rho^{td}_{i,j-1}) < 0 . \qquad (8-4.21)$$

We now need to find the quantities ρ^{max}_{ij} and ρ^{min}_{ij} in Eqs. (8–4.7) and (8–4.10). A good choice, as suggested by Zalesak, is

$$\rho^{a}_{i,j} = \max(\rho^{o}_{i,j}, \rho^{td}_{i,j}) \qquad (8-4.22)$$
$$\rho^{max}_{i,j} = \max(\rho^{a}_{i-1,j}, \rho^{a}_{i,j}, \rho^{a}_{i+1,j}, \rho^{a}_{i,j-1}, \rho^{a}_{i,j+1}) \qquad (8-4.23)$$
$$\rho^{b}_{i,j} = \min(\rho^{o}_{i,j}, \rho^{td}_{i,j}) \qquad (8-4.24)$$
$$\rho^{min}_{i,j} = \min(\rho^{b}_{i-1,j}, \rho^{b}_{i,j}, \rho^{b}_{i+1,j}, \rho^{b}_{i,j-1}, \rho^{b}_{i,j+1}) . \qquad (8-4.25)$$

The effect of Eqs. (8–4.14) – (8–4.21) is minimal, but if this refinement is used, it should be applied before Eqs. (8–4.3) – (8–4.13). Note that our search for $\rho^{max}_{i,j}$ and $\rho^{min}_{i,j}$ now extends over both coordinate directions.

This choice looks back to the previous timestep for upper and lower bounds on ρ^{n}_{ij}. There are many possible ways to determine ρ^{max}_{ij} and ρ^{min}_{ij}. The flux limiter described here allows any physically motivated upper and lower bound on ρ^{n}_{ij} and introduces an additional flexibility unavailable with the original flux limiter described in Section 8–3.2 on LCPFCT.

8–5. RESOLUTION AND REYNOLDS NUMBER LIMITATIONS

The nondimensional Reynolds number,

$$Re \equiv \frac{Lv}{\nu} , \qquad (8-5.1)$$

is defined as the ratio of the viscous (diffusive) decay time of a flow structure, with characteristic dimension L, to the transit time of fluid across that structure. Because a fluid flow often involves many scales, choosing the values that go into the definition of the Reynolds number can be somewhat ambiguous.

Usually the Reynolds number is a guide for making qualitative judgments about the flow, such as whether it is dominated by viscous damping, or whether turbulence can develop. As an example of when a Reynolds number might be confusing, consider flows that involve large velocities without variations or local gradients in the flow field. Then locally the fluid lacks short-wavelength structure and behaves as if the Reynolds number were very low. The long-wavelength components of the flow are not greatly damped because the viscosity is small, but the flow has persisted long enough that the short-wavelength components either are gone or never were present.

These considerations suggest that a complicated flow can be characterized by several different Reynolds numbers. In particular, two Reynolds numbers might be estimated, one for the short-wavelength structures and one for the long-wavelength structures. Because most interesting flows are complicated, this type of a separation can never be perfect, but it can be qualitatively informative.

The issues of resolution and numerical diffusion in simulating convective flows on an Eulerian grid are closely intertwined. The flow generally has structure on many different spatial scales simultaneously. For each application, there is often disagreement about how well the various scales have to be resolved in each region of the computational domain. In some cases, it is possible to resolve the range of scales needed and not be concerned about others. In other cases, the flow has important structures on any spatial scale we can afford to simulate numerically.

The least damped flow that a simple Eulerian method can describe can be characterized by a Reynolds number calculated as if the numerical diffusion were actually physical diffusion. We can evaluate this Reynolds number using the dissipation given in Eq. (8–3.1) and the amplification factor in Eq. (8–2.17),

$$\begin{aligned} |A|^2 &\approx 1 - 2|\epsilon|c(1-\cos\beta) - \epsilon^2\left[(c+\epsilon)^2 - 1\right](1-\cos\beta)^2 \\ &\approx \left[1 - |\epsilon|c(1-\cos\beta)\right]^2 . \end{aligned} \qquad (8-5.2)$$

Here c, the flux-limiting factor, is unity when a first-order method is used and zero when the net diffusion of the algorithm is zero. The amplification factor can also be identified with the exponential damping rate of the numerically computed solution

$$A \approx e^{-\gamma\,\Delta t} \approx 1 - \gamma\,\Delta t . \qquad (8-5.3)$$

From Eqs. (8–5.2) and (8–5.3),

$$\gamma\,\Delta t \approx |\epsilon|c(1-\cos\beta) \approx \frac{|\epsilon|}{2} c k^2\,\Delta x^2 . \qquad (8-5.4)$$

The characteristic length scale of the disturbance is $L \equiv 1/k$ and the characteristic damping time is $1/\gamma$. The quantity c is the average fraction of linear damping used by the composite flow algorithm, according to Eq. (8–3.1), to maintain positivity.

Letting Re_{nd} be the effective-dissipation Reynolds number due to numerical diffusion, we have

$$Re_{nd} \approx \frac{1/\gamma}{L/v} \approx \frac{2L}{c\,\Delta x}, \qquad (8-5.5)$$

which is independent of the timestep. This equation defines the *cell Reynolds number* for an Eulerian flow. It is roughly $2/c$ times the number of computational cells within a structure of characteristic size L.

There is a common misconception that the flow with the highest Reynolds number that can be calculated by an Eulerian method is roughly twice the number of cells across the system. This criterion, however, is based on the linear positivity condition, $c = 1$. It does not apply to nonlinear monotone methods such as FCT, which vary the flux-limiting factor c locally from cell to cell. The monotone methods require less overall dissipation and hence are potentially more accurate for simulating high Reynolds number flows than linear methods. If the average value of c is $1/10$, flows with Reynolds number 2×10^3 can be represented on a grid of 100 cells in each direction. In fact, the clipping factor generally decreases during the course of a calculation. This occurs because short-wavelength structures tend to be smoothed out. A balance is reached between their physical generation and numerical destruction by the nonlinear flux-correction process.

When short wavelengths are not dominant, the long wavelengths can be calculated much less dissipatively than is implied by the usual $Re_{nd} \approx 2L/\Delta x$ limit on the cell Reynolds number by using a monotone method such as FCT. In these cases, positivity is not a major problem and very small values of c suffice. Even when the short wavelengths have a large amplitude, the nonlinear monotone algorithms still propagate the longer wavelengths at a relatively high Reynolds number. Strong smoothing is only applied in a small region, and hence is applied preferentially to short wavelengths.

Multiple Reynolds numbers are generally required to describe flow on a computational grid. Specifically, a cell Reynolds number can be related to resolution and another related to the numerical damping of long wavelengths. It is useful to think of the long-wavelength Reynolds number as describing the viscous or numerical decay of the large scales, and the short-wavelength Reynolds number as describing the rate of numerical damping required to prevent truncation error, computational noise, and turbulent cascade from building up at the shortest wavelengths that can be represented. This short-wavelength numerical smoothing need not invade the long wavelengths appreciably if the phase errors of the intermediate and long wavelengths are small enough. The longer, slowly decaying wavelengths can be effectively calculated accurately in flows with much higher Reynolds numbers than the usual numerical cell Reynolds number suggests. Short-wavelength structures still exist in the calculation for a finite length of time. They are strongly damped, but they are not dominant and their numerical smoothing, because it is nonlinear, does not damp the long-wavelength structures proportionately.

8–6. APPROACHES TO MONOTONICITY

We summarize the monotonicity discussion with the statement first shown to be true by Godunov (1959): *there are no linear second-order or higher-order methods which guarantee monotonicity.* First-order methods, such as the Lax method or the donor-cell method, guarantee positivity but are extremely diffusive. Second-order linear methods, such as the Lax-Wendroff or MacCormack methods, are less diffusive but are susceptible to nonlinear instabilities and unphysical oscillations can occur in regions of high gradients. Enforced conservation can also cause a positive definite quantity to become negative.

Introducing diffusion into the algorithm is one traditional approach to dampen spurious oscillations in regions of large gradients. This is a compromise approach. Numerical dissipation is added into the method so that shocks and discontinuities are spread over distances which can be resolved on a practical computational mesh. If the oscillations are totally damped, the solution shows the same massive diffusion as the linear first-order methods. On the other hand, if the oscillations are not totally damped, then little has been gained. This is the fundamental dilemma of linear methods.

8–6.1 A Classification of Monotone Methods

We showed in Section 8–3 that one answer to this dilemma is to use nonlinear methods, and we discussed one nonlinear monotone method, Flux-Corrected Transport, in some detail. Here we summarize the major features of other approaches to maintaining monotonicity. The fundamental feature these methods have in common is that they use *flux-limiting* or *flux-correcting* in an attempt to maintain high resolution without the spurious oscillations associated with more classical methods. Different physical and mathematical interpretations are given to these limiting or correcting procedures, however. We divide monotone methods into three categories:

1. *Linear hybridization* methods
2. Methods based on Godunov's algorithm,
3. *Total Variational Diminishing* or *TVD* methods.

The first two categories are discussed by Woodward and Colella (1984a), and the third category was used by Baer and Gross (1986). A summary of some of the properties of selected methods is given by Woodward and Colella (1984a) and Rood (1987).

Linear Hybridization Methods

In linear hybridization, the results of a low-order method and a high-order method are blended to take advantage of the strong points of each. In smooth regions, the high-order linear method is used exclusively, assuring accuracy. Near regions of large gradients and extrema in the profile, however, the low-order linear method is averaged with the high-order method to an extent sufficient to maintain positivity and monotonicity. Though FCT was not originally cast in this form, Zalesak (1979) showed that it can be described in these terms. Currently there are a number of numerical methods based on this principle. In addition to FCT methods, the methods by Harten (1974), Chapman (1981), and Forester (1977) are in this category. Zalesak has shown that the switches used to perform this hybridization between two linear methods reduce to nonlinear formulas of the FCT flux-limiting type.

Godunov Methods

Godunov's method (1959) was developed to treat problems in which each discontinuity between piecewise constant cells is treated as a new Riemann problem. The discontinuity resolves itself in general into a contact discontinuity, a rarefaction, and a shock. A first-order method was used for the convection, but the exact local Riemann problem is solved. We return to this approach in Chapter 9.

Building this Riemann solution information into the solution allows shocks to be calculated much more accurately than with the classical methods. An important element in the Godunov approach is piecing together discontinuous solutions, instead of piecing together smooth, small-amplitude solutions. These discontinuous solutions are acceptable approximations to the smooth solutions when appropriate, and they do a much better job of approximating a solution with discontinuities. The intrinsic first-order smoothing of the underlying algorithm is its major drawback.

Higher-order methods based on the Godunov approach were developed by van Leer (1973, 1979). For example, the MUSCL code incorporates a Riemann solver, but when there is only linear advection such as in the square wave test problem, it reduces to a linear hybridization scheme giving results very similar to a second-order FCT. Besides the work of van Leer, Roe (1981, 1985) and Woodward and Colella (1984a, 1984b) are among the major contributors to this approach. The *PPM* or *Piecewise-Parabolic Method* of Woodward and Colella is an example of this technique.

Total Variational Diminishing Methods

The *TVD* methods represent an effort to place the study of monotone flux-correcting methods for solving hyperbolic systems of equations on a more rigorous mathematical foundation. A partial list of contributors to this effort includes Harten (1983), Davis (1984), Yee et al. (1983), Sweby (1984), and Chakravarthy and Osher (1985a,b). Many of the TVD methods use Riemann solvers and as such could also be classified as Godunov approaches. The work of Davis is analogous to the linear hybridization approach.

There are important aspects of TVD methods that are very similar to the two previously mentioned classes of monotone methods. This similarity is somewhat obscured by the different terminology and formalism. Zalesak (1987) has done extensive work to extract the basic ideas and the flux-limiting procedures in most of these methods, and has systematically compared them to a number of linear hybridization and Godunov methods.

8–6.2 Tests of Monotone Methods

In Chapter 4 we showed several examples of calculations of the convection of a square wave to test in various simple algorithms in flows with constant convection velocity. We return to this problem, adding two more linear tests: a sharp Gaussian profile and a half dome. These tests are important for several reasons. The Gaussian shows the effects of *clipping* the solution, and the half dome shows the effects of *terracing*.

The test problems are:

1. The square wave. A square wave is initially 20 cells wide. It is convected at a Courant number of 0.2, that is,

$$\frac{v\,\Delta t}{\Delta x} = 0.2 ,$$

   and we look at the profile after 800 steps. The maximum value of ρ is 2.0 and the minimum value is 0.5. This is a standard test problem used, for example, by Boris (1971) and Boris and Book (1973).
2. The Gaussian has half width $2\Delta x$. We convect this profile at a Courant number of 0.1 and examine the profile after 600 steps. The maximum

value of ρ is 1.6 and the minimum is 0.1. This is a standard test problem used by Forester (1977).

3. The half dome is a total of 30 cells wide. We convect the profile at a Courant number of 0.1 and examine the profile after 600 steps. The maximum value of ρ is 1.6 and the minimum is 0.1. This is a test problem suggested by B.E. MacDonald (see, for example, McDonald, 1985).

Figures 8–4 and 8–5 show the results of using different monotone algorithms to convect these profiles. The exact solution is the light solid line, and the computed solutions are the heavy dots. The methods shown are:

1. Second-order FCT (Zalesak, 1981)
2. Fourth-order FCT, similar to LCPFCT (Zalesak, 1981)
3. Sixteenth-order FCT (Zalesak, 1981)
4. PPM (Colella and Woodward, 1984)
5. HI-RES, a TVD method (Harten, 1983)
6. MUSCL (van Leer, 1979)
7. ETVD, "essentially" TVD (Chakravarthy et al., 1986)
8. SUPERBEE (Roe, 1985)

The results shown in these figures are quite similar. There is a little more diffusion in the HI-RES and the TVD methods. The second-order FCT algorithm shows a dispersive phase lag that slows short wavelengths, which is not a problem in the fourth-order FCT. In the Gaussian problem, the higher-order FCT methods do a little better than the others, but the differences are not enormous. The key to this high performance is the application of some form of nonlinear flux-limiting procedure. These monotone methods are all large improvements over the classical methods.

The methods which are less diffusive show the problem of terracing in the half dome solution. This is noticeable on the right or the left half of the dome, where the solution tends to set up terraces instead of smoothly decreasing. It is obvious in the PPM, low-order FCT, and SUPERBEE solutions. Adding diffusion selectively eliminates this effect, but then the solution is generally more diffusive, as in HI-RES, MUSCL, and ETVD, which is second order. The Gaussian profile is least clipped by the fourth-order and sixteenth-order FCT algorithms, followed closely by the PPM and SUPERBEE methods.

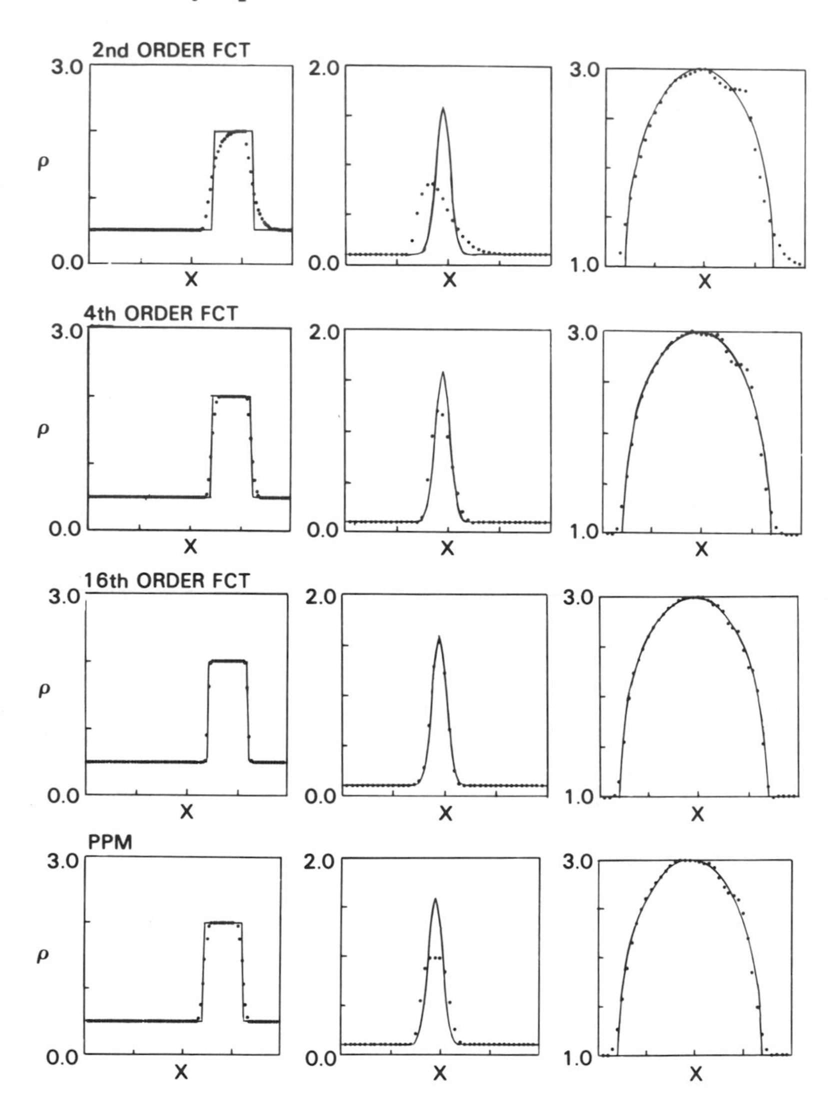

Figure 8–4. Tests of linear convection. (Figures are courtesy of S. Zalesak.)

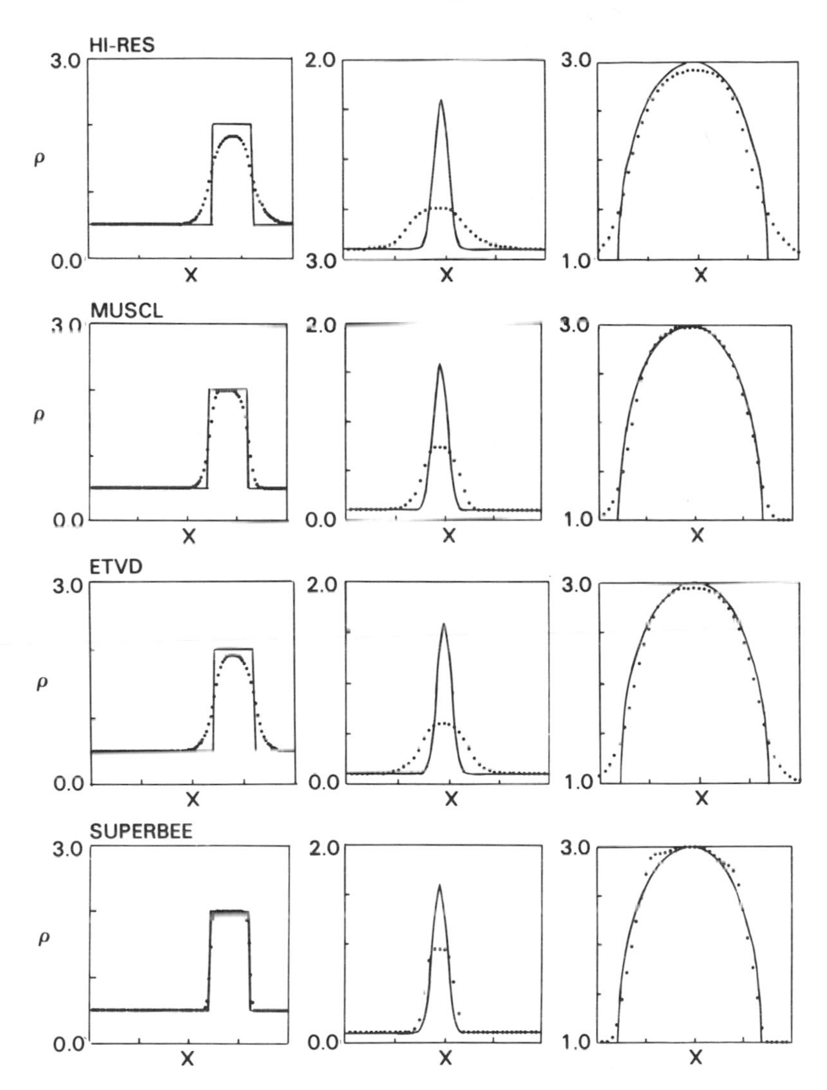

Figure 8–5. Tests of linear convection. (Figures are courtesy of S. Zalesak.)

8–6.3 Comments on Monotone Methods

The key to obtaining accurate solutions is maintaining conservation, positivity, and using a flux-correcting limiter to minimize numerical diffusion. There is little advantage to increasing the accuracy of the phases above fourth order, although improvements in going from second to fourth order are significant. The computational cost goes up quickly with order, but there is proportionately less gained. Formal order is advantageous in these linear convective problems, but it is not clear that it means much in nonlinear problems, in multidimensions, or with variably spaced grids.

Adding a Riemann solver is expensive. Using more knowledge about the underlying gas dynamics of the coupled continuity equation system can and does improve the solution at high Mach number. However, the Riemann solver is of little help in those cases for which the flow structure is not known ahead of time or when the flow is subsonic.

A lot of sound and fury is made over which algorithm is better. A subculture has grown up around comparisons of monotone algorithms, and the various methods are discussed with great agitation. Certainly advances have been made in the last 15 years, and some formal understanding has been gained of why this class of monotone methods works as well as it does. The comparisons on the preceding pages show that even the best monotone algorithms are imperfect, though much better than the linear algorithms of Chapter 4. These methods are the only practical way to solve many difficult convective flow problems where several processes are coupled together.

8–7. SPECTRAL AND FINITE-ELEMENT METHODS

Both finite-element and spectral methods use continuum expansions to solve the fluid equations. Both methods expand the fluid variables linearly in a set of basis functions whose coefficients are determined from integrating a complicated set of coupled usually nonlinear ODEs in time. For finite elements, the basis functions are usually piecewise polynomials localized to a few computational cells. In contrast, spectral methods use global basis functions.

Basic references for finite-element methods are Strang and Fix (1973), Zienkiewicz (1977), and Zienkiewicz and Morgan (1983). Basic references for spectral methods are Kreiss and Oliger (1973) and Gottlieb and Orszag (1977), which cover the principles of the method and introduce some applications. Voigt et al. (1984) is a collection of papers updating the theory and applications. Monotone, positivity-preserving techniques have been applied to finite elements by Löhner et al. (1986) and to spectral methods by Boris and Book (1976), Taylor et al. (1981), and Zalesak (1981).

This section is a brief overview of some of the features of these two methods, and of their similarities and differences. Neither method is presented in the detail with which we describe finite-difference methods. There are some problems with applying spectral methods to the reactive flow problems and these become apparent in the discussion below. Finite-element methods show promise, particularly in the adaptive gridding in compressible flows. However, the actual programming is complex and the costs are high. Both approaches are currently under active development and some of these difficulties may be overcome.

8–7.1 Continuous Representations

There are three basic techniques used in finite-element and spectral methods: Galerkin, tau, and collocation approximations. To understand how these are defined and differ, consider an initial value problem of the form

$$\frac{\partial u(x,t)}{\partial t} = \mathcal{L}(x,t)[u(x,t)] + f(x,t) \; . \qquad (8-7.1)$$

Here $\mathcal{L}$ is generally a nonlinear operator. For our current interests, u may be one of the dependent variables described by the continuity equation, such as ρ, ρv, or E, and $\mathcal{L}$ may be the spatial derivative operator in the continuity equation discussed in Section 8–1.

Assume that boundary conditions of the form

$$B_{\pm} u(x,t) = 0 \qquad (8-7.2)$$

are applied at the spatial boundaries $x = x_{\pm}$, where $B_{\pm}$ are linear operators and define the initial conditions

$$u(x,0) = u_o(x) \; . \qquad (8-7.3)$$

We can approximate the function u by a superposition of N basis functions $\{v_n(x)\}$, each satisfying the boundary condition in Eq. (8–7.2),

$$u(x,t) = \sum_{n=1}^{N} a_n(t) v_n(x) \; . \qquad (8-7.4)$$

If the expansion functions $\{v_n(x)\}$ form an orthonormal set, the coefficients $\{a_n\}$ satisfy

$$a_n(t) = \int_{x_-}^{x_+} dx \; v_n(x) u(x,t) w(x) \; , \qquad (8-7.5)$$

where $w(x)$ is a suitable weighting function assumed here to be unity. Equation (8–7.1) gives evolution equations for these coefficients,

$$\dot{a}_n \equiv \frac{da_n}{dt} = \sum_{m=1}^{N} \int_{x_-}^{x_+} dx\ v_n(x)\mathcal{L}v_m(x) + \eta_n(t) \ , \qquad (8-7.6)$$

where $\{\eta_n\}$, the expansion coefficients of the inhomogeneous term in Eq. (8–7.1), are chosen to satisfy

$$f(x,t) = \sum_{n=1}^{N} \eta_n(t) v_n(x) \ . \qquad (8-7.7)$$

Equation (8–7.6) is an explicit equation for $\dot{a}_n$. It has this particularly uncomplicated form because we have assumed that $\{v_n\}$ are orthonormal and the boundary condition operators $B_\pm$ do not change in time. In a more general case, it is necessary to solve N linear equations simultaneously,

$$\sum_{m=1}^{N} \left\{ \dot{a}_m \int_{x_-}^{x_+} dx\ v_n(x)v_m(x) - \int_{x_-}^{x_+} dx\ v_n(x)\mathcal{L}v_m(x) \right\}$$
$$= \sum_{n=1}^{N} \eta_m \int_{x_-}^{x_+} dx\ v_n v_m \ , \qquad n = 1,2,\ldots,N \ . \qquad (8-7.8)$$

Together, Eqs. (8–7.4) and (8–7.8) are the *Galerkin* approximation. A review of the application of Galerkin methods has been given by Fletcher (1984).

The *tau* approximation (Lanczos, 1956) also starts with Eq. (8–7.4), but the functions $v_n(x)$ do not have to satisfy the boundary conditions. Instead, there are two additional constraints,

$$\sum_{n=1}^{N} a_n B_\pm v_n(x) = 0 \ . \qquad (8-7.9)$$

The solution is found from these two equations along with the first $N-2$ of Eqs. (8–7.8) or (8–7.6).

In *collocation* methods, N points $\{x_n\}$ are chosen in the range $[x_-, x_+]$, such that

$$x_- < x_1 < \cdots < x_N < x_+ \ . \qquad (8-7.10)$$

These collocation points are effectively the grid points of the method. The solution for $u(x)$ in Eq. (8–7.1) is approximated by Eq. (8–7.4), so that the coefficients a_n satisfy

$$\sum_{m=1}^{N} a_m v_m(x_n) = u(x_n) \ , \qquad n = 1,2,\ldots,N \ . \qquad (8-7.11)$$

For example, if $\{x_n\}$ are uniformly distributed between x_- and x_+, and $\{v_n\}$ are functions such as sines and cosines, Eq. (8–7.11) is a discrete Fourier transform. The prescription for collocation methods is then to use Eq. (8–7.4) in Eq. (8–7.1), and find a set of N equations for $\{a_n\}$ by evaluating Eq. (8–7.1) at the points $\{x_n\}$. The distinguishing feature is that the continuum equation is satisfied exactly at only the grid points.

Note that in all three of these approximations, we have not yet specified the $\{v_n\}$ or supplied a prescription for carrying out the time integration. In each case, the formulation converts Eq. (8–7.1) into a set of ordinary differential equations.

8–7.2 Spectral Methods

In spectral methods, the dependent variables are written as linear superpositions of a set of basis functions of the independent variables, as in Eq. (8–7.4). The basis functions are chosen to have properties that simplify calculation of the series coefficients. For example, the functions should be analytic and orthogonal, should have some connection with the physical processes in the problem, and in some cases should satisfy boundary conditions.

Spectral methods and finite-element methods can be based on any of the three approximations described above: Galerkin, tau, or collocation. A number of different spectral approximations are possible, depending on which technique is adopted, the choice of basis functions, and the order of the approximation. All three types of spectral techniques can be generalized in a straightforward way to multidimensions. The time derivative is generally finite differenced and integrated either explicitly or implicitly. The most popular explicit integration is the leapfrog method, although Runge-Kutta methods are also used. Time differencing is discussed in Gottlieb and Orszag (1977).

Aliasing errors are often considered in conjunction with spectral methods. There is a very clear discussion of these in Roache (1982). The shortest-wavelength component of the expansion of $u(x,t)$ that can be resolved in a computational mesh has $\lambda = 2\Delta x$. Because there is generally more interest in accuracy at long wavelengths, this short-wavelength cutoff might not seem very important. However, short wavelengths are important if they interfere with the long wavelengths. In many problems of interest, the spectral components of u interact in such a way that energy appears in the long wavelengths that normally would appear in much shorter, unresolved wavelengths. On the coarse grid, these omitted short-wavelength components can appear as long *aliased* wavelengths and are subsequently treated that way in the calculations.

When the Galerkin approximation is used, the method is commonly called *spectral* or *fully spectral*. The basis sets used are generally orthonormal. When they are not, the integrals must be computed numerically. If the bases are orthonormal, and the operator $\mathcal{L}$ is linear, it is not necessary to solve a matrix for the $\{a_n\}$. If $\mathcal{L}$ is not linear, for some orthonormal bases there are extremely fast convolution processes which make the evaluation of the nonlinear terms relatively inexpensive. Such bases include, for example, Fourier and Chebyshev expansions, for which fast Fourier transforms and real cosine transforms exist. Aliasing errors do not appear in Galerkin and tau spectral methods. Galerkin methods throw away the terms with extraneous frequencies generated in the nonlinear terms. They eliminate aliasing by eliminating spurious frequencies from the configuration space interactions.

A *pseudospectral* method uses a collocation of points $\{x_i\}$ chosen to make the evaluation of Eq. (8–7.6) very efficient. For example, if Fourier series are used for the basis, the $\{x_i\}$ are equally spaced. If Chebyshev polynomials are used, the $\{x_i\}$ are roots of the polynomials. In these two representations, it is not necessary to solve a matrix, even when $\mathcal{L}$ is nonlinear. However, pseudospectral methods contain aliasing errors unless these are independently filtered. The aliasing errors are not large and can be neglected in many cases with enough dissipation in the calculation. Combining a pseudospectral method with suitable filtering essentially gives a spectral method.

For spectral methods to be competitive with finite-difference methods, very efficient algorithms are needed for performing the transform, Eq. (8–7.5), and its inverse, Eq.(8–7.4). Both of these operations generally require N^2 operations to perform, where N is the number of basis functions. Fortunately in Fourier analysis there are transform methods faster than N^2. An example of such a method is the fast Fourier transform or FFT whose cost scales as $N \log N$ rather than N^2 (Cooley and Tukey, 1965; Cooley, Lewis, and Welch, 1970; Temperton, 1983).

Consider an application of spectral methods to the set of two-dimensional incompressible Euler or Navier-Stokes equations (Gottlieb and Orszag, 1977). These are often solved in the vorticity and stream function formulation. The difficulties in solving the equations arise from nonlinear terms in the $\mathcal{L}$ operator in Eq. (8–7.6). The key is to use the fast transform to evaluate the derivative $\partial u/\partial x$, where now u stands for vorticity ω, and to evaluate the equation in conservation form in physical space. For complex exponential basis functions,

$$\frac{\partial u}{\partial x} = \frac{2\pi}{x_+ - x_-} \sum_{n=1}^{N} i\, n a_n \exp\left(\frac{2\pi i n \Delta x}{x_+ - x_-}\right) . \qquad (8-7.12)$$

The solution is then integrated forward in configuration space at the collocation points.

Galerkin methods are attractive because they are conservative and do not have aliasing errors. The Galerkin approximation formally enforces the condition that the error in the approximating function is orthogonal to the expansion functions. However, tau and Galerkin methods generally require twice the computational work of the collocation methods. Orszag (1972) showed that pseudospectral approximations can be roughly as accurate as Galerkin approximations at much less cost.

In summary, pseudospectral approximations have several advantages over Galerkin methods. They have fewer FFTs per step, or the same number on a smaller grid; they are simpler, and have less difficulty with complicated boundary conditions. Galerkin methods have no aliasing errors, however, and, as a wavenumber representation, spectra are readily and accurately available for diagnostic interpretations. Even when FFTs are used, over 80% of the time required by a spectral calculation is spent doing transforms whose cost can become excessive for physically complex, nonlinear problems.

When the problem has periodic boundary conditions, the basis should be composed of complex Fourier functions. However, if sinusoidal functions are used for problems without periodic boundary conditions, the answers can be relatively poor throughout the entire domain (Gottlieb and Orszag, 1977). This is understandable in terms of the global nature of spectral methods. All of the points are used to compute derivatives so an error at one location is felt immediately all over the computational region. When the boundary conditions are not periodic, the basis set should consist of, for example, Chebyshev or Legendre polynomials (Gottlieb and Orszag, 1977). Chebyshev polynomials are useful not only because they allow a variety of boundary conditions, but because FFTs (in particular, real cosine transforms) can be used very efficiently with them.

Spectral methods are best for problems with either square or circular boundaries. When the boundaries have other shapes, conformal mapping procedures can be used (Orszag, 1980). In principle this is not difficult, but each reformulation or motion that results in a change of boundary conditions or topology requires reprogramming the equations and changing aspects of the representation (Gottlieb and Orszag, 1977).

As long as there are no sharp gradients, spectral methods require less resolution than finite-difference methods as they have, in a sense, infinite order accuracy. Thus they are competitive and sometimes superior to finite-difference methods for incompressible flows. However, spectral methods do not guarantee positivity in the way that monotone finite-difference algorithms do. Therefore they are not as good for resolving moving discontinuities such as shock fronts and breaking internal waves. There have been several attempts to incorporate monotonicity in spectral methods through flux-limiting procedures, such as the initial effort of Boris and Book (1976), Taylor et al. (1981), Zalesak (1981), and McDonald et al. (1985).

There has been a recent effort to extend spectral methods to compressible flows, beginning with Gottlieb and Orszag (1977), and more recently by Gottlieb et al. (1984) and Hussaini (1984). There is a fundamental problem with this approach. The method is intrinsically global so the entire computational domain instantly senses what is happening everywhere else. This means that local phenomena such as shocks respond to what happens outside the Mach cone, a physically unrealistic situation. Implicit methods have this same problem with shocks, but it happens in both implicit and explicit spectral methods. Two examples of problems that arise are shocks with large oscillations from the Gibbs phenomena, and precursors in the material ahead of the shock. Possible remedies are to filter the unwanted information from the region around the discontinuity or to use shock-fitting techniques.

Spectral methods have become extremely popular with the development of fast transform methods and have been used extensively in numerical weather prediction and simulation of turbulent flows. They are best for idealized flows with only a few nonlinear terms where smoothly varying solutions must be very accurate. For idealized shock problems, filters can be found and the calculations appear satisfactory. However, for complicated problems with discontinuities and strong shocks, spectral methods are not optimal.

8–7.3 Finite-Element Methods

Finite-element techniques have been used for years to solve difficult, practical problems in structural engineering. Later they became interesting to mathematicians who formulated their properties in terms of broad classes of approximations. Since then, finite-element methods have been modified for a variety of problems in fluid dynamics and heat transfer, and most recently for time-dependent fluid dynamics. Their application to solutions of systems of continuity equations are of interest here. Good general references to finite-element methods include Strang and Fix (1973) and Zienkiewicz (1977). The recent book by Zienkiewicz and Morgan (1983) introduces the use of finite elements for solving partial differential equations. More recent collections of papers highlighting some of the latest methods for fluid dynamics are by Argyris et al. (1984) and Gallagher et al. (1985).

There are two ways of viewing finite-element methods for solving partial differential equations. The first is global. The independent variable u is expanded over the whole region $[x_-, x_+]$, as written in equation Eq. (8–7.4). Here N is the number of functions used in the expansion. One common choice for the basis functions v_n is called *tent*, *roof*, or *teepee* functions,

$$v_n(x) = \begin{cases} \dfrac{x - x_{n-1}}{x_n - x_{n-1}} & \text{if } x_{n-1} \le x \le x_n \ , \\[2ex] \dfrac{x_{n+1} - x}{x_{n+1} - x_n} & \text{if } x_n \le x \le x_{n+1} \ , \\[2ex] 0 & \text{otherwise} \ , \end{cases} \qquad (8-7.13)$$

where the $\{x_n\}$ satisfy Eq. (8–7.10). In this representation, each v_n consists of two straight lines. One of the lines has a positive slope between the *nodes* x_{n-1} and x_n and the other line has a negative slope between the nodes x_n and x_{n+1}. The region between two neighboring nodes is called an *element*. Equation (8–7.13) defines the *shape function* that spans the subregion $[x_{n-1}, x_{n+1}]$. This piecewise linear shape function has nonzero values over two elements, and is zero elsewhere, as discussed in Chapter 6.

Many choices of v_n are possible. However, the number of functions needed in an expansion such as Eq. (8–7.4) is related to the degree of the polynomial, the smoothness or continuity imposed on the nodes, and the number of elements in the domain.

The second approach is local. We expand each element in a polynomial

$$v_e(x) = \sum_{i=1}^{I} a_{e,i} x^i \qquad (8-7.14)$$

so that the coefficient $a_{e,i}$ refers now to the coefficient of the ith term of the polynomial for element e. In the global representation, continuity at the nodes is built in. With this local representation, added constraints ensure the proper degree of continuity of the shape functions at the nodes.

The next step in the development of the method is somewhat different from the spectral methods described above. We define an *error function* or *residual function* R_u, as a measure of how much our expansion fails to satisfy the original equation, Eq. (8–7.1),

$$R_u(\{v_e\}, t) = \frac{\partial u}{\partial t} - \mathcal{L}[u] - f \ . \qquad (8-7.15)$$

Then we ask that

$$\int_{x_-}^{x_+} W_l R_u \, dx \doteq 0 \ , \qquad l = 1, 2, \ldots, N \ , \qquad (8-7.16)$$

where $\{W_l\}$ is a set of independent *weighting functions*. Using the expression for u in Eq. (8–7.4) in Eq. (8–7.16) leads to a set of coupled ODEs of the form

$$\mathbf{C} \cdot \dot{\mathbf{a}} + \mathbf{K} \cdot \mathbf{a} - \mathbf{g} = 0 . \qquad (8-7.17)$$

The $\mathbf{C}$, $\mathbf{K}$, and $\mathbf{g}$ matrices are integrals over the entire region $[x_-, x_+]$ if we are using global functions and over a subregion if we use local functions,

$$\begin{aligned} C_{lm} &= \int W_l N_m d\Omega \\ K_{lm} &= -\int W_l \mathcal{L} N_m d\Omega \\ g_l &= -\int f W_l d\Omega . \end{aligned} \qquad (8-7.18)$$

This is called the *weighted residual approximation*. The functions $\{W_l\}$ may be chosen in a number of ways. The choice $W_l = v_l$ gives a Galerkin method. The choice $W_l = \delta(x - x_l)$ gives a point collocation method. Another weighting, $W_l = x^{l-1}$, is called the method of moments. Each of these choices has different strengths and weaknesses.

If time is advanced by an explicit finite-difference formula, it is necessary to solve a linear matrix problem at each step. When the time is advanced implicitly, there are additional problems. The cost in computer time for large matrix inversions at each timestep is often prohibitive, although some approaches circumvent this (Löhner, 1986; Löhner et al., 1986).

Finite-element methods in multidimensions require a great deal of computer storage. The cost of recomputing geometric and shape functions is often prohibitive for each physical variable transported. Thus these variables are usually stored, requiring dozens of quantities per node in two dimensions and many more in three dimensions.

There are compensating advantages to finite-element methods. It is relatively straightforward to generalize finite-element methods to multidimensional problems in several dependent variables. In particular, the shape of the domains do not have to be regular, so the method is well suited to problems with irregular boundaries. Often triangular elements are advantageous. The domain can also change in the course of time to allow improvements in resolution where needed.

We particularly want to mention two new finite-element approaches to solving time-dependent fluid dynamics problems. These are the *moving finite-element method* (see, for example, Miller and Miller, 1981; Gelinas et al., 1981; Djomehri et al., 1986), and the method developed by Löhner et al. (1984, 1985). Both of these are Galerkin methods using tent functions which are or can be made multidimensional. However, there are certain fundamental differences in the two approaches.

In the method by Löhner et al., time is advanced by an explicit second-order Lax-Wendroff discretization and has been extended to meshes that adapt in time. The degrees of freedom are the values of the dependent variables, represented by u, at a node. The nodes are subdivided in front of advancing gradients and structures according to known properties of the evolving system. This appears to be a rather straightforward method which combines the accuracy and flexible grid of finite elements with some of the potentially convenient features of finite-difference methods.

In the moving finite-element method, node positions are also treated as unknowns. Forming the weighted residual with this flexibility gives coupled ODEs for node positions and u. For example, in two dimensions, the ODEs contain $\dot{u}$, $\dot{x}$, and $\dot{y}$. The resulting equations are stiff and always implicit. The computational cost, therefore, can be substantial. However, in addition to the value of u, the method also produces "best" values for the positions of the nodes, given a certain number of nodes. Accordingly, nodes tend to migrate toward regions where there are abrupt changes in u. Thus problems characterized by advancing surfaces may be more efficiently solved by such methods because the number of nodes can be minimized, even though a relatively large amount of computation is required for each node.

An interesting question is whether finite-difference methods can be expressed as a special case of the weighted residual process with locally defined shape functions. In some cases, the simple finite-difference methods and finite-element methods result in the same algorithms. The answer is "generally yes" (see, for example, Zienkiewicz and Morgan, 1983), but it takes some effort to provide a proof for individual cases. The finite-difference expressions can be interpreted as a particular case of weighted residuals in which the weights are delta functions, $W_l = \delta(\mathbf{x} - \mathbf{x}_l)$, and the shape function represents the derivative in the same way as the finite-difference operator.

Finite-element methods have not been exploited for fluid calculations as fully as finite-difference and spectral methods. In general the number of operations per finite-element node exceeds that per mesh point in finite-difference or finite-volume methods, so finite-element methods must be more accurate if they are to be competitive. The recent developments by Löhner et al. (1986), which incorporate the monotone FCT algorithm into a finite-element approach, are competitive with finite-difference methods. An example of a typical problem of a shock over irregular object is shown in Figure 8–6. One drawback of this approach is the amount of computer memory per mesh point that it requires, a serious issue when memory is at a premium.

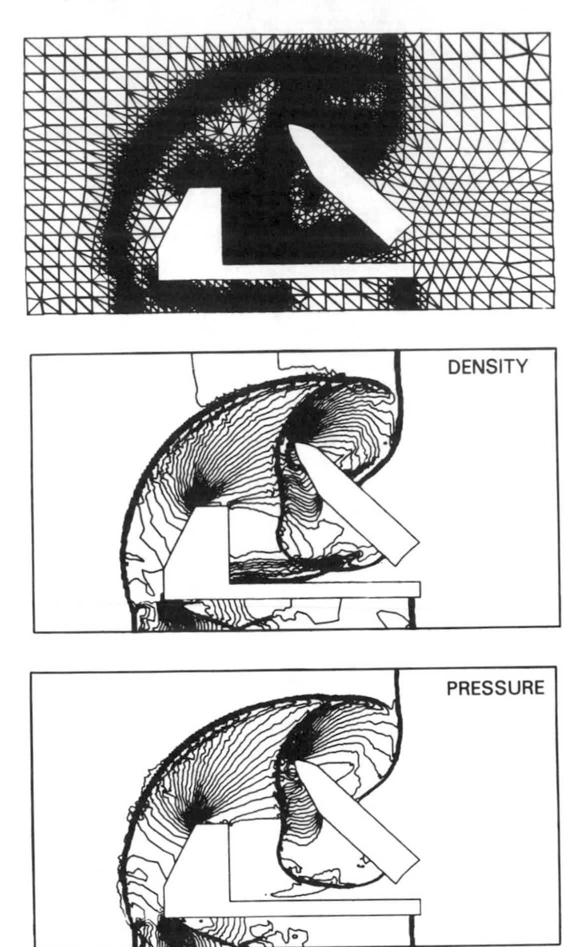

Figure 8–6. Finite-element FCT calculation of a shock over two irregularly shaped obstacles. (Figure is courtesy of R. Löhner.)

However, these methods are accurate, relatively fast, and allow very flexible, adaptive gridding. Extensions to three dimensions seem straightforward and potentially affordable.

References

Argyris, J.H., T.J.R. Hughes, and J.T. Oden, eds., 1984, *Comp. Meth. Appl. Mech. Eng.* 45: 1–362.

Baer, M.R., and R.J. Gross, 1986, *A Two-dimensional Flux-Corrected Transport Solver for Convectively Dominated Flows*, SAND85–0613, Sandia National Laboratories, Albuquerque, NM.

Boris, J.P., 1971, A Fluid Transport Algorithm that Works, in *Computing as a Language of Physics*, International Atomic Energy Agency, Vienna, 171–189.

Boris, J.P., 1976, Flux-Corrected Transport Modules for Generalized Continuity Equations, NRL Memorandom Report 3237, Naval Research Laboratory, Washington, DC.

Boris, J.P., and D.L. Book, 1973, Flux-Corrected Transport I: SHASTA — A Fluid Transport Algorithm that Works, *J. Comp. Phys.* 11: 38–69.

Boris, J.P., and D.L. Book, 1976, Solution of the Continuity Equation by the Method of Flux-Corrected Transport, *Methods in Computational Physics*, 16: 85–129.

Boris, J.P., J.H. Gardner, E.S. Oran, R.H. Guirguis, and G. Patnaik, 1987, LCPFCT — Flux-Corrected Transport for Generalized Continuity Equations, NRL Memorandum Report, Naval Research Laboratory, Washington, DC, to appear.

Chakravarthy, S.R., and S. Osher, 1985a, A New Class of High Accuracy TVD Schemes for Hyperbolic Conservation Laws, AIAA Paper No. 85-0363, AIAA, New York.

Chakravarthy, S.R., and S. Osher, 1985b, Application of a New Class of High Accuracy TVD Schemes to the Navier-Stokes Equations, AIAA Paper No. 85-0165, AIAA, New York.

Chakravarthy, S.R., A. Harten, and S. Osher, 1986, Essentially Non-Oscillatory Shock-Capturing Schemes of Arbitrarily-High Accuracy, AIAA Paper No. 86-0339, AIAA 24th Aerospace Sciences Meeting, AIAA, New York.

Chapman, M., 1981, FRAM — Nonlinear Damping Algorithms for the Continuity for Gas-dynamical Simulations, *J. Comp. Phys.* 44: 84–103.

Colella, P., and P.R. Woodward, 1984, The Piecewise Parabolic Method (PPM) for Gas-Dynamical Simulations, *J. Comp. Phys.* 54: 174–201.

Cooley, J.W. and J.W. Tukey, 1965, An Algorithm for the Machine Calculation of Complex Fourier Series, *Math. Comp.* 19: 297–301.

Cooley, J.W., P.A.W. Lewis, and P.D. Welch, 1970, The Fast Fourier Transform Algorithm: Programming Considerations in the Calculation of Sine, Cosine and Laplace Transforms, *J. Sound Vib.* 12: 315–337.

Courant, R., K.O. Friedrichs, and H. Lewy, 1928, Über die partiellen Differenzengleichungen der mathematischen Physik, *Math. Ann.* 100: 32–74. See translation in *IBM Journal*, March, 1967, 215–234.

Davis, S.F., 1984, *TVD Finite Difference Schemes and Artificial Viscosity*, ICASE Report No. 84–20, NASA, Langley Research Center, Hampton, VA.

Djomehri, M.J., S.K. Doss, R.J. Gelinas, and K. Miller, 1986, Applications of the Moving Finite Element Method for Systems in 2-D, *J. Comp. Phys.*, submitted.

Emery, A.F., 1968, An Evaluation of Several Differencing Methods for Inviscid Fluid Flow Problems, *J. Comp. Phys.*, 2: 306–331.

Fletcher, C., 1984, The Galerkin Method and Burgers' Equation, in J. Noye, ed., *Computational Techniques for Differential Equations*, North-Holland, New York, 355–476.

Forester, C.K., 1977, Higher Order Monotonic Convective Difference Schemes, *J. Comp. Phys.* 23: 1–22.

Gallagher, R.H., G.F. Carey, J.T. Oden, and O.C. Zienkiewicz, 1985, eds., *Finite Elements in Fluids*, Vol. VI, Wiley.

Gelinas, R.J., S.K. Doss, and K. Miller, 1981, The Moving Finite Element Method: Applications to General Partial Differential Equations with Multiple Large Gradients, *J. Comp. Phys.* 40: 202–249.

Gottlieb, D., and E. Turkel, 1976, Dissipative Two-Four Methods for Time Dependent Problems, *Math. Comp.* 30: 703–723.

Gottlieb, D., and S.A. Orszag, 1977, *Numerical Analysis of Spectral Methods: Theory and Applications*, SIAM, Philadelphia, PA.

Gottlieb, D., M.Y. Hussaini, and S.A. Orszag, 1984, in R.G. Voigt, D. Gottlieb, and M.Y. Hussaini, eds., *Spectral Methods for Partial Differential Equations*, SIAM, Philadelphia, PA, 1–55.

Godunov, S.K., 1959, Finite Difference methods for Numerical Computation of Discontinuous solutions of the Equations of Fluid Dynamics, *Mat. Sb.* 47: 271–306.

Harten, A., and G. Zwas, 1972, Self-Adjusting Hybrid Schemes for Shock Computations, *J. Comp. Phys.* 6: 568–583.

Harten, A., 1974, *The Method of Artificial Compression*, CIMS Rept. COO-3077-50, Courant Institute, New York University, New York.

Harten, A., 1983, High Resolution Schemes for Hyperbolic Conservation Laws, *J. Comp. Phys.* 49: 357–93.

Hussaini, M.Y., and T.A. Zang, 1984, in R.G. Voigt, D. Gottlieb, and M.Y. Hussaini, eds., *Spectral Methods for Partial Differential Equations*, SIAM, Philadelphia, PA, 119-140.

Kreiss, H. and J. Oliger, 1973, *Methods for the Approximate Solution of Time Dependent Problems*, Global Atmospheric Research Programme (GARP) Publication Series No. 10, World Meteorological Organization and International Council fo Scientific Unions, World Meteorological Organization, Geneva, Switzerland.

Kutler, P., and Lomax, H., 1971, The Computation of Supersonic Flow Fields about Wing-Body Combination by 'Shock Capturing' Finite Difference Techniques, in M. Hold, ed., *Proceedings of the Second International Conference on Numerical Methods in Fluid Dynamics, Lecture Notes in Physics*, vol. 8, Springer-Verlag, New York, 24–29.

Lanczos, C., 1956, *Applied Analysis*, Prentice-Hall, Englewood Cliffs, NJ.

Lax, P.D., and B. Wendroff, 1960, Systems of Conservation Laws, *Comm. Pure Appl. Math.*, 13: 217–237.

Lax, P.D., and B. Wendroff, 1964, Difference Schemes for Hyperbolic Equations with High Order of Accuracy, *Comm. Pure Appl. Math.*, 17: 381–398.

Löhner, R., 1986, An Adaptive Finite Element Schemo for Transient Problems in CFD, *Comp. Meth. Appl. Mech. Eng.*, submitted.

Löhner, R., K. Morgan, M. Vahdati, J.P. Boris, and D.L. Book, 1986, FEM-FCT: Combining Unstructured Grids with High Resolution, *J. Comp. Phys.*, submitted.

Löhner, R., K. Morgan, and O.C. Zienkiewicz, 1985, An Adaptive Finite Element Procedure for Compressible High Speed Flows, *Comp. Meth. Appl. Mech. Eng.*, 51: 441–465.

MacCormack, R.W., 1969, The Effect of Viscosity in Hypervelocity Impact Cratering, AIAA Paper No. 69-352, AIAA, New York.

MacCormack, R.W., 1971, Numerical Solution of the Interaction of a Shock Wave with a Laminar Boundary Layer, in M. Holt, ed., *Proceedings of the Second International Conference on Numerical Methods in Fluid Dynamics, Lecture Notes in Physics*, vol 8, Springer-Verlag, New York, 151-163.

McDonald, B.E., J. Ambrosiano, and S. Zalesak, 1985, The Pseudospectral Flux Correction (PSF) Method for Scalar Hyperbolic Problems, in R. Vichnevetsky, ed., *Proceedings of the Eleventh International Association for Mathematics and Computers in Simulation World Congress*, vol. 1, Rutgers University Press, New Brunswick, NJ, 67-70.

Miller, K., and R. Miller, 1981, Moving Finite Elements, Part I., *SIAM J. of Num. Anal.* 18: 1019–1032.

Orszag, S.A., 1972, Comparison of Pseudospectral and Spectral Approximations, *Stud. App. Math.* 51, 253–259.

Orszag, S., 1980, Spectral Methods for Problems in Complex Geometries, *J. Comp. Phys.* 37: 70–92.

Potter, D., 1973, *Computational Physics* Wiley, New York.

Richtmyer, R.D., 1963, *A Survey of Difference Methods for Nonsteady Fluid Dynamics*, NCAR Technical Note 63-2, Boulder, CO.

Richtmyer, R.D., and K.W. Morton, 1967, *Difference Methods for Initial-Value Problems*, Interscience, New York.

Roache, P.J., 1982, *Computational Fluid Dynamics*, Hermosa Publishers, Albuquerque, NM.

Roe, P.L., 1981, Approximate Riemann Solvers, Parameter Vectors, and Difference Schemes, *J. Comp. Phys.* 43: 357–372.

Roe, P.L., 1985, Some Contributions to the Modelling of Discontinuous Flows, *Lectures in Applied Mathematics* 22: 163–193.

Rood, R.B., 1987, Numerical Advection Algorithms and their Role in Atmospheric Transport and Chemistry Models, *Rev. Geophys.* 25: 71–100.

Strang, F., and G. Fix, 1973, *An Analysis of the Finite Element Method*, Printice-Hall, Engle Wood Cliffs, NJ.

Sweby, P.K., 1984, High Resolution Schemes using Flux Limiters for Hyperbolic Conservation Laws, *SIAM J. Numer. Anal.* 21: 995–1011.

Taylor, T.D., R.B. Myers, and J.H. Albert, 1981, Pseudo-Spectral Calculations of Shock Waves, Rarefaction Waves and Contact Surfaces, *Comp. Fluids* 4: 469–473.

Temperton, C. J., 1983, Self-Sorting Mixed-Radix Fast Fourier Transforms, *J. Comp. Phys.* 52: 1–23.

Turkel, E., 1980, Numerical Methods for Large-Scale Time-Dependent Partial Differential Equations, in W. Kollman, ed., *Computational Fluid Dynamics* 2: 127–262, Hemisphere Publishing Co., Washington, DC.

van Leer, B., 1973, Towards the Ultimate Conservative Difference Scheme. I. The Quest of Monotonicity, in H. Cabannes and R. Temam, eds., *Lecture Notes in Physics* 18, Springer-Verlag, Berlin, 163–168.

van Leer, B., 1979, Towards the Ultimate Conservative Difference Scheme. V. A Second-Order Sequal to Godunov's Method, *J. Comp. Phys.* 32: 101–136.

Voigt, R.G., D. Gottlieb, and M.Y. Hussaini, eds., 1984, *Spectral Methods for Partial Differential Equations*, SIAM, Philadelphia, PA.

von Neumann, J., and R.D. Richtmyer, 1950, A Method for the Numerical Calculation of Hydrodynamic Shocks, *J. App. Phys.* 21: 232–257.

Woodward, P., and P. Colella, 1984a, The Numerical Simulation of Two-Dimensional Fluid Flow with Strong Shocks, *J. Comp. Phys.* 54: 115–173.

Woodward, P., and P. Colella, 1984b, The Piecewise Parabolic Method (PPM) for Gas-Dynamical Simulations, *J. Comp. Phys.* 54: 174–201.

Yee, H.C., R.F. Warming, and A. Harten, 1983, Implicit Total Variation Diminishing (TVD) Schemes for Steady-state Calculations, AIAA Paper No. 83-1902, AIAA, New York.

Zalesak, S.T., 1979, Fully Multidimensional Flux-Corrected Transport Algorithms for Fluids, *J. Comp. Phys.* 31: 335–362.

Zalesak, S.T., 1981, Very High Order and pseudospectral Flux-Corrected Transport (FCT) Algorithms for Conservations Laws, in R. Vichnevetsky and R.S. Stepleman, *Advances in Computer Methods for Partial Differential Equations*, Vol. IV, International Association for Mathematics and Computers in Simulation (IMACS), Rutgers University, New Brunswick, NJ, 126–134.

Zalesak, S.T., 1984, A Physical Interpretation of the Richtmyer Two-Step Lax-Wendroff Scheme, and Its Generalization to Higher Spatial Order, in R. Vichnevetsky and R.S. Stepleman, *Advances in Computer Methods for Partial Differential Equations*, Vol. V, International Association for Mathematics and Computers in Simulation (IMACS), Rutgers University, New Brunswick, NJ, 491–496.

Zalesak, S.T., 1987, A Prelininary Comparison of Modern Shock-Capturing Schemes: Linear Advection, in R. Vichnevetsky and R.S. Stepelman, eds., *Advances in Computer Methods for Partial Differential Equations*, VI, IMACS, Rutgers University, New Brunswick, NJ, to appear.

Zienkiewicz, O.C., 1977, *The Finite Element Method*, McGraw-Hill, New York.

Zienkiewicz, O.C., and K. Morgan, and 1983, *Finite Elements and Approximation*, John Wiley, New York.

Chapter 9

COUPLED CONTINUITY EQUATIONS FOR FAST AND SLOW FLOWS

In Chapter 8 we described a number of ways to solve a single continuity equation numerically, and now we consider solving the coupled continuity equations of fluid dynamics. In this chapter we discuss Eulerian methods, and in Chapter 10 we discuss Lagrangian methods.

Table 2–2 listed five regimes of fluid speed used to characterize the behavior of fluids. These are, in order of increasing Mach number: incompressible, subsonic, transonic, supersonic, and hypersonic flows. The boundaries between these regimes loosely mark the onset or cessation of certain physical phenomena. Each of these five regimes have peculiar features that make certain solution procedures and algorithms more effective than others. In this chapter we describe methods for solving coupled flow problems in two regimes which are composites of those listed in Table 2–2: *slow flows*, whose velocities are subsonic and below, and *fast flows*, whose velocities are subsonic and above.

9–1. COUPLED CONTINUITY EQUATIONS

9–1.1. Equations of Incompressible and Compressible Flow

For a two- or three-dimensional *incompressible* flow, we can solve the *primitive equations*,

$$\frac{\partial \rho}{\partial t} + \mathbf{v} \cdot \nabla \rho = 0 , \qquad (9-1.1)$$

$$\nabla \cdot \mathbf{v} = 0 , \qquad (9-1.2)$$

$$\rho \left(\frac{\partial \mathbf{v}}{\partial t} + \mathbf{v} \cdot \nabla \mathbf{v} \right) + \nabla P = 0 . \qquad (9-1.3)$$

This representation uses the *primitive variables* ρ and $\mathbf{v}$ for density and velocity, respectively. When the fluid is incompressible, the equation for energy conservation is replaced by the constraint Eq. (9–1.2).

Alternately, Eqs. (9–1.2) and (9–1.3) can be replaced by an equivalent system, the *vorticity dynamics* equations,

$$\omega \equiv \nabla \times \mathbf{v} , \tag{9-1.4}$$

$$\mathbf{v} = \nabla \times \boldsymbol{\psi} , \tag{9-1.5}$$

$$\nabla^2 \boldsymbol{\psi} = \omega , \tag{9-1.6}$$

$$\frac{\partial \omega}{\partial t} + \nabla \cdot (\mathbf{v}\omega) = -\nabla \times \left(\frac{\nabla P}{\rho} \right) . \tag{9-1.7}$$

Here $\boldsymbol{\psi}$ is the vector stream function and ω is the vector vorticity. In two dimensions, both the stream function and vorticity are scalars, and Eqs. (9–1.5) – (9–1.7) become

$$\mathbf{v} = \hat{\mathbf{z}} \times \nabla \psi , \tag{9-1.8}$$

$$\nabla^2 \psi = \omega , \tag{9-1.9}$$

$$\frac{\partial \omega}{\partial t} + \nabla \cdot (\mathbf{v}\omega) = -\nabla \times \left(\frac{\nabla P}{\rho} \right) , \tag{9-1.10}$$

where $\hat{\mathbf{z}}$ is the unit vector in the z-direction. The pressure is found from an elliptic equation,

$$\nabla^2 P = \frac{1}{\rho} \nabla \rho \cdot \nabla P - \rho \nabla \mathbf{v} : \nabla \mathbf{v} , \tag{9-1.11}$$

obtained by taking the divergence of Eq. (9–1.3) and using Eq. (9–1.2). The density is advanced using Eq. (9–1.1). Note that for incompressible two-dimensional flows at constant density,

$$\frac{d\omega}{dt} \equiv \frac{\partial \omega}{\partial t} + \mathbf{v} \cdot \nabla \omega = 0 . \tag{9-1.12}$$

Both the primitive equation formulation and the vorticity dynamics formulation involve convective equations such as Eqs. (9–1.1), (9–1.3), (9–1.7), and (9–1.10). In addition, the vorticity formulation involves Poisson-like equations, such as Eq. (9–1.6), (9–1.9), and (9–1.11), whose numerical solutions are discussed in Chapter 11.

For compressible flows, we replace Eq. (9–1.1) by the continuity equation,

$$\frac{\partial \rho}{\partial t} + \mathbf{v} \cdot \nabla \rho + \rho \nabla \cdot \mathbf{v} = 0 , \tag{9-1.13}$$

and Eq. (9–1.2) either by an isentropic pressure equation,

$$\frac{\partial P}{\partial t} + \mathbf{v} \cdot \nabla P + \gamma P \nabla \cdot \mathbf{v} = 0 , \tag{9-1.14}$$

or by an equation for the energy E,

$$\frac{\partial E}{\partial t} + \nabla \cdot (\mathbf{v}(P + E)) = 0 . \qquad (9-1.15)$$

For an ideal gas,

$$E = \epsilon + \frac{1}{2}\rho v^2 . \qquad (9-1.16)$$

When $\nabla \cdot \mathbf{v} \neq 0$, the fluid velocity $\mathbf{v}$ is the sum of a vector and a scalar potential,

$$\mathbf{v} = \nabla \times \boldsymbol{\psi} + \nabla \phi . \qquad (9-1.17)$$

This is the same as Eq.(2–2.14) discussed in Chapter 2, except that we have omitted the particular solution ϕ_p. Both $\boldsymbol{\psi}$ and ϕ satisfy Poisson equations. The source of ϕ may be written

$$\sigma \equiv \nabla \cdot \mathbf{v} , \qquad (9-1.18)$$

where σ satisfies a convective equation

$$\frac{\partial \sigma}{\partial t} + \nabla \cdot (\mathbf{v}\sigma) = \sigma^2 - \nabla \mathbf{v} : \nabla \mathbf{v} + \frac{\nabla \rho \nabla P - \rho \nabla^2 P}{\rho^2} , \qquad (9-1.19)$$

which results from taking the divergence of Eq. (9–1.3). For compressible flow, there is generally no advantage in using the $\boldsymbol{\omega} - \sigma$ formulation.

Both Eqs. (9–1.14) and (9–1.15) are convective equations and can be solved by the same methods, but there is an important physical distinction between the two. Equation (9–1.14) can only be used to describe flows without shocks, that is, isentropic flows. Using the energy conservation equation, Eq. (9–1.15), brings an added difficulty, the need to extract the pressure P from Eq. (9–1.16). When the kinetic energy dominates the internal energy, that is, $P \ll E$, negative pressures can result from small truncation errors as differences of large numbers. These derived negative pressures can be made positive by using the isentropic pressure equation to provide a lower bound on the pressure, but this requires solving an additional continuity equation.

9–1.2. Time Integration

A fundamental decision in finite-difference approaches is what method to use for the time integration. This decision is dominated by the question of whether to choose an explicit or an implicit method. Consider an evolution formula of the general form

$$\rho^{n+1} = (1-\theta)F_e(\rho^n) + \theta F_i(\rho^{n+1}) \, , \qquad (9-1.20)$$

where ρ is evaluated at the new time, t^{n+1} in terms of values at the old time t^n and the F_e and F_i are functions of ρ. The quantity θ is the implicitness parameter: $\theta = 0$ for an explicit solution, $\theta = 1$ for an implicit solution, and $\theta = \frac{1}{2}$ for a time-centered or semi-implicit solution.

In the worst cases, implicit methods require solving complicated matrix equations at each timestep. The best cases require inverting a tridiagonal matrix system of equations. Though the cost per timestep can be quite high compared to explicit methods, the advantage of implicit methods is their ability to remain numerically stable with much longer timesteps.

Whether we want to use an implicit or an explicit approach depends on the answer to two questions about the problem:

1. What are the important time scales we need to resolve?
2. What is the relation of the timestep we would like to take to the stability limits of the algorithm?

Suppose we wish to resolve a particular flow structure in a convection problem and this structure varies appreciably on a time scale τ. No calculation with $\Delta t \gg \tau$ can accurately describe the detailed variation of this structure. This is completely analogous to the case of the finite spatial mesh. If the mesh is too coarse, the structure cannot be resolved. If the time scale τ is less than the timestep required by the Courant condition, an explicit method is needed. However, if τ is greater than the timestep required to satisfy the Courant condition, that is, if the structure varies slowly enough, it is worthwhile to use an implicit method but ensure $\Delta t < \tau$.

Consider a situation where the phenomena we wish to resolve is not the fastest effect in the problem. For example, consider the expanding flow which results as air is blown out of a whistle. The flow speed is slow, so that the small-scale, high-speed sound vibrations have little effect on the slower flow pattern. This is a case in which we would clearly like to use an implicit method, because the speed of sound is about two orders of magnitude larger than the flow speed. We do not really wish to resolve the sound waves as long as mistreating them does not seriously effect the accuracy of calculating the slower flow.

Unfortunately, this last sentence contains the crux of the problem. A major difficulty with implicit convection algorithms is that they tend to damp strongly the high-frequency modes that cannot be resolved adequately. The result is that sonic energy may be incorrectly deposited as heat near the whistle. A time-centered method, or one that is nearly centered, allows much of the sonic energy to propagate away as sound without heating the air. Again, the choice of method depends strongly on the specific problem.

9–1.3. Calculating Fast and Slow Flows

Modeling combustion systems requires the description of flows which are subsonic, such as flames, and flows which are supersonic, such as shocks and detonations. In the best of all possible worlds, both regimes of flow, coupled to chemical kinetics and diffusive transport effects, would be treated with one all-inclusive algorithm. Much work has gone into the search for such algorithms in one and multidimensions. These methods are implicit and include, for example, RICE (Rivard et al., 1975) and the implicit MacCormack method (MacCormack, 1981), both discussed in Section 9–3. There is generally a heavy price for being able to use a single algorithm for both fast and slow flows, a price which translates into many computer operations per timestep often spent in solving multiple and complicated matrix equations.

Fundamental conflicts arise in trying to construct such methods. For the method to be economical for slow flows, it should handle sound waves implicitly. This is generally done by solving the pressure equation instead of the energy equation and often requires artificial viscosity to be stable. Thus, for the benefit of being able to take long timesteps when the flow velocities are low, the spatial resolution is poor and shock fronts are unnecessarily thick when the flows are fast.

An implicit algorithm can be dangerous for supersonic flows because it transmits numerical precursors ahead of the shock with essentially infinite speed. The best explicit methods for supersonic flows contain numerical algorithms and models for shocks that degrade their performance in subsonic flows. The result is that the excess numerical damping, which stabilizes the nonlinear effects in supersonic flows, appears as rapid damping when the flow velocity is small compared to the sound speed.

We generally recommend different methods for very slow and very fast flows. Such methods are somewhat limited in generality, but offer advantages in accuracy, speed, and simplicity compared to existing global or composite implicit algorithms. For this reason, in the remainder of this chapter, we discuss two flow regimes separately: slow flows and fast flows.

9–2. METHODS FOR FAST FLOWS

Fast flows have characteristic velocities ranging from below the sound speed to far above it. As shown in Table 2–2, this includes flows that are subsonic, transonic, supersonic, and hypersonic.

In the subsonic flow regime, there are no shocks, but compressibility effects can be important. This is a borderline regime, which for some problems can be treated with the implicit or slow-flow methods described in Section 9–3, and for other problems must be treated by solutions of the coupled continuity equations that fully resolve the sound waves. For example, to simulate the acoustic-vortex interactions in subsonic combustion chambers, it is necessary to resolve the sound waves.

In the transonic regime, the finite sound speed as well as compressional effects become important and play a major role as the flow speed approaches the sound speed. The geometry becomes crucial near Mach one as shocks may or may not develop depending on subtle variations in geometry. Transonic problems are of particular importance to aerodynamic and missile technology.

The supersonic flow regime is generally characterized by strong shocks, often with complicated interacting flow structures. In numerical simulations, the shocks must be well localized and changes in the variables across them must be calculated accurately. Here the trade-off is between work invested in special purpose programming to resolve contact discontinuities and work spent on finely resolved calculations. In the last ten years a point of diminishing returns has been reached for supersonic flow algorithms. The most profitable way to get significantly better calculations is through adaptive gridding, not through research on improving Eulerian algorithms.

Because the fluid velocity in the hypersonic flow regime can become much larger than the sound speed, it can be difficult to calculate the temperature accurately and to conserve energy locally. Calculating the thermal energy, in a formalism that conserves energy, requires taking the difference of the total energy density and the kinetic energy density which are nearly equal. The result is that there are possible large errors from taking differences of large numbers. These errors are compounded by the high temperatures accompanying hypersonic flows. Atmospheric reentry problems, for example, fall in this hypersonic regime.

In all supersonic flows, the CFL condition (Chapter 8) restricts the computational timestep and the grid spacing according to

$$\left. \frac{c_s}{\Delta x} \right|_{max} < \frac{1}{\Delta t}, \qquad (9-2.1)$$

where c_s is the sound speed. This condition on the sound waves is similar to limiting the flow of material through the grid to one cell per timestep. Where there is a strong shock, the sound speed generally becomes comparable to the flow speed. In this flow regime, explicit, Eulerian methods are usually effective. There is usually no obvious advantage to using an implicit method. An implicit method could be useful, however, to avoid an even smaller timestep restriction imposed by, for example, a very fine grid needed to resolve a boundary layer elsewhere in the calculation.

Adopting an Eulerian or a Lagrangian formulation in the supersonic regime depends in many cases on the duration of the transient problem. When the calculation is over after the fast fluid or sound waves have crossed the system once or twice, a situation occurring in many explosion, detonation, and shock calculations, a Lagrangian grid may not become excessively distorted and so may be useful in tracking material interfaces in the flow. To date, almost all atmospheric reentry calculations have been calculations that "march" to the steady-state solution of the Navier-Stokes equations on fixed or adaptive Eulerian grids. Some effort has also been invested in developing Lagrangian particle dynamics models for the rarefied gas dynamics regime, where particle mean free paths are comparable to the body dimensions.

To understand the intrinsic ambiguity of using Lagrangian methods for compressible flows, consider Eq. (9–1.15), the equation for energy, which shows that the effective velocity at which energy is transported is $\mathbf{v}^* \equiv ([P + E]/E)\mathbf{v}$. In some regions of the flow, $\mathbf{v}^*$ can be as large as $\gamma\mathbf{v}$. Although a Lagrangian mesh moves with the fluid, almost as much energy and momentum flow across this moving grid as would if the mesh were fixed. For this reason, we emphasize Eulerian methods for supersonic flows.

Shock capturing methods attempt to represent the movement of a shock which is usually much thinner than one cell. For example, the thickness of a shock may be on the order of 50 – 100 Å in a gas, but the computational cell might be centimeters in size. Because of this fundamental disparity in scales, the propagation of shocks must be treated in special ways, some of which are described in this section.

9–2.1 Lax-Wendroff Methods

In Chapter 4 a simple Lax-Wendroff method was presented to solve the moving square-wave problem. Chapter 8 extended this method to solve a single general continuity equation in one dimension. Now we show how to solve systems of coupled continuity equations (Lax and Wendroff, 1960, 1964) which model compressible flows with shocks.

The fluid dynamic equations in one dimension may be written in vector form as

$$\frac{\partial \mathbf{u}}{\partial t} + \frac{\partial}{\partial x}\mathbf{f}(\mathbf{u}) = 0 . \qquad (9-2.2)$$

In Eq. (9–2.2), $\mathbf{u}$ and $\mathbf{f}$ are defined as vectors

$$\mathbf{u} = \begin{pmatrix} \rho \\ \rho v \\ E \end{pmatrix} \quad \text{and} \quad \mathbf{f} = \begin{pmatrix} \rho v \\ P + \rho v^2 \\ v(E+P) \end{pmatrix} . \qquad (9-2.3)$$

The Lax Wendroff method uses a Taylor series in time, as in Eq. (8–1.2),

$$\mathbf{u}_i^{n+1} = \mathbf{u}_i^n + \Delta t \left.\frac{\partial \mathbf{u}}{\partial t}\right|_i^n + \frac{\Delta t^2}{2}\left.\frac{\partial^2 \mathbf{u}}{\partial t^2}\right|_i^n + \cdots , \qquad (9-2.4)$$

and the expansion

$$\frac{\partial^2 \mathbf{u}}{\partial t^2} = -\frac{\partial}{\partial t}\frac{\partial \mathbf{f}}{\partial x} = -\frac{\partial}{\partial x}\frac{\partial \mathbf{f}}{\partial t} = -\frac{\partial}{\partial x}\left(\mathbf{A}\cdot\frac{\partial \mathbf{u}}{\partial t}\right) = \frac{\partial}{\partial x}\left(\mathbf{A}\cdot\frac{\partial \mathbf{f}}{\partial x}\right) . \qquad (9-2.5)$$

Here $\mathbf{A}(\mathbf{u})$ is the Jacobian of $\mathbf{f}(\mathbf{u})$ with respect to $\mathbf{u}$, that is

$$A_{kj} \equiv \frac{\partial f_k}{\partial u_j} . \qquad (9-2.6)$$

Writing the partial derivatives as finite differences, we can solve for $\mathbf{u}_i^{n+1}$,

$$\begin{aligned} \mathbf{u}_i^{n+1} &= \mathbf{u}_i^n - \frac{1}{2}\frac{\Delta t}{\Delta x}(\mathbf{f}_{i+1}^n - \mathbf{f}_{i-1}^n) \\ &\quad + \frac{1}{2}\left(\frac{\Delta t}{\Delta x}\right)^2 \left[\mathbf{A}_{i+\frac{1}{2}}^n \cdot (\mathbf{f}_{i+1}^n - \mathbf{f}_i^n) - \mathbf{A}_{i-\frac{1}{2}}^n \cdot (\mathbf{f}_i^n - \mathbf{f}_{i-1}^n)\right] \end{aligned} \qquad (9-2.7)$$

where

$$\mathbf{A}_{i+\frac{1}{2}}^n = \mathbf{A}\left(\frac{1}{2}\mathbf{u}_{i+1}^n + \frac{1}{2}\mathbf{u}_i^n\right) . \qquad (9-2.8)$$

Equations (9–2.7) and (9–2.8) generalize Eqs. (8–1.10) – (8–1.13) to a system of coupled continuity equations in one dimension.

Richtmyer (1963) has shown that these methods can be written in two-step forms in which the first step is a first-order Lax method and the second step is a centered leapfrog method (Section 8–1.2). The generalization of Eq. (8–1.15) is

$$\begin{aligned} \mathbf{u}_{i+\frac{1}{2}}^{n+\frac{1}{2}} &= \frac{1}{2}\left[\mathbf{u}_{i+1}^n + \mathbf{u}_i^n\right] - \frac{\Delta t}{2}\left[\frac{\mathbf{f}_{i+1}^n - \mathbf{f}_i^n}{\Delta x}\right] \\ \mathbf{u}_i^{n+1} &= \mathbf{u}_i^n - \Delta t \left[\frac{\mathbf{f}_{i+\frac{1}{2}}^{n+\frac{1}{2}} - \mathbf{f}_{i-\frac{1}{2}}^{n+\frac{1}{2}}}{\Delta x}\right] , \end{aligned} \qquad (9-2.9)$$

where the values of $\{\mathbf{f}_{i+\frac{1}{2}}^{n+\frac{1}{2}}\}$ in the second step are based on the values $\{\mathbf{u}_{i+\frac{1}{2}}^{n+\frac{1}{2}}\}$ in the first step.

This two-step method also can be extended to multidimensions. Consider a two-dimensional generalization of Eq. (9–2.2),

$$\frac{\partial \mathbf{u}}{\partial t} + \frac{\partial}{\partial x}\mathbf{f_x}(\mathbf{u}) + \frac{\partial}{\partial x}\mathbf{f_y}(\mathbf{u}) = 0 \,, \tag{9-2.10}$$

with the definitions

$$\mathbf{u} \equiv \begin{pmatrix} \rho \\ \rho v_x \\ \rho v_y \\ E \end{pmatrix} \quad \mathbf{f_x} \equiv \begin{pmatrix} \rho v_x \\ P + (\rho v_x)^2/\rho \\ \rho v_x \rho v_y/\rho \\ \rho v_x (E+P)/\rho \end{pmatrix} \quad \mathbf{f_y} \equiv \begin{pmatrix} \rho v_y \\ \rho v_x \rho v_y/\rho \\ P + (\rho v_y)^2/\rho \\ \rho v_y (E+P)/\rho \end{pmatrix} . \tag{9-2.11}$$

A two-step Lax-Wendroff method can be written as two one-dimensional Lax-method steps taking the solution to time $t^{n+\frac{1}{2}}$, and then a leapfrog step:

$$\begin{aligned}
\mathbf{u}_{i+\frac{1}{2},j}^{n+\frac{1}{2}} &= \frac{1}{2}\left[\mathbf{u}_{i,j} + \mathbf{u}_{i+1,j}\right]^n \\
&\quad - \frac{\Delta t}{2}\left[\frac{(\mathbf{f_x})_{i+1,j} - (\mathbf{f_x})_{i,j}}{\Delta x} + \frac{(\mathbf{f_y})_{i+\frac{1}{2},j+1} - (\mathbf{f_y})_{i+\frac{1}{2},j-1}}{2\Delta y}\right]^n \\
\mathbf{u}_{i,j+\frac{1}{2}}^{n+\frac{1}{2}} &= \frac{1}{2}\left[\mathbf{u}_{i,j} + \mathbf{u}_{i,j+1}\right]^n \\
&\quad - \frac{\Delta t}{2}\left[\frac{(\mathbf{f_x})_{i+1,j+\frac{1}{2}} - (\mathbf{f_x})_{i-1,j+\frac{1}{2}}}{2\Delta x} + \frac{(\mathbf{f_y})_{i,j+1} - (\mathbf{f_y})_{i,j}}{\Delta y}\right]^n \\
\mathbf{u}_{i,j}^{n+1} &= \mathbf{u}_{i,j}^n \\
&\quad - \Delta t\left[\frac{(\mathbf{f_x})_{i+\frac{1}{2},j} - (\mathbf{f_x})_{i-\frac{1}{2},j}}{\Delta x} + \frac{(\mathbf{f_y})_{i,j+\frac{1}{2}} - (\mathbf{f_y})_{i,j-\frac{1}{2}}}{\Delta y}\right]^{n+\frac{1}{2}} .
\end{aligned} \tag{9-2.12}$$

The half-step flux values can be estimated as described in Eqs. (8–1.17) and (8–1.18).

As written, Lax-Wendroff methods are second-order accurate. They give good results for smooth flows with no sharp gradients or discontinuities. Sometimes they can be used when discontinuities are present if optimal spatial resolution and high accuracy at the discontinuity are not required. For general use where discontinuities are present, it is necessary to add artificial viscosity, as discussed briefly at the end of Section 8–1. Richtmyer and Morton (1967) and Potter (1973) introduce this subject. Our general advice is that better ways now exist to treat such flows.

9–2.2 FCT for Coupled Continuity Equations

Section 8–3 described monotone FCT algorithms for integrating a single continuity equation. We now extend the approach to calculating coupled one-dimensional continuity equations. The discussion also applies to other one-dimensional monotone algorithms for solving continuity equations. Specifically, we want to solve the three gas-dynamic equations simultaneously,

$$\frac{\partial \rho}{\partial t} = -\nabla \cdot \rho \mathbf{v}, \qquad (9-2.13)$$

$$\frac{\partial \rho \mathbf{v}}{\partial t} = -\nabla \cdot (\rho \mathbf{v}\mathbf{v}) - \nabla P \qquad (9-2.14)$$

and

$$\frac{\partial E}{\partial t} = -\nabla \cdot E\mathbf{v} - \nabla \cdot (\mathbf{v}P) \,. \qquad (9-2.15)$$

We limit the discussion to the one-dimensional problem here, but conclude this section with a description of how to combine several one-dimensional calculations to make a multidimensional monotone calculation.

Solving these coupled equations is best done by determining the timestep, then integrating from the old time t^o for a half timestep to $t^o + \frac{\Delta t}{2}$, and then integrating from t^o to the full timestep $t^o + \Delta t$. The half-step approximation is used to evaluate centered spatial derivatives and fluxes. Assume that the cell-center values of all fluid quantities are known at t^o. The integration procedure is:

1. Half-step calculation to find the first-order time-centered variables.
   a. Calculate $\{v_i^o\}$ and $\{P_i^o\}$ using the old values of $\{\rho_i^o\}$, $\{\rho_i^o v_i^o\}$, and $\{E_i^o\}$.
   b. Convect $\{\rho_i^o\}$ a half timestep to $\{\rho_i^{\frac{1}{2}}\}$.
   c. Evaluate $-\nabla P^o$ for the momentum sources.
   d. Convect $\{\rho_i^o v_i^o\}$ to $\{\rho_i^{\frac{1}{2}} v_i^{\frac{1}{2}}\}$.
   e. Evaluate $-\nabla \cdot (P^o v^o)$ for the energy sources.
   f. Convect $\{E_i^o\}$ to $\{E_i^{\frac{1}{2}}\}$.

2. Whole-step calculation to find second-order results at the end of the timestep.
   a. Calculate $\{v_i^{\frac{1}{2}}\}$ and $\{P_i^{\frac{1}{2}}\}$ using the provisional values $\{\rho_i^{\frac{1}{2}}\}$, $\{\rho_i^{\frac{1}{2}} v_i^{\frac{1}{2}}\}$, and $\{E_i^{\frac{1}{2}}\}$.
   b. Convect $\{\rho_i^o\}$ to $\{\rho_i^1\}$.
   c. Evaluate $-\nabla P^{\frac{1}{2}}$ for the momentum sources.
   d. Convect $\{\rho_i^o v_i^o\}$ to $\{\rho_i^1 v_i^1\}$.
   e. Evaluate $-\nabla \cdot P^{\frac{1}{2}} v^{\frac{1}{2}}$ for the energy sources.
   f. Convect $\{E_i^o\}$ to $\{E_i^1\}$.
3. Do another timestep from t^1 to t^2, as above.

This two-step, second-order time integration greatly increases the accuracy of the calculations.

Often we want to couple species equations to Eqs. (9–2.13) – (9–2.15),

$$\frac{\partial n_i}{\partial t} = -\nabla \cdot n_i \mathbf{v} \,, \qquad i = 1, ..., N_s \,. \tag{9 – 2.16}$$

In general we find that we do not have to split the timestep for these variables provided that the half-step velocities are used in advancing $\{n_i^o\}$ to $\{n_i^1\}$. After integrating the fluid variables for the half and whole timestep, we convect these species the full timestep using the centered velocities $\{v_i^{\frac{1}{2}}\}$.

Even with the timestep-splitting procedure just outlined, there are still inaccuracies in the solution of the coupled continuity equations. Phase errors are not completely eliminated, and *synchronizing* the fluxes, so that the flux-correction formula works consistently on the three equations, is also of some concern. Because monotone methods are nonlinear, the amount of flux correction differs from one variable to another. Because a momentum jump may slightly lead the numerical jump in energy in the vicinity of a shock, there is still the possibility of undershoots and overshoots. There are many problem-dependent fixes for these remaining quantitative errors and inaccuracies. One simple fix that often works is to take a smaller timestep. Another is to put a lower limit on derived quantities such as pressure, if such a limit is known. These synchronization inaccuracies are most serious in a chemically reactive flow with a large energy release.

One approach to problems caused by unsynchronized fluxes is to try more complicated ways of limiting the fluxes. When we discussed one-dimensional FCT for a single continuity equation for the generic variable ρ, we described how ρ was flux limited. Now we have the freedom to limit ρ, ρv, and E differently and separately. Various combinations of limiting procedures can help the synchronization problem. Flux limiting and flux

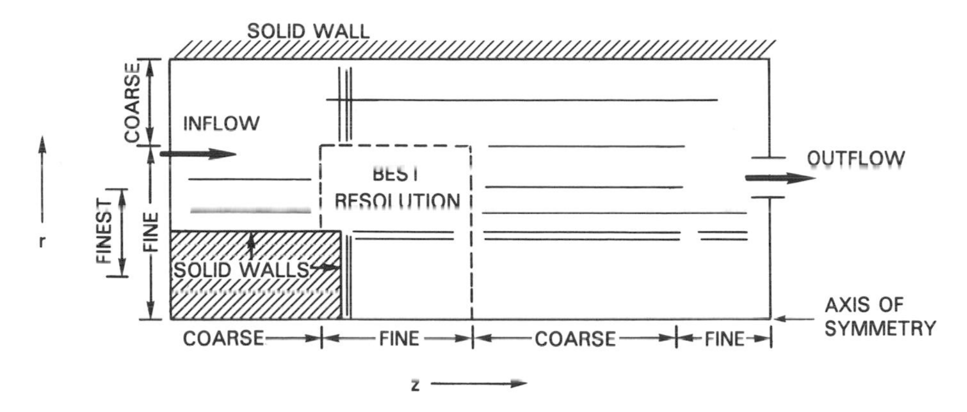

Figure 9–1. Schematic of the grid used in the two-dimensional axisymmetric flow problem.

synchronization have been discussed by Harten and Zwas (1972) and Zhmakin and Fursenko (1979), and recently by Löhner et al. (1986). How to limit the variables in a set of coupled equations is now the subject of avid research, and promises additional improvements. Note that the methods which incorporate Riemann solvers do not have this flux synchronization problem.

9–2.3 Multidimensions through Timestep Splitting

One-dimensional continuity equation solvers such as LCPFCT (Section 8–3) can be combined to construct a multidimensional program by timestep splitting in the several coordinate directions. This approach is straightforward when an orthogonal grid can be constructed with physical boundaries along segments of grid lines. The example shown schematically in Figure 9–1 is a cylindrically symmetric chamber with material entering through a nozzle in the center. The grid is orthogonal and lies along lines of constant r and z. The figure contains a region of high resolution formed by r and z grid lines clustered in the vicinity of the rim of the inlet. In this region, the smallest, most rapidly changing fluid motions arise which require the most spatial resolution and determine the timestep of the calculation. Other geometries, such as $(x - y)$, $(r - z)$, and the general orthogonal coordinates $(\eta - \xi)$ can also be used with timestep splitting.

The two-dimensional equations which describe ideal gas dynamics in axisymmetric $(r - z)$ geometry are

$$\begin{aligned}
\frac{\partial \rho}{\partial t} &= -\frac{\partial}{\partial z}(\rho v_z) - \frac{1}{r}\frac{\partial}{\partial r}(r\rho v_r) \\
\frac{\partial \rho v_r}{\partial t} &= -\frac{\partial}{\partial z}(\rho v_r v_z) - \frac{1}{r}\frac{\partial}{\partial r}(r\rho v_r v_r) - \frac{\partial P}{\partial r} \\
\frac{\partial \rho v_z}{\partial t} &= -\frac{\partial}{\partial z}(\rho v_z v_z) - \frac{1}{r}\frac{\partial}{\partial r}(r\rho v_z v_r) - \frac{\partial P}{\partial z} \\
\frac{\partial E}{\partial t} &= -\frac{\partial}{\partial z}[(E+P)v_z] - \frac{1}{r}\frac{\partial}{\partial r}[r(E+P)v_r] \, .
\end{aligned} \tag{9 - 2.17}$$

The pressure and energy are again related by

$$E = \epsilon + \frac{1}{2}\rho(v_r^2 + v_z^2) \, . \tag{9 - 2.18}$$

The right sides of Eqs. (9–2.17) are separated into two parts, the axial terms and the radial terms. This arrangement in each of the four equations separates the axial and the radial derivatives in the divergence and gradient terms into parts which can be treated sequentially by a general one-dimensional continuity equation solver.

Each axial row in the grid is integrated using the one-dimensional module to solve the four coupled continuity equations from time t to $t + \Delta t$. The axial split-step equations are written in the form of the general Eq. (8–3.11),

$$\begin{aligned}
\frac{\partial \rho}{\partial t} &= -\frac{\partial}{\partial z}(\rho v_z) \\
\frac{\partial \rho v_r}{\partial t} &= -\frac{\partial}{\partial z}(\rho v_r v_z) \\
\frac{\partial \rho v_z}{\partial t} &= -\frac{\partial}{\partial z}(\rho v_z v_z) - \frac{\partial P}{\partial z} \\
\frac{\partial E}{\partial t} &= -\frac{\partial}{\partial z}(E v_z) - \frac{\partial}{\partial z}(P v_z) \, ,
\end{aligned} \tag{9 - 2.19}$$

where α in Eq. (8–3.11) is set equal to one for planar geometry. Because the axial gradients and fluxes are being treated together, the one-dimensional integration connects those cells which are influencing each other through the axial component of convection.

The changes due to the derivatives in the radial direction must now be included. This is done in a second split step of one-dimensional integrations along each radial column,

$$\begin{aligned}
\frac{\partial \rho}{\partial t} &= -\frac{1}{r}\frac{\partial}{\partial r}(r\rho v_r) \\
\frac{\partial \rho v_r}{\partial t} &= -\frac{1}{r}\frac{\partial}{\partial r}(r\rho v_r v_r) - \frac{\partial P}{\partial r} \\
\frac{\partial \rho v_z}{\partial t} &= -\frac{1}{r}\frac{\partial}{\partial r}(r\rho v_z v_r) - \frac{\partial P}{\partial z} \\
\frac{\partial E}{\partial t} &= -\frac{1}{r}\frac{\partial}{\partial r}(rEv_r) - \frac{1}{r}\frac{\partial}{\partial r}(rPv_r)
\end{aligned} \tag{9 – 2.20}$$

where α in Eq. (8–3.11) has been set equal to two for cylindrical geometry. Again, the convection terms and source terms are written in the form of Eq. (8–3.11). The axial and radial integrations are alternated, each pair of sequential integrations constituting a full convection timestep. Thus a single highly optimized algorithm for a reasonably general continuity equation can be used to build up multidimensional fluid dynamics models.

There are obvious limitations to the use of this split-step approach. The timestep must be small enough that the distinct components of the fluxes do not change the properties of a cell appreciably during the timestep. We would like all the terms on the right sides of Eqs. (9–2.20) to be evaluated simultaneously. Their sequential calculation means there is an error. The fractional-step technique used here can be shown to be second-order accurate as long as the timestep is small and changed slowly enough, but there is still a bias built in depending on which direction is integrated first. To remove this bias, the results from two calculations for each timestep can be averaged. This is an expensive but effective solution.

9–2.4 The Riemann Problem and Riemann Methods

The solution to the one-dimensional Riemann problem describes the evolution of a single planar discontinuity separating two different but uniform fluid regions. For one-dimensional Euler equations, solutions may consist of shocks, contact discontinuities, and rarefactions. The discontinuity between the two regions might be in pressure or density, a situation which occurs in the standard shock diaphragm problem, but it could also be a material interface, a contact surface, or a rarefaction wave. In a shock, the density, energy, pressure, and velocity are all discontinuous. At a contact discontinuity, the density and energy are discontinuous, but the pressure and velocity are continuous. At these discontinuities, conservation laws give analytic jump conditions that can be used to determine how the discontinuity evolves in time. Even with ideal equations of state, the resulting equations are nonlinear and generally must be iterated to find solutions. Good descriptions of the Riemann problem and its use in convection problems are given by van Leer (1979) and Roe (1981).

Godunov first incorporated this analytic solution in a finite-difference method to improve accuracy at a discontinuity (Godunov, 1959; see also Richtmyer and Morton, 1967). His basic idea is to solve a Riemann problem cell-by-cell or region-by-region in the flow, and then to piece these local analytic solutions together. The method was initially designed to solve shock problems, and produced very encouraging results. It is first-order accurate, generally monotonic, and, because it calculates the transported fluxes for all three equations together, does not permit nonphysical decreases in entropy.

The Godunov method is basically a two-step method, first solving the Riemann problem to find provisional values and then using a first-order method to interpolate back onto the desired, usually fixed grid. The first step assumes that the solutions are initially piecewise constant in each computational cell. The method then solves a Riemann problem for the discontinuities at each cell interface to determine where the shocks, contact discontinuities, and rarefaction fans have moved after a time Δt. Figure 9–2 shows two cell interfaces along the horizontal axis with time on the vertical axis. At time t, there are discontinuities at the interfaces that later propagate outward as a function of time. It is important in this method that waves generated by the interaction of waves originating at the boundary do not get back to the cell boundary. This is assured by keeping Δt small enough, essentially the same timestep condition as in an explicit finite-difference method, and thus the conditions throughout each cell are known exactly. This process produces provisional values of v and P, which are used in the subsequent step.

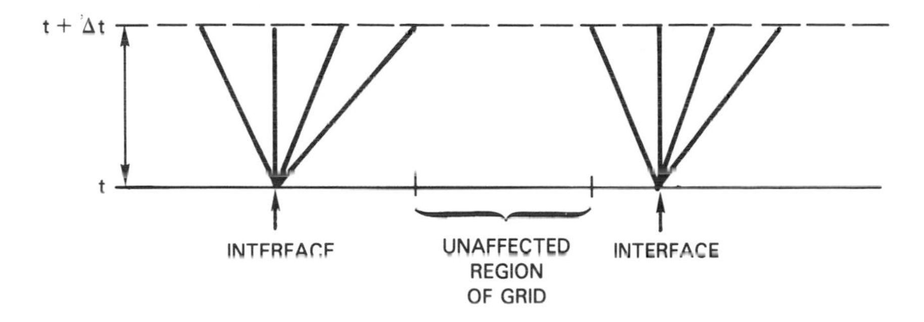

Figure 9–2. Solutions of the Riemann problem at cell interfaces.

Since Godunov's paper, the application of the solution of the Riemann problem has been used extensively to calculate accurate fluxes at interfaces in the mesh. In particular, in Chapter 8 we mentioned using the Riemann solution in a number of higher-order monotone methods to produce sharp discontinuities and more accurate representations of convection. For example, the PPM method combines a Riemann solver and a fourth-order convection algorithm (Colella and Woodward, 1984). Here the function in each computational cell is assumed to be piecewise parabolic, instead of piecewise constant. This naturally makes the Riemann problem more accurate, but also more costly and more difficult to solve, even using the fast, vectorizable, iterative Riemann solvers available. Riemann solvers are also included in the MUSCL method (van Leer, 1979), the TVD methods (see, for example, Sweby, 1984), and the SUPERBEE method (Roe, 1985), all briefly described in Section 8–6.

Upwind differencing methods are *one-sided* algorithms that attempt to discretize the equations using finite differences that are biased in the direction of the fluid flow. The donor-cell algorithm introduced in Section 4–4 to solve the square-wave convection problem is a one-sided, upwind method. For such methods, two models have been proposed to determine the interaction between computational cells. One is called *flux-difference splitting* (see, for example, Roe, 1981; Osher, 1981), in which neighboring cells interact through discrete, finite-amplitude waves. The properties of these waves, such as their propagation speed and amplitude, are determined by solving the Riemann problem for the discontinuity at the cell interface. The numerical method distinguishing between the influence of forward- and backward-moving waves is called *flux-difference splitting*. In the second approach, the neighboring cells interact through mixing *pseudo-particles* that move in and out of each cell with a given velocity distribution. This is the *Boltzmann approach*. The numerical method of distinguishing between the influence of the forward- and backward-moving particles is called *flux-vector splitting* or *flux splitting* (Sanders and Prendergast, 1974; Steger and Warming, 1981). Both approaches are described by van Leer (1982), Roe (1986), and Harten et al. (1983). Flux-difference splitting, based on the Riemann solution, is commonly used with the Riemann methods discussed in this chapter. Flux-vector splitting has been associated with the implicit methods described in Section 9–3.

Godunov approaches are among the more complex of the monotone methods. The major problems with such methods are the cost of solving the Riemann problem, the orientation of the discontinuities relative to the grid in multidimensions, and the extension to geometries other than Cartesian. In the last two problems, the analytic Riemann solution for assigning and following the system characteristics does not exist, so further (sometime unphysical) approximations must be made. To convert the problem formulation back to locally solvable Riemann problems, the problem must be timestep split so that

$$\frac{df}{dt} = \frac{df_x^R}{dt} + \frac{df_y^R}{dt} + g(f) \,, \qquad (9-2.21)$$

where df_x^R/dt and df_y^R/dt represent terms contributing to separable and solvable one-dimensional Riemann problems in the x- and y-directions, respectively. The remaining terms, $g(f)$, must contain all of the effects not representable as local Riemann problems. This includes geometric effects that occur even in one-dimensional cylindrical and spherical coordinate systems.

Riemann approaches can be expensive to use though they work well even in high Mach number flows. The general analytic solution of this particular specialized problem is definitely of great interest, but is of little use in improving the practical accuracy of an algorithm being applied to much more general problems. It is natural but misleading to pay careful attention to the parts of a problem that we can solve analytically, the df_x^R/dt and df_y^R/dt terms, for example, while paying little attention to the correction terms, $g(f)$, which can be just as important to the overall solution. Thus, many methods that are excellent for a specific one-dimensional Riemann test problem cannot be applied with confidence to more complicated flows. We note that Roe (1985), for example, incorporates an approximate Riemann solver which is less expensive without obvious deleterious effects. Significant effort has gone into transforming to local coordinate systems where the Rie-

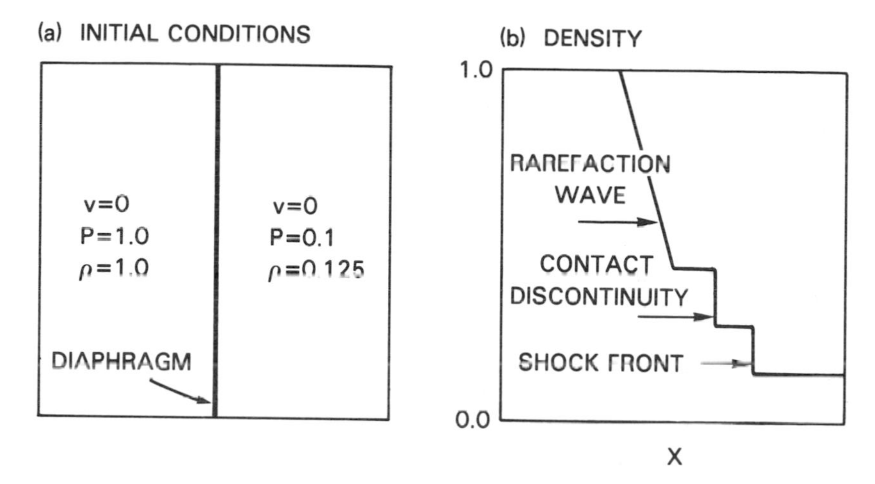

Figure 9–3. Riemann test problem. (a) Initial conditions. (b) Solution after some time.

mann problem is valid (Eidelman, 1986) but computational cost and global conservation remain issues.

There is another concern over the applicability of techniques based on the nonlinear Riemann solution: the importance of energy conservation in chemically reacting flows. Some techniques based on the Riemann solution have trouble guaranteeing the exact conservation of global mass, momentum, or energy. This is true of Glimm's method, discussed below, and algorithms that advance derived quantities such as temperature and entropy, rather than the conserved variables ρ, $\rho \mathbf{v}$, and E.

9–2.5 Comparison of Several Methods

In order to compare methods, consider a one-dimensional exploding-diaphragm problem. This is a fairly standard test problem that came to particular note after the review by Sod (1978). We also refer the reader to the review by Nittman (1982) who uses a shock interaction with a contact discontinuity to test three methods: a donor-cell method, a second-order version of FCT, and a flux-splitting method.

The initial conditions for the exploding-diaphragm test problem are shown in Figure 9–3a, and the general form of the solution after a short time is shown in Figure 9–3b. The calculations presented below used 100 evenly spaced cells and a fixed timestep chosen such that the Courant number was 0.5,

$$\frac{|c_s + v|\Delta t}{\Delta x} = 0.5 \ . \qquad (9-2.22)$$

Before we describe the results, a caveat concerning the evaluation of algorithms is in order. A "perfect" method does not exist. Different methods look best depending on the specific test problems. It is common for algorithms to be presented in ways that show their best features. For example, if one method is favored, the test problem on which it does best is highlighted. Therefore, we warn readers to reserve judgment about conclusions presented from one or even several test problems.

Figures 9–4 through 9–7 show the results of applying four methods to this exploding-diaphragm test problem:

a. Lax-Wendroff, in Figure 9–4. The algorithm is described in Section 9–2.1, and is used as described there, with no artifical viscosity. The application of this method to linear convection was shown in Figure 4–6 and corresponds to a standard version of the MacCormack algorithm.
b. LCPFCT, in Figure 9–5. This is the nonlinear monotone Flux-Corrected Transport algorithm described in Section 8–3. It is fourth order in phase, and second order in time. Application of this to linear convection was shown in Figure 4–6 and Figure 8–4. No auxiliary flux synchronization was attempted.
c. The Godunov method, in Figure 9–6. This is the linear first-order monotone method originally proposed by Godunov.
d. SUPERBEE, in Figure 9–7. This is the algorithm developed by Roe (1985) which includes an approximate Riemann solver.

Each of Figures 9–4 through 9–7 shows pressure, density, velocity, and temperature profiles at 63 steps after the diaphragm has burst and the leading shock has moved 25 cells downstream. On each frame, the theoretical solution to this Riemann problem is drawn as a solid line and the calculations are shown as dots. In the test shown, a very short time has elapsed since busting the diaphragm so that the shock and contact surface are only eight or nine cells apart. In longer tests, the two nonlinear monotone methods would appear even better relative to the Lax-Wendroff and Godunov methods.

The Lax-Wendroff method, shown in Figure 9–4, was programmed without introducing artificial viscosity. The absence of first-order damping shows large nonphysical fluctuations in the variables at almost all of the discontinuities. At the contact discontinuity where the velocity and pressure should be constant and continuous, the Lax-Wendroff solution has unacceptably

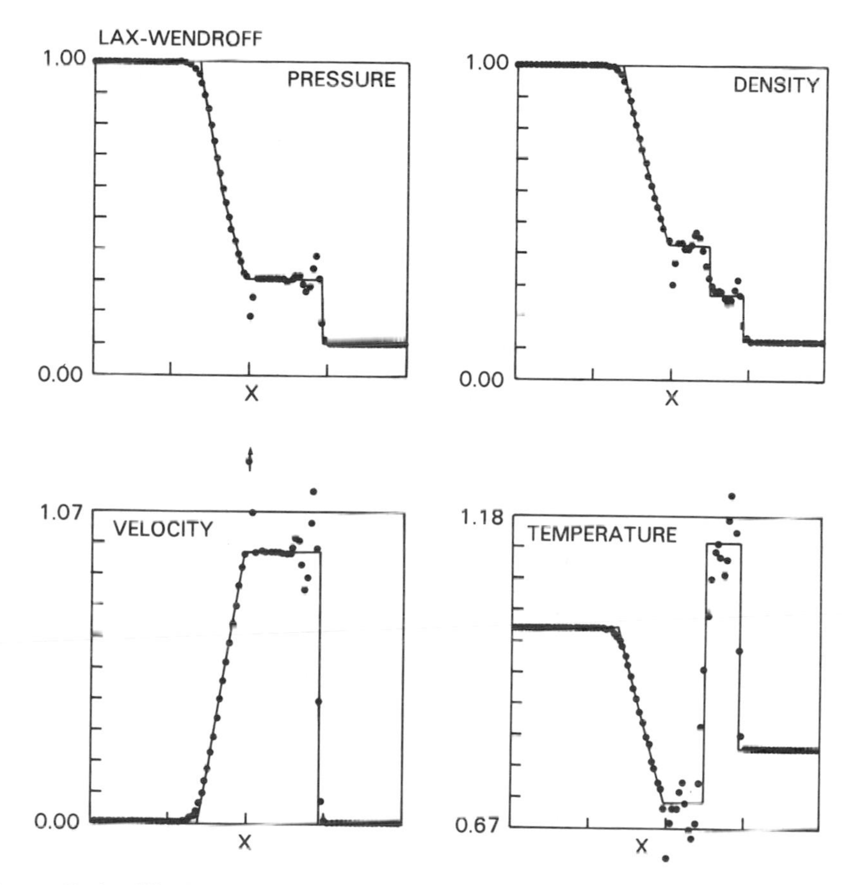

Figure 9–4. The Lax-Wendroff (MacCormack) method applied to the Riemann test problem. (Figure is courtesy of S. Zalesak.)

large oscillations. Because the method is second order, however, the locations of the discontinuities are rather well represented and the solution in the rarefaction is quite close to the theoretical solution.

Methods which make explicit use of physical and analytical insights into the behavior of the coupled system, in general, do better than methods that do not. The fourth-order Flux-Corrected Transport method used to calculate Figure 9–5 is basically a Lax-Wendroff method with the physical positivity condition enforced. This extra important piece of physics, when built directly into the method, improves the solution appreciably. Overshoots and undershoots are minimal, though there is a less than transient 4% error in the temperature peak behind the shock (which originates in the initial transients). The temperature and the velocity are derived quantities

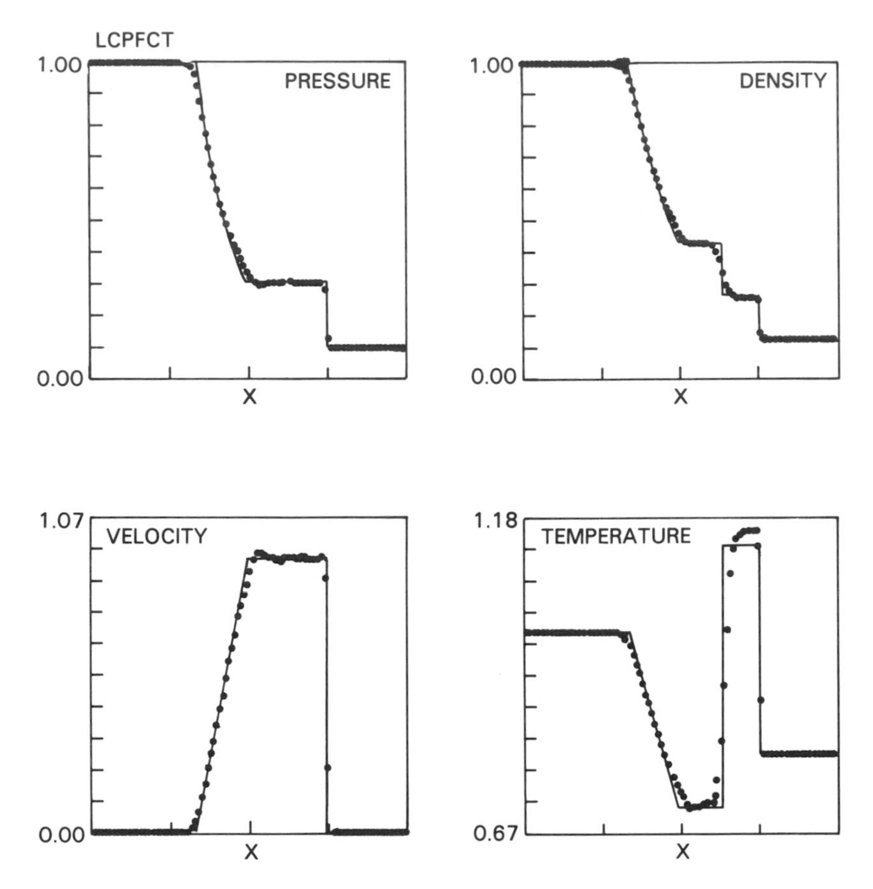

Figure 9–5. LCPFCT, a fourth-order version of Flux-Corrected Transport, applied to the Riemann test problem. (Figure is courtesy of S. Zalesak.)

so the monotone flux corrector does not work directly on these quantities. Löhner et al. (1986) have suggested synchronizing fluxes in ways that limit the composite solution nonlinearly and so decrease these errors. The errors also die out in long calculations.

Methods based on a Riemann solver use additional information about the actual solution relevant to this particular test problem, and such methods produce excellent results on this test problem. The Godunov method does particularly well on the Riemann test problem considering it is only a first-order method. If used on the linear convection problem shown in Figure 4–6, it would perform as the donor-cell calculation, which is really not very good at all. There are no wild oscillations in the Godunov calculation, in constrast to the Lax-Wendroff calculation in Figure 9–4, because the

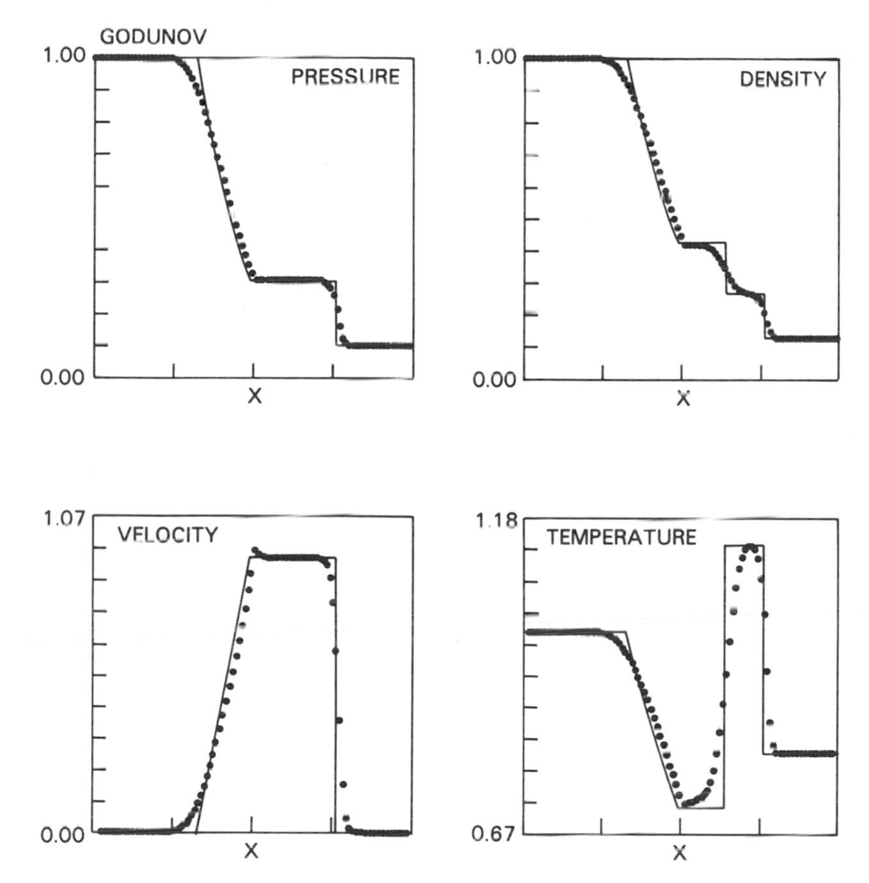

Figure 9–6. The Godunov method applied to the Riemann test problem. (Figure is courtesy of S. Eidelman and S. Zalesak).

Godunov method is monotone. The numerical diffusion is excessive in the region of the contact discontinuity and shock front. On a longer calculation of the same problem, the Godunov method shows appreciably more smearing at the contact discontinuity and rarefaction. However, the diffusion in the region of the shock front is limited by the natural tendency of shocks to steepen and can be reduced somewhat when the calculation is done at higher Courant number. A Courant number of 0.9 gives a reaonably accurate representation of the shock front.

The last example, Figure 9–7, uses the SUPERBEE method (Roe, 1985) which incorporates a solution of the Riemann problem, is a nonlinear monotone method with a flux-limiting procedure, and is second-order accurate. The quality of the solution is outstanding and the contact discontinuity is

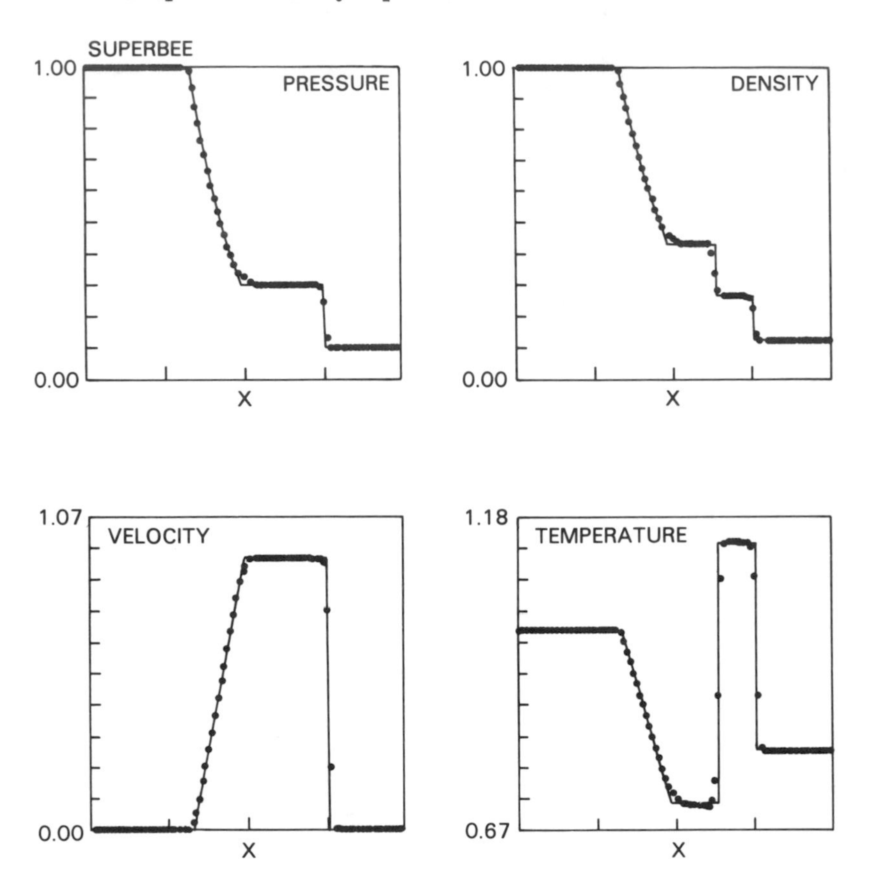

Figure 9–7. The SUPERBEE method applied to the Riemann test problem. (Figure is courtesy of S. Zalesak.)

better resolved than it is in Figure 9–5. The method was designed to solve Riemann problems accurately and succeeds. For pure convection, however, it does not do as well as a fourth-order monotone method. In addition, it *does not generalize to multidimensions or sliding grids*, and requires appreciable computation per timestep.

There is an advantage in any purely convection problem in using an algorithm that is high order in the spatial derivative. The advantage for the Riemann problem is including the Riemann solver. However, using a simple nonlinear monotone method produces good results at discontinuities in both cases.

9–2.6 Several Other Methods

Characteristic Methods

Characteristics were introduced in Chapter 2. The *characteristic* or *shock-fitting* methods for solving compressible flow problems are based on the idea of modeling an arbitrary flow as a series of isentropic regions separated by discontinuities. There are many good reviews of this method, for example, Liepmann and Roshko (1957) and the more recent discussion by Anderson et al. (1984). These methods were extremely popular when computers were small and slow, and even one-dimensional calculations were poorly resolved. These methods make maximum use of the limited data about the solution used in this representation, but have been generally replaced by more versatile, less expensive, and less complicated finite-difference methods. The characteristic methods described briefly below are not applicable to incompressible flows where the sound speed is effectively infinite. For such cases, characteristic techniques reduce to algorithms which are very similar to quasiparticle methods described in Chapter 11.

In the characteristic method, it is necessary to distinguish six different kinds of points in the flow (see, for example, Hoskin, 1964): points interior to isentropic regions, system boundaries, free surfaces, contact discontinuities, single shocks, and intersecting shocks. These points are labeled with coordinates x and t. Because time t, in general, varies from point to point, these methods do not work on a series of discrete time levels. To calculate the fluid properties at a new point on the intersection of characteristics running through two known points, only the physical quantities at the known points are needed. Thus it is necessary to store only an amount of information equivalent to describing one time level of a finite-difference scheme. In fact, some characteristic schemes use interpolation back onto a grid of points all at the same time level t. This permits the calculation to "march" from level to level.

Characteristic methods often can be formulated to take advantage of the physical content of a specific system of equations. For example, the equations of ideal magnetohydrodynamics (see, for example, Kulikovsky and Lyubimov, 1965) and ideal compressible fluid flow (Moretti and Abbett, 1966) can be formulated and solved by the method of characteristics. The characteristic solutions to these types of problems are often excellent because the discontinuities which arise naturally in these physical systems are followed individually and accurately. Also, it is relatively easy to ensure conservation and positivity in characteristic methods.

On the other hand, characteristic methods are not generally applicable.

The presence of even one diffusion term in the governing equations usually invalidates the use of characteristic methods because the characteristics of the composite model cease to exist! The diffusion also mixes regions of one entropy value with those of another, thereby requiring many more degrees of freedom to represent the smooth variation of the profile. A further drawback to characteristic methods is that they are quite complex and relatively inefficient for complicated and multidimensional flows. Special precautions are needed to describe a shock that forms from a steepening compression wave, a shock overtaking another shock, or a contact discontinuity forming when two shocks collide. All of these are situations in which discontinuities are introduced or removed. The logical complications of the characteristic method increase rapidly with the number of discontinuities present, and the advantages become overwhelmed by the difficulties of accounting for all of the special cases that can arise (Hoskin, 1964).

In two spatial dimensions, the characteristics are planar curves that intersect at moving points in x, y, t space. In three dimensions they are surfaces intersecting along moving curved lines. Thus in three dimensions, the method is even more complicated and the characteristics become progressively more and more difficult to follow as the flow proceeds in time.

The Method of Random Choice

The random choice method or Glimm's method, (Glimm, 1965) and the more recent variations (Sod, 1978; Chorin, 1976, 1977), attempt to avoid both the diffusion inherent in Eulerian donor-cell algorithms and the potential complexity of following characteristics in multidimensions. These methods discretize the location of discontinuities in the flow, but allow the heights of the discontinuities to be accurately modeled. The solution knows exactly how large the jumps are from one region to the next, but not exactly where they occur. This is just the opposite of more standard procedures in which the locations of characteristics are followed accurately, and then approximate strengths of the discontinuities are used to interpolate back onto a grid. This change of perspective is suggested by the desire to represent surfaces such as a contact discontinuity or a reactive species interface.

The price paid for precise knowledge of the strengths of shocks and other discontinuities is an imprecise knowledge of their location. The random or pseudo-random manner in which fluid values are assigned to the spatial grid leads to errors as large as those encountered in the more standard but diffusive formulations. Global conservation of mass, momentum, and energy cannot be guaranteed. A subtle form of the uncertainty principle seems to be acting here. Various types of errors can be traded off against each other, but not all reduced simultaneously. This uncertainty is just a reexpression of the necessarily finite resolution of a digital computation.

9–3. METHODS FOR SLOW FLOWS

We define the slow-flow regime as one in which the flow velocities of interest are much less than the speed of sound and thus the Mach number is much less than one. These flows are subsonic and sometimes incompressible. Any compressions or expansions that occur result from slow additions or subtractions of heat at essentially constant pressure.

A flame is an example of such a flow. The flame consumes combustible gases slowly enough for the pressure to stay essentially constant across the flame front. The density change across the flame can differ by factors of four to ten, however, due to heat released by chemical reactions. In flames neither the density change nor the actual fluid displacement associated with the expansion can be ignored. Many excellent numerical techniques are strained in this regime to allow compression and expansion while still integrating fast sound waves stably.

In subsonic flow the issue to resolve in the time integration algorithm is whether the sound waves are treated explicitly, implicitly, or asymptotically. Explicit algorithms, no matter how efficient computationally, become expensive when the maximum flow speed everywhere in the computational domain drops below 10% of the sound speed. Implicit methods usually involve expensive, iterated solutions of an elliptic equation. These additional computations may take as long as five or ten timesteps of an explicit algorithm and may require much more computer memory. Alternate approaches include the asymptotic slow-flow methods and methods that analytically rescale the properties of the fluid equations.

In incompressible flow, there are rotational and potential components in the velocity field, but the compressional component is identically zero. As much fluid volume must flow out of the system as flows into it. The fluid density does not have to be a constant in space during an incompressible flow, but it often is assumed to be. When the density is not constant everywhere, it is constant in the Lagrangian frame moving with each fluid element. Incompressible flow problems are frequently solved in the vorticity stream-function formulation as well as in the primitive-equation formulation.

When the solution of incompressible fluid dynamics problems uses a primitive-equation formulation, it is necessary to solve the Poisson equation for the fluid pressure. At each timestep, this formulation couples the solution implicitly over the whole grid to produce a solution that is stable and does not resolve sound waves. The correct pressure profile ensures that the time derivative of $\nabla \cdot \mathbf{v}$ remains zero. Within the Eulerian framework, there are a number of ways to develop an implicit algorithm and several of these are discussed below. In Chapter 10 we discuss Lagrangian methods appropriate for subsonic and incompressible flows.

9–3.1 Slow-Flow Methods

One class of algorithms for treating flows in the slow-flow regime are the *slow-flow methods*. These are asymptotic, not implicit, approaches to solving the coupled continuity equations for low-speed flows. These methods include the algorithms by Ramshaw and Trapp (1976), Jones and Boris (1977, 1979), Rehm and Baum (1978) and Rehm et al. (1982), and Paolucci (1982). The algorithm by Jones and Boris is used below as the basis for explaining the general idea.

This algorithm was designed to be used in multidimensional flame simulations in which the sound speed is large compared to the fluid velocity, and the pressure is essentially constant. The method solves the density equations,

$$\frac{\partial \rho}{\partial t} + \nabla \cdot \rho \mathbf{v} = 0 , \qquad (9-3.1)$$

$$\frac{\partial \rho_i}{\partial t} + \nabla \cdot \rho_i \mathbf{v} = -\nabla(\rho_i \mathbf{v}_{di}) + P_i - \rho_i L_i , \qquad i = 1, ..., N_s , \qquad (9-3.2)$$

the vorticity equation,

$$\frac{\partial \omega}{\partial t} = -\nabla \cdot (\mathbf{v}\omega) - \nabla \times \left(\frac{\nabla P}{\rho}\right) + \nabla \times \left(\frac{\nabla \cdot \nu \nabla \mathbf{v}}{\rho}\right) , \qquad (9-3.3)$$

the relations for the velocity potential and the stream function,

$$\nabla^2 \phi = \nabla \cdot \mathbf{v} , \qquad (9-3.4)$$

$$\nabla \times \nabla \psi = \nabla \times \mathbf{v} = \boldsymbol{\omega} , \qquad (9-3.5)$$

and the isentropic pressure equations with source terms,

$$\frac{\partial P}{\partial t} + \mathbf{v} \cdot \nabla P = -\gamma P \nabla \cdot \mathbf{v} + \left.\frac{\partial P}{\partial t}\right|_{\text{chem}} + (\gamma - 1)\left[\nabla \cdot \mathbf{q} - \mathbf{v} \cdot (\nabla \cdot \nu \nabla \mathbf{v}) + \sum_i h_i \rho_i \mathbf{v}_{di}\right] . \qquad (9-3.6)$$

We have included thermal conduction through the heat flux $\mathbf{q}$, thermochemistry through the enthalpies $\{h_i\}$, and the chemical reactions through the species production terms $\{P_i\}$ and loss terms $\{L_i\}$. The term $\partial P/\partial t|_{\text{chem}}$ arises from the change in enthalpies due to the chemical reactions and appears when we convert the energy equation to a pressure equation.

From Eq. (9–3.6), we can solve for the velocity divergence to lowest order in fluctuations of the pressure,

$$\begin{aligned}\gamma P \nabla \cdot \mathbf{v} \approx & \left\langle \left.\frac{\partial P}{\partial t}\right|_{\text{chem}} - \frac{\partial P}{\partial t} \right\rangle \\ & + (\gamma - 1)\left[\nabla \cdot \mathbf{q} - \mathbf{v} \cdot (\nabla \cdot \nu \nabla \mathbf{v}) + \sum_i h_i \rho_i \mathbf{v}_{di}\right] .\end{aligned} \tag{9 – 3.7}$$

This is a zeroth-order expression for $\nabla \cdot \mathbf{v}$ that neglects the $\mathbf{v} \cdot (\nabla P)$ term and averages over the time derivative. In certain types of self-consistent fluid dynamic problems, we must solve Eq. (9–3.7) to first order in pressure fluctuations as well as zeroth order to reproduce the vorticity generation.

One interesting feature of this equation is the first term on the right side, the average pressure change in the system from external energy addition or compression. This is written as the average of the $\partial P/\partial t$ terms in Eq. (9–3.6). A net addition of heat to the system through chemistry, thermal conduction, or external sources changes the average pressure with time. In a closed container, this average heating leads to no net velocity divergence and therefore is subtracted from Eq. (9–3.7) to yield the correct value of $\nabla \cdot \mathbf{v}$.

The solution to these slow-flow equations no longer includes sound waves and thus the formulation cannot treat shocks. However, the method still incorporates processes such as conduction, compression, and rarefaction as long as they evolve slowly compared to the propagation time of sound waves across the system. Two elliptic equations, one for ϕ and another for ψ, must be solved. To lowest order in the ratio of time scales, the pressure cannot vary in space, but it can vary globally on the time scale of the energy release. Higher-order corrections to the pressure can be calculated using the lower-order velocity field as source terms, but additional elliptic equations must be solved.

There is an important advantage to this type of method: the way in which Eqs. (9–3.1), (9–3.2), and (9–3.3) are advanced is left unspecified, so that any method can be chosen to solve these equations. For example, both Eulerian and Lagrangian finite-difference methods can be used. Jones and Boris used the FCT algorithm described in Section 9–2.

The algorithm is implemented on a staggered grid. For example, the density equation has ρ defined on the grid and $\mathbf{v}$ defined at intermediate locations. Then $\nabla \cdot \mathbf{v}$ is evaluated directly on the same grid as ρ. In solving for $\mathbf{v}$, the difference calculations of the velocity from ψ and ϕ, the stream function and velocity potential, is similarly straightforward if ψ and ϕ are defined at the right locations. When mass fluxes are required, such as for calculating diffusion velocities, the only quantities which need to be averaged are the mass densities at the half-cell positions. With this staggered grid, second-order accuracy is maintained in spatial integration for uniform or slowly varying cell sizes.

The quantities defined at grid points are $\nabla \cdot \mathbf{v}$, $\boldsymbol{\omega}$, ρ, ρ_i, T, P, and ϕ. The stream function ψ and velocity $\mathbf{v}$ are defined at the interfaces. The equations are integrated using a time-centered method. The general procedure to integrate one timestep is:

1. Estimate values for P, ρ, $\{\rho_i\}$, ω, $\nabla \cdot \mathbf{v}$, T, ψ, and ϕ at the new time t^n based on the values at t^o.
2. Find the source terms for Eqs. (9–3.1) – (9–3.5) based on centered values at time $t^o + \frac{1}{2}\Delta t$.
3. Integrate Eqs. (9–3.1) – (9–3.5) from t^o to $t^o + \Delta t$ using the source terms defined at the time- and space-centered positions.
4. Iterate steps 2 and 3 until the solution converges.
5. Integrate other terms, such as the chemical kinetics, from t^o to $t^o + \Delta t$.

Ramshaw and Trapp (1976) were interested in a multiphase flow problem in which they needed to represent the surface between a liquid and a gas. Their algorithm is similar to the slow-flow algorithm presented above with an important difference. They derive an expression for the velocity divergence, but do not decouple it from the sound waves in the system. Because it is not decoupled, they use an implicit formulation of the momentum and density equations which precludes using a positivity-preserving convection algorithm.

Rehm and Baum (1978) and Rehm et al. (1982) derived the heirarchy of slow-flow equations from a general perspective. They present a systematic expansion procedure where the only effects of compression allowed are the asymptotic long-time changes in density which result from local heating and cooling. The first-order term in their expansion of $\nabla \cdot \mathbf{v}$ is the expression given in Eq. (9–3.7). They originally presented their derivation in the context of primitive variables (1978), and later in the vorticity formulation (1982). Paolucci (1982) discussed the physical assumptions inherent in slow-flow methods, the scaling of the physical terms in the fluid equations, and the relation of the slow-flow equations to other approximations.

The asymptotic slow-flow approach works well when acoustic fluctuations are not important to the fluid dynamics, that is, for constant-pressure subsonic flows in which the divergence of the velocity is proportional to the local heat release through the energy equation. The generality and ease of use of this algorithm makes it appealing.

The slow-flow approximation breaks down when the kinetic energy of the flow becomes comparable to the internal energy. Because the equations are based on an asymptotic expansion, the sound waves are not even in the higher-order equations. Thus solving higher orders in this model cannot give solutions that merge continuously into the acoustic and compressional behavior of gas dynamics.

In many ablation and flame problems, pressure variations in the form of sound waves are important even though the fluid profiles themselves vary slowly. In these problems, it would be extremely expensive to resolve the solutions on the time scale of the sound speed by using an explicit integration algorithm, and yet the slow-flow approximation of constant pressure is not valid. Also, in atmospheric flows, in turbulence modeling for reactive systems, and in studies of Rayleigh-Taylor instabilities, the pressure fluctuations interact with the local density gradients to generate vorticity. In these cases, performing the acoustic calculation implicitly is valid as long as shocks are not present.

9–3.2 An Implicit Flux-Corrected Transport Algorithm

In cases where the slow-flow algorithm is too restrictive, some form of implicit treatment is useful. For example, the long-wavelength sound waves may need to be integrated accurately without a severe timestep restriction from the short wavelengths. At the same time, it is extremely useful to keep the advantages of a nonlinear, monotone convection method.

The Barely Implicit Correction to FCT, *BIC-FCT* (Patnaik et al., 1987), removes the timestep limit imposed by the sound speed by adding one implicit elliptic equation. The method is based on the analysis of Casulli and Greenspan (1984), who showed that it is not necessary to treat all terms in the gas dynamic equations implicitly to use longer timesteps than allowed by explicit stability limits. Only the two terms which force the numerical stability limit need to be coupled and solved implicitly.

The BIC-FCT method solves the compressible, gas dynamic conservation equations for density ρ, momentum density $\rho\mathbf{v}$, and total energy density E, where

$$E \equiv \epsilon + \frac{1}{2}\rho\mathbf{v}^2 \qquad (9-3.8)$$

and the equation of state is

$$P \equiv (\gamma - 1)\epsilon \, . \tag{9-3.9}$$

There are two stages to the algorithm. The first stage is an explicit predictor that determines the estimated new values $\tilde{\rho}$, $\tilde{\mathbf{v}}$,

$$\frac{\tilde{\rho} - \rho^o}{\Delta t} = -\nabla \cdot \rho^o \mathbf{v}^o \, , \tag{9-3.10}$$

$$\frac{\tilde{\rho}\tilde{\mathbf{v}} - \rho^o \mathbf{v}^o}{\Delta t} = -\nabla \cdot \rho^o \mathbf{v}^o \mathbf{v}^o - \nabla P^o \, . \tag{9-3.11}$$

The tilde denotes predictor values at the new time, and the superscripts o and n are used to denote the old time and new time, respectively. Only the time differencing in Eqs. (9–3.10) and (9–3.11) is fixed by these equations, not the spatial differencing. Thus any other convection algorithm could be substituted for the Flux-Corrected Transport module we use. The implicit forms of the momentum and energy density equations are

$$\frac{\rho^n \mathbf{v}^n - \rho^o \mathbf{v}^o}{\Delta t} = -\nabla \cdot \rho^o \mathbf{v}^o \mathbf{v}^o - \nabla[\theta P^n + (1-\theta)P^o] \, , \tag{9-3.12}$$

$$\frac{E^n - E^o}{\Delta t} = -\nabla \cdot (E^o + P^o)[\theta \mathbf{v}^n + (1-\theta)\mathbf{v}^o] \, . \tag{9-3.13}$$

When $\theta = 1$, the algorithm is fully implicit and reverts to the equations analyzed by Casulli and Greenspan.

This implicit system can be reduced to one scalar equation by eliminating $\mathbf{v}^n$ between Eqs. (9–3.12) and (9–3.13). Define a change in pressure, δP, as

$$\delta P \equiv \theta(P^n - P^o) \, . \tag{9-3.14}$$

The correction equation for momentum can be obtained in terms of δP by subtracting Eq. (9–3.11) from Eq. (9–3.12),

$$\frac{\rho^n \mathbf{v}^n - \tilde{\rho}\tilde{\mathbf{v}}}{\Delta t} = -\nabla \theta(P^n - P^o) = -\nabla \delta P \, . \tag{9-3.15}$$

The implicit pressure correction δP is just the additional pressure required to accelerate the fluid momentum from the explicit predictor values, $\tilde{\rho}\tilde{\mathbf{v}}$, to the final implicitly corrected values, $\rho^n \mathbf{v}^n$. Because this correction deals with the sound waves (which were improperly computed by the explicit monotone predictor step in this algorithm) and not with the convection, there is no interference caused by the FCT algorithm.

We obtain the new velocity by rearranging Eq. (9–3.15) and letting $\rho^n = \tilde{\rho}$, so that

$$\mathbf{v}^n = -\frac{\Delta t}{\tilde{\rho}} \nabla \delta P + \tilde{\mathbf{v}} \, . \qquad (9-3.16)$$

We obtain a correction equation for energy using the equation of state, Eq. (9–3.9), with γ constant,

$$\epsilon^n = \frac{\delta P}{(\gamma - 1)\theta} + \epsilon^o \, , \qquad (9-3.17)$$

where the θ appears through the definition of δP. We find δP by substituting Eqs. (9–3.16) and (9–3.17) into Eq. (9–3.13),

$$\begin{aligned} \frac{\tilde{\rho}\tilde{\mathbf{v}}^2 - \rho^o \mathbf{v}^{o2}}{2\Delta t} + \frac{\delta P}{(\gamma - 1)\theta \Delta t} &= \theta \Delta t \, \nabla \cdot \left(\frac{E^o + P^o}{\tilde{\rho}} \right) \nabla \delta P \\ &\quad - \theta \nabla \cdot (E^o + P^o) \tilde{\mathbf{v}} \\ &\quad - (1 - \theta) \nabla \cdot (E^o + P^o) \mathbf{v}^o \end{aligned} \qquad (9-3.18)$$

To clarify and simplify Eq. (9–3.18), Patnaik defines the explicitly determined corrector energy density $\overline{E}$,

$$\frac{\overline{E} - E^o}{\Delta t} \equiv -\nabla \cdot (E^o + P^o)(\theta \tilde{\mathbf{v}} + (1 - \theta)\mathbf{v}^o) \, . \qquad (9-3.19)$$

This allows Eq. (9–3.18) to be rewritten,

$$\frac{\delta P}{(\gamma - 1)\theta \Delta t} - \theta \Delta t \, \nabla \cdot \left(\frac{E^o + P^o}{\tilde{\rho}} \right) \nabla \delta P = \frac{\overline{E} - E^o}{\Delta t} - \frac{\tilde{\rho}\tilde{\mathbf{v}}^2 - \rho^o \mathbf{v}^{o2}}{2\Delta t} \, , \qquad (9-3.20)$$

a scalar elliptic equation for δP. Note that two equations with implicit terms have now been reduced to one equation, Eq. (9–3.20). The right hand side of Eq. (9–3.20) is evaluated explicitly using Eq. (9–3.19). After the elliptic equation is solved for δP, momentum and energy are corrected by Eqs. (9–3.15) and (9–3.17). From the form of Eqs. (9–3.15) and (9–3.11), we see that momentum is locally conserved throughout the algorithm. Furthermore, since the implicit correction appears as a gradient of a scalar, δP, using a monotone method for convection and an implicit stage for the acoustic waves does not interfere with the local conservation and convection of vorticity.

Solution Procedure

The derivation presented above does not prescribe a specific method for differencing the spatial derivatives but does require that the derivatives be evaluated at the appropriate time level indicated by the superscript. This allows flexibility in the choice of the differencing scheme for these terms. Therefore we integrate the explicit predictor equations, Eqs. (9–3.10), (9–3.11), and (9–3.19) with a fourth-order FCT to preserve steep species and vorticity gradients.

Each timestep is divided into the three stages:

1. Explicit predictor. The density, momentum, and energy are advanced explicitly as specified by Eqs. (9–3.10) and (9–3.11) using FCT. This produces the intermediate quantities, $\tilde{\rho}$, $\tilde{\rho}\tilde{\mathbf{v}}$, and $\tilde{E}$. The $\tilde{\mathbf{v}}$ is found from $\tilde{\rho}\tilde{\mathbf{v}}/\tilde{\rho}$. Then $\tilde{\mathbf{v}}$ and FCT are used to obtain the explicit convective energy corrector $\overline{E}$ given by Eq. (9–3.19).
2. Solution of Eq. (9–3.20) for δP. In one dimension, this requires solving a tridiagonal system of linear equations. In two dimensions, the solution requires solving an elliptic equation.
3. Momentum and energy corrections. These corrections are obtained from the implicit pressure change δP using Eqs. (9–3.15) and (9–3.17) or (9–3.13), respectively. These corrected values and the density obtained explicitly in the first stage are the starting conditions for the next timestep.

These three stages are repeated every timestep. The derivatives in the pressure difference equation, Eq. (9–3.20), are approximated by central differences. All physical quantities are calculated at cell centers, and those values needed at cell interfaces are obtained by averaging.

When BIC-FCT is applied in two dimensions, the same three-stage procedure is used (see Patnaik et al., 1987). Timestep splitting in the two spatial dimensions, but only with a single step, is needed instead of the half-step whole-step procedure described in Sections 9–2.2 and 9–2.3. For the method to work, however, the pressure correction equation must be solved in two dimensions, and its solution is a substantial part of the computational effort at each timestep. In one dimension, the finite-difference form of the pressure difference equation can be solved efficiently in $\mathcal{O}(N)$ operations, where N is the number of grid points, using standard tridiagonal methods. In two dimensions, it is important to use an efficient elliptic solver, preferably one that is not limited to specific types of problems with specific boundary conditions. Methods for solving tridiagonal matrices and elliptic equations are described briefly in Section 11–3.

Table 9–1 shows the time it takes to do one timestep of a two-dimensional calculation using BIC-FCT and LCPFCT. Both programs were

Table 9–1. Timings* per Step of BIC-FCT and LCPFCT

Grid	20 × 20	40 × 40	80 × 80
BIC-FCT			
explicit	6.8 ms	17.0 ms	54.1 ms
elliptic	3.8	8.4	22.5
other	2.7	5.9	17.1
total	13.3	31.3	93.7
per point-step†	33.3 μs	19.6 μs	14.6 μs
LCPFCT	13.6 ms	33.9 ms	108.1 ms
per point-step†	34.0 μs	21.2 μs	16.9 μs

* On CRAY XMP-12.
† Time per grid point per step.

fully optimized for a CRAY XMP-12 computer. The important point to note is that the cost of an implicit timestep is about the same as for an explicit timestep. As described in Section 9–2, the explicit calculations require both a half-step prediction and whole-step correction procedure each timestep. BIC-FCT does not split the timestep, but requires instead one matrix solution. If that solution is efficient enough, performing it and the auxiliary calculations take roughly the same computing time as the half timestep in the explicit FCT calculation.

Some General Comments

The BIC-FCT approach is more general than the slow-flow method, and therefore it is useful as a single accurate method in the range from nearly incompressible to subsonic flows. It also seems easier and better to use for faster flows because it allows acoustic modes. Patnaik has shown that when the timestep is reduced to the explicit limit determined by the sound speed, this algorithm does a reasonable calculation of the Riemann problem discussed in Section 9–2.

The implicitness parameter, θ, plays an important role whenever sound waves and pressure oscillations are important. Patnaik showed that the damping is negligible for long wavelengths and timesteps when $\theta = 0.5$ and the Courant number for the sound speed is less than two. Dispersion and damping become important for Courant numbers greater than five, irrespective of the value of θ. When sound waves are not important, θ can be set to unity.

9–3.3 Several Implicit Algorithms

The two methods described above, the slow-flow method and implicit adaptation of FCT, are only two of many approaches to solving fluid equations in the slow-flow regime. The slow-flow methods are really asymptotic methods, and implicit FCT is barely implicit in the sense that only two crucial terms are treated implicitly. Most of the methods in use now for the slow-flow regime, however, are fully implicit. Here it is appropriate to describe several of these. We recommend the book by Anderson et al. (1984) for detailed descriptions and extensive references to the block-implicit methods, and the article by Westbrook (1978) for a summary of the ICE and RICE methods.

Block-Implicit Methods

Block-implicit methods were developed by Lindemuth and Killeen (1973), McDonald and Briley (1975), Beam and Warming (1976, 1978), Warming and Beam (1977), Briley and McDonald (1975, 1977, 1980) for solving compressible, time-dependent Euler and Navier-Stokes equations. These methods are generally first or second order in time and very stable. Here we briefly describe the Beam and Warming method.

The starting point for the Beam and Warming method is the conservation equations for density, momentum, and energy written in matrix form in terms of the vector variable $\mathbf{u}$, as we did for the Lax-Wendroff algorithm in Section 9–1.1. Here the viscous terms are kept, giving the Navier-Stokes equations. The solution vector $\mathbf{u}$ is advanced in time by the formula

$$\begin{aligned}\Delta^n\mathbf{u} &\equiv \mathbf{u}^{n+1} - \mathbf{u}^n \\ &= \left(\frac{\theta_1 \Delta t}{1+\theta_2}\right)\frac{\partial}{\partial t}\Delta^n\mathbf{u} + \left(\frac{\Delta t}{1+\theta_2}\right)\frac{\partial}{\partial t}\mathbf{u} \\ &\quad + \left(\frac{\theta_2}{1+\theta_2}\right)\frac{\partial}{\partial t}\Delta^{n-1}\mathbf{u} + \mathcal{O}\left[(\theta_1 - \frac{1}{2} - \theta_2)(\Delta t)^2 + (\Delta t)^3\right] .\end{aligned} \tag{9 – 3.21}$$

The proper choice of θ_1 and θ_2 produces standard difference methods. For example, $\theta_1 = \theta_2 = 0$ gives an explicit first-order Euler method. The values $\theta_1 = 1$ and $\theta_2 = 0$ give an implicit Euler method, also first order in time. The most common method is $\theta_1 = 1$, $\theta_2 = \frac{1}{2}$, the three-point backward implicit method which is second order in time.

The procedure is to substitute the equations for $\mathbf{u}$ into Eq. (9–3.21) and linearize the nonlinear terms. For example, we find terms of the form

$$\Delta^n\mathbf{f}_y = \mathbf{A}_x\Delta^n\mathbf{u} + \mathcal{O}\left[(\Delta t)^2\right] , \tag{9 – 3.22}$$

where $\mathbf{f}_y$ is as defined in Section 9–2 and $\mathbf{A}$ is a Jacobian matrix. The resulting set of equations is a block-tridiagonal matrix system solved by ADI methods, discussed in Section 11–3.

Obayashi and Fujii (1985) developed a block-implicit method for the multidimensional Navier-Stokes equations using the flux-vector splitting method described in Section 9–2. They factored the fluxes into the right- and left-moving components, and treated each implicitly. Their equations resulted in bidiagonal instead of tridiagonal or pentadiagonal solutions in the appropriate directions. The different spatial dimensions are again treated by ADI methods. This method is first order and there is no monotonicity correction attempted, but the technique for simplifying the implicit calculations should be generally useful.

Implicit Methods with Riemann Solvers

A number of the explicit methods that include Riemann solvers have been made implicit. These include the MUSCL method, extended by Mulder and van Leer (1985), the PPM method, extended by Fryxell et al. (1986), and the implicit TVD method by Yee and Harten (1985) and Yee et al. (1985). These methods are stable, and first or second order. The implicit TVD methods have been applied to steady-state problems. The method described by Fryxell is particularly interesting in that it can transition smoothly from an explicit to an implicit method, as can BIC-FCT. These new methods are not described in detail here; we refer the interested reader to the references cited above.

Implicit MacCormack Method

The explicit MacCormack two-step method has a predictor step and a corrector step. The implicit version of this algorithm can be described as having two stages, although they can be combined in actually programming the method (MacCormack, 1981; see also Anderson et al., 1984). The first stage is the explicit MacCormack method and the second stage is implicit to take care of the stability restrictions. The resulting equations are upper or lower block *bi*diagonal and therefore easier to solve than tridiagonal. The method is unconditionally stable, and second order in space and time. The final solution depends on the timestep Δt. In practice, this is a problem. However, as Δt becomes small enough, the method converges.

ICE and RICE

The Implicit Continuous-fluid Eulerian method, *ICE*, developed by Harlow and Amsden (1971), has served as the basis for a number of models developed mainly at Los Alamos National Laboratory. These include *RICE*, (Rivard et al., 1975), which extended the method to reactive flows (Butler and O'Rourke, 1976), and to combustion (Westbrook, 1978). Another extension was the combination of ICE with the Arbitrary-Eulerian-Lagrangian method, *ALE* (Hirt et al., 1974), to form ICED-ALE (Amsden and Hirt, 1973). There are several other variations of the ICE method in current use.

In the original ICE method, the continuity and momentum equations are coupled together and solved simultaneously, using the equation of state to relate the time-advanced pressure to the time-advanced fluid density. After those equations are solved implicitly, the energy conservation equation is solved explicitly, decoupled from the continuity and momentum equations. The assumption is that changes in the fluid pressure in each computational cell are due mainly to density changes, and that pressure changes due to changes in the internal energy, temperature, or other physical processes are relatively small. This may be valid for problems in which the energy diffusion and deposition rates are small. If the change in internal energy is small, then the final correction to the pressure from the solution of the energy equation is also small. The pressure changes from density changes used in the solution of the fluid equations are then a good estimate of the true time-advanced pressure.

However, this assumption is not always valid. The analysis of Casulli and Greenspan (1984) shows that the energy and momentum equations should be coupled, rather than the density and momentum equations. The internal energy density can change as a result of rapid energy deposition in the fluid. In some cases, the ICE method can be adjusted appropriately to deal with this problem. For example, Ramshaw and Trapp (1976) used the energy equation, solved explicitly in time, to estimate the change in pressure over the timestep due to energy exchanges in two-phase flows. O'Rourke and Bracco (1979) used the rate of chemical reaction energy deposition, again evaluated explicitly in time, to help estimate the pressure changes in reacting flow problems.

In most combustion problems, the calculation of fluid dynamic properties of the reacting fluid is complicated by the fact that rapid exothermic chemical reactions take place in portions of the combustion volume. Because the mass and momentum flux terms in an implicit formulation depend on having an accurate estimate of the time-advanced pressure, it is essential to include all of the effects which significantly affect the pressure in the implicit, coupled fluid dynamics equations. For this reason, the original ICE method led to systematic errors.

Westbrook (1978) described a generalization that avoids these errors and therefore allows the method to be used more generally in combustion applications. The generalization takes account of the pressure dependence on other quantities, in addition to the density variation, and added operators, also decoupled from the continuity and momentum equations, to account for chemical kinetics rate equations and transport of chemical species.

SIMPLE

SIMPLE, the Semi-Implicit Method for Pressure-Linked Equations, was developed by Patankar and Spalding (1972). It is applicable to subsonic and incompressible flows and is relatively easy to apply. We recommend Anderson et al. (1984) for an explanation somewhat more detailed than given below, and also the book by Patankar (1980) for detailed discussions. The method, while far from the most accurate available, is very commonly used for industrial reactive flow problems.

The method is based on a series of predictor-corrector steps. First, we define the predicted quantites, P^o and $\mathbf{v}^o$, and the correction terms, P' and $\mathbf{v}'$,

$$\begin{aligned} P &= P^o + P' \\ \mathbf{v} &= \mathbf{v}^o + \mathbf{v}' \,. \end{aligned} \tag{9 – 3.23}$$

The velocity corrections are related to P' through the momentum equation,

$$\mathbf{v}' = -A \, \nabla P' \,, \tag{9 – 3.24}$$

where A is a time increment divided by density. Combining these two equations and substituting them into the continuity equation gives

$$\nabla^2 P' = \frac{1}{A} \nabla \cdot \mathbf{v}^o \,, \tag{9 – 3.25}$$

which is solved for P'.

The quantities are updated to find new predictor values for the next iteration through finite-difference formulas such as

$$\begin{aligned} P &= P^o + P' \\ v_x &= v_x^o - \frac{A}{2\Delta x}(P'_{i+1,j} - P'_{i-1,j}) \\ v_y &= v_y^o - \frac{A}{2\Delta y}(P'_{i,j+1} - P'_{i,j-1}) \,. \end{aligned} \tag{9 – 3.26}$$

The procedure is then:

1. Guess P^o on the grid.
2. Solve the momentum equations to find $\mathbf{v}_o$.
3. Solve the $\nabla \cdot P'$ equation to find P'.
4. Correct the pressure according to Eq. (9–3.26).
5. Replace the previous predictor values, P^o and $\mathbf{v}^o$, with the new values of P and $\mathbf{v}$, and return to step 2. The process is repeated until the solution converges.

There is often a convergence problem with this algorithm as presented above, especially when it is coupled to complex reactive flows. SIMPLER, where the R stands for "Revised," attempts to remedy this by allowing a relaxation parameter in the expression for the pressure,

$$P = P' + \alpha_p P' \ . \qquad (9-3.27)$$

Then $\mathbf{v}'$ is calulated the same way as in SIMPLE, and the P's may be closer to the correct value, depending on the choice of the relaxation parameter (Patankar, 1981). These methods are severely limited in their ability to preserve gradients.

9–3.4 Lagrangian Methods

This chapter dealt primarily with fluid dynamic algorithms, and generally ignored the grid on which they are implemented. Although it is possible to use a number of these algorithms in the Lagrangian framework, we concentrated primarily on Eulerian applications. In general, however, most of the algorithms described can be used on an arbitrary grid.

The algorithms described in the next chapter solve the same sets of equations given in Section 9–1, but are designed as Lagrangian algorithms. In one dimension and especially in multidimensions, the Lagrangian convection algorithms are intimately connected with the geometry of the problem and motion of the grid. Very often, as seen in Section 10–2, it is virtually impossible to separate them. Three of the methods described in the next chapter might have been described here in this section on slow-flow methods. These include the one-dimensional ADINC algorithm (Section 10–1), the general connectivity triangular grid approach (Section 10–2), and the vortex dynamics methods (Section 10–3). Quasiparticle methods (Section 10–4) are primarily for compressible flows. The Monotonic Lagrangian Grid (Section 10–5) is another Lagrangian data structure that has the potential to be a method for Lagrangian fluid dynamics.

References

Amsden, A.A., and C.W. Hirt, 1973, *YAQUI: An Arbitrary Lagrangian-Eulerian Computer Program for Fluid Flows at All Speeds*, Report LA-5100, Los Alamos Scientific Laboratory, Los Alamos, NM.

Anderson, D.A., J.C. Tannehill, and R.H. Pletcher, 1984, *Computational Fluid Mechanics and Heat Transfer*, Hemisphere Pub. Corp., Washington, DC.

Beam, R.M., and R.F. Warming, 1978, An Implicit Factored Scheme for the Compressible Navier-Stokes Equations, *AIAA J.* 16: 393–401.

Beam, R.M., and R.F. Warming, 1976, An Implicit Finite-Difference Algorithm for Hyperbolic Systems in Conservation-Law Form, *J. Comp. Phys.* 22: 87–110.

Briley, W.R., and H. McDonald, 1975, Solution of the Three-Dimensional Compressible Navier Stokes Equations by an Implicit Technique, in R.D. Richtmyer, ed., *Proceedings Fourth International Conference on Numerical Methods in Fluid Dynamics*, Springer-Verlag, New York, 105–110.

Briley, W.R., and H. McDonald, 1977, Solution of the Multi-Dimensional Compressible Navier-Stokes Equations by a Generalized Implicit Method, *J. Comp. Phys.* 24: 372–397.

Briley, W.R., and H. McDonald, 1980, On the Structure and Use of Linearized Block Implicit Schemes, *J. Comp. Phys.* 34: 54–73.

Butler, T.D., and J. O'Rourke, 1976, A Numerical Method for Two Dimensional Unsteady Reacting Flows, *Sixteenth Symposium (International) on Combustion*, The Combustion Institute, Pittsburgh, PA, 1503–1515.

Casulli, V., and D. Greenspan, 1984, Pressure Method for the Numerical Solution of Transient, Compressible Fluid Flows, *Int. J. Num. Methods Fluids* 4: 1001–1012.

Chorin, A.J., 1976, Random Choice Solution of Hyperbolic Systems, *J. Comp. Phys.* 22: 517–534.

Chorin, A.J., 1977, Random Choice Methods with Applications to Reacting Gas Flows, *J. Comp. Phys.* 25: 253–273.

Colella, P., and P.R. Woodward, 1984, The Piecewise Parabolic Method (PPM) for Gas-Dynamical Simulations, *J. Comp. Phys.* 54: 174–201.

Eidelman, S., 1986, Local Cell Orientation Method, *AIAA J.* 24: 530–531.

Fryxell, B.A., P.R. Woodward, P. Colella, and K.-H. Winkler, 1986, An Implicit-Explicit Hybrid Method for Lagrangian Hydrodynamics, to appear in *J. Comp. Phys.*.

Glimm, J., 1965, Solutions in the Large for Nonlinear Hyperbolic Systems of Equations, *Comm. Pure Appl. Math.* 18: 697–715.

Godunov, S.K., 1959, Finite Difference Methods for Numerical Computation of Discontinuous Solutions of the Equations of Fluid Dynamics, *Mat. Sb.* 47: 271–306.

Harten, A., and G. Zwas, 1972, Self-Adjusting Hybrid Schemes for Shock Computations, *J. Comp. Phys.* 9: 568–583.

Harten, A., P.D. Lax, and B. van Leer, 1983, On Upstream Differencing and Godunov-Type Schemes for Hyperbolic Conservation Laws, *SIAM Rev.* 25: 35–61.

Harlow, F.H., and A.A. Amsden, 1971, A Numerical Fluid Dynamics Method for All Flow Speeds, *J. Comp. Phys.* 8: 197–214.

Hirt, C.W., A.A. Amsden, and J.L. Cook, 1974, An Arbitrary Lagrangian Eulerian Computing Method for All Flow Speeds, *J. Comp. Phys.* 14: 227–254.

Hoskin, N.E., 1964, Solution by Characteristics of the Equations of One-Dimensional Unsteady Flow, in B. Adler, S. Fernback, and M. Rotenberg, eds., *Methods in Computational Physics*, vol. 3, Academic Press, New York, 265–293.

Jones, W.W., and J.P. Boris, 1977, Flame and Reactive Jet Studies Using a Self-Consistent Two-Dimensional Hydrocode, *J. Phys. Chem.* 81: 2532–2534.

Jones, W.W., and J.P. Boris, 1979, FLAME — A Slow-Flow Combustion Model, NRL Memorandum Report 3970, Washington, DC.

Kulikovsky, A.G., and G.A. Lyubimov, 1965, *Magnetohydrodynamics*, Addison-Wesley, Reading, MA.

Lax, P.D., and B. Wendroff, 1960, Systems of Conservation Laws, *Comm. Pure Appl. Math.* 13: 217–237.

Lax, P.D., and B. Wendroff, 1964, Difference Schemes for Hyperbolic Equations with High Order of Accuracy, *Comm. Pure Appl. Math.* 17: 381–398.

Liepmann, H.W., and A. Roshko, 1957, *Elements of Gasdynamics*, Wiley, New York.

Lindemuth, I., and J. Killeen, 1973, Alternating Direction Implicit Techniques for Two Dimensional Magnetohydrodynamics Calculations, *J. Comp. Phys.* 13: 181–208.

Löhner, R., K. Morgan, M. Vahdati, J.P. Boris, and D.L. Book, 1986, FEM-FCT: Combining High Resolution with Unstructured Grids, to appear in *J. Comp. Phys.*

MacCormack, R.W., 1981, *A Numerical Method for Solving the Equations of Compressible Viscous Flow*, AIAA Paper 81-0110, AIAA, New York, NY.

McDonald, H., and W.R. Briley, 1975, Three-Dimensional Supersonic Flow of a Viscous or Inviscid Gas, *J. Comp. Phys.* 19: 150–178.

Moretti, G., and M. Abbett, 1966, A Time-Dependent Computational Method for Blunt Body Flows, *AIAA J.* 4: 2136–2141.

Mulder, W.A., and B. van Leer, 1985, Experiments with Impicit Upwind Methods for the Euler Equations, *J. Comp. Phys.* 59: 232–246.

Nittman, J., 1982, Donor Cell, FCT-SHASTA and Flux Splitting Method: Three Finite Difference Techniques Applied to Astrophysical Shock-Cloud Interactions, in K.W. Morton and M.J. Baines, eds., *Numerical Methods for Fluid Dynamics*, Academic Press, New York, 497–517.

Obayashi, S., and K. Fujii, 1985, Computation of Three-Dimensional Viscous Transonic Flows with the LU Factored Scheme, *AIAA 7th Computational Fluid Dynamics Conference*, AIAA, New York, 192–202.

O'Rourke, P.J., and F.V. Bracco, 1979, Two Scaling Transformations for Numerical Computation of Multidimensional Unsteady Laminar Flames, *J. Comp. Phys.* 33: 185–203.

Osher, S., 1981, Numerical Solutions of Singular Perturbation Problems and Hyperbolic Systems of Conservation Laws, in O. Axelsson, L.S. Frank, and A. van der Sluis, eds., *Mathematical Studies* 47: 179–204. North-Holland, New York.

Paolucci, S., 1982, *On the Filtering of Sound from the Navier-Stokes Equations*, Sandia Report SAND82-8257, Sandia National Laboratories, Albuquerque, NM.

Patankar, S.V., 1980, *Numerical Heat Transfer and Fluid Flow*, Hemisphere, Washington, DC.

Patankar, S.V., 1981, A Calculation Procedure for Two-Dimensional Elliptic Situations, *Numer. Heat Transfer* 4: 409–425.

Patankar, S.V., and D.B. Spalding, 1972, A Calculation Procedure for Heat, Mass and Momentum Transfer in Three-Dimensional Parabolic Flows, *Int. J. Heat Mass Transfer* 15: 1787–1806.

Patnaik, G., R.H. Guirguis, J.P. Boris, and E.S. Oran, 1987, A Barely Implicit Correction for Flux-Corrected Transport, to appear in *J. Comp. Phys.*

Potter, D., 1973, *Computational Physics*, Wiley, New York.

Ramshaw, J.D., and J.A. Trapp, 1976, A Numerical Technique for Low-Speed Homogeneous Two-Phase Flow with Sharp Interfaces, *J. Comp. Phys.* 21: 438–453.

Rehm, R.G., and H.R. Baum, 1978, The Equations of Motion for Thermally Driven, Buoyant Flows, *J. Res. NBS* 83: 297–308.

Rehm, R.G., H.R. Baum, and P.D. Darcy, 1982, Buoyant Convection Computed in a Voriticity, Stream-Function Formulation, *J. Res. NBS* 87: 165–185.

Richtmyer, R.D., 1963, *A Survey of Difference Methods for Nonsteady Fluid Dynamics*, NCAR Technical Note 63-2, Boulder, CO.

Richtmyer, R.D., and K.W. Morton, 1967, *Difference Methods for Initial-Value Problems*, Interscience, New York.

Rivard, W.C., O.A. Farmer, and T.D. Butler, 1975, *RICE: A Computer Program for Multi-component Chemically Reactive Flows at All Speeds*, Los Alamos Scientific Laboratory Report LA-5812, Los Alamos Scientific Laboratory, Los Alamos, NM.

Roe, P.L., 1981, Approximate Riemann Solvers, Parameter Vectors, and Difference Schemes, *J. Comp. Phys.* 43: 357–372.

Roe, P.L., 1985, Some Contributions to the Modelling of Discontinuous Flows, *Lect. Appl. Math.* 22: 163–193.

Roe, P.L., 1986, Characteristic-Based Schemes for the Euler Equations, *Ann. Rev. Fluid Mech.* 18: 337–365.

Sanders, R.H., and K.H. Prendergast, 1974, On the Origin of the 3 Kiloparsec Arm, *Astro. J.* 188: 489–500.

Sod, G.A., 1978, A Survey of Several Finite Difference Methods of Systems of Nonlinear Hyperbolic Conservation Laws, *J. Comp. Phys.* 27: 1–31.

Steger, J.L., and R.F. Warming, 1981, Flux-Vector Splitting of the Inviscid Gas Dynamic Equations with Applications to Finite-Difference Methods, *J. Comp. Phys.* 40: 263–293.

Sweby, P.K., 1984, High Resolution Schemes Using Flux Limiters for Hyperbolic Conservation Laws, *SIAM J. Numer. Anal.* 21: 995–1011.

van Leer, B., 1979, Towards the Ultimate Conservative Difference Scheme. V. A Second-Order Sequel to Godunov's Method, *J. Comp. Phys.* 32: 101–136.

van Leer, B., 1982, Flux-Vector Splitting for the Euler Equations, in E. Krause, ed., *Eighth International Conference on Numerical Methods in Fluid Dynamics, Lecture Notes in Physics*, 170, Springer-Verlag, New York, 507–512.

Warming, R.F., and R.M. Beam, 1977, On the Construction and Application of Implicit Factored Schemes for Conservation Laws, *Symposium on Computational Fluid Dynamics*, SIAM-AMS Proceedings, Vol. 11, pp. 89–129, SIAM, Philadelphia.

Westbrook, C.K., 1978, A Generalized ICE Method for Chemically Reactive Flows in Combustion Systems, *J. Comp. Phys.* 29: 67–80.

Yee, H.C., and A. Harten, 1985, Implicit TVD Schemes for Hyperbolic Conservation Laws in Curvilinear Coordinates, *AIAA 7th Computational*

Fluid Dynamics Conference AIAA, New York, 228–241.

Yee, H.C., R.F. Warming, and A. Harten, 1985, Implicit Total Variation Diminishing (TVD) Schemes for Steady-State Calculations, *J. Comp. Phys.* 57: 327–360.

Zhmakin, A.I., and A.A. Fursenko, 1979, *On a Class of Monotonic Shock-Capturing Difference Schemes*, Ioffe Physicotechnical Institute, USSR Adaccmy of Sciences, Leningrad, USSR.

Chapter 10

LAGRANGIAN METHODS

Lagrangian methods appear to offer substantial gains over Eulerian methods. Their primary advantage is to eliminate numerical diffusion from the solution of a continuity equation. In a reference frame moving at the local velocity $\mathbf{v}$, the advection term does not appear directly in the equations. Writing a single continuity equation in Lagrangian form,

$$\frac{d\rho}{dt} = \frac{\partial \rho}{\partial t} + \mathbf{v} \cdot \nabla \rho = -\rho \nabla \cdot \mathbf{v} , \qquad (10-0.1)$$

the troublesome advection term does not have to be differenced, and the equation appears simpler. This Lagrangian continuity equation becomes simpler still when the flow is incompressible, $\nabla \cdot \mathbf{v} = 0$.

Thus a Lagrangian formulation of fluid dynamics equations is attractive for numerical calculations. Each discretized fluid element evolves through local interactions with its changing environment and with external forces. There is no nonphysical numerical diffusion from the flow of fluid across cell boundaries. In addition, the paths of the fluid elements are themselves a flow visualization. Lagrangian methods are also a natural choice for treating fluid dynamics with free surfaces, interfaces, or sharp boundaries.

In practice, Lagrangian methods in numerical simulations have generally been restricted to "well behaved" flows. Shear, fluid separation, or even large amplitude motions produce severe distortions in a Lagrangian grid which arise because Lagrangian grid points can move far enough that their nearest neighbors change in the course of a calculation. When differential operators are approximated by finite differences over a highly distorted mesh, the approximate derivatives become inaccurate. It is possible to regain some of the accuracy by interpolating the physical quantities onto a new, more regular grid, but this interpolation introduces unwanted numerical diffusion back into the calculation.

Another difficulty arises when Lagrangian methods are applied to compressible flows. The simplified energy equation, originally given as Eq. (9–1.13), is

$$\frac{\partial E}{\partial t} = -\nabla \cdot E\mathbf{v} - \nabla \cdot P\mathbf{v} , \qquad (10-0.2)$$

where for an ideal gas,

$$E = \frac{1}{2}\rho v^2 + \frac{P}{\gamma - 1} . \qquad (10-0.3)$$

By defining a new velocity,

$$\mathbf{v}^* \equiv \mathbf{v}\left(\frac{E+P}{E}\right) , \qquad (10-0.4)$$

which is always parallel to $\mathbf{v}$, Eq. (10–0.2) becomes a simple continuity equation,

$$\frac{\partial E}{\partial t} = -\nabla \cdot E\mathbf{v}^* . \qquad (10-0.5)$$

This has the same form as the mass conservation equation, Eq. (10–0.1), but $\mathbf{v}^* \neq \mathbf{v}$. A grid moving at velocity $\mathbf{v}$ is Lagrangian for density but not for energy. A grid moving at $\mathbf{v}^*$ is Lagrangian for energy but not for density. Although a Lagrangian mesh moves with the fluid, energy and momentum still convect through the grid much as they would if the mesh were fixed. Thus they are subject to numerical diffusion even when the grid is Lagrangian. In compressible reactive flows, the full set of equations is usually solved simultaneously, so that it is not always possible to avoid the Eulerian aspects of the problem completely.

We discuss a number of Lagrangian methods in this chapter. One- and two-dimensional finite-difference methods based on equations in primitive variable form, Eqs. (9–1.1) – (9–1.3) are as described first. The next methods considered are quasiparticles and vortex dynamics techniques. Both of these are based on moving representative particles. Vortex dynamics uses the the vorticity formulation, given in Eqs. (9–1.4) – (9–1.6). The chapter concludes with a brief description of the basic ideas of the Monotonic Lagrangian Grid, a potentially useful data structure for Lagrangian representations.

10–1. A ONE-DIMENSIONAL LAGRANGIAN METHOD

Here we describe a Lagrangian algorithm called ADINC, for *AD*iabatic and *INC*ompressible flows (Boris, 1979), designed to solve the mass, momentum, and energy equations for general one-dimensional fluid systems. Nonadiabatic processes are added to the basic module by timestep-splitting methods. The primary motivation was to have a very accurate convection algorithm with no numerical diffusion for simulating the time-dependent behavior of one-dimensional flames (see, for example, Kailasanath et al., 1982a,b). The convection algorithm, however, is not limited to combustion problems and has been applied to problems in controlled thermonuclear fusion (Boris, 1979; Cooper et al., 1980).

In some systems, the physical phenomena of interest vary slowly compared to the time it takes sound waves to cross the system. Nevertheless, substantial compressions and expansions occur that make the incompressibility assumption invalid. This class of problems requires an implicit treatment of sound waves, as discussed in Chapter 9. In other systems, rather sharp interfaces between regions with different compositions or different temperatures must be maintained, yet the fluids interact across the interface. This requires a Lagrangian description so that the exact interface location is available at any time. In an Eulerian calculation, the interface becomes spread over several zones due to the numerical diffusion. While we can tolerate nonphysical smearing of gradients in some problems, in others, such as flame propagation, it is not acceptable.

The implicit Lagrangian algorithm in ADINC deals with these two problems and allows a general equation of state and complicated boundary conditions. Large amplitude acoustic waves are treated and nonideal effects can be introduced through timestep splitting. Chapter 13 describes a one-dimensional laminar flame model that uses ADINC.

10–1.1 The Basic Equations

The density, momentum, and energy in one dimension satisfy the coupled, Lagrangian equations

$$\frac{d\rho}{dt} = -\rho\nabla\cdot\mathbf{v} \, , \qquad (10-1.1)$$

$$\frac{ds}{dt} = 0 \, , \qquad (10-1.2)$$

$$\rho\frac{d\mathbf{v}}{dt} = -\nabla P \, . \qquad (10-1.3)$$

The energy equation is eliminated by using an isentropic equation of state. Explicitly missing in this restricted set of fluid equations is viscous heating for shocks. Equation (10–1.2) indicates that the entropy S is constant in each fluid element bounded by Lagrangian interfaces. Nonisentropic processes, such as external heating, thermal conduction, or chemical energy release, may be added using timestep-splitting methods, as shown in Chapter 13.

The default equation of state in each computational cell is assumed to be

$$\rho(P,S,...) = \rho_c + \left(\frac{P}{S}\right)^{1/\gamma} \, . \qquad (10-1.4)$$

When $\rho_c = 0$, Eq. (10–1.4) is correct for adiabatic compression and expansion in an ideal gas and generally $1.2 \leq \gamma \leq 1.7$. Also, $\rho_c \neq 0$ adequately represents a mildly compressible liquid. Water, for example, has

$\rho_c = 1$ g/cm^3 and $S = 2.5 \times 10^{11}$cm-s^{-2}. Thus in this crude model, a pressure of 250 Kbar, or 2.5×10^{11} dynes/cm^2, causes a substantial increase in the compression. More complicated equations of state for solids and liquids may replace Eq. (10–1.4).

During a timestep, ρ_c, γ, and S are treated as constants and only the variation of ρ with P is considered. Rather than the usual approach of using ρ to find P, the ADINC algorithm calculates ρ given an approximation to P and does not involve the temperature. The density from the equation of state is compared to the density derived from Eq. (10–1.1), and the difference is forced to zero by adjusting the locations of the cell interfaces using a quadratically convergent iteration.

The equation of state appears in the form $\rho(P, S, ...)$ for two reasons. First, the density is a much less sensitive function of the pressure for liquids and solids than the pressure is of the density. During the iteration process, finite errors in pressure and density are expected. In the form, $P(\rho, S, ...)$, the errors in density appear as large fluctuations in the pressure. For gases and plasmas, the two forms of the equations of state have about the same accuracy. The second reason is that the algorithm can be used for looking at material interfaces such as a gas-solid interface. At such an interface, the pressure is continuous but the density does not have to be. Therefore, finite differences in the pressure are more accurate computationally than mathematically equivalent differences in the density.

The ADINC algorithm uses a quadratically convergent implicit iteration of Eqs. (10–1.1), (10–1.2), and (10–1.3) that produces an improved pressure approximation. The iteration uses the derivative $d\Lambda/dP$

$$\frac{1}{\Lambda}\frac{d\Lambda}{dP} = -\frac{1}{\gamma \rho P}\left(\frac{P}{S}\right)^{\frac{1}{\gamma}} , \qquad (10-1.5)$$

where Λ is the volume of a computational cell.

To complete the set of equations, the geometry of the calculation must be given. ADINC incorporates four one-dimensional geometries: Cartesian, cylindrical, spherical, and power series, and "nozzle" coordinates. Figure 10–1 shows a general one-dimensional region of variable cross-sectional area $A(r)$. The volume of the region is

$$V(r) = \int_{r_L}^{r} A(r')dr' . \qquad (10-1.6)$$

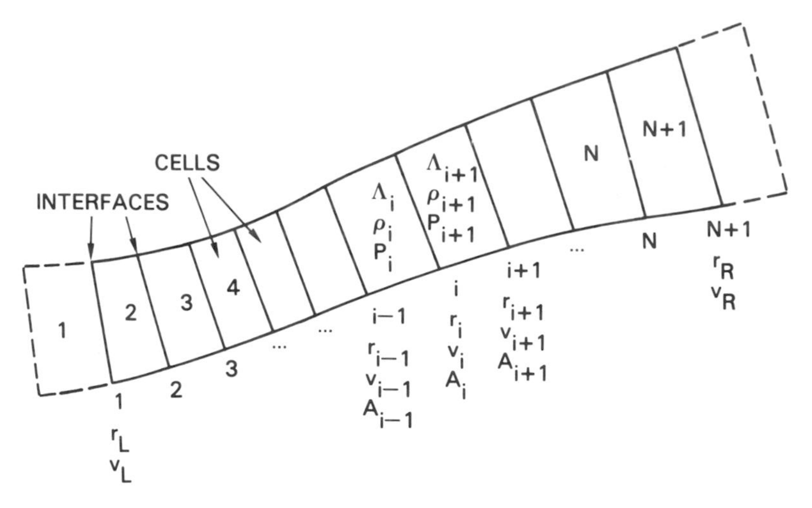

Figure 10–1. A general one-dimensional region of variable cross-sectional area $A(r)$ as used in the Lagrangian algorithm ADINC.

10–1.2 The Numerical Algorithm

A typical computational domain for the ADINC algorithm, shown in Figure 10–1, consists of N cells of volume $\{\Lambda_i\}$, $i = 2, 3, ..., N+1$, bounded by $N+1$ interfaces of area $\{A_i\}$, $i = 1, ..., N+1$. The interfaces are located at $\{r_i\}$, $i = 1, ..., N+1$, so that $A_i \equiv A(r_i)$, where $A(r)$ is determined by the chosen geometry. The cell volumes, $\Lambda_i \equiv V(r_i) - V(r_{i-1})$, are the difference of the volume integral from Eq. (10–1.6) at the two cell-bounding interface locations. The locations of the cell centers can also be defined,

$$R_i \equiv \frac{A_i r_{i-1} + A_{i-1} r_i}{A_i + A_{i-1}} , \qquad (10-1.7)$$

which lie between r_{i-1} and r_i.

The first physical cell in the computational domain is $i = 2$, and lies between interfaces r_1 and r_2. The last physical cell is $i = N+1$ and lies between intefaces r_N and r_{N+1}. Interface 1 is the left boundary, $r_1 \equiv r_L$, and interface $N+1$ is the right boundary, $r_{N+1} \equiv r_R$. The interfaces are fully Lagrangian, therefore the interface velocities $\{v_i\}$ and the interface positions $\{r_i\}$ must be advanced by the algorithm in conjunction with the changing cell pressures and densities. The velocity at the left boundary is $v_L \equiv v_1$, and the velocity at the right boundary is $v_R \equiv v_{N+1}$. The interior interface locations and velocities are integrated from time t to $t + \Delta t$ and their values at the system boundaries are given.

The cell interface positions $\{r_i\}$ satisfy

$$\frac{dr_i}{dt} \equiv v_i \,, \qquad (10-1.8)$$

which is discretized to

$$r_i^n = r_i^o + \Delta t[\epsilon_r v_i^o + (1-\epsilon_r)v_i^n] \,. \qquad (10-1.9)$$

The superscript n indicates variables at the new time, $t + \Delta t$, and the superscript o indicates variables at the old time, t. The quantity ϵ_r is the explicitness parameter for the interface positions, $0 \le \epsilon_r \le 1$. When $\epsilon_r < 1$, the method is implicit. When $\epsilon_r = \frac{1}{2}$, the method is centered and nominally most accurate. If we want to take long timesteps, we must use $\epsilon_r \le \frac{1}{2}$ to keep the algorithm stable. When $\epsilon_r = 0$, the calculation is fully implicit. This is the most stable algorithm, but it is only first-order accurate. It is possible to vary ϵ_r from cell to cell and from timestep to timestep.

The momentum equation, Eq. (10–1.3), for an interface velocity is

$$\frac{dv_i}{dt} \equiv \frac{-1}{\rho_{\text{interface } i}} \left.\frac{\partial P}{\partial r}\right|_{\text{interface } i} \qquad (10-1.10)$$

where the density and pressure gradient must be estimated at cell interfaces. The discretization used is

$$v_i^n = v_i^o - \frac{\Delta t \; \epsilon_v}{\langle \rho \Delta r \rangle_{i+\frac{1}{2}}}(P_{i+1}^o - P_i^o) - \frac{\Delta t(1-\epsilon_v)}{\langle \rho \Delta r \rangle_{i+\frac{1}{2}}}(P_{i+1}^n - P_i^n) \,. \quad (10-1.11)$$

Here ϵ_v is the explicitness parameter for the interface velocity and has the same properties as ϵ_r. Although the ϵ_v and ϵ_r can be distinct, we generally use the same value. The interface averaged mass, indicated $\langle \rho \Delta r \rangle_{i+\frac{1}{2}}$ is both a spatial and a temporal average. The average $\langle \rho \Delta r \rangle_{i+\frac{1}{2}}$ is defined so that the discretization in Eq. (10–1.11) is insensitive to numerical errors arising from large density discontinuities that can occur at material interfaces.

Figure 10–2 shows two cells i and $i+1$ that straddle interface i. The pressures P_i and P_{i+1} are defined at cell centers R_i and R_{i+1} as shown, and the densities ρ_i and ρ_{i+1} are assumed constant throughout their respective cells. Because ρ_i and ρ_{i+1} differ, using a straight line to model the pressure gradient from R_i to R_{i+1} means that the local acceleration is different for fluid to the right and to the left of interface i. If the fluid were allowed to move according to these different accelerations, the fluids would either overlap or separate near the interface i after a short time. To prevent this, a fictional pressure P_i^* is calculated for interface i, defining two straight line

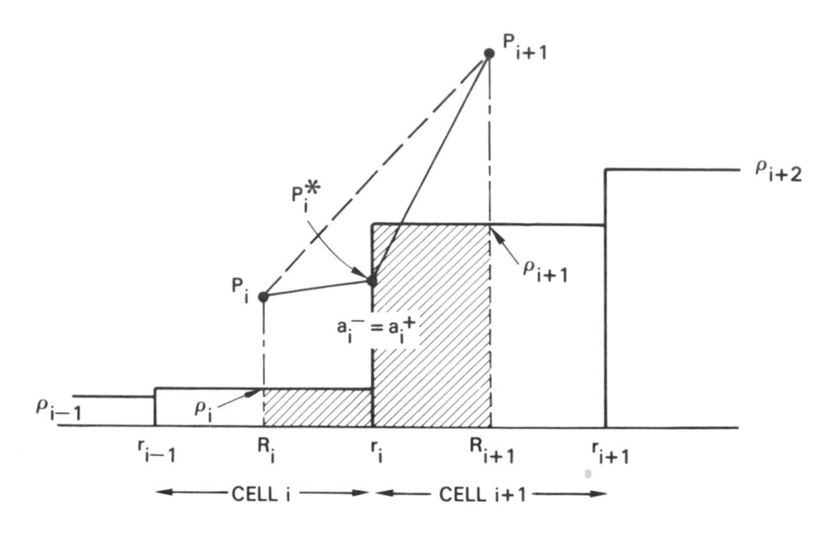

Figure 10–2. The interfaces between two cells, i and $i+1$, that straddle interface i. The diagram schematically shows the interface acceleration algorithm used in the one-dimensional algorithm ADINC.

pressure profiles, such that the fluid acceleration at the interface calculated from the left equals the acceleration calculated from the right. That is,

$$P_i^* \equiv \frac{P_{i+1} f_{i+1}^- + P_i f_i^+}{f_{i+1}^- + f_i^+} , \tag{10 – 1.12}$$

where

$$f_i^+ \equiv \frac{1}{\rho_i (r_i - R_i)} \qquad f_i^- \equiv \frac{1}{\rho_{i+1}(R_{i+1} - r_i)} . \tag{10 – 1.13}$$

Using the correct average for $\langle \rho \Delta r \rangle_{i+\frac{1}{2}}$, we can eliminate the interface pressures $\{P_i^*\}$ from further consideration. The average that accomplishes this acceleration matching is

$$\langle \rho \Delta r \rangle_{i+\frac{1}{2}} \equiv \rho_{i+1}(R_{i+1} - r_i) + \rho_i (r_i - R_i) \tag{10 – 1.14}$$

and defines the spatial part of the average.

The suitable temporal average in Eq. (10–1.14) has yet to be evaluated. ADINC uses

$$\langle \rho \Delta r \rangle_{i+\frac{1}{2}} \equiv [\rho_{i+1}^p (R_{i+1}^h - r_i^h) + \rho_i^p (r_i^h - R_i^h)] \tag{10 – 1.15}$$

where the superscript p (for "previous") indicates the latest iterated approximation to the new value of the variable, in this case ρ_i^n. The superscript h is used to indicate the exact "half-time" average. In Eq. (10–1.15),

$$r_i^h \equiv \frac{1}{2}(r_i^o + r_i^p) \quad \text{and} \quad R_i^h \equiv \frac{1}{2}(R_i^o + R_i^p) \, . \tag{10-1.16}$$

These may not be the best averages. The freedom remaining in this part of the calculation might be used to further improve the accuracy of the algorithm. However, no problems arising from these particular choices have yet been identified in calculations.

We now derive a tridiagonal equation for P_i^n. The momentum equation, Eq. (10–1.11), can be simplified to

$$v_i^n = \alpha_i - \beta_i(P_{i+1}^n - P_i^n) \, , \qquad i = 1, ..., N \, , \tag{10-1.17}$$

by using the auxiliary definitions

$$\begin{aligned} \alpha_i &\equiv v_i^o - \frac{\Delta t \ \epsilon_v}{\langle \rho \Delta r \rangle_{i+\frac{1}{2}}}(P_{i+1}^o - P_i^o) \, , \\ \beta_i &\equiv \frac{\Delta t(1-\epsilon_v)}{\langle \rho \Delta r \rangle_{i+\frac{1}{2}}} \, . \end{aligned} \tag{10-1.18}$$

The new cell volume, computed from the equation of state using the new values of pressure, $\Lambda_i^{n(\text{eos})}$, must equal the new cell volume computed from the fluid dynamics, $\Lambda_i^{n(\text{fd})}$, based on how much these same new implicitly coupled values of pressure move the cell interfaces. At successive iterations, indexed by p, the difference

$$\Delta\Lambda_i^p \equiv \Lambda_i^{p(\text{eos})}(P_i^p, S_i, ...) - \Lambda_i^{p(\text{fd})}(\{r_i^p\}) \, , \tag{10-1.19}$$

should approach zero. Changing P_i^p to P_i^n varies both terms in Eq. (10–1.19). The quantity r_i^p converges to r_i^n as a function of the pressure through Eq. (10–1.17) and Eq. (10–1.9). The Newton-Raphson algorithm gives a quadratically convergent iteration to obtain the desired solution at time $t + \Delta t$,

$$\Lambda_i^{n(\text{eos})}(P_i^n, S_i, ...) = \Lambda_i^{n(\text{fd})}(\{r_i^n\}) \, . \tag{10-1.20}$$

The difference between $\Lambda_i^{p(\text{fd})}$, which is known at each iteration, and the desired $\Lambda_i^{n(\text{fd})} = \Lambda_i^{n(\text{eos})} = \Lambda_i^n$ can be written in terms of the cell interface areas and the desired new fluid velocities of the Lagrangian cell interfaces,

$$\Lambda_i^n - \Lambda_i^{p(\text{fd})} \approx (1-\epsilon_r)\Delta t\left[A_i^h(v_i^n - v_i^p) - A_{i-1}^h(v_{i-1}^n - v_{i-1}^p)\right] \, . \tag{10-1.21}$$

The same treatment of the equation of state gives

$$\Lambda_i^n - \Lambda_i^{p(\text{eos})} \approx (P_i^n - P_i^p) \left.\frac{\partial \Lambda}{\partial P}\right|^{p(\text{eos})} . \qquad (10-1.22)$$

Letting

$$d_i^+ \equiv -(1-\epsilon_r)\Delta t \, A_i^h , \qquad d_i^- \equiv -(1-\epsilon_r)\Delta t \, A_{i-1}^h , \qquad (10-1.23)$$

then Eq. (10-1.21) becomes

$$\Lambda_i^n - \Lambda^{p(\text{fd})} \approx -d_i^+(v_i^n - v_i^p) + d_i^-(v_{i-1}^n - v_{i-1}^p) . \qquad (10-1.24)$$

Equating Λ_i^n in Eqs. (10–1.21) and (10–1.22) gives

$$\Delta\Lambda_i^p + (P_i^n - P_i^p) \left.\frac{\partial \Lambda}{\partial P}\right|_i^{p(\text{eos})} \approx -d_i^+(v_i^n - v_i^p) + d_i^-(v_{i-1}^n - v_{i-1}^p) . \qquad (10-1.25)$$

Then define

$$c_i \equiv P_i^p \left.\frac{\partial \Lambda}{\partial P}\right|_i^{p(\text{eos})} + d_i^+ v_i^p - d_i^- v_{i-1}^p - \Delta\Lambda_i^p , \qquad (10-1.26)$$

so that Eq. (10–1.25) becomes

$$P_i^n \left.\frac{\partial \Lambda}{\partial P}\right|_i^{p(\text{eos})} + d_i^+ v_i^n - d_i^- v_{i-1}^n \approx c_i . \qquad (10-1.27)$$

This is the implicit tridiagonal, linear equation that must be solved for the estimated new pressures $\{P_i^n\}$. Equation (10–1.17) is used to eliminate $\{v_i^n\}$ in terms of $\{P_i^n\}$. Expanding Eq. (10–1.27) gives

$$P_i^n \left.\frac{\partial \Lambda}{\partial P}\right|_i^{p(\text{eos})} + d_i^+ \left[\alpha_i - \beta_i(P_{i+1}^n - P_i^n)\right] - d_i^- \left[\alpha_{i-1} - \beta_{i-1}(P_i^n - P_{i-1}^n)\right] \approx c_i . \qquad (10-1.28)$$

Equation (10–1.28) is the basic tridiagonal equation solved by ADINC. Iteration is necessary because Eqs. (10–1.24) and (10–1.22) are only approximate equalities. The iteration is quadratically convergent, however, because only second-order terms are neglected at each stage of the iteration.

When integrating the fluid dynamic equations, this algorithm assumes that each interface moves in a fully Lagrangian manner according to Eq. (10–1.8). The change in density from one timestep to the next in a cell is therefore given simply by the change in cell volume according to the mass conservation equation,

$$\rho_i^n \Lambda_i^n \equiv \Delta M_i \equiv \rho_i^o \Lambda_i^o . \qquad (10-1.29)$$

When individual species number densities must be advanced, they are also found using Eq. (10–1.29).

10–1.3 Some General Comments

The accuracy and stability of this Lagrangian algorithm is based on the assumption that the interface positions $\{r_i\}$ increase monotonically with increasing i, and that the interface areas $\{A_i\}$ are positive. The largest potential source of instability in ADINC is nonphysical crossing of cell interfaces. This can occur for large timesteps even though the implicit algorithm given above is nominally stable for sound waves at arbitrary timestep.

To prevent interface crossings, a Courant condition must still be satisfied for the flow velocities $\{v_i\}$, even though $|v| \ll \sqrt{P/\rho^\gamma}$ throughout the gas. If $\{v_i^n\}$ becomes large enough, $\{r_i^n\}$ can cross adjacent interfaces even though the original timestep estimate should prevent this. The maximum timestep is limited by

$$\Delta t \;\leq\; \frac{1}{2}\min\left\{\left(\frac{r_{i+1}-r_i}{|v_i|+\delta}\right),\ \left(\frac{r_i-r_{i-1}}{|v_i|+\delta}\right)\right\}\,, \qquad i \;=\; 2,...,N\,. \tag{10 – 1.30}$$

A very small number, δ, is added to $|v_i|$ to prevent dividing by zero in Eq. (10–1.30). This equation is a conservative estimate and usually longer timesteps may be taken.

The cell interface positions are advanced using an average of the new and old velocities in Eq. (10–1.9). Because the timestep has to be estimated at the beginning of a cycle, it is possible for $\{v_i^o\}$ to be small or zero while $\{v_i^n\}$ can be large. A user-supplied maximum timestep usually provides a suitable limit.

The software package which implements ADINC is designed to be used as many ordinary differential equation packages are used. The program internally adjusts the timestep and the number of iterations at each step to maximize speed and maintain accuracy. Using the algorithm given above for estimating the timestep, ADINC subcycles the fluid dynamics as often as necessary to integrate over the time interval specified.

Within each timestep (or subcycle) ADINC performs a convergence calculation on the new iterated tridiagonal pressure solution. The convergence condition used for each cell is

$$\frac{|P_i^n - P_i^p|}{P_{\max}} \;<\; 10^{-9}\sqrt{\frac{M_{\max}}{M_{\min}}}\,, \tag{10 – 1.31}$$

where $P_{\max}$ is the largest pressure in the system, $M_{\max}$ is the largest cell mass in the system, and $M_{\min}$ is the smallest cell mass in the system. Equation (10–1.31) is heuristic, but works well. Using the timestep condition given above, convergence in $\{P_i^n\}$ is often obtained in two iterations and almost always in three or four. The maximum number of interations is limited to six, and this generally gives double-precision accuracy.

ADINC has been used for a number of one-dimensional, subsonic flows. Because it is Lagrangian, it should be more accurate than the Eulerian BIC-FCT algorithm described in Chapter 9. However, BIC-FCT can also handle shocks with short timesteps, conserves energy, and can also be generalized easily to multidimensions. As shown next, this is not the case for a fully Lagrangian algorithm such as ADINC.

10–2. GENERAL CONNECTIVITY TRIANGULAR GRIDS

This section describes a fluid dynamics algorithm based on the Lagrangian triangular grid mentioned briefly in Chapter 6 for solving the equations of incompressible fluid motion in two spatial dimensions. The important feature of this method is the implementation of a *general connectivity* or *free-Lagrangian* mesh, which permits the irregularly connected grid to change its connectivity during a calculation. The algorithms described in this section were developed by Fritts and Boris (1979) and Fyfe et al. (1986). The proceedings edited by Fritts et al. (1985) summarizes results of a number of free-Lagrange methods and their related problems. Although the algorithm described here is incompressible, the basic principles of the slow-flow algorithm described in Chapter 9 also can be incorporated to make this two-dimensional Lagrangian method valid for subsonic flows in which sound waves do not have to be explicitly treated.

10–2.1. The Triangular Grid

Consider a two-dimensional domain divided into triangular cells. Figure 10–3, shows a section of this mesh including an interface between fluid of type I and fluid of type II. In Figure 10–3a, the vertices, V_1, V_2, and V_3, of triangle j are connected by sides S_1, S_2, and S_3. The direction of increasing vertex and side indices around each triangle is counterclockwise and the z-axis is directed out of the page. Since the mesh can be irregularly connected, an arbitrary number of triangles can meet at each vertex.

We define a cell surrounding a vertex by the shaded region surrounding V_3, shown in Figure 10–3b. The *vertex-centered cells*, or *vertex cells*, are bordered by line segments joining the centroid of each triangle with the midpoints of the two triangle sides connected to the vertex, for all triangles surrounding that vertex. This definition of a vertex cell equally apportions the area of a triangle to each of its three vertices and provides a straightforward way to evaluate the finite-difference operators. Other definitions generally require additional calculations to determine cell intersection points, making it more expensive to integrate cell quantities.

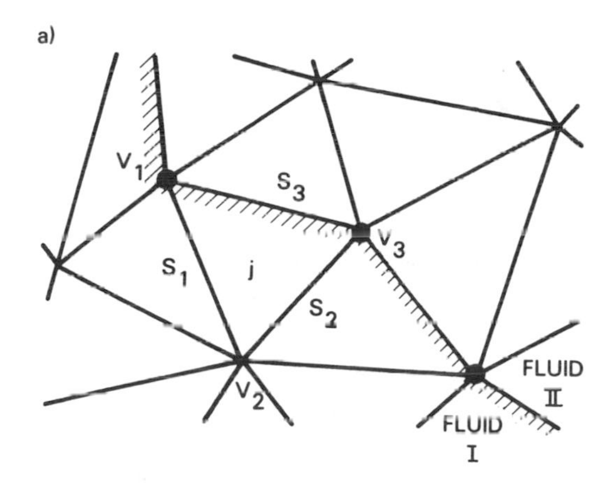

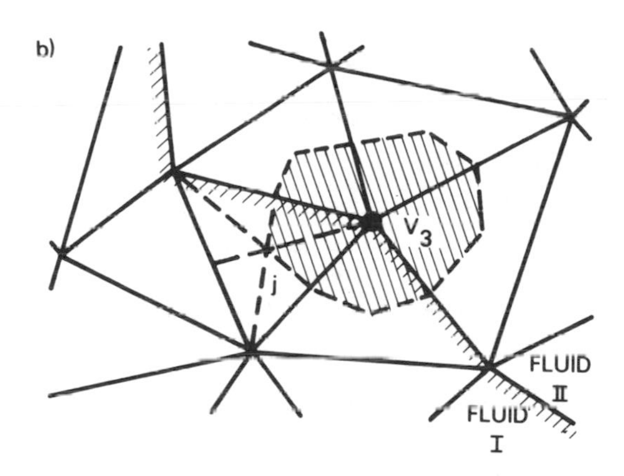

Figure 10–3. A segment of a two-dimensional domain composed of triangles. (a) The interface between materials of type I and II is identified by the line with hatching on the left. A typical triangle is labeled j, its vertices are labeled by V, and its sides by S. (b) The hatched region is the vertex cell surrounding vertex V_3. This is formed from line segments joining the centroid of each triangle with the midpoints of the two triangle sides connected to the vertex, for all triangles surrounding that vertex.

An alternate definition of a vertex cell is obtained by bisecting the side of each triangle. The perpendicular bisections of the sides intersect at a point, the center of the circle inscribed through the three triangle vertices. Then the cells are constructed as before with the intersection point used instead of the triangle centroid. This definition apportions the triangle area so that the area associated with each vertex contains all points within the triangle which are closer to it than to the other two vertices. It is intuitively reasonable, because the closest material is connected as part of the vertex cell. It has the disadvantage that the intersection could be outside of the triangle itself.

The cell definitions above give a *Voronoi mesh* if each vertex cell contains all those points in the domain that are closer to its central vertex than to any other vertex in the mesh. If the grid reconnection method presented below is used, the proper mesh connectivity automatically produces the cell-centered Voronoi mesh (Fritts, 1979). The triangular mesh is then the dual of the Voronoi mesh. Either definition of a vertex cell uses the same mesh connectivity and the same vertices in its numerical templates.

10–2.2 Finite Differences on a Triangular Grid

The finite-difference approximations for derivatives on the triangular grid are derived from the expressions for the integral of the gradient of a scalar funtion, f, and the divergence and curl of a vector field, $\mathbf{v}$. In two Cartesian dimensions,

$$\int_A \nabla f \, dA = \oint_C f \; d\mathbf{l} \times \hat{z} \, , \qquad (10-2.1)$$

$$\int_A \nabla \cdot \mathbf{v} \, dA = \oint_C \mathbf{v} \cdot (d\mathbf{l} \times \hat{z}) \, , \qquad (10-2.2)$$

$$\int_A \nabla \times \mathbf{v} \, dA = \oint_C \mathbf{v} \cdot d\mathbf{l} \;\; \hat{z} \, , \qquad (10-2.3)$$

where A is the region enclosed by the curve C, $d\mathbf{l}$ is the vector arc length around C in the counterclockwise direction, and $\hat{z}$ is a unit vector in the z-direction. By using these definitions in a finite-volume conservation equation approach, the definitions for spatial derivatives given below have also been extended to two-dimensional axisymmetric geometry (Emery et al., 1981).

A triangle-centered quantity is assumed to be piecewise constant over the triangles with discontinuities at the triangle sides. A vertex-centered quantity is piecewise linear over the triangles and thus continuous everywhere. A triangle-centered derivative uses the sides of the triangle to form the curve C in Eqs. (10–2.1) – (10–2.3). The area integral is approximated by the area of the triangle times the average value of the derivative on the triangle, and the line integral is approximated using the trapezoidal rule along each side of the triangle. For example, the gradient of a scalar function f, defined at the vertices, is a triangle-centered quantity, $(\nabla f)_j$, given by

$$A_j(\nabla f)_j = \frac{1}{2}\sum_{i(j)} f_i \ (\mathbf{r}_{i-1} - \mathbf{r}_{i+1}) \times \hat{z} \ . \qquad (10-2.4)$$

Here $\mathbf{r}_i = (x_i, y_i)$ is the coordinate vector for vertex i, A_j is the area of triangle j, and the $\sum_{i(j)}$ is over vertices i of triangle j. In the material presented below, the index i designates vertex-centered quantities and the index j designates triangle-centered quantities.

To form a vertex-centered derivative, the vertex-centered cell is used as the area A. The area integral on the left side of Eq. (10–2.1) – (10–2.3) is approximated by the area of the vertex-centered cell times the value of the derivative at the vertex. The line integral is approximated using the value on each triangle and the appropriate vector length through the triangle. For example, the curl of the vector field $\mathbf{v}$ at a vertex c is

$$A_c(\nabla \times \mathbf{v})_c = \frac{1}{2}\sum_{i(c)} \mathbf{v}_{i+\frac{1}{2}} \cdot (\mathbf{r}_{i+1} - \mathbf{r}_i) \ \hat{z} \ . \qquad (10-2.5)$$

Here $A_c = \frac{1}{3}\sum_{j(c)} A_j$ is the vertex-centered cell area, $\sum_{j(c)}$ is a sum over the triangles around the central vertex c, $\sum_{i(c)}$ is a sum over the vertices around vertex c, and $\mathbf{v}_{i+\frac{1}{2}}$ is the value of the vector field $\mathbf{v}$ on the triangle having vertices c, i, $i+1$. Similarly, the divergence of the vector field $\mathbf{v}$ at a vertex is approximated by

$$A_c(\nabla \cdot \mathbf{v})_c = \frac{1}{2}\sum_{i(c)} [\mathbf{v}_{i+\frac{1}{2}} \times (\mathbf{r}_{i+1} - \mathbf{r}_i)] \cdot \hat{z}. \qquad (10-2.6)$$

10–2.3 Equations for Incompressible, Inviscid Flow

The Lagrangian equations for inviscid, incompressible fluid dynamics are

$$\frac{d\rho}{dt} = 0, \qquad (10-2.7)$$

$$\nabla \cdot \mathbf{v} = 0, \qquad (10-2.8)$$

$$\rho\frac{d\mathbf{v}}{dt} + \nabla P = \mathbf{f}_e. \qquad (10-2.9)$$

In two dimensions the fluid density ρ, pressure P, and velocity $\mathbf{v}$ are assumed to vary with x, y, and t. The term $\mathbf{f}_e$ represents external forces applied to the fluid, for example, gravity. Equation (10–2.8), the condition for fluid incompressibility, removes the sound waves from the system.

Judiciously choosing which of the physical variables, ρ, $\mathbf{v}$, and P, should be defined as vertex-centered quantities and which should be defined as triangle-centered quantities helps to maintain the system conservation properties. In an incompressible, invisid flow, vorticity should be conserved. Numerically this is implemented by conserving circulation, the area-integral of vorticity defined in Eq. (2–2.9). Then vorticity is conserved as long as the area of each fluid element is maintained accurately. Defining velocities as triangle-centered quantities makes the conservation of circulation straightforward, and prescribing the densities on triangles and pressures at vertices allows conservation of vertex-cell areas. These choices also allow triangle sides to be used as interfaces between different materials or phases.

The velocities are integrated using a reversible split-step algorithm. First, the half-timestep triangle velocities are computed from

$$\mathbf{v}_j^{\frac{1}{2}} = \mathbf{v}_j^o - \frac{\Delta t}{2\rho_j}(\nabla P)_j^o + \frac{\Delta t}{2\rho_j}\mathbf{f}_e, \qquad (10-2.10)$$

where the superscript o designates the values at the old timestep. The initial guess for the new triangle velocities is

$$\mathbf{v}_j^{n,(0)} = \mathbf{v}_j^{\frac{1}{2}}$$

which is used to iterate

$$\mathbf{v}_i^{\frac{1}{2},(k)} = \frac{1}{2}(\mathbf{v}_i^o + \mathbf{v}_i^{n,(k-1)}), \qquad (10-2.11)$$

$$\mathbf{x}_i^{n,(k)} = \mathbf{x}_i^o + \Delta t\, \mathbf{v}_i^{\frac{1}{2},(k)}, \qquad (10-2.12)$$

$$\tilde{\mathbf{v}}_j^{\frac{1}{2},(k)} = \mathbf{R}(\{\mathbf{x}_j^o\}, \{\mathbf{x}_j^{n,(k)}\}) \cdot \mathbf{v}_j^{\frac{1}{2}}, \qquad (10-2.13)$$

$$\mathbf{v}_j^{n,(k)} = \tilde{\mathbf{v}}_j^{\frac{1}{2},(k)} - \frac{\Delta t}{2\rho_j}(\nabla P)_j^{n,(k)} + \frac{\Delta t}{2\rho_j}\mathbf{f}_e, \qquad (10-2.14)$$

where k is the index for the iteration. The vertex velocity $\mathbf{v}_i^{n,(k)}$ in Eq. (10–2.11) is obtained from the area-weighted $\mathbf{v}_j^{n,(k)}$ determined in the previous iteration,

$$\mathbf{v}_i^n = \frac{\sum_{j(i)} w_j \mathbf{v}_j^n}{\sum_{j(i)} w_j}, \qquad (10-2.15)$$

where $w_j = \theta_j \rho_j A_j$ and θ_j is the angle (in radians) of triangle j at vertex i divided by π. The transformation $\mathbf{R}$ results from the requirement of conservation of circulation, and is discussed below.

The pressures $\{P_i^{n,(k)}\}$ in Eq. (10–2.14) are derived from the condition that the new velocities $\{\mathbf{v}_j^{n,(k)}\}$ should be divergence-free at the new timestep, satisfying Eq. (10–2.8). The Poisson equation for the pressure is derived from Eq. (10–2.14) by setting $\nabla \cdot \mathbf{v}_j^{n,(k)} = 0$ to obtain a pressure $P_i^{n,(k)}$, such that

$$\left(\nabla \cdot \frac{\Delta t}{2\rho_j}(\nabla P)_j^{n,(k)}\right)_i = (\nabla^{n,(k)} \cdot \mathbf{v}_j^{\frac{1}{2},(k)})_i + \left(\nabla^{n,(k)} \cdot \frac{\Delta t}{2\rho_j}\mathbf{f}_e\right)_i . \quad (10-2.16)$$

Both terms in Eq. (10–2.16) are easy to evaluate because the divergence is evaluated on triangle-centered quantities. Two features of the Poisson equation, Eq. (10–2.16), are noteworthy. First, it is derived from $\nabla^2\phi = \nabla \cdot \nabla\phi$, as in the continuum case. Second, the left hand side results in familiar second-order accurate templates for the Laplacians when applied to homogeneous fluids and regular mesh geometries.

10–2.4 Conservation of Circulation

The approach we have been outlining is basically a *control volume* approach using an integral formulation to derive the difference algorithms. Equation (10–2.13), which ensures conservation of circulation over vertex-cell volumes, is a consequence of this approach. It reflects the fact that the triangle velocities must be altered for geometric consistency as the grid rotates and stretches. The transformation $\mathbf{R}$ is derived by considering the circulation about each vertex. Because triangle velocities are constant over the triangle, the circulation taken about the boundary of the vertex cell can be calculated from Eq. (10–2.5). The conservation of vorticity then takes the form of the operator $\mathbf{R}$ which preserves the value of the circulation about each vertex as the Lagrangian motion of the vertices causes the grid to change.

Conservation of circulation requires that a corrected set of triangle velocities $\{\mathbf{v}_{i+\frac{1}{2}}\}$ be calculated at each timestep, such that for each vertex, c,

$$\sum_{i(c)} \tilde{\mathbf{v}}_{i+\frac{1}{2}}^{\frac{1}{2},(k)} \cdot (\mathbf{r}_{i+1}^{n,(k)} - \mathbf{r}_i^{n,(k)}) = \sum_{i(c)} \mathbf{v}_{i+\frac{1}{2}}^{\frac{1}{2}} \cdot (\mathbf{r}_{i+1}^{o} - \mathbf{r}_i^{o}) . \quad (10-2.17)$$

Because this equation is linear in the unknowns $\{\tilde{\mathbf{v}}_i^{\frac{1}{2},(k)}\}$, we can obtain the change in triangle velocities by considering the change produced by the movement of a single vertex c, with coordinates $\mathbf{r}_c^{\frac{1}{2},(k)}$, and sum the resultant expression over all vertices. It is reasonable to assume that the rotator should change only the velocities of the triangles which have c as a vertex. The final result for the required change in triangle velocity from $\mathbf{v}_{i+\frac{1}{2}}^{\frac{1}{2},(k)}$ to $\tilde{\mathbf{v}}_{i+\frac{1}{2}}^{\frac{1}{2},(k)}$ is

$$\Delta\mathbf{v}_{i+\frac{1}{2}}^{\frac{1}{2},(k)} = \frac{\hat{z}\times\Delta\mathbf{r}_{i+\frac{1}{2}}^{n,(k)}}{A_{i+\frac{1}{2}}^{n,(k)}}\left[\frac{b - \sum\limits_{l(c)}\dfrac{c_{l+\frac{1}{2}}|\Delta\mathbf{r}_{l+\frac{1}{2}}^{n,(k)}|}{A_{l+\frac{1}{2}}^{n,(k)}}(\mathbf{v}_{l+\frac{1}{2}}^{\frac{1}{2}} - \mathbf{v}_{i+\frac{1}{2}}^{\frac{1}{2}})\cdot\Delta\mathbf{r}_c^{n,(k)}}{\sum\limits_{l(c)}\dfrac{c_{l+\frac{1}{2}}|\Delta\mathbf{r}_{l+\frac{1}{2}}^{n,(k)}|}{A_{l+\frac{1}{2}}^{n,(k)}}}\right] \tag{10 - 2.18}$$

where $\Delta\mathbf{r}_{i+\frac{1}{2}} = \frac{1}{2}(\mathbf{r}_{i+1} - \mathbf{r}_i)$. Several alternatives are possible for the definitions of b and c. For example, if we conserve divergence about the vertex c, then $c_{i+\frac{1}{2}} = |\mathbf{r}_{i+1} - \mathbf{r}_i|$, and $b = 0$.

The entire algorithm advances vertex positions and velocities reversibly, as it maintains the correct circulation about every interior vertex. The transformation **R** prescribed by Eq. (10–2.18) is time-reversible, hence Eqs. (10–2.10) – (10–2.14) are also reversible. This technique is unique for Lagrangian methods, which usually either ignore conservation of circulation completely or conserve circulation through an iteration performed simultaneously with the pressure iteration. With this method, circulation is conserved exactly regardless of whether the pressures and new velocities have been iterated to their final values. This makes the Lagrangian triangular grid method equivalent to a vortex dynamics method (discussed in Section 10–4) where the order N^2 velocity summation is replaced by an order $N \log N$ Poisson equation solution on a Lagrangian grid determined by the vortex locations.

10–2.5 Viscous Flows

Viscosity modifies Eq. (10–2.9), so that now

$$\rho\frac{d\mathbf{v}}{dt} + \nabla P = \mathbf{f}_e + \mu\nabla^2\mathbf{v}. \tag{10 - 2.19}$$

The additional term in the momentum equation is discretized using the same approach used for other terms. Because the velocity is a triangle-centered quantity, the viscosity requires a discrete vertex-centered gradient operator and a discrete triangle-centered divergence operator. We then have

$$A_c(\nabla f)_c = \frac{1}{2}\sum_{i(c)} f_{i+\frac{1}{2}}(\mathbf{r}_{i+1} - \mathbf{r}_i)\times\hat{z}, \tag{10 - 2.20}$$

and

$$A_j(\nabla \cdot \mathbf{v})_j = \frac{1}{2}\sum_{i(j)}[\mathbf{v}_i \times (\mathbf{r}_{i+1} - \mathbf{r}_{i-1})] \cdot \hat{z}. \qquad (10-2.21)$$

The Laplacian is found by taking the divergence of the gradient.

The finite-difference equations, Eqs. (10–2.10) and (10–2.14), can be modified to account for the additional viscous term in the momentum equation by

$$\mathbf{v}_j^{\frac{1}{2}} = \mathbf{v}_j^o - \frac{\Delta t}{2\rho_j}(\nabla P)_j^o + \frac{\Delta t}{2\rho_j}\mathbf{f}_e + \frac{\mu_j \Delta t}{2\rho_j}(\nabla^2 \mathbf{v})_j^o, \qquad (10-2.22)$$

$$\mathbf{v}_j^{n,(k)} = \tilde{\mathbf{v}}_j^{\frac{1}{2},(k)} - \frac{\Delta t}{2\rho_j}(\nabla P)_j^{n,(k)} + \frac{\Delta t}{2\rho_j}\mathbf{f}_e + \frac{\mu_j \Delta t}{2\rho_j}(\nabla^2 \mathbf{v})_j^{n,(k)} . \qquad (10-2.23)$$

These equations are implicit in the velocities and so must be iterated.

10–2.6 Conservation of Vertex-Cell Areas

Equations (10–2.10) – (10–2.14) are implicit in the triangle velocities $\{\mathbf{v}_j^n\}$. Because these equations must be solved iteratively to produce a divergence-free velocity field, residual error usually remains. In addition, vertex velocities are derived from the divergence-free triangle velocities. In practice this means that vertex-cell areas may not be exactly conserved even though the flow is divergence-free. Furthermore, as the flow progresses, the triangle sides should distort even though we assume straight triangle sides at each timestep. This does not produce the equivalent cell area about any given vertex that would result from "correctly" distorted triangle sides. However, because we know what the area should be, it is possible to correct the known error iteratively. This correction operation is performed at the end of each timestep as we move the vertex positions. The curl-conserving rotator ensures that this correction does not change the circulation about each of the vertices due to numerical errors. The circulation is still free to change due to physical processes.

To expand or contract a vertex-cell area, the surrounding triangle areas must expand or contract. To expand a triangle j with area A_j and vertex coordinates $\mathbf{r}_i$ by an amount ΔA_j, move each vertex $\mathbf{r}_i$ an amount

$$\mathbf{r}_i^{new} - \mathbf{r}_i = \Delta \mathbf{r}_i = d_j \left[\hat{z} \times (\mathbf{r}_{i-1} - \mathbf{r}_{i+1})\right] . \qquad (10-2.24)$$

The vertices are moved normal to the opposite side a distance prescribed by the triangle expansion factor, d_j. If d_j is positive, the triangle area increases. Using the vector definition for the area of a triangle, we have

$$\begin{aligned}
2\Delta A_j &= 2A_j^{\text{new}} - 2A_j \\
&= [(\mathbf{r}_{i+1}^{\text{new}} - \mathbf{r}_i^{\text{new}}) \times (\mathbf{r}_{i-1}^{\text{new}} - \mathbf{r}_{i+1}^{\text{new}})] \cdot \hat{z} \\
&\quad - [(\mathbf{r}_{i+1} - \mathbf{r}_i) \times (\mathbf{r}_i - \mathbf{r}_{i-1})] \cdot \hat{z} \\
&= [(\mathbf{r}_{i+1} - \mathbf{r}_i) \times (\Delta\mathbf{r}_{i-1} - \Delta\mathbf{r}_{i+1})] \cdot \hat{z} \\
&\quad + [(\Delta\mathbf{r}_{i+1} - \Delta\mathbf{r}_i) \times (\mathbf{r}_{i-1} - \mathbf{r}_{i+1})] \cdot \hat{z} \\
&\quad + [(\Delta\mathbf{r}_{i+1} - \Delta\mathbf{r}_i) \times (\Delta\mathbf{r}_{i-1} - \Delta\mathbf{r}_{i+1})] \cdot \hat{z} \\
&= \beta^2 d_j + 6A_j d_j^2 \ ,
\end{aligned} \tag{10-2.25}$$

where β^2 is the sum of the squares of the sides of the triangle. This quadratic in the expansion factor, d_j, can be solved to yield

$$d_j = \frac{-\beta^2 + \sqrt{\beta^4 + 48A_j\Delta A_j}}{12A_j} \ . \tag{10-2.26}$$

The sign in front of the square root was chosen to ensure that d_j has the same sign as ΔA_j.

The change in triangle area, ΔA_j, is related to the conservation of vertex-cell areas through

$$\Delta A_j = \frac{A_j}{3} \sum_{i(j)} \frac{A_i^c - A_i}{A_i} = \sum_{i(j)} \frac{A_j}{3A_i} \Delta A_i \ . \tag{10-2.27}$$

Here the sum is over the three vertices of the triangle, A_i is the current area about vertex i, and A_i^c is the original or "correct" cell area about vertex i. The change in vertex-cell areas is apportioned to each contributing triangle according to that triangle's contribution to the vertex-cell area.

Although this area correction is a small numerical effect, the ability to expand triangles is useful for modeling other physical processes. In a compressible problem involving slow energy release, this algorithm produces the required expansion of the vertex cells.

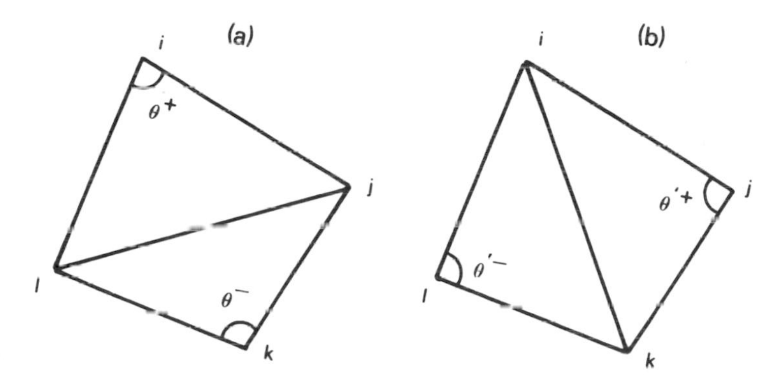

Figure 10–4. Automatic vertex reconnection proceeds according to conditions on the triangle angles shown in these two figures.

10–2.7 Grid Restructuring

In multidimensional Lagrangian calculations, the grid can become extremely distorted. Unless this grid is restructured, the calculation on the distorted grid becomes extremely inaccurate and eventually unstable. This problem is handled by reconnecting the vertices to reduce the local grid distortion and by adding new vertices and deleting old ones. There are many ways to formulate a reconnection algorithm. The one we use is based on solving the Poisson equation for the pressure. Because this equation is solved iteratively, we want the iteration to converge as rapidly as possible.

The reconnection algorithm is chosen to ensure that the off-diagonal elements in the finite-difference form of the Poisson equation are positive. The off-diagonal coefficient relating vertex l to vertex j is

$$a = \frac{1}{2}(\cot\theta^- + \cot\theta^+) \ . \tag{10 – 2.28}$$

Here θ^+ and θ^- are the angles opposite the line from the vertex j to the vertex l as shown in Figure 10–4a. For positive area triangles, θ^+ and θ^- both range between 0° and 180°. Thus each term in Eq. (10–2.28) is negative only when $\theta^+ + \theta^- > 180°$, because

$$a = \frac{\sin(\theta^+ + \theta^-)}{2\sin\theta^+ \sin\theta^-} \ . \tag{10 – 2.29}$$

If $\theta^+ + \theta^-$ is greater than 180°, the grid line is reconnected as shown in Figure 10–4b. The new angles, θ'^+ and θ'^-, sum to less than 180° because $(\theta^+ + \theta^- + \theta'^+ + \theta'^-)$, the sum of the interior quadrilateral angles, must be 360°. This reconnection algorithm preferentially eliminates large angles in triangles, because the diagonal is chosen which divides the largest opposing angles.

Vertex addition algorithms are needed where the flow naturally depletes vertices where new flow structure is developing, and at material interfaces that should not reconnect. For vertex addition, satisfying conservation integrals is straightforward. The vertex added at the centroid of a triangle subdivides that triangle into three smaller triangles. A vertex added to the midpoint of a side subdivides the two adjacent triangles into four smaller triangles. If the new triangle velocities are all the same as the velocity of the subdivided triangles, all conservation laws are satisfied. Because the reconnection algorithm is also conservative, subsequent reconnections to other vertices ensure that the only effect of vertex addition is to increase resolution.

The procedure for vertex deletion is not as straightforward. Reconnections can be used to surround any interior vertex within a triangle. Although some of the angle constraints may not be satisfied by completely enclosing the vertex, this is not important because the vertex is then removed. The new larger triangle is assigned a velocity that is the area-weighted sum of the old velocities,

$$A_l \mathbf{v}_l = A_i \mathbf{v}_i + A_j \mathbf{v}_j + A_k \mathbf{v}_k \, . \qquad (10-2.30)$$

Such a substitution redistributes circulation with a corresponding loss of local resolution. Figure 10–5 shows the triangles before and after vertex removal. If ς_4 is the vorticity about vertex 4 before it is removed, then the vorticity about each of the other three vertices is increased by an amount ς_i' given by

$$\begin{aligned} \varsigma_1' &= A_j \varsigma_4 / A_l \\ \varsigma_2' &= A_k \varsigma_4 / A_l \\ \varsigma_3' &= A_i \varsigma_4 / A_l \, . \end{aligned} \qquad (10-2.31)$$

Vorticity is conserved because

$$\varsigma_1' + \varsigma_2' + \varsigma_3' = \varsigma_4$$

and

$$A_i + A_j + A_k = A_l \, .$$

The total vorticity is conserved but is redistributed by vertex deletion which therefore has smoothing or diffusive properties.

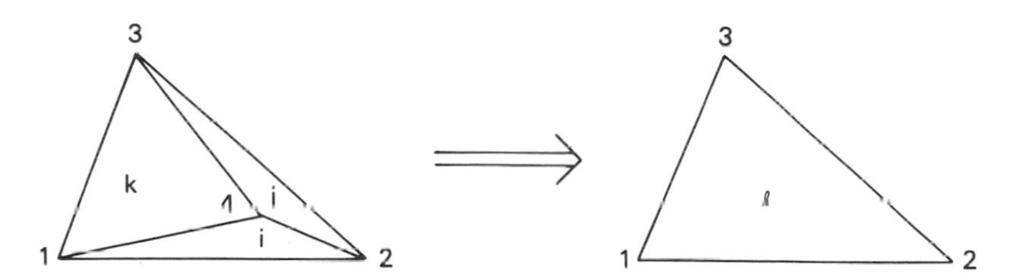

Figure 10–5. The vertex deletion process to remove vertex 4.

10–2.8 Some General Comments

Multidimensional Lagrangian algorithms are not easy algorithms to program or use. The logic of the grid restructuring algorithms in this case, or the grid remapping in other methods, is an essential part of a Lagrangian algorithm and is quite complicated. Many of the algorithms that seem obvious have required time-consuming efforts to find the best implementation. These algorithms are incorporated in the SPLISH modules and have been used for a number of different types of calculations including flows over obstacles (Miner et al., 1983), Rayleigh-Taylor instabilities (Emery et al., 1982), and Couette and Taylor-vortex flows (Emery et al., 1981). We consider a particular application of SPLISH in Chapter 13, calculating droplet distortion and breakup in a multiphase flow.

10–3. QUASIPARTICLE METHODS

Quasiparticle methods are attempts at solving convection problems by simulating, at some level of approximation, the behavior of the particles which make up the fluid. The collective, continuum behavior of the fluid is extracted as suitable averages over the quasiparticles although it is never possible to represent the actual number of particles in the system. It is typical to simulate a galaxy, consisting of 10^{11} stars with 10^5 quasiparticles, or 1 cm^3 of water, consisting of about $\sim 10^{22}$ particles, with 10^6 quasiparticles.

If a large number N of quasiparticles are initialized at positions $\{\mathbf{x}_j(0)\}$ at time $t = 0$, their later positions can be found at time t by integrating the (Lagrangian) trajectory equations

$$\frac{d\mathbf{x}_j}{dt} = \mathbf{v}_j(t) \,, \qquad (10-3.1)$$

where the velocity $\mathbf{v}_j$ of each quasiparticle j is found by integrating a force law

$$\frac{d\mathbf{v}_j}{dt} = \mathbf{f}_j(\{\mathbf{x}_k\}, \{\mathbf{v}_k\}, t) \ . \qquad (10-3.2)$$

Here the force $\mathbf{f}_j$ on the jth particle can depend on the positions and velocities of all of the other particles. The fluid mass density is found by summing the masses m_j of all the particles j at or near the ith location,

$$\rho(x_i) = \left\langle \sum_i m_i \delta(\mathbf{x}_j - \mathbf{x}_i) \right\rangle \ . \qquad (10-3.3)$$

The fluid momentum density is found by summing the particle momenta,

$$\rho \, \mathbf{v}(x_i) = \left\langle \sum_j m_j \mathbf{v}_j \delta(\mathbf{x}_j - \mathbf{x}_i) \right\rangle \ . \qquad (10-3.4)$$

The brackets indicate an averaging procedure that smoothes out the discreteness, generally some form of area- or volume-weighting over finite-sized particles among several cells.

When the number of quasiparticles becomes very large, and a long-range interaction law is used, it is no longer practical to treat the forces by calculating direct interactions. A potential ϕ is introduced,

$$\mathbf{f}_j = -\nabla\phi(\mathbf{x}_j) \ . \qquad (10-3.5)$$

This potential satisfies a Poisson equation relating the source terms, such as mass, and the corresponding long-range potential fields ϕ. This field ϕ is found by methods discussed in Section 11–3. Equations (10–3.1) and (10–3.2) are ordinary differential equations, for which many solution methods were described in Chapter 5.

When each quasiparticle has a velocity that is determined by the particle position at any time,

$$\mathbf{v}_j = \mathcal{V}\left(\mathbf{x}_j(t), \ t\right) , \qquad (10-3.6)$$

such as in the case of vorticity transport being considered here, the quasiparticles cannot pass through each other. When they do pass each other, it is a result of errors due to taking finite timesteps or singularities in the form of $\mathcal{V}(\mathbf{x}, t)$. The model is only appropriate for equilibrium, collisional fluids.

The incompressibility condition, $\nabla \cdot \mathbf{v} = 0$, is difficult to enforce rigorously in quasiparticle models. Incompressibility really only arises as a statistical concept in real fluids. Because there are so many fewer quasiparticles in the model than real particles in the fluid, the statistical fluctuations about constant density can be quite large. When several quasiparticles can occupy essentially the same location in space, and yet move with significantly different velocities, the model can be used to describe *collisionless* systems, such as plasmas on the microscopic scale or stars in self-gravitating systems. Collisionless quasiparticle methods are reviewed by Birdsall and Fuss (1969), Langdon and Lasinski (1976), Birdsall and Langdon (1985) and in the *Proceedings of the Fourth Conference on Numerical Simulation of Plasma*, edited by Boris and Shanny (1970). The original algorithms stem from work by Hockney (1965, 1966) and Buneman (1967).

Several of the original collisional quasiparticle methods developed include the *Particle-in-Cell* method (PIC) (Evans and Harlow, 1957), the *Marker-and-Cell* method (MAC) (Harlow and Welch, 1965), the *Grid-and-Particle* method (GAP) (Marder, 1975), and the *VORTEX* method (Christiansen, 1973). All of these methods use a finite-difference grid to obtain certain derived continuum quantities, such as the pressure, while using averaged particle values of the mass, momentum, and energy to preserve the Lagrangian features of the flow.

For example, PIC (Harlow, 1964) uses a rectilinear Eulerian mesh. A mean velocity, density, internal energy, and pressure are defined within each computational cell. In the first stage of the calculation the particles are regarded as fixed. The density $\rho_{\alpha i}$ at a computational cell i, for a species α, is the mass m_α times the number of particles of species α within the given cell, divided by the volume associated with the cell. The pressure is then calculated from the partial pressures $P_{\alpha i}$ derived from $\rho_{\alpha i}$ using a suitable equation of state. Then Eqs. (9–1.3), (9–1.13), or an analogous equation for the internal energy density ϵ, are solved without the advection terms to obtain intermediate values of $\mathbf{v}$ and E (or ϵ) on the grid. The equations are differenced to conserve momentum and energy. The energy is obtained by using the mean of the old and the intermediate velocities in the compression term. The particles are then moved in a Lagrangian manner using the area-weighted intermediate velocities. Some of them cross cell boundaries and thus transfer mass, momentum, and energy from one cell to another. This results in updated velocities, energy densities, and mass densities.

The PIC method, as well as other quasiparticle methods, have a problem with "graininess" when the number of particles of given species per computational cell becomes small. In addition, the Eulerian differencing scheme in PIC is unstable. The method works only because there is an effective viscosity, $\bar{\mu} \approx \frac{1}{2}\rho\bar{v}\,\Delta x$, in the calculation associated with particles crossing grid boundaries. This is effectively first-order donor-cell smoothing. Near stagnation regions where $\mathbf{v}$ approaches zero, the results are inaccurate. There are also problems associated with extreme expansions and weak shocks, where an additional artificial viscosity is needed.

The MAC method (Harlow and Welch, 1965) is essentially an Eulerian finite-difference method in which marker particles are advected passively by the computed flow field. The markers participate in the dynamics only in determining the position of a surface or interface. The MAC method is used extensively in surface-wave problems, and we return to it in Chapter 11 in the discussion of interface tracking.

The GAP method (Marder, 1975) differs from PIC in that the particles carry fluid properties, including specific volume. Area weighting is used to determine P and ρ on the grid and then to calculate from them the forces on the particles. GAP treats shocks of any strength without needing artificial viscosity and has better resolution and stability properties than PIC.

The quasiparticle methods are generally stable and conservative. They can also guarantee positivity and can be made reasonably general and flexible. Their drawback is a relative lack of efficiency and accuracy. Because the statistics, and hence the smoothness, of the computed solutions depend on the number of quasiparticles which the user can afford, many applications of interest are inaccessible to the particle approach. For example, three-dimensional problems are generally unacceptably expensive.

10–4. VORTEX DYNAMICS METHODS

Vortex dynamics methods are based on the vorticity formulation of the incompressible fluid equations, given by Eqs. (9–1.4) – (9–1.6). They are an alternate Lagrangian representation to the usual pressure-velocity formulation. In vortex dynamics, the flow is simulated by representing a continuous vorticity field by many discrete parcels of circulation and then tracking this discretization in a Lagrangian model. Local velocities are calculated using the discrete vorticity as a source term. The result is a set of nonlinear ordinary differential equations giving the time evolution of the discrete vortex locations. Excellent reviews of this method have been given by Leonard (1980, 1985) and they form the basis for much of the discussion of vortex methods presented here.

10–4.1 Point Vortex Methods

The earliest vortex method was presented by Rosenhead (1931). He considered the problem of the time-dependent behavior of a two-dimensional vortex sheet, which he approximated by following the movement of a system of point vortices. A scalar vorticity field was written as a sum

$$\omega(\mathbf{r},t) = \sum_{i=1}^{N} \Gamma_i \, \delta(\mathbf{r} - \mathbf{r}_i(t)) \, , \tag{10-4.1}$$

where δ is the two-dimensional Dirac delta function, the vectors

$$\mathbf{r}_i = (x_i, y_i) \tag{10-4.2}$$

are the locations of N singular point vortices, and $\{\Gamma_i\}$ are the circulations, or strengths, of each vortex, defined as

$$\Gamma_S = \oint_S \omega \, dS \, . \tag{10-4.3}$$

To satisfy the inviscid vorticity transport equation,

$$\frac{d\omega}{dt} = \frac{\partial \omega}{\partial t} + (\mathbf{v} \cdot \nabla)\omega = 0 \, , \tag{10-4.4}$$

these discrete vortices must be convected conservatively with the flow. Thus the velocity of each vortex is the value of the velocity field at the vortex location,

$$\frac{d\mathbf{r}_i}{dt} = \mathbf{v}(\mathbf{r}_i, t) \, . \tag{10-4.5}$$

The velocity equation is computed as the solution to the Poisson equation

$$\nabla^2 \mathbf{v} = -\nabla \times \omega \hat{z} \, , \tag{10-4.6}$$

where $\hat{z}$ is the unit vector in the z-direction. If the two-dimensional flow field has no interior boundaries and the fluid is at rest at infinity, the solution to Eq. (10–4.6) may be written as a Biot-Savart integral. Then using Eq. (10–4.1) for the vorticity, the $\{\mathbf{r}_i\}$ are the solution to the following set of nonlinear ordinary differential equations,

$$\frac{d\mathbf{r}_i}{dt} = -\frac{1}{2\pi} \sum_{j \neq i}^{N} \frac{(\mathbf{r}_i - \mathbf{r}_j) \times \hat{z}\Gamma_j}{|\mathbf{r}_i - \mathbf{r}_j|^2} \, . \tag{10-4.7}$$

The point vortex method is not stable and does not generally converge, though a solution can be obtained for some finite time with the method.

Using an increased number of point vortices of reduced strength with more accurate integration methods does not produce a converged solution. The best results are obtained when only a few point vortices are used and with a fairly diffusive time-integration scheme. The problem is that treating points instead of distributed vorticity produces singularities in the solution after a finite time.

10–4.2 Distributed-Core Vortex and Filament Methods

The *distributed-core* or *vortex blob* method is an attempt to remedy the problems with point vortices. In this approach, vortices have finite cores (Chorin, 1973; Chorin and Bernard, 1973) and vorticity (in two dimensions) is now represented by

$$\omega(\mathbf{r},t) = \sum_{i=1}^{N} \Gamma_i \; \gamma_i(\mathbf{r} - \mathbf{r}_i(t)) \; . \qquad (10-4.8)$$

Here γ_i, the vorticity distribution of the vortex at $\mathbf{r}_i$, is a function of $(\mathbf{r} - \mathbf{r}_i)$ and is normalized such that

$$\int \gamma_i(\mathbf{r}) d\mathbf{r} = 1 \; , \qquad (10-4.9)$$

where A is an area element. Also the characteristic radii of the distributions differ among the vortices depending on a parameter, σ_i. Thus γ_i is given by

$$\gamma_i(\mathbf{r} - \mathbf{r}_i) = \frac{1}{\sigma_i^2} \, f\left(\frac{|\mathbf{r} - \mathbf{r}_i|}{\sigma_i}\right) \qquad (10-4.10)$$

where the *shape function* f is taken common to all vortices. The parameter σ_i is a measure of the size of the *core* of *vortex element* i. Using this definition of γ_i gives a smoother representation of the vorticity and the induced velocities in and near vortex elements are now bounded.

Various types of core distribution functions have been used. One choice is a Gaussian function that identically satifies the viscous part of the vorticity transport equation for viscous flows. Thus σ can be programmed to increase in time according to a well understood similarity solution. Forms of γ proportional to $|r|^{-1}$ have been used to obtain a constant velocity within the vortex core (Chorin, 1973). It has been shown that using γ containing both signs of vorticity can increase the accuracy (Hald, 1979).

There are two ways of calculating the velocity in a distributed-core vortex. One way is to evaluate the velocity field at the center of the blob,

$$\frac{d\mathbf{r}_i}{dt} = \mathbf{v}(\mathbf{r}_i,t) = -\frac{1}{2\pi}\sum_{j=1}^{N}\frac{(\mathbf{r}_i-\mathbf{r}_j)\times\hat{z}\,\Gamma_j\,G\left(\frac{|\mathbf{r}_i-\mathbf{r}_j|}{\sigma_i}\right)}{|\mathbf{r}_i-\mathbf{r}_j|^2}, \qquad (10-4.11)$$

with

$$G(a) = 2\pi\int_0^a f(z)z\,da\,. \qquad (10-4.12)$$

Another approach is to use a vorticity-weighted average of the velocity over the blob,

$$\begin{aligned}\frac{d\mathbf{r}_i}{dt} &= \int \gamma_i(\mathbf{r}_i-\mathbf{r}')\mathbf{v}(\mathbf{r}',t)dA' \\ &= -\frac{1}{2\pi}\sum_{j=1}^{N}\frac{(\mathbf{r}_i-\mathbf{r}_j)\times\hat{z}\,\Gamma_j\,H\left(|\mathbf{r}_i-\mathbf{r}_j|,\sigma_i,\sigma_j\right)}{|\mathbf{r}_i-\mathbf{r}_j|^2},\end{aligned} \qquad (10-4.13)$$

where

$$H\left(|\mathbf{r}_i-\mathbf{r}_j|,\sigma_i,\sigma_j\right) \equiv (2\pi)^2\int_o^\infty\int_o^\infty dx\,dy\,f(x)f(y)\,J\left(|\mathbf{r}_i-\mathbf{r}_j|,\sigma_i x,\sigma_j y\right), \qquad (10-4.14)$$

where

$$J(b,s,t) = \begin{cases} 1, & x+t\le b, \\ \frac{1}{\pi}\cos^{-1}\left(\frac{s^2+t^2-b^2}{2st}\right), & |s-t|\le b\le s+t, \\ 0, & b\le|s-t|\,. \end{cases} \qquad (10-4.15)$$

Again, these give a set of nonlinear coupled ordinary differential equations.

Problems with spatial accuracy in the distributed-core vortex method are associated with the assumption of a constant shape function for the computational elements at all times. A real fluid element with a fixed amount of circulation is strained and distorted. One way to test the accuracy of this approximation is to see if the method maintains integral constraints on the motion (see Leonard, 1980). For example, for two-dimensional flow at rest at infinity with no interior boundaries, there are constraints satisfying conservation of total circulation and linear impulse. There are additional constraints for inviscid flow related to conservation of angular impulse, total kinetic energy, and the fact that the flow is incompressible and the vorticity of each blob is constant. It has been found that the vorticity-weighted average of the velocity over vortices satisfies these constraints best although it is more expensive to compute. Also, the Gaussian core is second-order accurate in σ. Higher-order accuracy can be obtained by specially constructing the core distribution (Hald and Del Prete, 1978).

It is important to consider how to add viscosity to these methods. Viscosity appears in the vorticity diffusion term, $\mu\nabla^2\omega$, in the vorticity transport Eq. (10–4.4). Diffusion of vorticity is treated by allowing the Gaussian vortex cores to increase in size according to

$$\frac{d\sigma^2}{dt} = 4\nu \, . \tag{10 – 4.16}$$

Another approach proposed by Chorin (1973) is to add a random walk each timestep, with a steplength proportional to $(\nu\Delta t)^{\frac{1}{2}}$.

Various ways have been proposed to account for vorticity generation at boundaries. One is to model the effect phenomenologically using data from experiments or other calculations. Another approach is to create vortices at the boundary to maintain the no-slip condition at the surface. The vortices created close to the surface then simulate the evolving boundary layer. A great number of point vortices are needed near the layer to resolve the short scale lengths, so this approach can be expensive. Another approach is to use elongated blobs, which mirrors the fact that derivatives parallel to the surface are much smaller than derivatives in the normal direction. This idea comes from Eulerian methods which often use high-aspect-ratio cells near a solid boundary.

Vortex filaments are the generalization of distributed-core vortices to three dimensions. It is assumed that the vorticity field is represented by a collection of N filaments. In analogy to the equations given above for inviscid flow, the circulation of a filament i is given by

$$\Gamma_i = \oint_{s_i} \omega \cdot d\mathbf{a} \, , \tag{10 – 4.17}$$

where $\mathbf{a}$ is the surface area in a cross section of filament i. The vorticity can be written as

$$\omega(\mathbf{r},t) = \sum_{i=1}^{N} \Gamma_i \int \gamma_i\big(\mathbf{r} - \mathbf{r}_i(s',t)\big) \frac{\partial r_i}{\partial s'} ds' \, , \tag{10 – 4.18}$$

where the curve $\mathbf{r}_i(s',t)$ describes the configuration of filament i at time t and s is a parameter along the curve. The vorticity distribution profile γ_i is normalized as in Eq. (10–4.9) and σ_i, the radius of the core of a filament i, varies with s along the filament. Because a collection of vortex lines are moved as a single entity according to Eq. (10–4.18), an average velocity must be chosen. Again, there are several ways to do this in analogy with the two-dimensional distributed-core vortex methods. For viscous effects, as in the two-dimensional case, γ_i can be modified to show a viscous flattening of the core. Leonard (1980) points out that for low Reynolds number (large ν) this should work well. For flows with a high Reynolds number, the curve $\mathbf{r}_i$ can be distorted greatly in space due to large velocity gradients. These turbulent effects can be modeled by introducing an eddy viscosity approximation or a more sophisticated model.

10–4.3 Contour Dynamics

In *contour dynamics* (Zabusky et al., 1979), regions of piecewise-constant vorticity are enclosed by contours. If the vorticity in a region is initially constant, it stays that way for inviscid motion, but each region is distorted as it is strained by the velocity field. The contour dynamics method calculates the evolving boundaries of each region of vorticity. An evolution equation for the planar curves that delineate boundaries of the vorticity region can be written as a line integral along the curve.

To track these curves, Zabusky et al. define points or *nodes* on the curves, and follow these points. When boundary curves form singularities, cusps, or filaments, or when the total length of the curves increase in time, it becomes necessary to redistribute the points along the curves. The result is a loss of small-scale features, somewhat analogous to a structure passing into a coarser grid in a finite-difference calculation. The expectation is that this process does not influence the large-scale structures appreciably.

10–4.4 Vortex-in-Cell Methods

Evaluation of the velocity integrals required in distributed-core vortex and filament methods calls for $\mathcal{O}(N^2)$ operations when there are N elements, either two-dimensional blobs or three-dimensional filaments. The idea of vortex-in-cell methods is to integrate the vorticity in Lagrangian coordinates to eliminate numerical diffusion, but to solve the Poisson equation for the velocity on a mesh of M points. This Poisson solution requires $O(M \log M)$ operations. Mesh values for the vorticity must be built up from the Lagrangian locations, and the corresponding velocities must be interpolated from the mesh at the locations of the Lagrangian vortices. The result is a total operation count of $M + M \log M$. Thus, there is a large gain if M is kept small enough. In two dimensions, this has been implemented in the cloud-in-cell technique by Christiansen (1973), and in three dimensions by Couët et al. (1981).

10–4.5 Some Comments on Vortex Methods

An excellent summary of vortex methods has been given by Leonard (1980, 1985). Because vortices are needed only in rotational parts of the flow, only a minimal description of the flow field is needed. However, the number of operations per timestep is $\mathcal{O}(N^2)$, and this can be expensive. Using a Lagrangian formulation eliminates the need to treat convective derivatives explicitly. However, there are errors generated in representing smooth flows and it is difficult to treat viscous effects. Vortex methods can automatically allow small-scale structure to develop by locally concentrating points, but these often have to be consolidated into fewer larger vortices because keeping all of them would require too many points in the calculation. There are methods for treating boundary conditions at infinity and for outflow in vortex methods, but no-slip boundary conditions at a solid wall are difficult to treat. Leonard also describes and references a number of applications of these methods, such as simulations of incompressible turbulence and certain boundary-layer flows.

A point of current popular discussion is the application of these methods to reactive flows, particularly to combustion. Combustion is a variable density flow, even for slow flames. Therefore, it is necessary to resolve expansions, compressions, and even the effects of sound waves properly. This cannot be done in a pure vortex method which is basically incompressible. For a complete solution, it is necessary to introduce a grid and solve a Poisson equation for the pressure, as done with the vortex-in-cell methods described above. Combustion simulations that do not add this second, compressible component to a rotational vortex flow are qualitative, at best. However, vortex dynamics methods have been combined with an interface tracking method for the flame front and used to simulate the behavior of flames (see, for example, Ashurst and Barr, 1983; Barr and Ashurst, 1984; Ghoneim, 1986).

10–5. THE MONOTONIC LAGRANGIAN GRID

Lagrangian fluid dynamics shares a number of problems with many-body dynamics. For example, both need to continually reevaluate which points are near neighbors in a large set of nodes that often seem to be moving randomly. Both need to compute several different types of interactions. Both are expensive and the program performance must be optimized. This expense arises because each node can potentially interact with any of $N-1$ other nodes in the system. This is called the N^2 problem because the N^2 interactions between pairs of nodes are potentially important.

In Chapter 6 we noted that the optimum Lagrangian method actually has no grid, but is only a collection of moving locations. However, as indicated in the discussion of vortex-in-cell methods, some kind of grid is necessary to beat the N^2 problem. In the approach described here, a Lagrangian grid has its nodes arranged in computer memory in a way that requires minimal data reshuffling and reindexing to change the grid. We call this data structure a Monotonic Lagrangian Grid, or *MLG* (Boris, 1986; Lambrakos and Boris, 1987). This method has applications in particle calculations, data base structuring, and fluid dynamics.

The MLG is really a data structure for storing the positions and other data needed to describe N moving nodes. These N nodes could, for example, represent fluid elements, atomic particles, droplets, or dust particles that must be tracked in a multiphase combustion calculation. A node with three spatial coordinates has three indices in the MLG data arrays. The data relating to each node are stored in the memory locations indicated by these indices and ensure that nodes which are close to each other in real space are always near neighbors in the MLG data arrays. This simplifies and speeds most interaction and correlation calculations. A computer program based on the MLG data structure does not need to check $N-1$ possible distances to find which nodes are close to a particular node. The indices of the neighboring nodes are automatically known because the MLG node indices vary monotonically in all directions with the Lagrangian node coordinates. The cost of the algorithm in practical calculations is dominated by the calculation of the interactions of nodes with their near neighbors, and the timing thus scales as N.

Figure 10–6 shows an example of a small two-dimensional MLG. Because spatial coordinates define a natural ordering of positions, it is always possible to associate two grid indices with a set of random locations in a two-dimensional space. The MLG shown is one of the many possible 4×4 MLGs linking 16 nodes distributed nonuniformly in a two-dimensional region. This freedom is illustrated in Figure 10–7, which shows three different MLGs passing through the same set of nodes.

The two-dimensional MLGs shown satisfy the following indexing constraints:

$$x(i,j) \leq x(i+1,j) \qquad i = 1,...,N-1; \qquad j = 1,...,N \qquad (10-5.1a)$$

$$y(i,j) \leq y(i,j+1) \qquad i = 1,...,N; \qquad j = 1,...,N-1\,. \qquad (10-5.1b)$$

The nodes are shown at their irregular spatial locations, but they are indexed regularly in the MLG by a monotonic mapping between the grid indices and

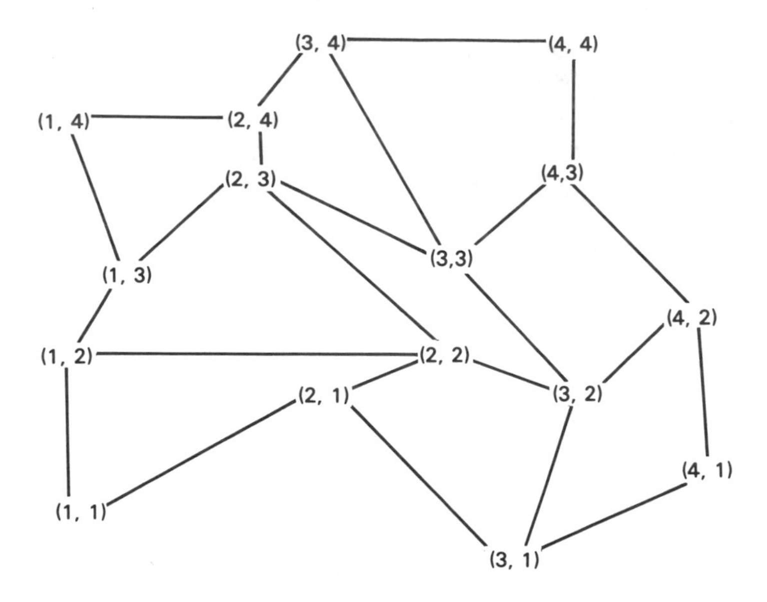

Figure 10–6. An example of a two-dimensional Monotonic Lagrangian Grid.

Figure 10–7. Three Monotonic Lagrangian Grids passing through the same set of nodes.

the locations. Each grid line in each spatial direction is forced to be a monotone index mapping. Nodes stored in the data memory according to this prescription are at most two or three nodes away from the neighboring nodes which can affect them. Thus for gradient, derivative, or force calculations, only a fraction of the possible node-node interactions need to be considered. It is not necessary to search for near neighbors and the necessary logic and computation is ideal for parallel or multiprocessing methods.

Using an MLG to index the positions and physical properties describing a node in computer memory, a Lagrangian neighborhood in the MLG can be determined by a maximum index offset, N_c, rather than a short-range cutoff radius R_c. The quantity N_c is the MLG integer analog of the radius R_c. Computations for the interactions of a particular node are only considered with those nodes whose MLG indices are offset from the index of the particular node by an amount less than or equal to N_c. Thus computations of interactions are only made between nodes located in a small contiguous portion of computer memory spanned by $2N_c$. Although this approach results in computing interactions for some distant nodes, it provides a substantial reduction in computational cost. The computations can be vectorized efficiently because close-by nodes are indexed through contiguous memory.

A construction algorithm that scales as $N \log N$ can be used to build an MLG from randomly located nodes. An MLG can always be found for arbitrary data although it is not necessarily unique. Further, when node motions in real space destroy some of the monotonicity conditions, another faster $N \log N$ algorithm exists to iteratively swap the node data between adjacent cells to restore the MLG order. The swapping iterations continue until a new MLG is found that satisfies the monotonicity conditions and any additional constraints relating to special boundary conditions. The swapping iteration converges rapidly even for large deviations from MLG ordering. Swapping is a discrete, local, and reversible operation.

Research efforts are underway to use the MLG data structure and algorithms for compressible and incompressible Lagrangian fluid dynamics. The method combines features of both the Lagrangian and the Eulerian representations. Some of the nodes can be located at interfaces and move with them, so these interfaces are resolved in a Lagrangian manner. The grid has a simple global structure which does not change, however, so the efficiency of the method approaches the efficiency of Eulerian algorithms. The local structure of an MLG is not quite as good as with Lagrangian triangular grids or Voronoi meshes described in Chapter 6 and in Section 10–2. More nodes are needed in a finite-difference grid using an MLG, typically a factor or two, to get the same accuracy as with locally adapting Lagrangian triangular grids. The potential gain with the MLG, however, comes from the parallel computing it allows.

References

Ashhurst, W.T., and P.K. Barr, 1983, Stochastic Calculation of Laminar Wrinkled Flame Propagation via Vortex Dynamics, *Comb. Sci. Tech.* 34: 227–256.

Barr, P.K., and W.T. Ashurst, 1984, *An Interface Scheme for Turbulent Flame Propagation*, Sandia Report SAND82-8773, Sandia National Laboratories, Albuquerque, NM.

Birdsall, C.K., and D. Fuss, 1969, Clouds-in-Clouds, Clouds-in-Cells Physics for Many-Body Plasma Simulation, *J. Comp. Phys.* 3: 494–511.

Birdsall, C.K., and A.B. Langdon, 1985, *Plasma Physics via Computer Simulation*, McGraw-Hill, New York.

Boris, J.P., 1979, *ADINC: An Implicit Lagrangian Hydrodynamics Code*, NRL Memorandum Report No. 4022, Naval Research Laboratory, Washington, DC.

Boris, J.P., 1986, A Vectorized "Near Neighbors" Algorithm of Order *N* Using a Monotonic Logical Grid, *J. Comp. Phys.* 66: 1–20.

Boris, J.P., and R.A. Shanny, eds., 1970, *Proceedings of the Fourth Conference on Numerical Simulation of Plasmas*, U.S. GPO, Washington, DC.

Buneman, O., 1967, Time-Reversible Difference Procedures, *J. Comp. Phys.* 1: 517–535.

Chorin, A.J., 1973, Numerical Study of Slightly Viscous Flow, *J. Fluid Mech.* 57: 785–796.

Chorin, A.J., and P.S. Bernard, 1973, Discretization of a Vortex Sheet, with an Example of Roll-up, *J. Comp. Phys.* 13: 423–429.

Christiansen, J.P., 1973, Numerical Simulation of Hydrodynamics by the Method of Point Vortices, *J. Comp. Phys.* 13: 363–379.

Cooper, A., J.M. Pierre, P.J. Turchi, J.P. Boris, and R.L. Burton, 1980, Modeling of LINUS-Type Stabilized Liner Implosions, in P.J. Turchi, ed., *Megagauss Physics and Technology*, Plenum, New York, 447–460.

Couët, B., O. Buneman, and A. Leonard, 1981, Simulation of Three-Dimensional Incompressible Flows with a Vortex-in-Cell Method, *J. Comp. Phys.* 39: 305–328.

Emery, M.H., M.J. Fritts, and R C. Shockley, 1981, *Lagrangian Simulation of Taylor-Couette Flow*, NRL Memorandum Report No. 4569, Naval Research Laboratory, Washington, DC.

Emery, M.H., S.E. Bodner, J.P. Boris, D.G. Colombant, A.L. Cooper, and M.J. Fritts, 1982, *Stability and Symmetry in Inertial Confinement Fusion*, NRL Memorandum Report No. 4947, Naval Research Laboratory, Washington, DC.

Evans, M.W., and F.H. Harlow, 1957, *The Particle-in-Cell Method for Hydrodynamic Calculations*, LASL Report No. LA-2139, Los Alamos National Laboratory, Los Alamos, NM.

Fritts, M.J., and J.P. Boris, 1979, The Lagrangian Solution of Transient Problems in Hydrodynamics Using a Triangular Mesh, *J. Comp. Phys.* 31: 173–215.

Fritts, M.J., 1979, Numerical Approximations on Distorted Lagrangian Grids, in R. Vichnevetsky and R.S. Stepleman, eds., *Advances in Computer Methods for Partial Differential Equations — III. Proceedings of the Third IMACS International Symposium of Computer Methods for Partial Differential Equations*, IMACS, New Brunswick, NJ, 137–142.

Fritts, M.J., W.P. Crowley, and H. Trease, Eds., 1985, *The Free-Lagrange Method*, Springer-Verlag, New York.

Fyfe, D.E., E.S. Oran, and M.J. Fritts, 1986, Surface Tension and Viscosity with Lagrangian Hydrodynamics on a Triangular Mesh, *J. Comp. Phys.*, to appear.

Ghoneim, A.F., 1986, Effect of Large Scale Structures on Turbulent Flame Propagation, *Comb. Flame* 64: 321–336.

Hald, O.H., 1979, Convergence of Vortex Methods for Euler's Equations, II. *SIAM J. Numer. Anal.* 16: 726–755.

Hald, O.H., and V.M. Del Prete, 1978, Convergence of Vortex Methods for Euler's Equations, *Math. Comput.* 32: 791–809.

Harlow, F.H., 1964, The Particle-in-Cell Computing Method for Fluid Dynamics, in B. Adler, S. Fernback, and M. Rotenberg, eds., *Methods in Computational Physics* 3, Academic Press, New York, 319–343.

Harlow, F.H., and J.F. Welch, 1965, Numerical Calculation of Time-Dependent Viscous Incompressible Flow of Fluid with Free Surface, *Phys. Fluids* 8: 2182–2189.

Hockney, R.W., 1965, A Fast Direct Solution of Poisson's Equation Using Fourier Analysis, *J. Comm. Assoc. Comp. Mach.* 12: 95–113.

Hockney, R.W., 1966, *Further Computer Experimentation on Anomalous Diffusion*, SUIPR Report No. 202, Institute for Plasma Research, Stanford University, Stanford, CA.

Kailasanath, K., E.S. Oran, J.P. Boris, and T.R. Young, 1982a, Time-Dependent Simulation of Flames in Hydrogen-Oxygen-Nitrogen Mixtures, in N. Peters and J. Warnatz, eds., *Numerical Methods in Laminar Flame Propagation*, Friedr. Wieweg, Wiesbaden, West Germany, 152–166.

Kailasanath, K., E.S. Oran, and J.P. Boris, 1982b, A Theoretical Study of the Ignition of Premixed Gases, *Comb. Flame* 47: 173–190.

Lambrakos, S., and J.P. Boris, 1987, Geometric Properties of the Monotonic Lagrangian Grid Algorithm for Near Neighbors Calculations, *J. Comp. Phys.*, to appear.

Langdon, A.B., and B.J. Lasinski, 1976, Electromagnetic and Relativistic Plasma Simulations Models, in B. Adler, S. Fernbach, and J. Killeen, eds., *Methods in Computational Physics* 16, Academic Press, New York, 327–366.

Leonard, A., 1980, Vortex Methods for Flow Simulation, *J. Comp. Phys.* 37: 389–335.

Leonard, A., 1985, Computing Three-Dimensional Incompressible Flows with Vortex Elements, *Ann. Rev. Fluid Mech.* 17: 523–559.

Marder, B.M., 1975, GAP — A PIC-Type Fluid Code, *Math. Comp.* 24: 434–436.

Miner, E.W., M.J. Fritts, O.M. Griffin, and S.E. Ramberg, 1983, Free Surface Wave Motions and Interactions, *Int. J. Num. Meth. Fluids* 3: 399–424.

Rosenhead, L., 1931, The Formation of Vortices from a Surface of Discontinuity, *Proc. Roy. Soc. London* A134: 170–192.

Zabusky, N.J., M.H. Hughes, and K.V. Roberts, 1979, Contour Dynamics for the Euler Equations in Two Dimensions, *J. Comp. Phys.* 30: 96–106.

Chapter 11

BOUNDARIES, INTERFACES AND MATRIX ALGORITHMS

Developing appropriate boundary conditions to simulate an infinitely large region of space or to resolve boundary layers near walls is an economically important and technically complicated issue with wide latitude for conceptual and programming errors. It is necessary to determine the correct conditions to apply, how they should be implemented numerically, and whether there are inconsistencies between the boundary conditions and the description of the physical system within the computational domain. Although it is not always necessary to understand the solution completely to model the interior of a computational domain, implementing realistic boundary conditions requires some understanding of the important interior and exterior phenomena and how they interact.

Interfaces are internal boundaries that have a structure and can move with the flow. When interfaces are present, they greatly increase the complexity of the simulation. Additional physical processes that do not occur within the flow field, such as surface tension, evaporation, condensation, or catalytic surface reactions, must be incorporated at these interfaces. Often the behavior of the interface has to be modeled phenomenologically as part of a much larger overall problem. This occurs when there are orders of magnitude difference in the gradients perpendicular to and parallel to the interface. A shock, for example, may be curved on radii of centimeters or meters, but it may only be a micron thick. The layer in which ice melts and sublimates is only fractions of a millimeter thick. Flame thicknesses are typically on the order of a millimeter or less. In these cases, the interface can be treated as a discontinuity on macroscopic length scales.

This chapter also discusses matrix algebra for finite differences. We address the problem of how to solve certain types of systems of algebraic equations that arise naturally in the numerical solution methods described in the previous chapters. Because the methods described are subjects of entire books, we present a short summary discussion and suggestions. Details are left to the references.

11–1. BOUNDARY CONDITIONS

Boundary conditions are needed to model confined and unconfined computational domains. When the system is effectively unconfined, an infinite volume must be represented with only a finite number of degrees of freedom. Depending on the physical modes in the system, the treatment of these boundary conditions differs greatly. In hyperbolic systems that simulate convection and acoustic phenomena, waves may travel much faster than the fluid flows through which the waves propagate. These waves carry information through the medium in all directions and thus information can enter the computational grid as well as leave it. In parabolic systems, the information flow is generally one sided so the solution can be advanced preferentially in one direction. Diffusion is characterized by a flux in only one direction at each point. Some supersonic, compressible flows have all characteristics moving in one direction so that one-sided or direction-biased flow equations can be integratated along streamlines.

Because boundary conditions are usually developed for a specific problem, relatively little of a general nature has been written about the edges of the computational domain compared to the volumes that have been written about the numerical techniques used in the interior. Ramshaw and Dukowicz (1979), Hyman (1979), and the review article by Turkel (1980) discuss methods of applying boundary conditions to a finite-difference representation. Oliger and Sundström (1978) discuss general problems with boundary conditions. We especially recommend the introductory article by Bayliss and Turkel (1982) and the proceedings of the symposium *Numerical Boundary Condition Procedures* (Kutler, 1982) for a collection and overview of selected recent work in boundary conditions in computational fluid dynamics.

There are several ways to implement boundary conditions in numerical models:

1. Expand the continuum fluid variables in a linear superposition of expansion functions with the boundary conditions built into each of them, so that any combination automatically satisfies the boundary conditions. Although expansions are used in many methods, this approach cannot be applied systematically to most problems or to most numerical solution procedures.
2. Develop special finite-difference formulas for the boundary cells which reflect the formulas used in the interior of the mesh, but use auxiliary relations to replace grid variables outside the computational domain.
3. Develop extrapolations from the interior to cells outside the computational domain that continue the numerical mesh a finite distance beyond the boundary. Boundary cells are then treated as interior cells, and for-

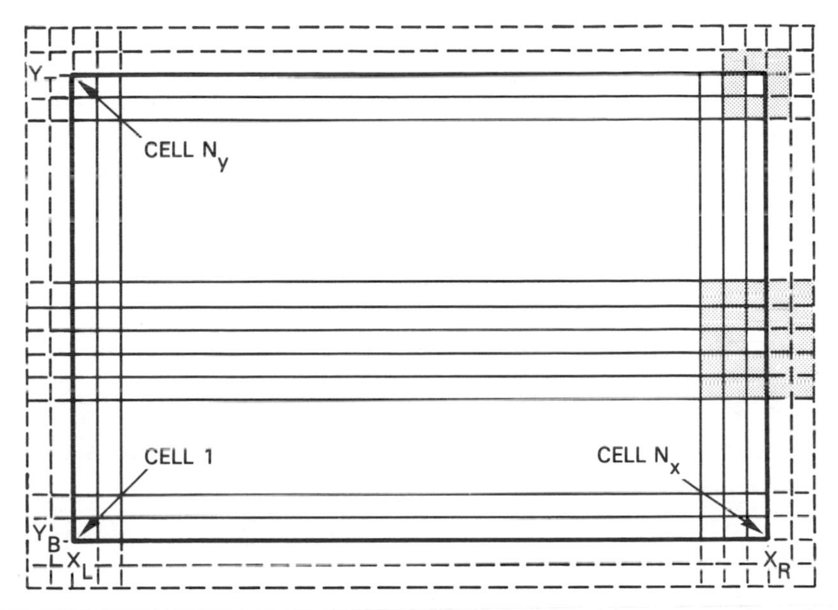

Figure 11–1. A two-dimensional computational domain with two rows of guard cells. Values of variables must be specifed in these cells to model various boundary conditions.

mulas such as those developed for the previous method can be used to define the values in the outside cells.

4. Develop an analytic formulation for the boundary variables that uses information about the behavior of the system as it approaches infinity.

Of these methods, the most easily implemented is the use of guard cells. Figure 11–1 shows a two-dimensional uniformly spaced grid where the boundary of the computational domain has been outlined by thicker lines. The grid has N_x cells in the x-direction, extending from x_L to x_R, and N_y cells in the y-direction, extending from y_B to y_T. The two rows of cells in dashed lines which surround the computational grid are called *guard cells* or *ghost cells.* By assigning appropriate values to the guard cells, the fluid elements in the domain interior can be made to satisfy an external condition as if there were no boundary at all. Usually only one or two layers of guard cells are needed. When the values of variables assigned to guard cells are stored in the same arrays as the interior variables, the calculation can be advanced on the entire grid using a single set of finite-difference equations.

Using special finite-difference formulas near boundaries or guard cells to define boundary values is often a matter of preference. Usually it is possible to define values at guard cells that are equivalent to special formulas applied at the boundaries. Conversely, separate formulas can be written to incorporate the values in the guard cells. Which of these approaches to use is determined by the computational representation, the complexity of the problem, the availability of fast computer memory, the specific numerical algorithms used, and by personal preferences.

To reduce the amount of extra memory required in three-dimensional simulations, guard cells are sometimes used for only one plane at a time. For example, a $20 \times 20 \times 20$ grid uses 8000 memory locations per variable. If this grid is extended two guard cells in all directions, the resulting $24 \times 24 \times 24$ grid has 13,824 cells, almost a factor of two increase with no added resolution! Because guard cells provide such a convenient approach to boundary conditions, it is useful to have the additional computer memory available. As the fast memory commonly available in personal computers and supercomputers increases, an extra factor of two in memory requirements will not be as important as it has been in the past.

11–1.1 Boundary Conditions for Confined Domains

Ideal Symmetry or Nonboundary Conditions

The easiest way to treat boundary conditions accurately and consistently is to eliminate them. This can be done in regions where a symmetry condition exists. Boundary points can be treated as interior points by adding guard cells whose values are known exactly from the symmetric values at interior locations. In these cases, guard cells simplify programming because only one set of difference formulas applies to all of the cells, whether or not they are near the boundary.

Symmetry conditions can often be applied in the interior of a system as well as at a boundary. A system may have a natural symmetry plane or line, such as the axis of an axisymmetric flow. Often a symmetry plane is a good approximation to the three-dimensional system, as in the case of two equal and opposite impinging jets. Finding simple symmetry and reflection boundary approximations can be extremely important to the success of a simulation.

Figures 11–2 show the right boundary and the last few cells of two one-dimensional grids. The edge of the grid is shaded to indicate a wall or mirror showing an image system. Several guard cells are indicated beyond the boundary. The upper figure depicts the symmetry boundary at the outer interface of the last cell. This is called a *cell-interface boundary*. The lower figure shows a *cell-centered boundary*, in which the symmetry passes through the center of the last cell. Similar figures could also be drawn

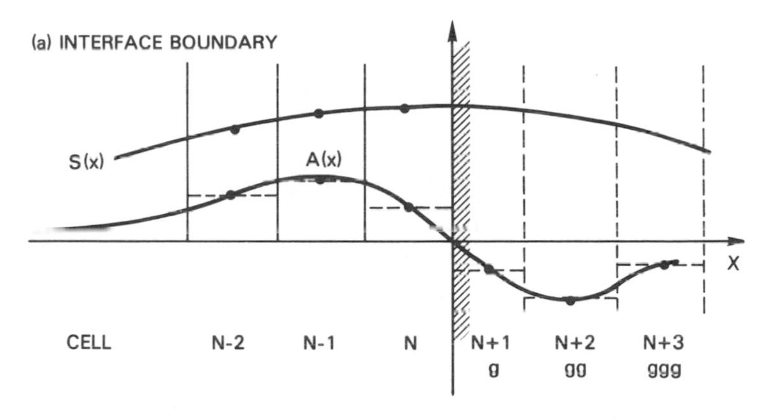

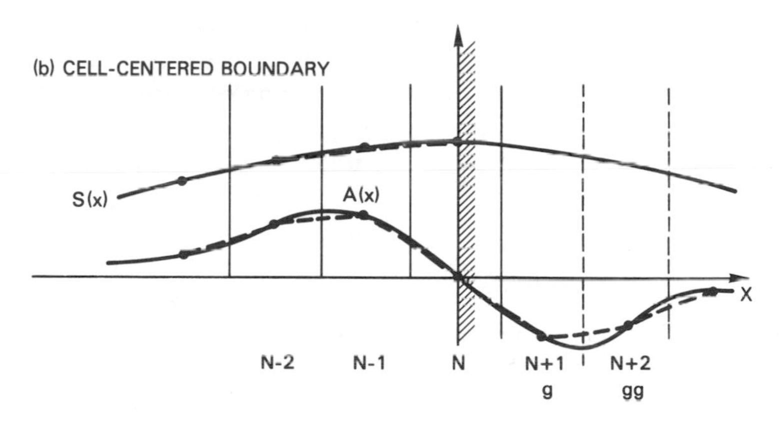

Figure 11–2. Different applications of symmetry boundary conditions with guard cells. The interior of the computational domain is on the left of the shaded band, and the guard cells are on the right. $S(x)$ is a symmetric function and $A(x)$ is an antisymmetric function. Both $S(x)$ and $A(x)$ have zero value on the x-axis. (a) Piecewise constant representation with a cell-interface boundary. (b) Piecewise linear representation with a cell-centered boundary.

for the left boundary and the first few cells of the grid. Figures 11–2 also show two different interpretations of the discretization: a piecewise constant representation in the top panel and a piecewise linear representation in the bottom panel.

Choosing between the two options shown in Figures 11–2 is generally a matter of taste. When the grid locations define the cell centers, the piecewise linear representation is natural because the discrete values of the fluid variables can be interpreted as the specific values at those points. In this interpretation, the conservation sum described by Eq. (6–2.7) always has a different, half-cell term at the boundary. By placing the symmetry plane at the known cell-center location, there is no ambiguity about its location. The last half cell, which appears in the mass integral formulation, occurs naturally in the geometry of the simulation boundary.

When the grid locations specify the cell interfaces, it is natural to use the piecewise constant interpretation even though there are apparent discontinuities at the cell interfaces. In this representation, the boundary cells are complete, as are the interior cells, so it is natural to end the computed domain at the edge rather than the middle of the last cell. Boundary fluxes enter and exit the system at interfaces where the geometric areas as well as the variable values must be known. Conservation is ensured by controlling the fluxes at these cell interfaces.

In a piecewise linear representation, the cell-interface locations are generally interpolated from the cell-center locations. When the cell-center and interface locations are constantly shifting, ensuring consistency in these interpolations makes the algorithm and programming more complex. The piecewise linear format looks smoother, which is a major advantage when the time comes to show results. In fact, most contour-plotting routines interpret the given function as values at the specified grid of points with a linear approximation made between the points. The resulting contours are relatively smooth. At multidimensional cell interfaces they show slight discontinuities in direction but not in value. Piecewise constant representations are often presented graphically as if they were piecewise linear. The piecewise constant format is ideally suited for presentation as pixel plots where a rectangle of color is filled in for each cell of a variable.

Each panel in Figure 11–2 shows two fluid variables, $S(x)$ and $A(x)$. $S(x)$ is symmetric and $A(x)$ is antisymmetric with respect to the boundary plane. In addition to variables $A(x)$ and $S(x)$, we also consider a periodic function $P(x)$. Periodic boundary conditions are variants of the symmetry conditions in which interior values somewhere are used to set guard-cell values elsewhere. For example, the values of the variables at cell $N+1$ are equal to those at cell 1 in a periodic system. Periodic boundary conditions arise in circular systems, such as stacks of turbine blades, cylindrical systems, or spherical systems. These periodic cases are relatively easy to treat numerically.

We generally assume that the guard cells are the same size as the corresponding cells just inside the boundary. Table 11–1 lists simple guard-cell formulas for symmetric, antisymmetric, and periodic variables, for both interface-centered boundary systems and cell-centered boundary systems. The entries in this table show another difference in the two representations.

Table 11–1. Guard-Cell Values*: Symmetric (S), Antisymmetric (A), Periodic (P)

Left Boundary	Right Boundary
Interface-Centered Boundaries	
$S_g = S_o = S_1$	$S_g = S_{N+1} = S_N$
$S_{gg} = S_{-1} = S_2$	$S_{gg} = S_{N+2} = S_{N-1}$
$A_g = A_o = -A_1$	$A_g = A_{N+1} = -A_N$
$A_{gg} = A_{-1} = -A_2$	$A_{gg} = A_{N+2} = -A_{N-1}$
$P_g = P_o = P_N$	$P_g = P_{N+1} = P_1$
$P_{gg} = P_{-1} = P_{N-1}$	$P_{gg} = P_{N+2} = P_2$
Cell-Centered Boundaries	
$S_g = S_o = S_2$	$S_g = S_{N+1} = S_{N-1}$
$S_{gg} = S_{-1} = S_3$	$S_{gg} = S_{N+2} = S_{N-2}$
$A_g = A_o = -A_2$	$A_g = A_{N+1} = -A_{N-1}$
$A_{gg} = A_{-1} = -A_3$	$A_{gg} = A_{N+2} = -A_{N-2}$
$A_1 = 0$	$A_N = 0$
$P_g = P_o = P_{N-1}$	$P_g = P_{N+1} = P_2$
$P_{gg} = P_{-1} = P_{N-2}$	$P_{gg} = P_{N+2} = P_3$
$P_1 = P_N$	$P_N = P_1$

* Subscript g indicates the first guard cell and gg indicates the second guard cell.

With a cell-centered boundary, the two half-cell boundary elements have the same value. Thus there is a constraint on the interior values as well as the guard-cell values. In the case of the periodic function, the values of the variable just outside of the left boundary are the same as the values just inside on the right.

A slightly more general formulation that includes the symmetry, antisymmetry, and periodicity conditions is given by

$$f_g^n \;=\; bf_{sc}^n + c + pf_{pc}^n \,, \tag{11 – 1.1}$$

where f_g^n is the value of $f(x)$ at the guard cell at timestep n, f_{sc}^n is the current value of f at the symmetry cell (sc) point inside the domain, f_{pc}^n is the periodic cell value of f at timestep n, b and p are boundary condition factors, and c is an additive constant. All of the cases in Table 11–1 can be obtained by setting $c = 0$ and making b and p appropriate combinations of $+1$, 0, and -1. Equation 11–1.1 also allows more complicated and realistic boundary conditions, including the inflow and outflow conditions described below.

Table 11–2. Free-Slip Flow along an Insulating Hard Wall*

Boundary at Cell Interface	Boundary at Cell Center
Density, Temperature, and Pressure:	
$\rho_g = \rho_s$	$\rho_g = \rho_s$
$T_g = T_s$	$T_g = T_s$
$P_g = P_s$	$P_g = P_s$
	ρ_{wall}, T_{wall}, P_{wall} calculated
Tangential Velocity:	
$v_{\parallel g} = v_{\parallel s}$	$v_{\parallel g} = v_{\parallel s}$
	v_{wall} calculated
Normal Velocity:	
$v_{\perp g} = -v_{\perp s}$	$v_{\perp g} = -v_{\perp s}$
	$v_{\perp \text{wall}} = 0$
Species Number Densities:	
$n_{i,g} = n_{i,s}$	$n_{i,g} = n_{i,s}$
	$n_{i,\text{wall}}$ calculated

* Subscript s indicates value taken at the symmetry cell.

When a flow with both normal and tangential components involves an ideal wall, the physical conditions in the guard cells can be found from the values in the nearby interior cells. This is important for bounded Euler flows and for high Reynolds number viscous flows, in which the boundary layers are approximated by additional phenomenologies. Both symmetry and antisymmetry conditions must be applied at such a wall. The lowest-order condition is symmetric, as the density, temperature, and pressure have zero slope at the wall. The tangential velocity for *free-slip* conditions is symmetric but the normal velocity is antisymmetric. The guard-cell values of variables for flow involving a wall are given in Table 11–2.

Sometimes values of the physical variables on the boundary cannot be determined by directly applying finite-difference equations with symmetry conditions. It is often necessary to develop modified boundary formulas that depend on the information from the interior. This allows representation of more physically complex situations such as viscous and turbulent boundary layers. The extrapolations generally use one-sided formulas.

Diffusion or Edge-Flux Boundary Conditions

Diffusion is one of the easiest of the nonlocal phenomena to simulate accurately. Similarly, formulating boundary conditions for diffusion problems is usually straightforward. For example, consider the finite-difference diffusion formula given in Chapter 4,

$$\frac{\rho_j^n - \rho_j^{n-1}}{\Delta t} = \frac{\nu\theta}{(\Delta x)^2}\left[\rho_{j+1}^n - 2\rho_j^n + \rho_{j-1}^n\right] + \frac{\nu(1-\theta)}{(\Delta x)^2}\left[\rho_{j+1}^{n-1} - 2\rho_j^{n-1} + \rho_{j-1}^{n-1}\right], \qquad (11-1.2)$$

where θ is the implicitness parameter. New values ρ_1^n and ρ_N^n need to be specified in terms of fictitious values at guard cells. These values can be used to control the flux of ρ entering or leaving the computational domain during a timestep, generally by specifying the gradient of ρ at the boundaries.

In some circumstances, the external guard-cell value, ρ_{N+1}^n, is known at all timesteps n. This situation occurs when a constant temperature wall bounds a system described by a thermal conduction equation. Using the known boundary values allows the solution of Eq. (11–1.2) to be advanced from step $n-1$ to step n.

Sometimes the flux of the diffused quantity is known and thus the gradient of the quantity being diffused is given at the boundary. In this case, the boundary value, ρ_N^n, and the guard-cell value, ρ_g^n, change together so that the gradient,

$$G^n = \frac{\rho_{N+1}^n - \rho_N^n}{\Delta x}, \qquad (11-1.3)$$

is fixed. Again, Eq. (11–1.2) can be solved numerically because ρ_{N+1}^n can be eliminated from Eq. (11–1.2) using Eq. (11–1.3).

Boundary Layers and Surface Phenomenologies

In a numerical simulation, boundaries often model interfaces between two different phases or two materials. A solid container with a gas inside may be chemically inert, thermally insulating, absolutely rigid, and perfectly smooth. If these approximations are acceptable, simplified symmetry and guard-cell algorithms can be used, as indicated above. Many phenomena, however, depend on the nonideal nature of the boundaries.

Analysis of numerical methods for boundary layers and surface phenomenologies could fill several books and could be enlarged further to include subgrid phenomenologies for representing other types of physics not generally resolved in the simulations. For example, catalytic reactions at walls and thermal boundary layers can be modeled by surface phenomenologies. As another example, when heat transfer to or from a wall is high enough, condensation, evaporation, or ablation can occur.

Developing such surface phenomenologies requires satisfying the conservation equations at and through the boundaries. This means that fluxes of mass, momentum, energy, and enthalpy, which enter or leave the computational domain through the boundaries must exactly equal the fluxes to the exterior world. Causality and the conservation laws provide valuable constraints in these problems. For example, if a thermal boundary layer forms in a gas next to a cold metal wall, the thermal energy moving from the last cell into the wall should not exceed the thermal capacity of that cell. If the temperature scale lengths in the gas are resolved and can be estimated in the metal wall, simple energy interchange approximations between the interior cells and the exterior are adequate. At the conclusion of this section, we derive a surface phenomenology that includes models for the spatial and temporal variations of a thermal boundary layer.

If catalytic surface reactions are occurring, the corresponding boundary condition must estimate the amount of the reactant that encounters the boundary and evaluate the probabilities of each possible reaction pathway. When the grid is finely resolved at the wall so that molecular diffusion scale lengths can be resolved, treating the wall interaction is relatively straightforward. When the cells are too large, however, the reacting species seem to be spread throughout the last cells even though they are actually concentrated near the wall. If subsequent reactions depend on the volumetric density of these reactants, the model underestimates the volumetric rates at the wall and greatly overestimates them one or two cells from the wall. Surface effects appear as volume averages leading to spurious numerical rates.

A similar effect arises from viscosity. The usual macroscopic treatment of the Navier-Stokes equation assumes the *zero-slip condition*, which means that the tangential and normal velocity components at the wall are zero. On a microscopic scale, molecules rebounding from a rigid wall are assumed to lose any memory of their collective drifts as they collide with the wall. Thus they have equal probability of scattering forward or backward relative to the flow of the fluid far from the wall. A thin laminar boundary layer forms near the wall. From the macroscopic fluid point of view, this boundary layer develops on a very small spatial scale which is very expensive to resolve. Again, a subgrid phenomenology is often used to satisfy the physical conditions at this rigid surface. When the computational cells are large, the tangential fluid momentum deficit due to this boundary layer is small even in the cells adjacent to the wall, so that very little fluid is slowed down. However, some

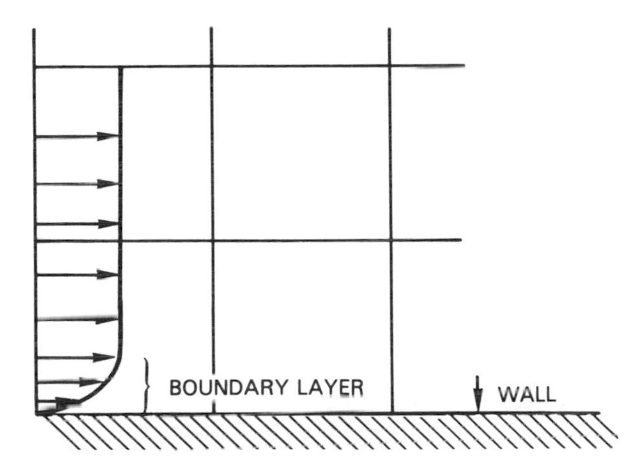

Figure 11–3. Schematic of a grid where the physical boundary layer is substantially smaller than the computational cell adjacent to the wall.

fluid is moving at almost zero velocity, as shown in Figure 11–3.

The normal component of velocity must vanish at the wall on both the macroscopic and the microscopic scale. This condition is easily satisfied in most continuum formulations and provides the necessary and sufficient condition for solving the Poisson equation. The additional tangential zero-slip condition is a complicating factor arising from the inhomogeneous contributions to the incompressible flow associated with the vorticity. In vortex dynamic formulations, the converse difficulty occurs. It becomes difficult to ensure the normal velocity is zero as the numerically generated vortex blobs are shed from the boundary.

In the primitive variable formulation, the correct drag, or "momentum deficit" has to be subtracted from the fluid near the boundary for the velocity at the wall to fall to zero. Relatively small cells normal to the wall are needed to resolve this boundary layer. These small cells impose a severe timestep restriction unless the equations are integrated implicitly. In such models, the velocity at the wall is rigorously zero, and viscosity diffuses the momentum deficit from the boundary layer into the free flow further from the wall.

11–1.2 Boundary Conditions for Unconfined Domains

Simulating an unconfined flow requires representing an effectively infinite region as a finite computational domain. In this case, the boundary conditions must transmit information to and from the entire outside world, properly absorbing any signals coming from the computational domain. Gustafsson and Kreiss (1979) have shown that systems coupled to an exterior domain can be rigorously computed on a bounded domain only when the variables and coefficients in the problem become constant at infinity. When the simplifying constant-coefficient conditions are not met, approximations are always needed, and there will be an accuracy limit having nothing to do with the solution methods interior to the domain.

One approach is to map the infinite region into a finite domain by analytically redefining the independent spatial variables. The problem with this approach is that finite-wavelength components are not resolved properly near the edges of the transformed grid. The spatial resolution of the grid becomes inadequate to propagate information at wavelengths of interest (see, for example, Grosch and Orszag, 1977). Another approach is to truncate the simulated domain at a finite distance and analytically model the influence of the exterior on the boundaries. The shorter wavelengths can now propagate up to the boundaries but are partially reflected in a nonphysical way if an exact analytic condition is not available

We recommend a combination of these approaches. First, the cells should be made progressively larger away from the central region of interest, thereby pushing the computational problems far away. This is shown in a number of the panels in Figures 6–4 and 6–5, in which the upper and lower boundaries may be viewed as the frame of a window outlining a portion of an infinite medium. By stretching the cells near the edges, or equivalently, by making them small only in regions of interest, the computational domain can be quite large without a corresponding increase in computer storage. This pushes the computational boundary far away. Errors still arise from lack of knowledge of the solution in the exterior region, but these affect the solution only weakly, and only after a delay for the numerical boundary condition influences to reach the central region.

Cell stretching should be limited to 10–20% per cell in any direction to control inaccuracies that arise from varying cell sizes. Some of the gridding techniques shown in Figures 6–4 and 6–5 allow this gradual variation. When the methods do not allow stretched cells, special formulas must be generated to interlace coarse and fine mesh solutions together at their boundaries. Such internal boundaries, which may have to be changed as the solution evolves, are a major difficulty associated with adaptive embedded grids.

In a problem with an open boundary, analytic or phenomenological models can be used to approximate the values of the simulation variables in the region outside of the computed domain. This is done using computed values near the edge of the domain and auxiliary information about the exterior behavior of the solution. In this way, the potentially unstable numerical extrapolation beyond the computational domain is replaced by a more stable interpolation.

In many cases the behavior of the system is closely coupled to the external world. This coupling often appears as conserved integrals over the discretized domain which are governed by interactions with parts of the fluid outside. In other cases this means that the value of some important physical property of the interior system is known a priori, rather than being determined by the simulation. For example, a flame in open air burns essentially at constant pressure, so the pressure at infinity becomes a constraint on the dynamics of the flow. The rate of expansion in a small burning element of a premixed gas is governed by the rate of exothermic chemical reactions. The density decreases exactly as fast as the temperature increases to keep pressure essentially constant. There is a constraint on the fluxes into or out of the computational domain, and this constraint represents feedback from the outside world which must be included if the answers are to be quantitatively correct.

In unconfined systems, boundary conditions for waves and convection are more difficult than for diffusion. Because waves are often weakly damped as they propagate, smoothing from diffusion cannot be relied on to mask errors at boundaries in the same way it does for diffusive phenomena. The most difficult boundary conditions have always been continuitive boundary conditions to describe a fluid flowing into and out of a computational mesh with a superimposed wave field. The complexities of the fluid motion essentially guarantee that the coefficients in the equations are time varying, and no analytic solution exists.

Continuitive or Inflow-Outflow Boundary Conditions

Considerations based on the characteristics and wave systems of the flow have been discussed by a number of authors. These determine how much information can be specified independently at a boundary for certain types of flows. In particular, we recommend the articles by Pulliam (1982), Oliger and Sundström (1978), and Yee (1981). The incremental changes in the evolution of a fluid system can be separated locally into a number of well-defined characteristic behaviors, generally in terms of the normal modes of the locally linearized system. In compressible gas dynamics, there are typically three characteristics in one dimension, the entropy mode, that describes convecting density and temperature variations, and the right- and left-moving acoustic modes. By separating the solution near the boundary into its characteristics, those parts of the solution moving off the grid can be extrapolated from the interior, and those parts moving onto the grid can be specified externally. In an incompressible flow, the entropy mode remains, and vorticity becomes another conserved characteristic quantity.

The difficulty with outflow boundary conditions is nonphysical reflection of the flow characteristics moving off the boundary, back into the fluid. In all but the simplest linear problems, the boundary algorithms can only be approximate because inward and outward propagating waves and pulses become locally indistinguishable in the nonlinear fluid dynamic equations. Monotone convection algorithms (Chapter 8) extrapolate the flow parameters off the edge of the system by setting the guard-cell values based on values just inside the mesh. In two- or three-dimensional flows, this extrapolation should be done along the flow lines as a better approximation to analyzing the characteristics.

Such extrapolation is often unstable in linear convection algorithms for which positivity is not guaranteed. One of the major advantages of monotone methods is their ability to operate stably and reasonably accurately using such simplifications of characteristic analysis. By separating the fluid disturbance into its constituent characteristics, however, and extrapolating each of these out to the guard cells, more stable, approximate outflow conditions are obtained for any but the most finicky convection algorithm.

Table 11–3 describes boundary conditions for a fluid flowing off the edge of a finite computational domain. Used with a monotone method, these formulas for the guard-cell values take into account the continuity of flow in the vicinity of the boundary. The lowest-order extrapolation for guard-cell values uses the adjacent cell values, that is, $b = 1$ and $\tau = \infty$. The next higher-order extrapolation uses the two cells just inside the boundary to extrapolate linearly to the guard cells. The values of the variables at or near infinity, such as ρ_∞, feed information about the external domain into the computational domain. Without these terms, that is, with $\tau = \infty$, a simulation cannot relax to the asymptotic pressure, and this leads to growing errors called *secular errors*.

As an example, the grid shown in Figure 6–10b was used to calculate an unstable shear flow generated as a fast planar jet flows across a thin splitter plate into a chamber. The instability results in vortical structures, shown in Figure 6–10a, triggered by pressure fluctuations at the lip of the plate, shown at 2 cm on the horizontal axis. In this system, both the inflow and

Table 11–3. Interpolation of Flow Off a Grid Boundary

Boundary at Cell Interface or Cell Center*
Density, Temperature, and Pressure:
$\rho_g = b\rho_s + c_\rho$
where $b = [1 - \frac{\Delta t}{\tau}]$,
$c_\rho = \frac{\Delta t}{\tau}\rho_\infty$,
τ is time constant for the relaxation of boundary conditions to the far field value.
$T_g = bT_s + c_T$
$P_g = bP_s + c_P$
Different values of τ for temperature, density, and pressure may be appropriate for subsonic flows.
Tangential Velocity:
$v_{\parallel g} = v_{\parallel s}$
Normal Velocity:
$v_{\perp g} = v_{\perp g}$ (same sign to extrapolate off the boundary)
Species Number Densities:
$n_{i,g} = bn_{i,s} + c_{n_i}$

* Subscript g refers to the guard cell and subscript s refers to the symmetry cell inside the computational domain for the particular type of grid chosen.

the outflow conditions are influenced by feedback effects from the events occuring in the computational domain (see, for example, Grinstein et al., 1986). The boundary conditions use guard cells to extrapolate the pressure to the ambient pressure at the outflow. The pressure at the inflow is allowed to vary, through a zero-slope symmetry condition (assuming subsonic inflow), to reflect acoustic perturbations from downstream back into the system. While the inflow pressure varies slightly, two properties of the incoming fluid can be specified in one dimension, typically the entropy of the gas and its mass flux. Extensive tests by Grinstein have shown that the calculations are far more accurate if some of the the splitter plate itself is included in the computational domain.

Radiative or Wave-Transmission Boundary Conditions

If there is essentially no convection at the boundaries but the problem has wavelike solutions which radiate or propagate from the local region of interest into the unresolved exterior world, the boundary conditions must simulate this radiation. If these radiation conditions are specified incorrectly, waves again are reflected nonphysically at the edge of the computational mesh and appear in the simulation as energy propagating in from infinity. This at least confuses what is going on inside the grid and may force incorrect conclusions or seriously degrade the expected accuracy.

In linear problems, it is sometimes possible to isolate the incoming and outgoing wave systems at a boundary analytically. When this can be done, nonreflective boundary algorithms, sometimes called *radiative boundary conditions*, can be developed by zeroing the incoming wave system and analytically extrapolating the outgoing waves. However, this process is very different from the local analysis of characteristics discussed above. Determining the wave components usually involves a Fourier or equivalent nonlocal analysis along the entire boundary for cells two or three layers deep. In effect, the outgoing wave solutions at infinity are built into the representation so detailed resolution of a particular scale is not required. These outer solutions must be matched to the simulated inner solutions at the boundaries of the computational domain. For example, matching spherical harmonics on a constant radius onto outgoing solutions that are calculated numerically inside this radius allows the bounding sphere to be brought close to the near field. In cylindrical coordinates, a circle can be matched with Bessel function solutions for outgoing waves. In ducts or semi-infinite geometries, Cartesian plane waves can be used.

In almost all cases the far field will be linearized so the individual components of the outwardly propagating acoustic or electromagnetic radiation can be combined additively. Once the functional form of the asymptotic expansion is known, we can derive differential relations that are satisfied by any function having the desired form. These relations are then used as boundary conditions to match the computed solution to the asymptotic expansion valid near infinity. These radiation boundary conditions become increasingly accurate as they match the solution to more terms in the exterior expansion and as the boundaries are taken far enough from the source of waves that their amplitudes become small.

A different approach is proposed by Engquist and Majda (1977, 1979), who constructed a pseudodifferential operator to exactly annihilate outgoing waves. This pseudodifferential operator is a global boundary operator. In order to derive local boundary conditions, they expanded the pseudodifferential operators in the deviation of the wave direction from some preferred direction of propagation. In this manner they construct local boundary conditions to absorb waves in a progressively larger band around a given propagation direction.

As a more efficient but usually less general alternative, an artificial dissipative region may be set up in the first few layers of cells adjacent to the boundary. When fluid flow is involved, this region is often called a Rayleigh viscous zone. As a wave disturbance propagates into this dissipative region, it encounters a medium of progressively greater damping and becomes trapped with minimal reflection until it can no longer be seen. Because there is some signal reflection off the gradient in the damping coefficient, the optimum absorbing boundary is obtained when the spatial variation of the boundary damping coefficient is gradual enough. Then the waves become trapped before the unphysically reflected wave component becomes appreciable.

This damping technique has been used for electromagnetic waves, sound waves, and gravity waves, although it does not work for bulk convection. It is particularly simple to apply because it can be added to a model generated by more idealized methods. Different wavelengths can be treated differently by this approach, but the maximum damping can usually be arranged to coincide with the strongest outgoing wavelengths. Although the absorption is not perfect, four or five cells are often enough to reduce the amplitude of the spuriously reflected wave by over an order of magnitude. Ten cells should be enough to guarantee 98–99% absorption.

A Thermal Boundary Layer

Determining phenomenological boundary conditions to represent a thermal boundary layer is a challenging problem that occurs in many reactive flows and is usually not treated as physically as it should be. Here we use this problem as an example of coupling a boundary phenomenology to a simulation model to simulate complicated phenomena relatively inexpensively.

Consider a shock in a chemically reacting gas medium that reflects from a smooth, cold metal wall. Shortly after the shock reflects, the fluid velocity near the wall returns essentially to zero. The shock heats the gas, and compresses it adjacent to the wall. The temperature in the wall begins to rise immediately and the gas temperature adjacent to the wall begins to drop. All the energy taken out of the fluid enters the wall, first as a thin boundary layer and later in a broad diffusion profile. Because the heat capacity is much lower for the gas than for the metal, the gas undergoes a larger temperature change than the metal.

We are primarily interested in boundary conditions for the gas, not the details of the temperature in the metal as a function of time and depth. To formulate a suitable phenomenological model of the temperature in the metal from knowledge of its time evolution at the surface, several assumptions are necessary. Here we assume the wall is rigid, and its thermal conductivity and heat capacity are finite.

Figure 11–4 shows the temperature profile in the vicinity of the wall at two times after the shock has reflected. The exposed surface of the wall is at $x = 0$, and the temperature there is $T_W(t)$. The temperatures calculated in the two cells adjacent to the wall are T_N and T_{N-1}. These values are calculated at discrete timesteps and must be coupled to the changing wall temperatures. At later times, a rarefaction wave may reach the wall, reducing the gas temperature below the peak temperature of the wall. This leaves a complicated temperature profile in the wall, as shown in Figure 11–4b. The energy flux out of the wall reheats the gas, as indicated by the positive slope of the temperature profiles for T_N and T_{N-1}.

The correct solution in the wall is a complicated time-dependent profile that we would have to simulate in detail to know accurately. However, the first step in constructing a suitable phenomenology is to assume a shape for the spatial profile of the temperature in the wall with a few free parameters adjusted to ensure that energy is conserved and is conducted away from the fluid as fast as it is conducted into the wall. Taking a Gaussian profile for the shape, as shown in Figure 11–4, the free parameters are: $x_m(t)$, the distance from the surface to the peak of the Gaussian; $k(t)$, the characteristic inverse scale length of the Gaussian; $T_W(t)$, the temperature at the wall; and T_∞, the wall temperature far from the surface.

The assumed functional form of the temperature in the wall is then

$$T(x,t) \;=\; T_\infty + T_m(t)e^{-k^2(t)(x-x_m(t))^2} \,, \qquad (11-1.4)$$

where

$$T_m(t) \;=\; (T_W(t) - T_\infty)e^{\eta^2(t)} \,, \qquad (11-1.5)$$

with

$$\eta(t) \;\equiv\; -k(t)x_m(t) \,. \qquad (11-1.6)$$

Here $\eta(t)$ is positive when $T_N > T_W$, that is, when the peak of the Gaussian would be out in the fluid. When η is negative, shown in Figure 11–4b, the peak of the Gaussian is inside the wall.

The equations satisfied by the temperature inside the wall and inside the fluid are

$$\frac{\partial T}{\partial t} \;=\; \lambda_W \frac{\partial^2 T(x,t)}{\partial x^2} \qquad \text{for } x \,>\, 0 \,, \qquad (11-1.7a)$$

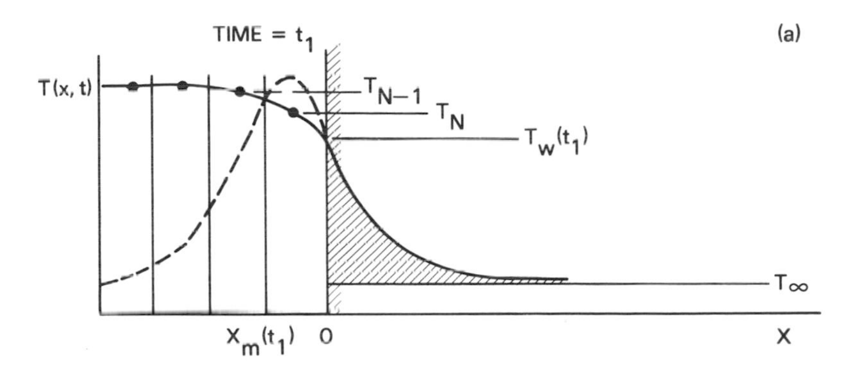

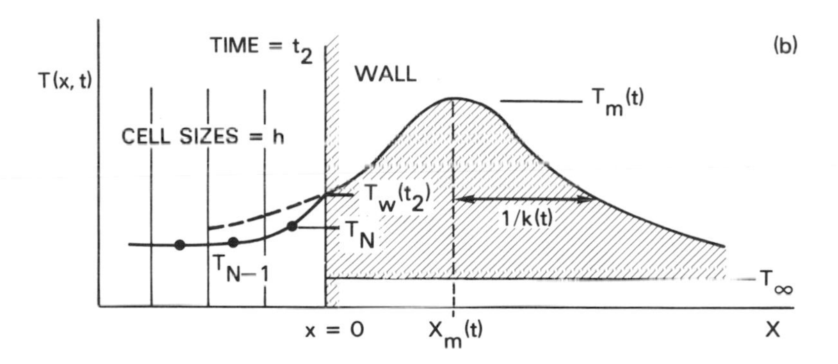

Figure 11–4. Temperature profile shown at two times after a strong shock has reflected from a cold metal wall.

$$\frac{\partial T}{\partial t} = \lambda_N \frac{\partial^2 T(x,t)}{\partial x^2} + \dot{T} \qquad \text{for } x < 0 . \qquad (11-1.7b)$$

Here $\dot{T}$ indicates changes in the temperature of the fluid due to processes in the fluid itself, such as convection and chemical energy release. We have subsumed the heat capacities into the thermal conduction coefficients, λ_N and λ_W (which in cgs units have units of cm^2/s).

Energy conservation inside the wall relates $T_W(t)$ to $k(t)$. The integrated energy entering the wall from the fluid is

$$E(t) = \int_0^\infty (T(x',t) - T_\infty)dx' = \frac{T_m(t)\mathcal{E}(\eta)}{k(t)} , \qquad (11-1.8)$$

where $\mathcal{E}(\eta)$, related to the error function, is given by

$$\mathcal{E}(\eta) = \int_0^\infty e^{-(\eta'+\eta)^2} d\eta' . \qquad (11-1.9)$$

The second relation connecting T_W, k, and η is derived by equating the fluxes used in Eqs. (11–1.7) and $\lambda \nabla T$ at the interface,

$$\lambda_N \frac{\partial T_N}{\partial x}\bigg|_W = \lambda_W \frac{\partial T}{\partial x}\bigg|_W . \qquad (11-1.10)$$

In the fluid we write a finite-difference approximation for the left side of Eq. (11–1.10) and use the analytic form, Eq. (11–1.4), to evaluate the right side. The result is

$$\frac{T_W - T_N}{T_W - T_\infty} = -\left(\frac{\lambda_W \eta}{\lambda_N}\right) kh . \qquad (11-1.11)$$

When the initial conditions are a half Gaussian in the wall with $\eta = 0$, and the fluid temperature in the gas varies in a way that keeps the temperature gradient at the wall zero, the solution inside the wall should evolve like the correct Gaussian decay profile. In this case the behavior of the inverse scale length $k(t)$ is known,

$$k(t) = \sqrt{4\lambda_W (t - t_o)} . \qquad (11-1.12)$$

The adjustable parameter t_o is the time when the Gaussian was singular. In a general case, we try to choose an average, representative value for $k(t)$ that reduces to the desired result, Eq. (11–1.12), when all the energy is deposited at once. One way to do this is to choose a suitable value for t_o which averages over the time-varying energy input, and is biased in favor of the most recent changes in total energy. For example, one choice is

$$t_o(t) = \frac{\displaystyle\int_{-\infty}^{t} t' \left|\frac{dE(t')}{dt}\right|^4 e^{t'/\tau} dt'}{\displaystyle\int_{-\infty}^{t} \left|\frac{dE(t')}{dt}\right|^4 e^{t'/\tau} dt'} . \qquad (11-1.13)$$

It is straightforward to verify that t_o is the correct constant when the energy deposition is a delta function.

There is another limiting case for which we know how the phenomenology should behave. When the energy input is exponential, growing as

$$\left|\frac{dE(t)}{dt}\right| \approx e^{\Gamma t} , \qquad (11-1.14)$$

the temperature profile should become exponential with a scale length k which does not change in time. This scale length is related to the exponential energy addition rate Γ by

$$k = \sqrt{\Gamma / \lambda_W} . \qquad (11-1.15)$$

Using the fourth power of the energy addition rate in Eq. (11–1.13) ensures that this scale length is recovered when the integrals are evaluated. The answer obtained for this scale length is

$$k = \sqrt{\left(\Gamma + \frac{1}{4}\tau\right) / \lambda_W}\,, \qquad (11-1.16)$$

which agrees with the theoretical result when the transient heating is fast compared to the long memory time, τ. The adjustable time τ is included in Eq. (11–1.13) to bias the determination of t_o in the calculation of $k(t)$ to energy addition occurring in the recent past.

As the transient heating rate increases, we expect the exponential solution to be valid and indeed the model becomes quite accurate. When the transient heating decreases, the exponential solution is less valid and the model changes smoothly to an averaged decaying Gaussian that again evolves in the analytically prescribed manner.

The model just described is only a phenomenology with adjustable parameters, even though physical constraints (conservation, continuity, and two limiting cases) are built in. Further improvements are possible. The value of τ chosen for the simulation should be comparable to a memory time expected of the interior thermal solution. To implement this model, the fluid cell value T_N can be advanced from t to $t + \Delta t$ using Eq. (11–1.11) with all of the variables solved implicitly. The energy is advanced by calculating the flux that leaves the fluid and enters the wall and by increasing the accumulated energy integral $E(t)$ by the correct amount. Then using Eq. (11–1.8) and evaluating Eq. (11–1.13), all the parameters can be determined at the new time, $t + \Delta t$. The integrals in Eq. (11–1.13) are accumulated from the energy exchanged between the fluid and the wall each timestep. The implicit solution approach requires an iteration to solve the nonlinear transcendental equations that arise. This is not a problem because it is done only at a boundary, involves only a few variables, and has much less stringent stability conditions than the numerical model used for the fluid.

11–2. INTERFACES AND DISCONTINUITIES

An interface is an internal boundary separating two different materials or two different states of matter. We identify two types of interfaces here. A *passive (resolved) interface* is inert in the sense that it is only a line or contour that models the physical behavior of the discontinuity and moves with the fluid. An *active (unresolved) interface* has thickness and properties of its own that neither of the bordering materials have and does not simply advect with the local flow velocity.

Whether an interface is active or passive is sometimes ambiguous and depends on how accurately we choose to represent it. An example of a passive interface could be the border between a solid, inert material and a gas flowing past it. However, this same interface can be active in a molecular dynamics calculation of the interaction between the molecules on the solid surface and the gas molecules impinging upon it.

Calculating resolved interfaces is called *interface capturing*. We attempt to describe the properties of the interface with detailed models of the individual contributing physical processes with adequate numerical resolution and accurate numerical methods. Usually when we resolve an interface, it is an essential aspect of the calculation. When simulating passive interfaces, we refer to the procedure as *interface tracking*, and use special algorithms to represent and determine its location in space accurately. There are four general methods used to track interfaces numerically: moving-grid methods, surface-tracking methods, volume-tracking methods, and gradient methods. Hyman (1984) and Laskey et al. (1987) review some of these methods, and Hirt and Nichols (1981) present a useful introduction.

11–2.1 Resolving Active Interfaces

The only way to capture an active interface is to model the controlling physical processes and to resolve the interface adequately. Grid refinement methods, such as those discussed in Chapter 6, can be used to provide the necessary resolution. Here we give several examples of resolved, captured interfaces.

First, consider a laminar flame front moving through a mixture of hydrogen and oxygen gas, as shown in Figure 11–5 (Oran and Boris, 1981). The program that generated this figure, described in Section 13–3, modeled the details of the chemistry, thermal conduction, molecular diffusion, and convection with enough resolution to simulate the detailed structure in the flame front. It is necessary to model and calculate the individual processes very accurately in order to produce quantitatively correct values for the properties of the flame front. To do this requires accurate input values for chemical reaction rates and the various diffusion coefficients. In this example, much more resolution is needed around the interface, here a flame front, than elsewhere in the system.

Next consider a calculation of a shock front propagating through a grid. The real shock structure is on the scale of a few mean free paths, and is orders of magnitude too small to be resolved in a macroscopic fluid calculation. Nevertheless, most algorithms for simulating shocks are shock-capturing algorithms. Numerical diffusion, flux limiters, or artificial viscosity play the role on the macroscopic scale that molecular viscosity plays on the microscopic scale. Generally at least a few cells are used to represent a shock profile. The shock is captured in the sense that the actual physical shock discontinuity is near the middle of the numerical gradient. The shock is not modeled in the same detailed sense as the resolved flame front described above, and yet the conservation form of the equations generally ensures that the correct jump conditions across the shock are maintained. This global property of satisfying the jump conditions also ensures that shocks and contact surfaces move through or with the fluid at the correct local speed. Captured shocks are shown in Figures 9–4 through 9–7 which solve a Riemann test problem by four different methods.

Another example of a resolved, active interface is the calculation of the structure of a detonation front propagating in liquid nitromethane, as shown in Figure 11–6 (Guirguis et al., 1986). This calculation was performed with the two-dimensional detonation model described in Section 13–2. The input to this model is a nitromethane equation of state and a model of the chemical reactions that includes measured chemical induction times and energy release times. Very fine resolution is required around the detonation front to calculate the complicated, interacting shock structure and the reaction zones that follow the leading shock.

The final example is another kind of reactive flow interface: an ablation layer caused by the deposition of laser energy on a solid target. One such calculation is shown in Figure 11–7, taken from the work of Emery et al. (1982). The interface resolved is the receding surface of the solid target, on the left of the two figures shown. As energy from the laser is deposited on the surface of the target, a high-pressure plasma layer is formed. The surface of the target is unstable to the Rayleigh-Taylor instability as a result of the acceleration from this low-density, high-pressure layer. This instability is generally analyzed in terms of a heavy fluid sitting on top of a light fluid in the presence of gravity. The FAST2D models that performed the calcualtions are described further in Section 13–2. The difference between the models is that here algorithms are included for strong electron thermal

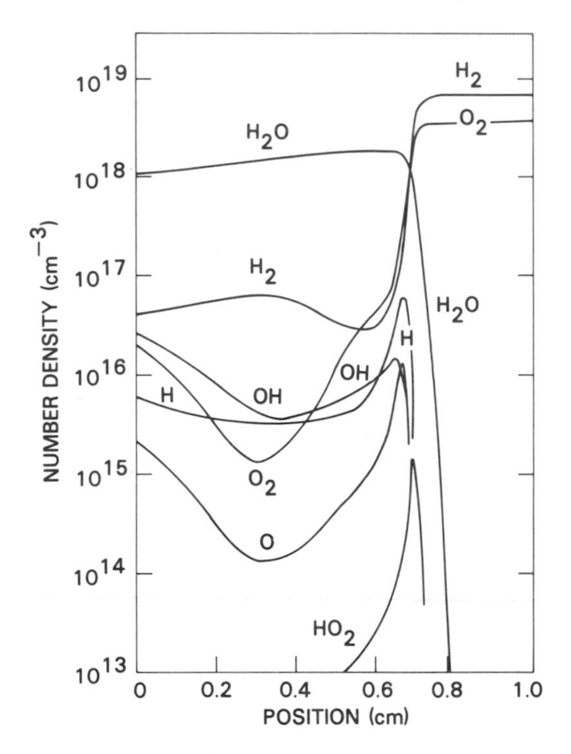

Figure 11–5. Calculation of the chemical structure of a flame propagating in a mixture of hydrogen and oxygen (Oran and Boris, 1981).

Figure 11–6. Pressure and temperature profiles for a two-dimensional calculation of a detonation propagating in liquid nitromethane (Guirguis et al., 1986).

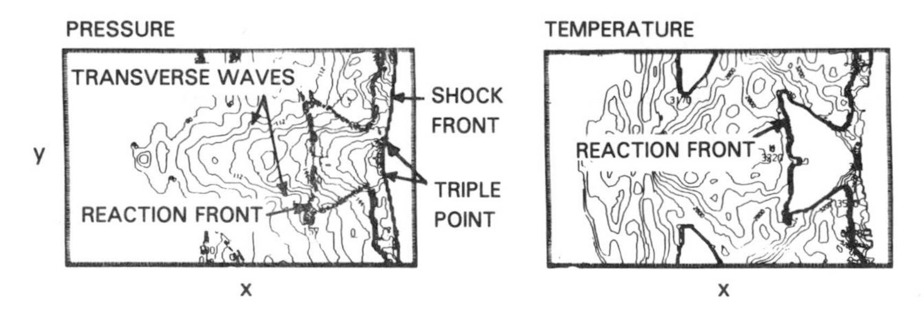

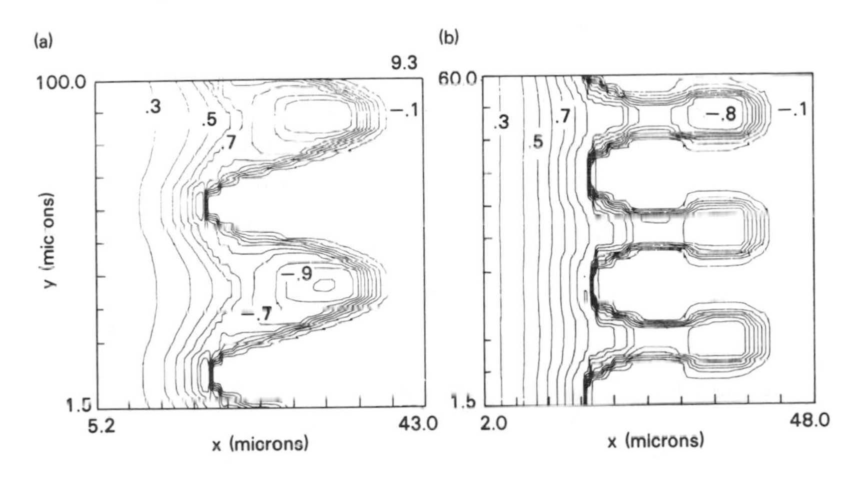

Figure 11–7. Calculation of the ablation of material from a plastic target irradiated by a laser. The Rayleigh-Taylor instability is initiated by small density perturbations at the surface. The figures show the results of two different wavelength perturbations along the surface (Emery et al., 1982).

conduction and inverse bremsstrahlung. The two panels show the nonlinear evolution of two different wavelength perturbations at the interface.

11–2.2 Moving-Grid Methods

Moving-grid methods adjust the grid so that the interface is always located on well-defined cell boundaries. Maintaining a cell boundary between different fluids controls numerical smearing that can occur at the interface as it moves. There are several approaches to implement this idea. One is to maintain a grid of distorted quadrilaterals. Another approach is to use generalized orthogonal grids that fit the form of the interface. Yet another approach is to use a Lagrangian representation with triangular cells. We might also try combination methods that use a limited Lagrangian grid to represent the interface as it moves through a rectangular Eulerian grid on which the overall fluid problem is solved.

All of these approaches have advantages and disadvantages. The common advantages are obvious: a potentially good representation of the interface. However, using a quadrilateral moving grid can cause problems when there are motions in the flow that severely distort the grid, as shown in Figure 11–8. One remedy requires defining a new grid and interpolating the physical variable onto that grid. This interpolation reintroduces numerical diffusion into the calculation. Some triangular grid approaches do not have this problem, but instead have significant bookkeeping and computer storage requirements.

Combining both an Eulerian grid and a superimposed Lagrangian grid has the versatility of representing the interface on an adaptable grid and still keeping the convenient features of a rectangular grid for the fluid dynamics. In the CEL program (Noh, 1964), a series of straight-line segments represents the interface. For example, an isolated pocket of fluid is bounded by an irregular polygon. The calculation for one timestep proceeds in several stages. First, the Lagrangian grids defining interfaces are advected through the Eulerian domain. Then, the calculations for the various fluids are performed on the Eulerian grid using the newly calculated Lagrangian positions of the interface. Finally, the velocity and pressure fields from the Eulerian calculation are used to calculate the Lagrangian positions at the start of the next timestep. An initial CEL grid for a sample problem with a number of interfaces is shown in Figure 11–9. Although methods such as CEL isolate the various fluids from each other, the problem of highly distorted Lagrangian grids is still present. In addition, this method requires storing information for both the Eulerian and Lagrangian grids, as well as rather expensive logic and computation to interpolate back and forth between the two computational grids.

A fully Lagrangian method seems to be the most natural way to track many interfaces. In Chapter 10 we introduced the Lagrangian triangular grid designed to avoid problems related to rectangular cells. When space is divided into quadrilateral cells, the connectivity of grid points is inflexible: each vertex is common to four quadrilaterials. When we divide space into triangles, a vertex may be common to any number of triangles. Thus the distortion problem can be avoided without diffusion by locally restructuring the grid. For example, an elongated cell arising from shear flow in the fluid can be replaced by one that is more symmetric, as shown in Figure 11–10.

Figure 11–11 shows a calculation of the Rayleigh-Taylor instability at the interface between two fluids using the Lagrangian algorithm described in Section 10–2. The initial configuration is a heavy fluid on top of a light fluid in the presence of a vertical gravitational field. Given a small perturbation at the interface between the fluids, the interface is unstable and the heavy fluid tends to penetrate the light fluid as it falls. By the last panel in the figure, the heavy fluid has penetrated the light fluid significantly and mixing is well underway. A number of new cells have been added to keep the interface well resolved as it rolls up and there are now isolated patches of one kind of fluid mixed in with another. The interface between fluids is a series of connected triangle sides, so that the border is composed of straight-

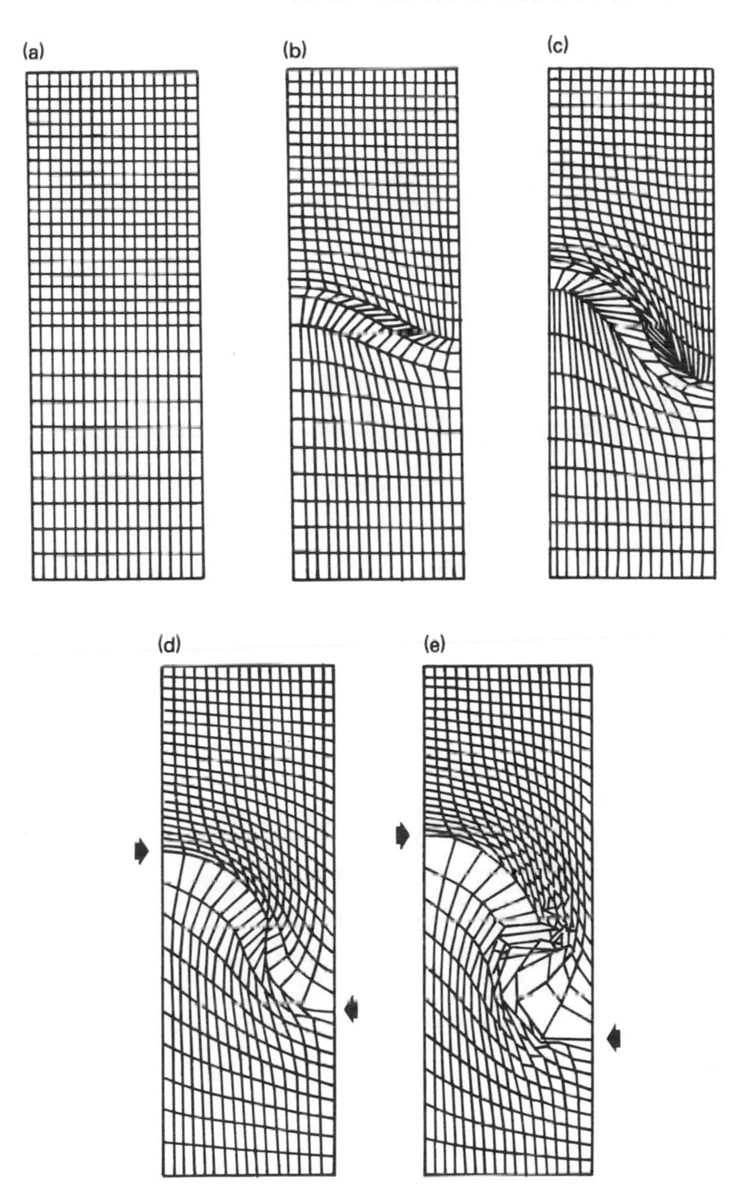

Figure 11–8. A Lagrangian calculation of a Rayleigh-Taylor instability on a quadrilaterial grid. A heavy fluid is on top of the light fluid, and the effects of gravity are included. A small perturbation at the interface at the beginning of the calculation initiates the instability. The arrows on (d) and (e) indicate the location of the interface on each side of the figures. By (e), the calculation cannot proceed because grid lines have crossed. To continue the calculations, some regridding procedure must be carried out. (Figures are courtesy of M. Fritts.)

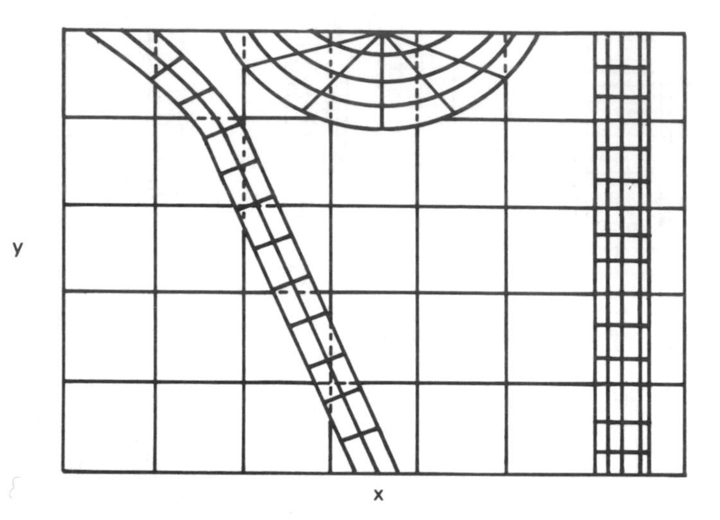

Figure 11–9. A possible initialization for a problem with multiple interfaces using the CEL method (W. Noh, 1964).

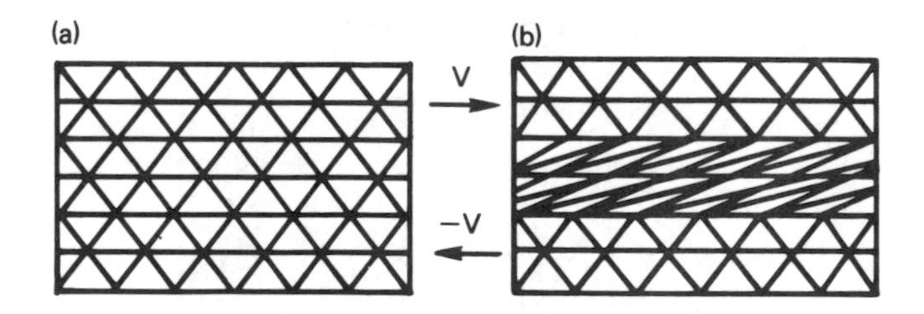

Figure 11–10. The initial conditions for a shear flow calculation using a Lagrangian triangular grid are shown in (a). The velocity of the upper fluid is v and the lower fluid is $-v$. After some time, the grid can distort, as shown in (b). Then the grid vertices can be reconnected again to produce the grid (a).

line segments. Within the limitations of a given resolution, minimum and maximum length triangle sides are permitted, so the interface can be tracked quite well even though the surface is represented as straight lines.

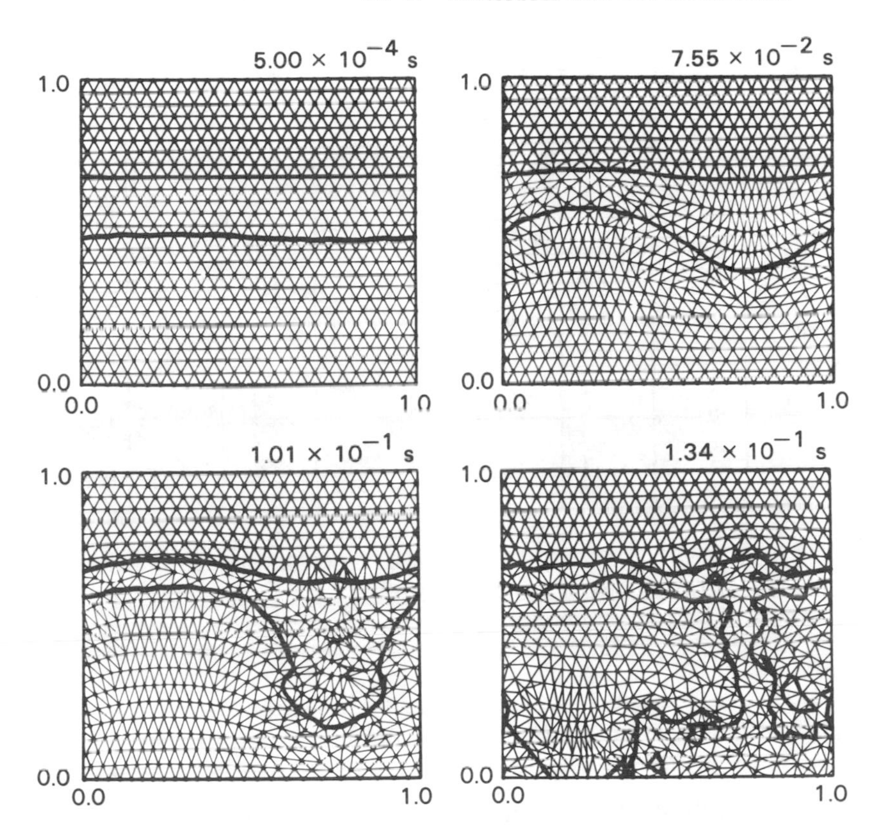

Figure 11–11. Sequence showing the evolution of a Rayleigh-Taylor instability calculated with a reconnecting Lagrangian grid of triangles. (Figure is courtesy of M. Fritts.)

11–2.3 Surface-Tracking Methods

Surface-tracking methods represent an interface as a connected series of curves interpolated through points on the interface. At each timestep, the points are advected with the flow field and the sequence in which they are connected is saved. The points can also move to simulate processes other than convection, such as chemical reactions, phase changes, or ablation. Surface-tracking methods have been used by Glimm et al. (1983) and Chern et al. (1986). The contour dynamics and contour integral methods described in Section 10–4 for certain types of fluid problems are closely related to the surface-tracking algorithms described here.

In simple surface-tracking methods for two dimensions, the points are saved as a sequence of heights above a given reference line, as shown in Figure 11–12a. Two curves are shown bounding the top and the bottom of

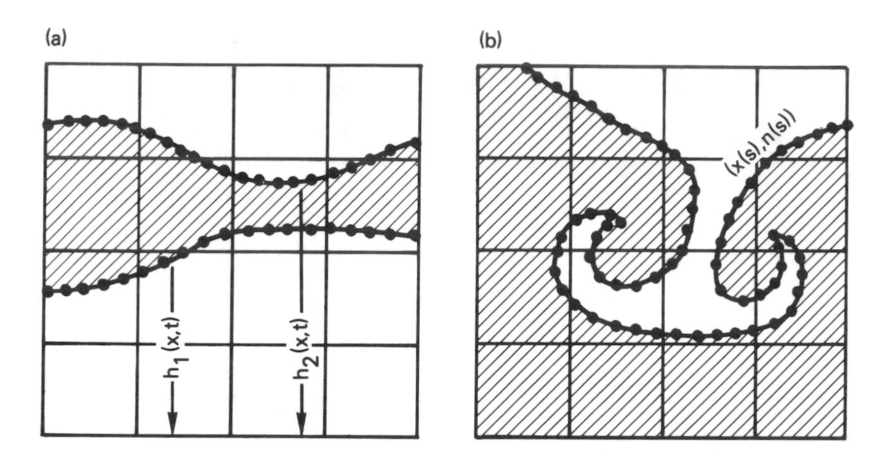

Figure 11–12. The interface may be determined by (a) a sequence of heights above a reference line, or (b) a series of points parametrized by a distance along the interface. (J.M. Hyman, 1984.)

a region of fluid. This approach fails if the curve becomes multivalued or does not extend all the way across the region. However, this failure may be avoided if the points are parametrically represented, as that shown in Figure 11–12b. The formulation is more complex, but it can represent fine detail in the interface if enough points are used. An interesting feature is that surface-tracking methods can resolve features of the interface that are smaller than the cells of the Eulerian grid on which the curves are overlaid. There is naturally a price paid for storing this additional information and the timestep is limited by how much change can be tolerated each timestep on the smallest resolved space scale.

There are still two major difficulties in surface-tracking methods. First, it is difficult to handle merging interfaces or to join a part of an interface to itself. This requires detecting the crossing, reordering the interface points (which could involve significant computational bookkeeping), and possibly tracking additional interfaces. Second, the points can accumulate along one segment of the interface leaving other segments without enough resolution. For good accuracy, it is best to limit the distance between neighboring points to less than the minimum size of the local grid (Hirt and Nichols, 1981). Interface areas typically increase continually in complex flows. Thus it is necessary to add points along the interface automatically. Conversely, points should be deleted where there are too many. When points must be added or deleted, the best way to interpolate new points and the best way to represent and manipulate contours with changing lengths are also major computational issues.

Development of surface-tracking methods continues, but the problems of changing topology from simply connected to multiply connected regions, of merging fronts, of eliminating weakened fronts, and generating interfaces at new fronts have yet to be solved. Such methods have generally been used in one- and two-dimensional calculations for interfaces that do not interact. The complexity in specifying an interface and treating its three-dimensional interactions makes surface tracking in three dimensions an area requiring substantial development.

11–2.4 Volume-Tracking Methods

Unlike surface-tracking methods that store a representation of the interface at each timestep, volume-tracking methods reconstruct the interface whenever it is needed. The reconstruction is done cell by cell and is based on the location of a marker within the cell or on some other volumetric progress variable. Whereas the surface-tracking methods represent the interface by a continuous curve, the interface generated by volume tracking consists of a set of disconnected line segments in the cells containing parts of the interface.

The earliest volume-tracking methods used marker particles whose number density in each cell indicated the density of the material. This method, first proposed by Harlow (1955), was called the Particle-in-Cell or PIC method. In the Marker-and-Cell or MAC method (Harlow and Welch, 1965; Welch et al., 1966), the particles are tracers, marker particles with no mass.

As an example, consider some of the features of the PIC method implemented by Amsden (1966) using an Eulerian grid in which velocity, internal energy, and total cell mass are defined at cell centers. The different fluids are represented by Lagrangian mass points, the marker particles, that move through the Eulerian grid. The marker particles have a constant mass, a specific internal energy, and a recorded location in the grid which is updated with the local cell velocity. The particle mass, momentum, and specific internal energy are transported from one cell to its neighbor when the marker particle crosses the cell boundary. Cells containing marker particles of two fluids contain an interface. Because the interface can be reconstructed locally at any time, the problems associated with interacting interfaces and large fluid distortions are reduced and the method generalizes to any number of fluids.

Unlike surface-tracking, marker-particle methods cannot resolve details of the interface which are smaller than the mesh size. They are also more expensive in terms of computer time and memory, typically requiring many words of information per Eulerian cell. As with the surface-tracking methods, particles may accumulate in portions of the grid, thus leaving other portions poorly resolved. Unacceptable statistical fluctuations arise when there are not enough marker particles causing the variations of the accumulated marker-particle attributes to be too large.

Many volume-tracking methods now use the fraction of a cell volume occupied by one of the materials to reconstruct the interface. If this fraction is zero for a given cell, there is no interface present. Conversely, if the fraction is one, the cell is completely occupied by the material and again there is no interface present. An interface is present only if the fractional marker volume is between zero and one. This representation, like the gradient method described below, makes optimal use of computer memory.

The Simple Line Interface Calculation, or SLIC, algorithms were first proposed by Noh and Woodward (1976). Each cell is partitioned by a horizontal or vertical line such that the resulting partial volume equals the fractional marker volume. The orientation of the line is chosen so that, if possible, similar types of fluid in neighboring cells are adjacent. The orientation is chosen to give lines through cells normal to the direction of flow. This minimizes numerical diffusion by having all of one fluid move across a cell boundary into a region of like fluid, before the other type of fluid can enter a cell that previously had a single fluid. The method assumes timestep splitting so extensions to two or three dimensions are straightforward. For the two-dimensional case, the interface is constructed cell by cell before advection in the x-direction. After the x-integration, the interface is reconstructed and the y-integration is done. Line segments normal to the direction of flow result in different representations of the interface for the x- and y-sweeps. To avoid such a directional bias, the order of x- and y-integrations is changed every timestep. Samples of SLIC interface approximations are shown in Figures 11–13 and 11–14.

Chorin (1980) improved the resolution of the original SLIC algorithm by adding a corner interface structure to the straight horizontal and vertical lines and using the same fractional cell volume. An example of this SLIC interface is shown in Figure 11–15b. Chorin used the vortex dynamics method described in Section 10–4 to calculate the vorticity field. The vorticity gives the velocity field and these velocities are used to advect the interface. Advection is done in the same time-split manner as in the original SLIC algorithm.

To simulate flame propagation, that is, the active evolution of a propagating interface, Chorin uses an algorithm based on Huygens principle.

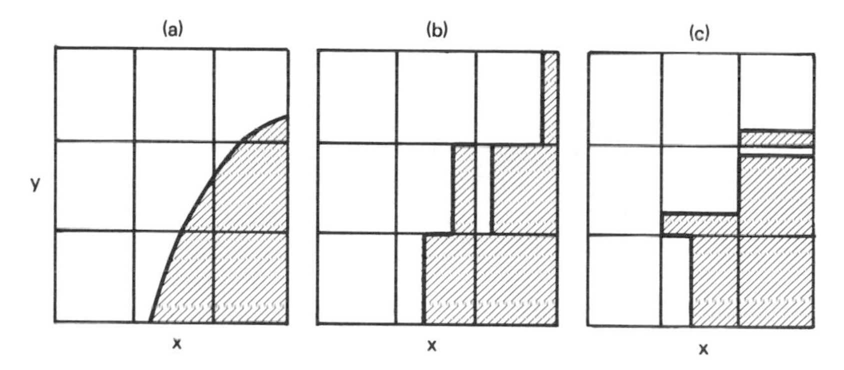

Figure 11–13. (a) The actual interface. (b) The SLIC representation on the x-pass. (c) The SLIC representation on the y-pass. (Noh and Woodward, 1976.)

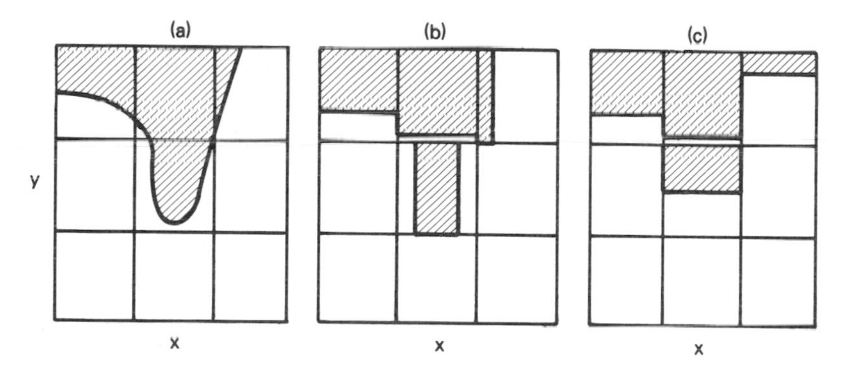

Figure 11–14. (a) The actual interface. (b) The SLIC representation on the x-pass. (c) The SLIC representation on the y-pass. (Noh and Woodward, 1976.)

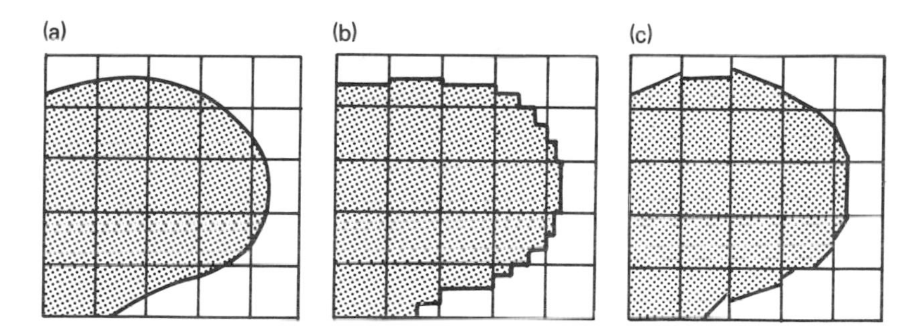

Figure 11–15. (a) The actual interface. (b) Reconstructed SLIC interfaces for convection (Chorin, 1980). (c) Reconstructed interface for boundary conditions using VOF. (Barr and Ashurst, 1984.)

Given a number of different directions, the interface is propagated at a known speed in each of the directions separately and the effect on the fractions of volume of interest in each direction is stored. The final interface position is the position that assigns the largest increase in the burned volume to each cell. Eight directions were chosen at angles $\alpha_l = (l-1)\pi/4$, $l = 1, \ldots, 8$. Each propagation is carried out using the same timestep-splitting process as for the advection. In the straightforward application of this idea, it is possible for circles to evolve into octagons. This problem, due to preferential propagation in the eight directions, may be avoided in several ways, for example, by using random starting angles. This interface construction, advection, and propagation scheme appears in combustion studies by Ghoniem et al. (1981, 1986) and Sethian (1984).

The VOF method of Hirt and Nichols (1981) also represents the interface within a cell by a straight line, but the line may have any slope. A numerical estimate of the x-direction and y-direction derivatives of the volume fraction (the progress variable) are obtained for a given cell. If the magnitude of the x-direction derivative is smaller than the magnitude of the y-direction derivative, the interface is more nearly horizontal and its slope is the value of the x-direction derivative. A more nearly vertical interface has a slope equal to the value of the y-direction derivative. If the x-direction derivative gives the slope of the interface, then the sign of the y-direction derivative determines whether the fluid is above (positive) or below (negative) the interface. For a more vertical line, the sign of the x-direction derivative determines whether the location of the marker fluid is to the right (positive) or the left (negative) of the interface. Given the slope of the interface and the side of the interface on which the fluid is located, the position of the interface within the cell is set. These calculations are done for every cell with an occupied volume between zero and one. Hirt and Nichols use values of the fractional volume of fluid averaged over several cells to calculate the derivatives.

Other implementations of VOF use simple central differences (Barr and Ashurst, 1984). The VOF method depends on the ability to advect the volume fraction through the grid accurately without smearing from numerical diffusion. Hirt and Nichols have described a "donor-acceptor" method that ensures that only the appropriate constituent fluid moves to or from a cell containing an interface. This helps to avoid the cell averaging that results in numerical diffusion.

To analyze combustion problems, Barr and Ashurst (1984) use both VOF and the Chorin version of SLIC in a method called SLIC-VOF. The SLIC algorithm is used to define and advect interfaces and VOF is used to define the normal direction and to give a smoother interface for the flame propagation phase. Figure 11–15 shows how SLIC and VOF approximate the same curved interface. For a two-dimensional flame propagation problem, SLIC-VOF performs the x-integration of the SLIC interface, the y-integration of the SLIC interface, the x-direction flame propagation of the VOF interface, and then the y-direction flame propagation of the VOF interface. The interfaces are reconstructed following each advection or propagation calculation. At each full timestep, the order of the x- and y-sweeps are interchanged, but convection always precedes burning. As in the Chorin implementation, flow velocities are calculated from a vorticity field. The flame propagation speed in the x- and y-directions is the sum of the flow velocity and the flame speed directed normal to the interface defined by VOF.

Barr and Ashurst (1984) give a detailed discussion of shortcomings in SLIC algorithms. In brief, curved surfaces may be flattened or even indented. This distortion depends on the Courant number of the flow and the interface geometry. In some cases, this distortion appears as the interface first moves across a cell and then does not increase. In other cases, the distortion may continue to grow. Including a model for propagating chemical reaction fronts apparently decreases this distortion because the propagation step smooths short wavelength wrinkles.

11–2.5 The Gradient Method

The gradient method (Laskey et al., 1987) does not define the exact location of the interface within a cell, but represents the interface as a relatively continuous gradient over several cells. Keeping the resolution of the interface within the limits of the numerical convection algorithm reduces the computer storage requirements and also the cost of interface tracking. Laskey has applied this method to flame fronts and it may also be useful for other types of interfaces. We briefly describe the elements of flame-front tracking using the gradient method because of its potential use in reactive flow simulations.

In a system of gases reacting to form a product whose number density is n_p, we must solve an equation of the form

$$\frac{\partial n_p}{\partial t} + \nabla \cdot (\mathrm{v} n_p) \;=\; w \;, \qquad (11-2.1)$$

where n_p is the number density of product molecules and w is the production term. The left side can be solved by any method for solving continuity equations. The production term is added to the convection by timestep splitting methods so that the rate of progress of the flame is controlled.

Conservation among the reactants and the product determines the amount of energy released in each cell.

The reaction fronts are identified by the region where the gradient, ∇n_p, is large. The integral of the gradient from the front to the back of the extended interface, the change in n_p in crossing the flame, is known. Choosing the local energy release rate proportional to the local gradient guarantees the correct overall energy release along the convoluted front. The amount of new product formed is

$$\Delta n_p = |\nabla n_p| l , \qquad (11-2.2)$$

where Δn_p is the change in n_p and l is the distance the front moves normal to itself locally during the time interval of interest. The direction of the normal to the reaction front is the same as the direction of the gradient. The speed of the front is the local flame speed. Therefore l is given by

$$l = v_f \Delta t , \qquad (11-2.3)$$

where v_f is the known local flame speed normal to the interface. The amount of product formed per unit volume in the time interval Δt is

$$w \equiv \Delta n_p = v_f \Delta t |\nabla n_p| . \qquad (11-2.4)$$

The calculation requires determining $|\nabla n_p|$. Laskey et al. (1987) have found that a one-sided maximum difference for evaluating the gradient, although technically only first-order accurate, gives the best results. In addition, there are several tests which must be made on the gradient and the values of n_p in the vicinity to determine if the numerical estimate of the gradient is a valid quantity. Finally, some modification is required for a nonuniform pressure field. The flame speed v_f becomes a function of position and the number densities considered must be scaled to a reference temperature. An example of a gradient method calculation of two round flame fronts that grow and merge is shown in Figure 11–16 for a 40×40 finite-difference grid.

There are several beneficial features of the gradient method. It treats the effects of merging interfaces with relatively little difficulty. Adding other interface processes, and eventually ignoring weakened interfaces results naturally from the formulation. Because no additional variables are needed, computer memory requirements are modest. The algorithm as implemented is fully vectorized. Finally, it is straightforward to extend the two-dimensional formulation to three dimensions.

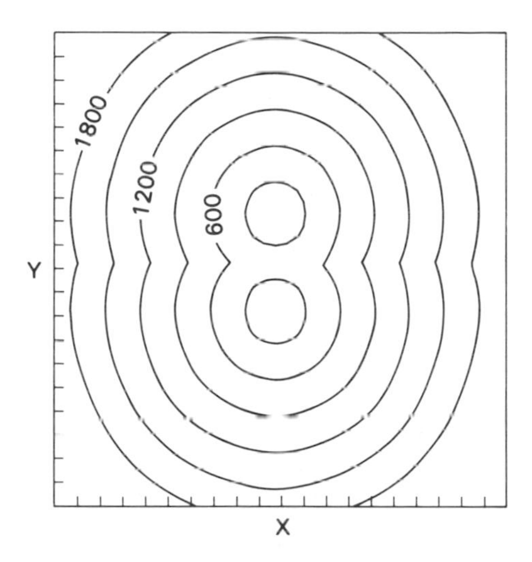

Figure 11–16. Calculation of two intersecting flame fronts by the gradient method. Successive contours moving out from the center indicate successive times. (Figure is courtesy of K. Laskey.)

The price of this flexibility is that we do not know the exact location of the interface. The gradient method ensures that the right amount of reaction takes place in the vicinity of the gradient, as defined by the macroscopic grid. This method works well for the types of problems for which it was designed, for example, flame fronts propagating in complicated flow fields. However, it is not suitable for simulations which must rigorously separate two materials and track the curvature at the surface on scales comparable to or smaller than the grid spacing. An example of a problem for which the gradient method is not suitable is the dynamics of a droplet in which it is important to know the surface curvature very accurately to calculate the effects of surface tension. The method could locate interfaces accurately enough for extracting information about flame curvature and would also be useful for describing evaporation at a droplet surface and the spread of vaporized material into the background gas.

In the gradient method, the gradients should remain quite large and the interface should advance through the fluid. Because the gradients should be well resolved and localized, we recommend use of one of the better monotone methods described in Chapters 8 and 9.

11–3. MATRIX ALGEBRA FOR FINITE DIFFERENCES

Solving the equations for reactive flows numerically often leads to sets of matrix algebraic equations. Generally the matrices are *sparse*, which means that most of the elements off the matrix diagonals are zero. Using general matrix inversion methods is almost never required or recommended for solving sparse matrices. The most general matrix inversion algorithms scale as N_c^3, where N_c is the number of computational cells in the problem. In multidimensional calculations, N_c is very large, typically 10^4 to 10^6. By taking advantage of the fact that the matrices are sparse, we can use algorithms that scale as N_c^2. For certain problems it is also possible to invert matrices with algorithms that scale as $N_c \ln N_c$ or even N_c. Below we use N to mean the dimension of the vector, which means N_c for the cases discussed here (and therefore N and N_c are interchangeable).

In this section we outline solution techniques applicable to the matrix equations that arise in parabolic equations describing fluid dynamics and heat transport in reactive flow calculations. It is not our intent to give the details of the techniques, which are the subject of many volumes written at many levels. Good general references which may not appear in the numerical matrix algebra literature include Hockney (1970), Potter (1973), Roache (1982), and Anderson et al. (1984).

11–3.1 Statement of the Problem

In many discretized representations of the reactive flow equations, the partial differential equations becomes finite algebraic matrix equations. In explicit formulations, the equations usually degenerate into relations between diagonal matrices. However, in implicit formulations, matrix equations must often be inverted.

In the general case, simultaneous linear algebraic equations relate a vector of unknown quantites, $\mathbf{u}$, to a vector of known source quantites, $\mathbf{s}$, through a matrix $\mathbf{M}$,

$$\mathbf{M} \cdot \mathbf{u} = \mathbf{s} . \qquad (11-3.1)$$

Assuming that there are N equations and N unknowns, $\mathbf{M}$ is a square matrix. In typical problems, N is the number of cells, N_c, in the discretization of a spatial domain for a continuous fluid representation.

The one-dimensional diffusion equation treated in Section 4–3 provides a good example,

$$\frac{d\rho}{dt} = D\frac{d\rho^2}{dx^2} + s. \qquad (11-3.2)$$

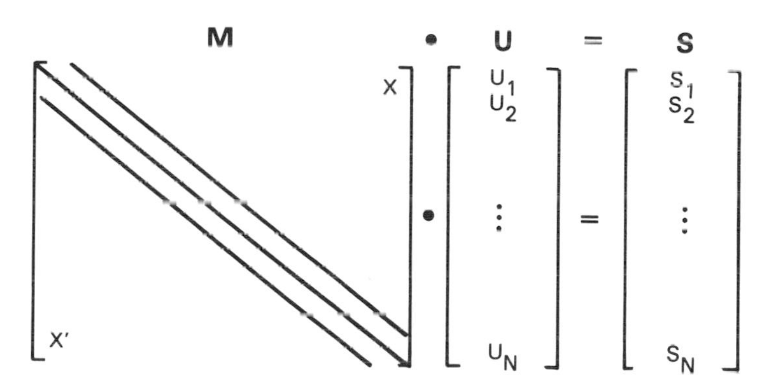

Figure 11–17. Generic form of a tridiagonal matrix. The slanted solid lines indicate nonzero coefficients, and the blank areas are assumed to be zero. The elements X and X' are zero unless the problem is periodic.

With an evenly spaced grid with cell size Δ and a constant diffusion coefficient D, Eq. (11–3.2) (see Eq. (4–3.8)) can be differenced as

$$\begin{aligned}\frac{\rho_j^n - \rho_j^{n-1}}{\Delta t} &= \frac{D\theta}{(\Delta)^2}\left[\rho_{j+1} - 2\rho_j + \rho_{j-1}\right]^n \\ &+ \frac{D(1-\theta)}{(\Delta)^2}\left[\rho_{j+1} - 2\rho_j + \rho_{j-1}\right]^{n-1} + s_j^{n-1} .\end{aligned} \qquad (11-3.3)$$

Here j indexes each of the N_c computational cell values and the quantities in square brackets are evaluated either at the new time, n, or the old time, $n-1$. Equation (11–3.3) for the new values $\{\rho_j^n\}$ can be written symbolically as

$$\mathbf{M}^n \cdot \boldsymbol{\rho}^n = \mathbf{M}^{n-1} \cdot \boldsymbol{\rho}^{n-1} + \mathbf{s}^{n-1} , \qquad (11-3.4)$$

where $\mathbf{M}^n$ and $\mathbf{M}^{n-1}$ are tridiagonal matrices. The generic form of a tridiagonal matrix is shown in Figure 11–17. Only the elements on the diagonal and adjacent to it are nonzero. In a periodic one-dimensional problem of the form of Eq. (11–3.3), the corner matrix elements X and X' are also nonzero.

The one-dimensional Poisson equation,

$$\frac{d^2\phi}{dx^2} = -s(x) , \qquad (11-3.5)$$

also results in a tridiagonal matrix. In fact, any equation that can be put in the form

$$f\frac{d^2\rho}{dx^2} + g\frac{d\rho}{dx} + h\rho = s , \qquad (11-3.6)$$

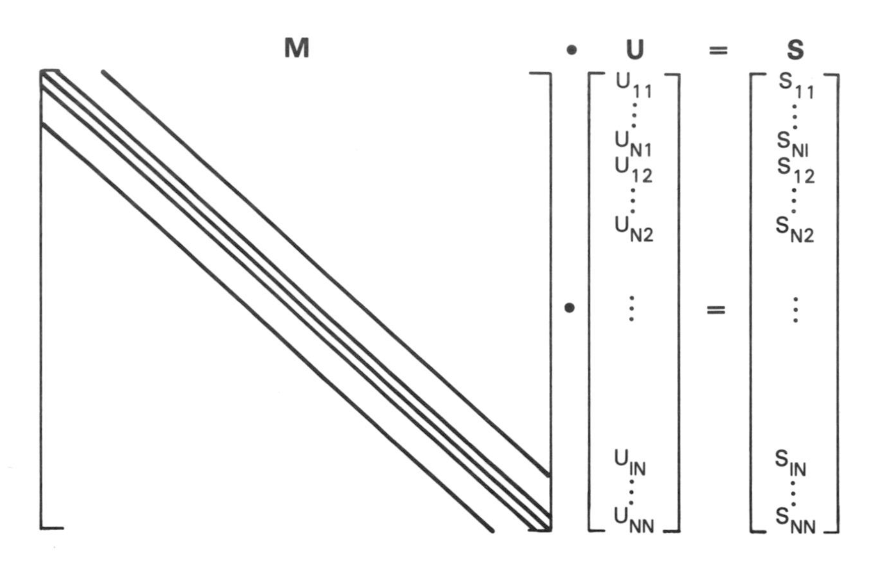

Figure 11–18. Generic form of a pentadiagonal matrix. The slanted solid lines indicate nonzero coefficients and the blank areas assume zero values in the matrix **M**.

where $\rho(x)$ is the unknown and $f(x)$, $g(x)$, $h(x)$, and $s(x)$ are known functions of x, gives a tridiagonal matrix equation when using algorithms coupling only nearest computational cells.

In the two-dimensional generalization of Eq. (11–3.2),

$$\frac{d\rho}{dt} = D\frac{d^2\rho}{dx^2} + D\frac{d^2\rho}{dy^2} + s(x,y,t) . \qquad (11-3.7)$$

With $\Delta \equiv \Delta x = \Delta y$, a simple finite-difference approximation is

$$\begin{aligned} \frac{\rho^n - \rho^{n-1}}{\Delta t} &= \frac{D\theta}{\Delta^2}\left[(\rho_{i+1,j} - 2\rho_{ij} + \rho_{i-1,j}) + (\rho_{i,j+1} - 2\rho_{ij} + \rho_{i,j-1})\right]^n \\ &\quad + \frac{D(\theta-1)}{\Delta^2} \times \left[(\rho_{i+1,j} - 2\rho_{ij} + \rho_{i-1,j}) \right. \\ &\quad \left. + (\rho_{i,j+1} - 2\rho_{ij} + \rho_{i,j-1})\right]^{n-1} . \end{aligned} \qquad (11-3.8)$$

Now $\mathbf{M}^n$ and $\mathbf{M}^{n-1}$ are no longer tridiagonal, but have the pentadiagonal form shown schematically in Figure 11–18. The four off-diagonal rows of coefficients are not contiguous although the matrix is still sparse. In fact, rather general partial differential equations can be approximated in this form as long as they can be approximated using the five-point template depicted in Figures 7–7a. Equation (11–3.1) encompasses all of these cases, but also disguises the underlying structure which can usually be exploited to optimize the solution numerically.

11–3.2 Exact Solution of the Matrix Equation

Equation (11–3.1) has the formal solution

$$\mathbf{u} = \mathbf{M}^{-1} \cdot \mathbf{s} , \qquad (11-3.9)$$

where the matrix $\mathbf{M}^{-1}$ is the $N \times N$ inverse of $\mathbf{M}$. Further, we are interested in very large matrices. A two-dimensional problem on a 100×100 grid gives matrices of dimension 10,000 with 10^8 elements, so we cannot use the direct application of Cramer's rule.

Matrices derived from finite-difference algorithms usually have special properties, and it can be valuable to take advantage of these properties in solving the equations. Although the matrices are often very large, they are usually very sparse. Physically this occurs because nearby cells are coupled directly but distant cells are coupled only indirectly. When matrices are sparse, there are relatively fast iterative methods which produce reasonable solutions. As in all iterations, however, the rate of convergence and the pathological cases determine how useful the method really is.

An exact solution of Eq. (11–3.1) to find $\mathbf{M}^{-1}$ that does not take advantage of any special properties of the matrix scales as N^3 and therefore is expensive. Most general methods that fall into this category are based on the method of *Gaussian elimination*. The basic idea of Gaussian elimination can best be seen when Eq. (11–3.1) is written in component form

$$\sum_j m_{ij} u_j = s_i , \qquad i = 1, ..., N_c . \qquad (11-3.10)$$

The first equation can be used to define u_1 in terms of the other $(N-1)$ variables. Then, using this equation, u_1 is successively eliminated from the $(N-1)$ equations by multiplying each of the $(N-1)$ other equations by m_{11}/m_{i1} and subtracting the first equation. This leaves the partially reduced matrix equation

$$\begin{aligned} m_{11}u_1 + m_{12}u_2 + \cdots + m_{1N}u_N &= s_1 \\ m'_{22}u_2 + \cdots + m'_{2N}u_N &= s'_2 \\ m'_{32}u_2 + \cdots + m'_{3N}u_N &= s'_3 \\ \vdots \qquad\qquad \vdots \quad &= \vdots \\ m'_{N2}u_2 + \cdots + m'_{NN}u_N &= s'_N . \end{aligned} \qquad (11-3.11)$$

The same process eliminates all of the other variables from successively smaller series of equations, ending up with an upper triangular form with only nonzero elements below the diagonal,

$$\begin{aligned} m_{11}u_1 + m_{12}u_2 + m_{13}u_3 + \cdots + m_{1N}u_N &= s_1 \\ m'_{22}u_2 + m'_{23}u_3 + \cdots + m'_{2N}u_N &= s'_2 \\ m''_{33}u_3 + \cdots + m''_{3N}u_N &= s''_3 \\ &\vdots \\ m_{NN}^{N-1}u_N &= s_N^{(N-1)} \ . \end{aligned} \qquad (11-3.12)$$

This process requires several times N^3 operations to eliminate the elements below the diagonal, and great care to control errors that can occur because of large differences in the matrix elements.

The numerical process which completes the solution is called *back substitution*. First we evaluate u_N from the last of Eqs. (11–3.12), and then u_{N-1}, and so on, and finally u_1. This requires of order N^2 operations.

Because Gaussian elimination scales as N^3, it is too expensive to be done repeatedly in an evolving simulation. Thus other algorithms must be found that use specific properties of the pattern of sparseness in the matrix or relationships between the coefficients to reduce the scaling at least to N^2. Fortunately, there are fast direct methods and fast iterative methods. In this context, the term "fast" generally means approximately $N \ln N$ scaling of the computational cost with the system size.

11–3.3 Tridiagonal Matrices

We wish to solve a set of equations of tridiagonal form that can be written as

$$\alpha_j u_{j-1} + \beta_j u_j + \gamma_j u_{j+1} = s_j \ , \qquad (11-3.13)$$

for $1 < j < N$ with the appropriate boundary conditions at u_1 and u_N. Here the problems are usually one dimensional, $N = N_c = N_x$, for example. Given u_j, we want a relation that gives us u_{j+1}. All easily solvable tridiagonal systems replace u_o and u_{N+1} in Eq. (11–3.13) with known values in terms of the bounding values of u in cells 1, 2, $N-1$, and N. For example,

$$u_o = s_o + \alpha_o u_1 + \beta_o u_2 + \gamma_o u_N \ , \qquad (11-3.14)$$

and

$$u_{N+1} = s_{N+1} + \alpha_{N+1} u_1 + \beta_{N+1} u_{N-1} + \gamma_{N+1} u_N \ , \qquad (11-3.15)$$

can be used to define u_o and u_{N+1} in Eq. (11–3.13). When the α_{N+1} and γ_o terms are nonzero, Eqs. (11–3.14) and (11–3.15) correspond to periodic boundary conditions for which Gaussian elimination can be used.

One basically scalar approach to solving the tridiagonal matrix equation, Eq. (11–3.13), results in recursion relations. Consider u_{j+1} such that

$$u_{j+1} = x_j u_j + y_j \; . \qquad (11-3.16)$$

Putting this into Eq. (11–3.13) and rearranging terms gives

$$u_j = -\frac{\alpha_j}{\gamma_j x_j + \beta_j} u_{j-1} + \frac{s_j - \gamma_j y_j}{\gamma_j x_j + \beta_j} \; . \qquad (11-3.17)$$

We can now equate the terms in Eq. (11–3.16) and Eq. (11–3.17) to give

$$\begin{aligned} x_{j-1} &= -\frac{\alpha_j}{\gamma_j x_j + \beta_j} \\ y_{j-1} &= \frac{s_j - \gamma_j y_j}{\gamma_j x_j + \beta_j} \, , \end{aligned} \qquad (11-3.18)$$

for all of the values of $\{x_j\}$ and $\{y_j\}$.

The procedure is then to start at N, and work down to $j = 1$ to find the values of x_j and y_j. Then to go from $j = 1$ to $j = N$ using Eq. (11–3.16) to get the u_j. The result is an easily programmed double-recursion relation essentially the same algorithm as obtained by Gaussian elimination. This algorithm is generally accurate, stable, and scales as N. It is about minimal in terms of operation count. It is not easily vectorized or set up for parallel processing unless a number of similar but separate tridiagonal systems can be solved simultaneously.

The $N \ln N$ folding algorithms for solving tridiagonal matrices are all quite similar to each other, and somewhat more complex than the double-sweep scalar recursion algorithms. The equations relating u_j for j even to u_{j+1} and u_{j-1} are used to eliminate u_j from the j odd equations, giving half as many equations. New, composite coefficients then relate alternate unknowns. This new system is still tridiagonal and thus can be folded again. When the repeatedly folded system becomes short enough, the scalar algorithm is used to complete the solution of the short system. Using the (now known) odd j half of the variables at each stage of the unfolding, the j even unknowns can be found. Algorithms for all of these cases and corresponding test problems have been assembled by Boris (1976). As parallelism becomes more and more important, the $N \ln N$ cylic reduction algorithm or its variants that can be adapted for parallel processing will become standard.

11–3.4 Direct Solutions of the Poisson Equation

The simplest case of a multidimensional, sparse matrix problem with pentadiagonal form arises from a local five-point finite-difference approximation to the Poisson equation on a rectangular grid. Even when the cells are nonuniform in size and shape, we retain the regular sparse matrix structure as long as the grid is topologically rectangular and each vertex has four adjacent cells. By taking advantage of the fact that the nonzero elements of the matrix are localized in the diagonal and adjacent blocks, the pattern of the entries allows us to reduce the operation count to $\mathcal{O}(N^2)$ because most of the blocks are entirely zero. However, we would like algorithms for inverting the two-dimensional five-point Poisson operator in $\mathcal{O}(N \ln N)$ operations, where $N = N_c = N_x \times N_y$ is the number of distinct potential values sought. On a 100×100 mesh, an $\mathcal{O}(N_c \ln N_c)$ solution reduces the solution time by about a factor of 15.

General elliptic equations in two or three dimensions are difficult and expensive to solve, particularly when antisymmetric cross terms are comparable in magnitude to the symmetric terms. Symmetry in the form of the second-order derivatives in the operator,

$$\nabla^2 = \frac{\partial^2}{\partial x^2} + \frac{\partial^2}{\partial y^2} , \qquad (11-3.19)$$

makes possible several direct, fast solution algorithms that do not apply in the general case. Because of these symmetries, the eigenfunctions of Eq. (11–3.19) can be written as the product of separate eigenfunctions along each of the grid directions. The result is that multidimensional problems can sometimes be separated into a series of one-dimensional problems, each of which is tridiagonal. Some direct methods are described briefly in the following paragraphs.

A number of well tested packages exist to solve elliptic and Poisson equations in multidimensions. Before writing your own, check the software library on your computer system or ask what is available at a few major computer centers. There are a number of public domain software libraries that can be accessed. The journal *Computer Physics Communications* (published by North Holland) is devoted to documenting and disseminating this kind of software. There is also a report (Boris, 1976) containing algorithms and programs for a number of optimized tridiagonal solvers.

Multiple Fourier Analysis

This method involves performing a complete Fourier analysis along each spatial dimension (Hockney, 1970; Boris and Roberts, 1969). In two dimensions this requires of order $\mathcal{O}(N_c \ln N_c)$ arithmetic operations and can be accomplished in the same operation count as a number of one-dimensional Fourier transforms or as a single two-dimensional transform. One of the attractions of this method is the ability to use standard, highly optimized Fast Fourier Transform (FFT) algorithms. The FFT is one of the best-studied numerical algorithms and usually is one of the first to be optimized for a new computer system. When the double Fourier harmonics of the potential ϕ are also eigenfunctions of the two-point gradient and three- and five-point second-derivative difference operators under many circumstances, the Fourier decomposition can often be performed right through the finite-difference operators. This decomposes the problem to be solved into a large number of simple equations for the separate harmonics.

Multiple Fourier analysis algorithms scale as $N_c \ln N_c$ when the system size is suitably chosen and can be used for three-dimensional problems. Typically FFTs are readily available when the number of cells in the system is a power of two or four. They are also efficient for all system lengths that can be expressed as the products of small prime integers. Because of their use in spectral and pseudospectral methods (Chapter 9), the best place to learn how to use FFTs under nonideal situations is in that literature. To use FFTs in a relatively straightforward way, a uniformly spaced grid and constant coefficients in the elliptic or Poisson equations are necessary. Strictly periodic, symmetric (reflecting), and antisymmetric boundary conditions are also straightforward to apply.

The operation count is $N \ln N$ using FFTs in one dimension, somewhat more than the $\mathcal{O}(N)$ scaling of tridiagonal algorithms but about the same as folding or cyclic reduction algorithms which also scale as $N \ln N$. Fourier transform algorithms are generally more accurate when they can be used because each harmonic contains information from everywhere else in the grid. Higher-order errors introduced by finite-difference derivative operators elsewhere in the calculation, for example, by differencing the velocity potential to obtain the velocity, can be corrected by adjusting the harmonic amplitudes. Birdsall and Langdon (1985) review these considerations in a plasma simulation context.

Fourier Analysis Decoupling Tridiagonal Systems

This hybrid method involves Fourier analysis along one spatial direction to decouple the Fourier harmonics. It gives separate tridiagonal matrix equations in the transverse direction for each of the harmonics. This technique was introduced by Hockney (1970) who used a cyclic reduction (folding) technique on the tridiagonal equations. In two dimensions, this requires of order $N_x N_y \ln N_x N_y$ arithmetic operations. It is a very fast method and applicable to axisymmetric or more complex geometries because the Fourier analysis is only done in one dimension.

A variant uses a scalar double-sweep algorithm for separated tridiagonal equations. Vectorization is possible because a number of tridiagonal systems of identical length are solved at once. In three dimensions, these hybrid techniques can be used if Fourier transforms in two directions can decouple the problem into a system of one-dimensional tridiagonal matrices.

Double Cyclic Reduction

This method involves cyclic reduction along each spatial dimension and was introduced by Buneman (1969). When the coefficients in each of the two spatial directions are constant, an analytic method eliminates half of the equations in each row using the basic five-point difference formulas to remove the alternate variables from the equations. Although cyclic reduction can be used on tridiagonal matrices with rather general coefficients, the need to perform the reduction procedure in two directions at once constrains the coefficients appreciably.

Cyclic reduction is used repeatedly until the matrix is small enough to solve directly. The boundary and grid limitations are roughly similar to those needed to make Fourier transforms convenient. However, the resulting optimized program can be faster because the cyclic reduction technique is somewhat more efficient that Fourier transforms. The double cyclic reduction method lacks the flexibility of FFTs to introduce higher-order corrections to the finite-difference operators by modifying the Fourier coefficients.

11–3.5 Iterative Solutions of Elliptic Equations

In a few special cases, direct solutions of a complete elliptic system are possible in order $N_c \ln N_c$ operations, where N_c is $N_x \times N_y$ cells in two dimensions and $N_x \times N_y \times N_z$ in three dimensions. These special cases do not, however, allow general coefficients which arise, for example, from variable rectilinear gridding, a temperature-dependent diffusion coefficient, or a variable sound speed. It takes at least of order N_c^2 operations to solve Eq. (11–3.1) using Gaussian elimination or an equivalent method. For the large matrices of interest here, this operation count is not acceptable. It is particularly odious in a time-dependent problem, where the matrix equation would have to be solved once at every timestep. Here we consider iterative methods to use when fast, direct methods cannot be used. Extensions and generalization of the discussion given below can be found in Potter (1973) and Varga (1962).

The inexact iterative methods start with a guess at an initial value of $\mathbf{u}$, called $\mathbf{u}^{(o)}$. They proceed through p iterations, arriving at $\mathbf{u}^{(p)}$, which should be an improved approximation. The iteration formula is

$$\mathbf{u}^{(p+1)} = \mathbf{P} \cdot \mathbf{u}^{(p)} + \mathbf{c} , \qquad (11-3.20)$$

where $\mathbf{P}$ is the iteration matrix related to $\mathbf{M}^{-1}$ and $\mathbf{c}$ is related to $\mathbf{s}$ and previous approximate solutions. Each improved solution vector $\mathbf{u}^{(p+1)}$ may be obtained explicitly from $\mathbf{u}^{(p)}$. The methods described below all fit into this format, except for the last, the alternating-direction implicit method, which is an implicit, two-step iterative procedure. The computational issues are how many iterations are needed and what is the best form of $\mathbf{P}$. Simply using a matrix multiply operation as implied by Eq. (11–3.20) requires $\mathcal{O}(N_c^2)$ operations per iteration. The important point is that $\mathbf{P} \cdot \mathbf{u}^{(p)}$ should require fewer operations than $\mathcal{O}(N_c^2)$ when $\mathbf{P}$ is suitably chosen.

Here we illustrate the methods by applying them to the Poisson equation, rewritten in a more graphic notation,

$$u_c - \frac{1}{4}(u_n + u_s + u_e + u_n) = \frac{1}{4}s_c\Delta^2 = w_c, \qquad (11-3.21)$$

where the (i,j)th point on the computational mesh is denoted c and the surrounding points are denoted n, s, e, and w for north, south, east, and west, respectively, as shown in Figure 11–19. For iteration procedures applied to Eq. (11–3.21), we write

$$u_c^{(p+1)} = (1-\omega)u_c^{(p)} + \frac{1}{4}\omega(u_n^{(p)} + u_s^{(p)} + u_e^{(p)} + u_w^{(p)}) + \omega w_c , \quad (11-3.22)$$

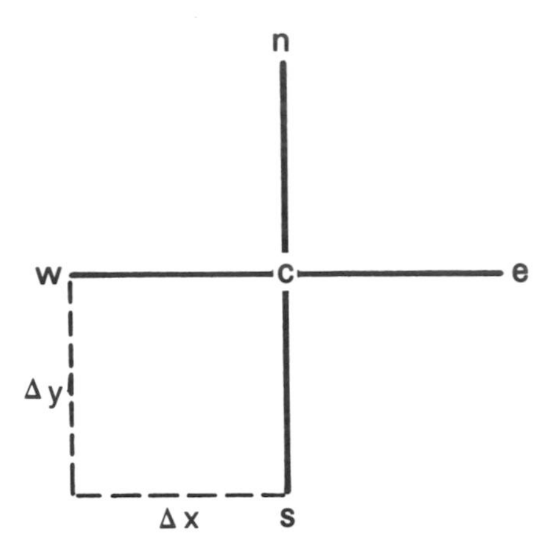

Figure 11–19. Schematic showing a particular grid point in a computational mesh, c, surrounded by north, south, east, and west points, designated n, s, e, and w, respectively.

where ω is an adjustable relaxation parameter. For convergence in two dimensions, $0 \leq \omega \leq 1$. When ω is zero, the new iteration is the same as the previous one. When ω is greater than one, the solution is being extrapolated beyond the previous iteration.

We have also found it convenient to use another notation that is more physically motivated and better suited for optimized programming than either the usual matrix subscripting or the n, s, e, and w notation of Figure 11–19. For example, when dealing with arrays of cell quantities such as $\{\rho_{ij}\}$ for the cell values of the density $\rho(x, y)$ in two dimensions, or E_{ijk} for the cell values of the energy density $E(x, y, z)$ in three dimensions, the first index i in our notation increases monotonically with x, the second index j increases monotonically with y, and the third index k increases monotonically with z. In two dimensions, this notation inverts the usual matrix notation in which the second index scans horizontally along the row indicated by the first index. This physical notation also has the second index increasing upward, so the sense of j and y are the same, whereas the row index in matrix notation increases downward. This notation ensures that adjacent x values are stored in adjacent memory locations, and more directly coincides with usual computer language notation.

Jacobi Iteration

This is Eq. (11–3.22) with $\omega = 1$, equivalent to averaging the four adjacent nodes to find the next iteration in Laplace's equation. The method converges slowly and is not commonly used.

Gauss-Seidel Iteration

The trick in this method is to note that each component of $\mathbf{u}^{(p+1)}$ is obtained in sequence. For example, when c is reached in Poisson's equation, we already have $u_s^{(p+1)}$ and $u_w^{(p+1)}$ and can use them as improved values in the iteration. If $\omega = 1$, then

$$u_c^{(p+1)} = \frac{1}{4}(u_n^{(p)} + u_s^{(p+1)} + u_e^{(p)} + u_w^{(p+1)}) + w_c \,. \qquad (11-3.23)$$

This method is somewhat faster than the Jacobi iteration. However, it is really only suitable for scalar computation because one computation must be completed before another can be started, that is, u_c^{p+1} depends on u_w^{p+1}.

The Successive Over-Relaxation (SOR) Method

This method uses the Gauss-Seidel formulation but optimizes the value of ω. For example, for the Poisson problem with large values of N_x and N_y, the best value of ω, denoted ω_b, is

$$\omega_b = \frac{2}{1 + \sqrt{1 - \mu_m^2}} \,, \qquad (11-3.24)$$

where

$$\mu_m = \frac{1}{2}\cos\frac{\pi}{N_x} + \frac{1}{2}\cos\frac{\pi}{N_y} \,. \qquad (11-3.25)$$

Therefore

$$\omega_b \approx \frac{2}{1 + \pi\sqrt{\frac{1}{2N_x^2} + \frac{1}{2N_y^2}}} \,, \qquad (11-3.26)$$

which goes to the value $\omega = 2$ in the limit of an infinitely fine mesh. This method converges faster than the other methods described so far. However, convergence is still slow in the early stages of iteration because the error can only propagate away from a cell at a rate of one cell per iteration. Further, the error during these stages can even grow from step to step.

A variation on this is a "double" or "cyclic" iteration. This takes advantage of the fact that variables at even points are coupled directly only to variables at odd points, and vice versa. The idea is to improve all variables on even points first, and then use these to improve all the variables on odd points. Sometimes this is called "red-black" ordering, recalling the red and black squares on a chess board (see, for example, Adams and Jordan, 1984). This iteration also generalizes to three dimensions.

Certainly SOR is the simplest algorithm. It is often best to use the SOR method to test various portions of a composite model in the development phases of large, complex models. However, the price for this conceptual and programming simplicity is its slow convergence and high cost.

It is possible to extrapolate the starting iteration $\mathbf{u}^{(o)}$ from the previous two timesteps to get a better starting value for the SOR iteration. This extrapolation generally works better than might be expected because it is most accurate at long wavelengths where the iteration converges most slowly. Relatively large errors introduced by the extrapolation in the short-wavelength components of the solution decay away quickly while the starting error in the long wavelengths is often orders of magnitude smaller. The result, in terms of time advancement from one timestep to the next, is that the starting error is much smaller in the initial guess and many fewer iterations are generally required to obtain a given accuracy.

Cyclic Chebyshev Method

This is similar to the previous method, except that the early stages of convergence are improved by varying ω from step to step using a specific sequence $\{\omega^{(p)}\}$. For the Poisson problem we are considering,

$$\begin{aligned} \omega^{(0)} &= 1 \\ \omega^{(1)} &= \frac{1}{1 - \frac{1}{2}\mu_m^2} \\ &\vdots \\ \omega^{(p)} &= \frac{1}{1 - \frac{1}{4}\mu_m^2 \omega^{(p-1)}} \end{aligned} \qquad (11-3.27)$$

where μ_m is defined in Eq. (11–3.25). Note that ω increases with each step, and in the limit, $\omega(\infty) = \omega_b$ from Eq. (11–3.24).

Alternating-Direction Implicit Method (ADI)

This method involves successive iterated implicit steps in each spatial directions. Each step involves the solution of a linear tridiagonal system. For the Poisson equation in two dimensions, the method may be written

$$\begin{aligned} u_c^{(p+\frac{1}{2})} - u_c^{(p)} &= \frac{1}{2} w_c \Big[(u_n - 2u_c + u_s)^{(p+\frac{1}{2})} \\ &\qquad + (u_e - 2u_c + u_w)^{(p)} \Big] \\ u_c^{(p+1)} - u_c^{(p+\frac{1}{2})} &= \frac{1}{2} w_c \Big[(u_n - 2u_c + u_s)^{(p+\frac{1}{2})} \\ &\qquad + (u_e - 2u_c + u_w)^{(p+1)} \Big] . \end{aligned} \tag{11 - 3.28}$$

It can be shown the the asymptotic rate of convergence is the same as the SOR or Chebyshev methods. Extensions to three dimension are described in Richtmyer and Morton (1967).

The block implicit approach, applied in an ADI framework, is one way to calculate subsonic multidimensional flows (see, for example, Section 9–3; also Lindemuth and Killeen, 1973; Briley and McDonald, 1975, 1977, 1980; McDonald and Briley, 1975; Beam and Warming, 1976, 1978; McDonald, 1979). In this method, calculations are performed implicitly for one direction using a tridiagonal solver with the fluid properties from the other direction defined explicitly using the values of the physical variables from the previous timestep or iteration. The directions are then reversed and the ADI procedure is repeated. A number of iterations are required to couple the whole calculation accurately. Including chemical reactions and molecular diffusion complicates this approach because the matrices which form each block can become prohibitively large.

The iterative methods described above all use the structure of the finite-difference grid and the sparsity of the matrix to reduce the operation count. Each iteration requires of order N_c operation, but at least $\max(N_x, N_y, N_z)$ iterations are required to share information over the entire grid. In two dimensions, a good approximate solution is obtained in $\mathcal{O}(N_c^{3/2})$ operations, and in three dimensions the scaling is $\mathcal{O}(N_c^{4/3})$ although N_c itself is usually very large. Even though these approximate scaling laws are an improvement over the N_c^2 scaling, $N_c \ln N_c$ scaling is better if it can be obtained.

Incomplete Cholesky – Conjugate Gradient, ICCG Iteration

The *Cholesky* method is a form of Gaussian elimination for *symmetric* linear systems. The *Conjugate Gradient* method is an iterative method that uses successive approximations to reach the final solution. These two methods are often used together, in the *Incomplete Cholesky – Conjugate Gradient* method, or *ICCG*, to solve systems of linear equations for physical problems in which the matrix is positive definite and nearly symmetric. Good descriptions of the ICCG method are given by Kershaw (1978) and Meijerink and van der Vorst (1977).

The Conjugate Gradient method is iterative and involves successive approximations, each of which is a projection of the solution on a progressively larger subspace of the vector space in which the solution lies. Each successive iteration searches for a correction to the previous iteration in a direction normal to the subspace of the most recent approximation. This search continues until the entire vector space is searched. When the matrix is symmetric and positive definite, convergence to the exact solution is assured in N_c iterations. Unfortunately, N_c is quite large and, when the matrix is not symmetric, convergence is not assured.

Conjugate Gradient methods are often improved with some kind of "preconditioning" of the matrix. If convergence is only required to some finite value, then it is possible to precondition the matrix in such a way that the solution converges to within that finite value in considerably fewer than N_c iterations. The Incomplete Cholesky method is one way to do this.

The ICCG method works best for those physical problems in which the matrix is almost symmetric. This is especially true for problems for which viscous or diffusive effects are dominant. These include slow viscous fluid flows, magnetohydrodynamic problems in which resistivity or viscosity dominate, or Fokker-Planck equations in which the collision operator is a diffusion term. The method is also a good possible choice for the Poisson equation of the SPLISH algorithm described in Chapter 10, in which the matrix is very large and symmetric.

The multigrid, described next, and the ICCG method are now being used as alternatives to the simpler, but relatively expensive SOR and ADI iterations. As these methods become as well tuned and exhaustively tested, they will approach matrix methods in general availability and ease of use. They are also similar to matrix inversion techniques in that multigrid and ICCG methods use a relatively large amount of storage. Twenty to fifty times as much memory is often required as there are cells in the two- or three-dimensional finite-difference mesh being solved.

Multigrid Iteration Methods

During the last decade, there has been considerable effort devoted to multigrid methods for solving many types of numerical problems. The general principles of these methods apply equally well to finite-difference and finite-element approaches to partial differential equations. We recommend the book edited by Hackbusch and Trottenberg (1983) and the articles by Brandt (1977, 1981). The basic idea of multigrid methods was originally presented by Federenko (1961).

The essence of the multigrid approach is to set up a discretization of the computational domain, and to define a sequence of smaller, perhaps nested, auxiliary spaces in this domain. The solutions in these smaller (less expensive) spaces can be used to approximate the true solution in the largest space. The solution in any space can be improved by combining relaxation iterations in that space, which smooth the fine-scale errors, with correction iterations using the smaller spaces, which reduce the coarse-scale errors. The smaller spaces have many fewer unknowns, and so require far less computational effort than the iterations on the largest space. The final result can be a substantial savings in the work required to solve the problem and an $N_c \ln N_c$ scaling.

Here we are interested in multigrid algorithms for elliptic problems. These have been proposed and analyzed by a number of authors, for example, Brandt (1977), Bank and Dupont (1981), and Douglas (1984). Douglas found an efficient multigrid algorithm for elliptic problems and this was subsequently converted to vectorized form for use on high-speed computers by DeVore (1984).

Suppose we have an elliptic equation to solve in the (x, y) plane with an $N_c = N_x \times N_y$ grid with uniform spacing h in each direction. After finite-differencing the differential equation and incorporating the boundary conditions, the problem is reduced to solving the matrix equation,

$$\mathbf{M}^h \cdot \mathbf{u}^h = \mathbf{s}^h , \qquad (11-3.29)$$

where $\mathbf{M}^h$ is the matrix of coefficients, $\mathbf{u}^h$ is the desired solution, and $\mathbf{s}^h$ is the inhomogeneous source vector. We could attempt to solve Eq. (11–3.29) by a direct method, such as Gaussian elimination, or by an iterative method, such as a Gauss-Seidel relaxation. For the direct approach, the operation count is $\mathcal{O}(N_c^2)$ to find the N_c unknowns in Eq. (11–3.29). This rapidly becomes too costly for large problems. For the iterative approach, the operation count per iteration is proportional to the number of unknowns. However, due to the slow relaxation of the coarse-scale errors in the solution, many iterations must be performed to get an accurate result, and again the operation count is very large because the coefficient of N_c is large. Multigrid methods combine direct and iterative methods. Thus they take advantage of the reduction in effort required by direct methods as the number of unknowns in reduced, and of the rapid relaxation of fine-scale errors in the solution by iterative methods.

We now describe the basic elements of a multigrid algorithm for solving this elliptic problem. Choose N to be even, and lay two uniform grids on the rectangle, one with $N_x \times N_y$ spacing h, and the other $\frac{N_x}{2} \times \frac{N_y}{2}$ with mesh spacing $2h$. Equation (11–3.29) holds on the fine grid, and on the coarse grid

$$\mathbf{M}^{2h} \cdot \mathbf{u}^{2h} = \mathbf{s}^{2h} . \qquad (11-3.30)$$

We begin by solving Eq. (11–3.30) using a direct method. Because the number of unknowns is smaller by a factor of four, this requires a factor of eight less work than the same task on the fine grid. We then interpolate the solution onto the fine grid. The result has fine-scale errors due to the lack of fine-scale information, in addition to any errors introduced by the interpolation process. We therefore do a small number of relaxation iterations on the new solution to smooth the fine-scale errors, and this process produces the first approximation, $\mathbf{u}^{h(1)}$. We can improve the approximation and find $\mathbf{u}^{h(2)}$ by solving a correction problem based on the residual vector. This step requires a projection step in which values from the fine grid are projected onto the coarse grid, usually by summing or averaging.

This process describes a 2-level, 2-cycle algorithm. It could be extended to use additional coarse grids with spacings $4h$, $8h$, etc., or to solve additional correction cycles, or both, leading to an $N_c \ln N_c$ algorithm. Various multilevel algorithms have been proposed by Brandt (1977) and Douglas (1982).

11–3.6 Summary

Transforming the partial differential equations of reactive flow into a finite number of algebraic relations produces a set of equations that can be written, manipulated, and analyzed in matrix form. In some situations, specifically in an explicit formulation, these equations can be solved directly with minimal effort. Very often, however, implicit solution algorithms are needed. In implicit time-dependent problems, a different elliptic equation must be solved at each timestep.

Iterative techniques such as SOR and ADI methods are useful and easy to program but are computationally expensive relative to faster direct solution methods. ADI methods are intermediate in complexity between full implicit matrix methods and explicit SOR techniques. One of the attractions of simply solving the big matrices, an expensive procedure, is the possibility of using a well tested, documented, optimized package that does the job.

Fast direct methods have been considered extensively by Hockney (1970) and Hockney and Eastwood (1981) and are very attractive when they can be used. For these methods, the cost in floating-point operations in multidimensional applications scales as $\mathcal{O}(N_c \ln N_c)$. When the finite-difference coefficients are constant in any direction, that is, in periodic or reflecting Cartesian geometries or the azimuthal direction in cylindrical coordinates, Fourier transform and cylic reduction (folding) techniques can be used to decouple the solution into smaller distinct parts, thus reducing the effective dimensionality of the problem. When a problem is reduced to one dimension, a direct solution method exists using fast tridiagonal algorithms which again scale as $N_c \ln N_c$.

When the coefficients are variable in the elliptic equation, whether for physical or geometric reasons, the $N_c \times N_c$ matrix is initially very sparse but rapidly fills in. Direct solutions are still possible but generally much more expensive, scaling as $\mathcal{O}(N_c^3)$ or $\mathcal{O}(N_c^2)$. In these situations, some form of iteration using an approximate inverse is usually used to speed up the solution. The ICCG and multigrid methods and a number of other variants all recover the $N_c \ln N_c$ scaling of a fast direct method, but generally with a larger numerical factor in front. The efficiency of these iterative algorithms is higher for a required accuracy than using methods based on ADI or SOR iteration, but many iterations are still needed in some situations. A direct method discussed by Madala (1978) is based on an error sweepout technique and this has been used for difficult problems in weather modeling. Unfortunately no generalization of the efficient direct solutions of Buneman (1969) or Hockney (1970) to the general coefficient elliptic problem on a rectilinear grid has been discovered.

References

Adams, L.M., and H.F. Jordan, 1984, *Is SOR Color-Blind?*, ICASE Report No. 84-14, NASA Langley Research Center, Hampton, VA.

Amsden, A.A., 1966, *The Particle-in-Cell Method for the Calculation of the Dynamics of Compressible Fluids*, LA-3466, Los Alamos Scientific Laboratory, Los Alamos, NM.

Anderson, D.A., J.C. Tannehill, and R.H. Pletcher, 1984, *Computational Fluid Mechanics and Heat Transfer*, Hemisphere Pub. Corp., Washington, DC.

Bank, R.E., and T. Dupont, 1981, An Optimal Order Process for Solving Elliptic Finite Element Equations, *Math. Comp.* 36: 35–51.

Barr, P.K., and W.T. Ashurst, 1984, *An Interface Scheme for Turbulent Flame Propagation*, SAND82-8773, Sandia National Laboratory, Livermore, CA.

Bayliss, A., and E. Turkel, 1982, Far Field Boundary Conditions for Compressible Flows, in P. Kutler, ed., *Numerical Boundary Condition Procedures*, NASA Conference Publication 2201, NASA Ames Research Center, Moffett Field, CA.

Beam, R.M., and R.F. Warming, 1976, An Implicit Finite-Difference Algorithm for Hyperbolic Systems in Conservation-Law Form, *J. Comp. Phys.* 22: 87–110.

Beam, R.M., and R.F. Warming, 1978, An Implicit Scheme for the Compressible Navier-Stokes Equations, *AIAA J.* 16: 393–402.

Birdsall, C.K., and A.B. Langdon, 1985, *Plasma Physics via Computer Simulation*, McGraw-Hill, New York.

Boris, J.P., 1976, *Vectorized Tridiagonal Solvers*, NRL Memorandum Report No. 3408, Naval Research Laboratory, Washington, DC.

Boris, J.P., and K.V. Roberts, 1969, The Optimization of Particle Calculations in Two and Three Dimensions, *J. Comp. Phys.* 4: 552–571.

Brandt, A., 1977, Multi-Level Adaptive Solutions to Boundary-Value Problems, *Math. Comp.* 31: 333–390.

Brandt, A., 1981, Multigrid Solvers on Parallel Computers, in M.H. Shultz, ed., *Elliptic Problem Solvers*, Academic Press, New York, 39–83.

Briley, W.R., and H. McDonald, 1975, Solution of the Three-Dimensional Compressible Navier Stokes Equations by an Implicit Technique, in R.D. Richtmyer, ed., *Proceedings Fourth International Conference on Numerical Methods in Fluid Dynamics*, Springer-Verlag, NY, 105–110.

Briley, W.R., and H. McDonald, 1977, Solution of the Multi-Dimensional Compressible Navier-Stokes Equations by a Generalized Implicit Method, *J. Comp. Phys.* 24: 372–397.

Briley, W.R., and H. McDonald, 1980, On the Structure and Use of Linearized Block Implicit Schemes, *J. Comp. Phys.* 34: 54–73.

Buneman, O., 1969, *A Compact Non-Iterative Poisson Solver*, Stanford University Institute for Plasma Research, Report No. 294, Stanford University, Palo Alto, CA.

Chern, I-L., J. Glimm, O. McBryan, B. Plohr, and S. Yaniv, 1986, Front Tracking for Gas Dynamics, *J. Comp. Phys.* 62: 83–110.

Chorin, A.J., 1980, Flame Advection and Propagation Algorithms, *J. Comp. Phys.* 35: 1–11.

Chorin, A.J., 1985, Curvature and Solidification, *J. Comp. Phys.* 58: 472–490.

DeVore, C.R., 1984, *Vectorization and Implementation of an Efficient Multigrid Algorithm for the Solution of Elliptic Partial Differential Equations*, NRL Memorandum Report 5504, Naval Research Laboratory, Washington, DC.

Douglas, C.C., 1984, Multi-Grid Algorithms with Applications to Elliptic Boundary-Value Problems, *SIAM J. Numer. Anal.* 21: 236–254.

Douglas, C.C., 1982, *Multi-Grid Algorithms for Elliptic Boundary-Value Problems,* Yale University Technical Report No. 223, Yale University, New Haven, CT.

Emery, M.H., J.H. Gardner, and J.P. Boris, 1982, Nonlinear Aspects of Hydrodynamic Instabilities in Laser Ablation, *Appl. Phys. Lett.* 41: 808–810.

Engquist, B., and A. Majda, 1977, Absorbing Boundary Conditions for the Numerical Simulation of Waves, *Math. Comp.* 31: 629-651.

Engquist, B. and A. Majda, 1979, Radiation Boundary Conditions for Acoustic and Elastic Wave Calculations, *Comm. Pure Appl. Math.* 32: 312-358.

Federenko, R.P., 1961, A Relaxation Method for Solving Elliptic Difference Equations, *Z. Vycisl. Mat. i. Mat. Fiz.* 1: 922–927.

Ghoniem, A.F., A.J. Chorin, and A.K. Oppenheim, 1981, Numerical Modeling of Turbulent Combustion in Premixed Gases, *Eighteenth Symposium (International) on Symposium on Combustion*, The Combustion Institute, Pittsburgh, PA, 1375–1383.

Ghoniem, A.F., D.Y. Chen, and A.K. Oppenheim, 1986, Formation and Inflammation of a Turbulent Jet, *AIAA J.* 24: 224–229.

Glimm, J., B. Lindquist, O. McBryan, B. Plohr, and S. Yaniv, 1983, Front Tracking for Petroleum Reservoir Simulation, *Proceedings of the Seventh SPE Symposium on Petroleum Reservoir Simulation*, SPE-12238, Society of Petroleum Engineers, Dallas, TX, 41–49.

Grinstein, F.F., E.S. Oran, and J.P. Boris, 1986, Numerical Simulations of Asymmetric Mixing in Planar Shear Flows, *J. Fluid Mech.* 165: 201–220.

Grosch, C.E., and S.A. Orszag, 1977, Numerical Solution of Problems in Unbounded Regions: Coordinate Transformations, *J. Comp. Phys.* 25: 273–295.

Guirguis, R., E.S. Oran, and K. Kailasanath, 1986, Numerical Simulations of the Cellular Structure of Detonations in Liquid Nitromethane — Regularity of the Cell Structure, *Comb. Flame* 65: 339–365.

Gustafsson, B., and H.-O. Kreiss, 1979, Boundary Conditions for Time Dependent Problems with an Artificial Boundary, *J. Comp. Phys.* 30: 333–351.

Hackbusch, W., and U. Trottenberg, Eds., 1983, *Multigrid Methods*, Lecture Notes in Mathematics, Nr. 960, Springer-Verlag, New York.

Harlow, F.H., 1955, *A Machine Calculation Method for Hydrodynamic Problems*, LAMS-1956, Los Alamos Scientific Laboratory, Los Alamos, NM.

Harlow, F.H., and J.F. Welch, 1965, Numerical Calculation of Time-Dependent Viscous Incompressible Flow of Fluid with Free Surface, *Phys. Fluids* 8: 2182–2189.

Hirt, C.W., and B.D. Nichols, 1981, Volume of Fluid (VOF) Method for the Dynamics of Free Boundaries, *J. Comp. Phys.* 39: 201–225.

Hackbusch, W., and U. Trottenberg, Eds., 1983, *Multigrid Methods*, Lecture Notes in Mathematics Nr. 960, Springer-Verlag, NY.

Hockney, R.W., 1970, The Potential Calculation and Some Applications, in B. Adler, S. Fernbach, and M. Rotenberg, eds., *Methods in Computational Physics* 9: 135-211, Academic Press, New York.

Hockney, R.W., and J.W. Eastwood, 1981, *Computer Simulation Using Particles*, McGraw-Hill, New York.

Hyman, J.M., 1984, Numerical Methods for Tracking Interfaces, *Physica* 12D: 396–407.

Hyman, J.M., 1979, A Method of Lines Approach to the Numerical Solution of Conservation Laws, in R. Vichnevetski and R.S. Stepleman, eds., *Third IMACS International Symposium on Computer Methods for Partial Differential Equations*, IMACS, Rutgers University, New Brunswick, NJ, 313–321.

Kershaw, D.S., 1978, The Incompete Cholesky – Conjugate Gradient Method for the Solution of Systems of Linear Equations, *J. Comp. Phys.* 26: 43–65.

Kutler, P., Ed., 1982, *Numerical Boundary Condition Procedures*, NASA Conference Publication 2201, NASA Ames Research Center, Moffett Field, CA.

Laskey K.J., E.S. Oran, and J.P. Boris, 1986, *Approaches to Resolving and Tracking Interfaces and Discontinuities*, to appear as Naval Research Laboratory Memorandum Report.

Laskey, K.J., E.S. Oran, and J.P. Boris, 1987, The Gradient Method for Interfacing Tracking, to appear in *J. Comp. Phys.*.

Lindemuth, I., and J. Killeen, 1973, Alternating Direction Implicit Techniques for Two-Dimensional Magnetohydrodynamic Calculations, *J. Comp. Phys.* 13: 181–208.

Madala, R.V., 1978, An Efficient Direct Solver for Separable and Non-Separable Elliptic Equations, *Monthly Weather Rev.* 106: 1735–1741.

McDonald, H., 1979, Combustion Modelling in Two and Three Dimensions: Some Numerical Considerations, *Prog. Energy Comb. Sci.* 5: 97–122.

McDonald, H., and W.R. Briley, 1975, Three-Dimensional Supersonic Flow of a Viscous or Inviscid Gas, *J. Comp. Phys.* 19: 150–178.

Meijerink, J.A., and H.A. van der Vorst, 1977, An Iterative Solution Method for Linear Systems of which the Coefficient Matrix is a Symmetric M-Matrix, *Math. Comp.* 31: 148–162.

Noh, W.F., and P. Woodward, 1976, SLIC (Simple Line Interface Method), *Proceedings of the Fifth International Conference on Numerical Methods in Fluid Dynamics*, in A.I. van de Vooren and P.J. Zandbergen, eds., *Lecture Notes in Physics*, vol. 59, Springer-Verlag, New York, 330–340.

Noh, W.F., 1964, *CEL: A Time-Dependent, Two-Space-Dimensional, Coupled Eulerian-Lagrange Code*, in B. Adler, S. Fernbach, and M. Rotenberg, eds., *Methods in Computational Physics*, vol. 3, 117–179.

Oliger, J., and A. Sundström, 1978, Theoretical and Practical Aspects of Some Initial Boundary Value Problems in Fluid Dynamics, *SIAM J. Appl. Math.* 35: 419–446.

Oran, E.S., and J.P. Boris, 1981, Theoretical and Computational Approach to Modeling Flame Ignition, in J.R. Bowen, N. Manson, A.K. Oppenheim, and R.I. Soloukhim, eds., *Combustion in Reactive Systems*, vol. 76, *Progress in Astronautics and Aeronautics*, AIAA, New York, 154–171.

Potter, D., 1973, *Computational Physics*, Wiley, New York.

Pulliam, T.H., 1982, Characteristic Boundary Conditions for the Euler Equations, in P. Kutler, ed., *Numerical Boundary Condition Procedures*, NASA Conference Publication 2201, NASA Ames Research Center, Moffett Field, CA.

Ramshaw, J.D., and J.K. Dukowicz, 1979, *APACHE: A Generalized-Mesh Eulerian Computer Code for Multicomponent Chemically Reactive Flows*, Report LA-7427, Los Alamos Scientific Laboratory, Los Alamos, NM.

Richtmyer, R.D., and K.W. Morton, 1967, *Difference Methods for Initial-Value Problems*, Interscience, New York.

Roache, P.J., 1982, *Computational Fluid Dynamics*, Hermosa Publishers, Albuquerque, NM.

Sethian, J., 1984, Turbulent Combustion in Open and Closed Vessels, *J. Comp. Phys.* 54: 425–452.

Turkel, E., 1980, Numerical Methods for Large-Scale Time-Dependent Partial Differential Equations, in W. Kollmann, ed., *Computational Fluid Dynamics*, vol. 2, Hemisphere, Washington, DC, 127–262.

Varga, R.S., 1962, *Matrix Iterative Analysis*, Prentice-Hall, Englewood Cliffs, NJ.

Welch, J.E., F.H. Harlow, J.P. Shannon, and B.J. Daly, 1966, *The MAC Method: A Computing Technique for Solving Viscous, Incompressible, Transient Fluid Flow Problems Involving Free Surfaces*, LA-3425, Los Alamos Scientific Laboratory, Los Alamos, NM.

Yee, H.C., 1981, *Numerical Approximation of Boundary Conditions with Applications to Inviscid Equations of Gas Dynamics*, NASA Technical Memorandum 81265, Ames Research Center, Moffett Field, CA.

Chapter 12

DIFFICULT TOPICS

Three topics are covered in this chapter: turbulence, radiation transport, and multiphase flows. These are extremely difficult in most physical contexts, and each really deserves a multivolume discussion. The equations generally used to represent these phenomena are extensions or generalizations of the multicomponent gas phase equations given in Chapter 2. Here we briefly describe some of the more common and useful models used, and refer to previous chapters for appropriate numerical methods and algorithms.

12–1. TURBULENCE

Turbulence persists as one of the major theoretical and computational problems of fluid dynamics. If it were possible to solve the full set of time-dependent, three-dimensional fluid dynamic equations for mass, momentum, and energy for a large enough range of time and space scales, turbulence would appear naturally in the solutions. Because the range of important scales is continuous and extensive, there is no clear way to separate scales. Modeling approaches to turbulent flows are generally inadequate if they are limited to a small range of scales or if they assume that the important large and small scales are widely separated.

We begin this section by describing *detailed simulations* that attempt to resolve the time-dependent structure of the turbulent flows. Unfortunately, the fastest and largest computers are neither fast nor large enough to solve most turbulent flow problems from first principles, even though the fundamental set of equations are generally agreed to be adequate. In some idealized low Reynolds number flows, simulations have been able to resolve the full range of scales down to the *Kolmogorov scale*, the small scale at which turbulent structures are dissipated into heat. In higher Reynolds number or less idealized flows, only the large scales can be calculated in detail, and the small scales have to be modeled phenomenologically.

There have been many approaches to modeling turbulence. Statistical theories divide the solution into a mean, time-averaged flow and a component that fluctuates in time. These models generally involve sets of conservation equations with a few extra terms that represent the fluctuations in the system. Models relating these extra terms to known quantities must be postulated to *close* the set of equations. Once such statistical models are postulated, the equations can be solved by the techniques described in previous chapters. Introducing more phenomenologies, as we go from models of nonreacting incompressible flows to models of highly compressible turbulent flows with chemical reactions, makes the validity of the calculations even more questionable.

12–1.1 Detailed Simulations of Turbulence

The term *direct simulation* refers to calculations in which the full range of relevant scales is resolved, down to the Kolmogorov scale. The basic idea behind these calculations is to use the power of the computer to resolve the time-dependent behavior of a complicated turbulent or transitional flow. The computational region must be big enough to include the largest scales, and the mesh spacing must be fine enough to resolve the smallest scales of interest. Limitations on computer speed and memory have meant that direct simulations can only be done for a limited class of low Reynolds number problems in idealized configurations. For example, direct simulations have been done recently using $128 \times 128 \times 128$ grids for Reynolds numbers of a few hundred, and no doubt these numbers will increase steadily in the future. Such calculations will certainly be expanded as computers with very large memories become available.

Practical flows in complex geometries generally have a wide range of important spatial scales. Because of this, it is often assumed that the flow can be decomposed into two parts. The first part contains the large scales. This part depends on the boundary conditions. The second part contains the small scales, and this is modeled using *subgrid turbulence models.* The large scales driving the flow are assumed to contain a significant part of the turbulent energy. The small scales are assumed to follow given decay laws. The explicit assumption is that there are no coherent small-scale sources of turbulence or vorticity which feed back to the large scales. Using a subgrid model, we can truncate the representation at scales considerably larger than the Kolmogorov scale. Simulations which make this distinction in scales are referred to as *large-eddy simulations.* We refer to both large-eddy simulations and direct simulations as *detailed simulations* of turbulent flows.

Three time-dependent numerical approaches are used for detailed simulations of turbulent and transitional flows: spectral methods, finite-difference methods, and vortex-dynamics methods. Sometimes computations use a combination of these methods. The types of problems simulated have included the evolution and decay of homogeneous turbulence, turbulent flow in a channel, the evolution of mixing layers in shear flows, and flow past obstacles. We complete this discussion with several examples of detailed

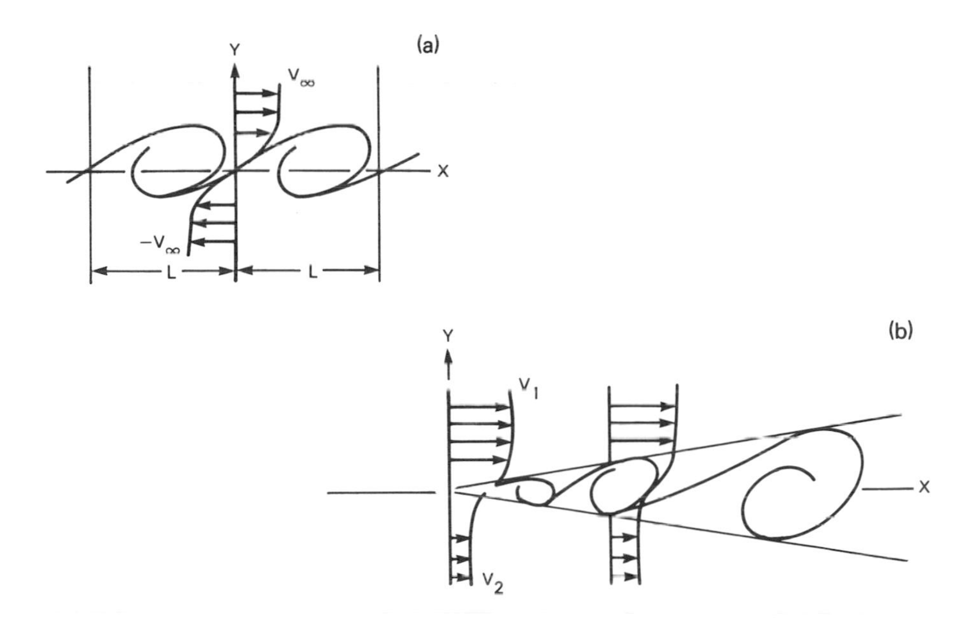

Figure 12–1. Schematic diagram of shear layers. (a) A temporally evolving shear layer with periodic boundaries on the left and right. (b) A spatially evolving shear with inflow on the left and outflow on the right. Velocity profiles across the shear layer are indicated by horizontal lines with arrows.

simulations of the evolution of a mixing layer at a shear interface. Such flows are common in nature, and their effects are observed over space and time scales that vary from centimeters to megaparsecs. They are observed as galaxies eject material to form astrophysical jets, and they occur in engines when a stream of conbustible material enters a region of slower or faster flow.

The simulations that have been performed of planar shear flows have been either *temporally evolving* or *spatially evolving*. Figure 12–1 qualitatively shows the difference in the two types of flows. In both situations, two streams move at different velocities with the shear at the common interface. In a spatially evolving flow, the vortical structures form and grow by merging with each other as they move away from their point of origin. In temporally evolving flows, two parallel streams of equal speed move in opposite directions, and the boundary conditions in the direction of the shear flow, the streamwise direction, are periodic.

There is no precise transformation between the temporally and spatially developing mixing layers, though there are some similarities between them. Although natural flows are spatially evolving, it is easier to cal-

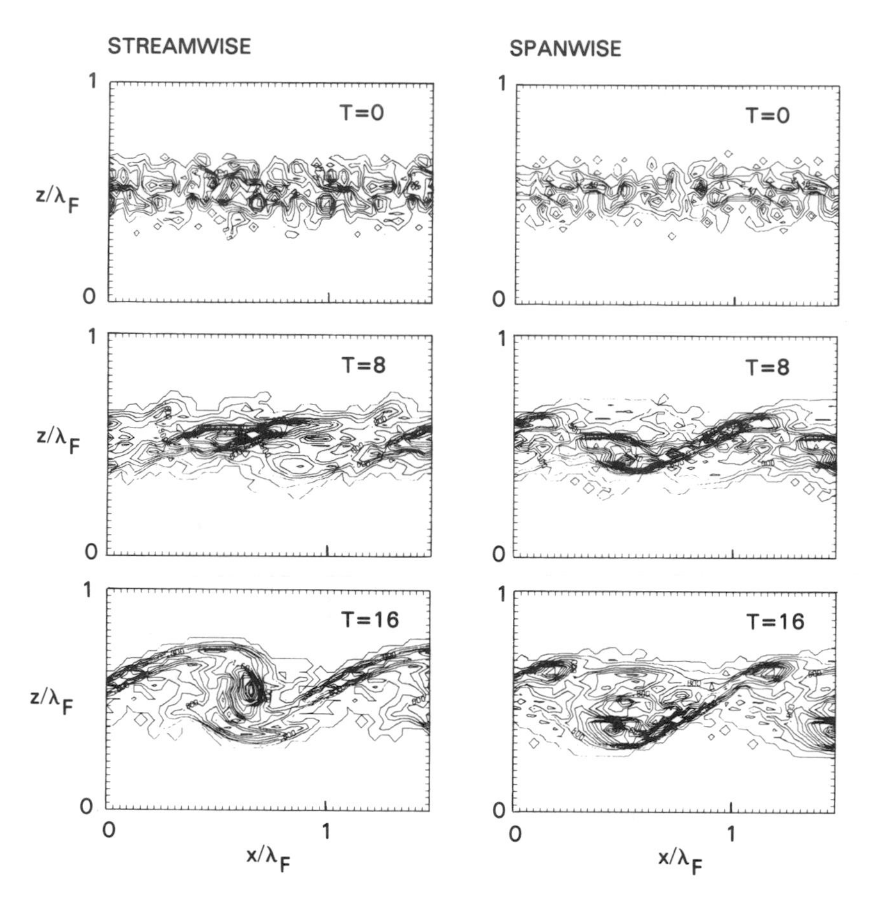

Figure 12–2. Calculation of a temporally evolving shear layer using a pseudo-spectral method and showing vorticity contours in both streamwise and spanwise directions (Riley and Metcalfe, 1980).

culate temporally evolving flows because the calculations do not require inflow and outflow boundary conditions. It is hoped that the temporal evolution of a flow with periodic boundary conditions adequately describes some of the important physical features of the natural flow. The calculations of two-dimensional shear flows described below use a spectral method to calculate a three-dimensional incompressible temporally evolving flow, a vortex-dynamics method to calculate two-dimensional incompressible spatially evolving flows, and a finite-difference method to calculate a compressible axisymmetric flow in a more complex confined geometry.

A Spectral-Method Calculation

Riley and Metcalfe (1980) have carried out two- and three-dimensional pseudospectral simulations (Section 8–7) of the evolution of a temporally evolving planar mixing layer. They solve the Navier-Stokes equations with a fairly large kinematic viscosity. This keeps the Reynolds number low enough so that the full range of scales, down to the Kolmogorov scale, can be resolved and there is no need to use a subgrid model.

Figure 12–2 shows the evolution of the vorticity contours for a three-dimensional calculation in which the initial energy spectrum and mean flow field were specified. The turbulent mixing layer was allowed to evolve from these conditions. In the streamwise direction, a large-scale structure emerges from the initial spectrum. However, in the spanwise direction, the structures are not nearly as distinct. Additional calculations have shown the result of imposing a frequency on this system (Riley and Metcalfe, 1980). More recent work describes the formation of riblike structures in the spanwise direction in three-dimensional flows (Metcalfe et al., 1987).

A Vortex-Dynamics Calculation

Two- and three-dimensional calculations of mixing layers using vortex-dynamics methods have been reported, for example, by Acton (1976), Ashurst (1979), Aref and Siggia (1980), and Couet and Leonard (1981). Figure 12–3, taken from Ghoniem and Ng (1986a,b), is part of a study of confined, two-dimensional splitter-plate simulations of the effects of upstream forcing on the development of structures and the rate of downstream entrainment. The figure shows the vorticity fields at various time intervals. Each vortex element is represented by a circle, and its velocity with respect to the mean velocity is shown by a line vector starting at the center. The results show the formation of eddies and their successive pairings. Similar to the spectral methods in general use, the vortex-dynamics methods are incompressible and so do not incorporate sound waves in the flow. They are thus most applicable to slow flow velocities in constant density liquids, for which compressibility effects are substantially smaller than for gases.

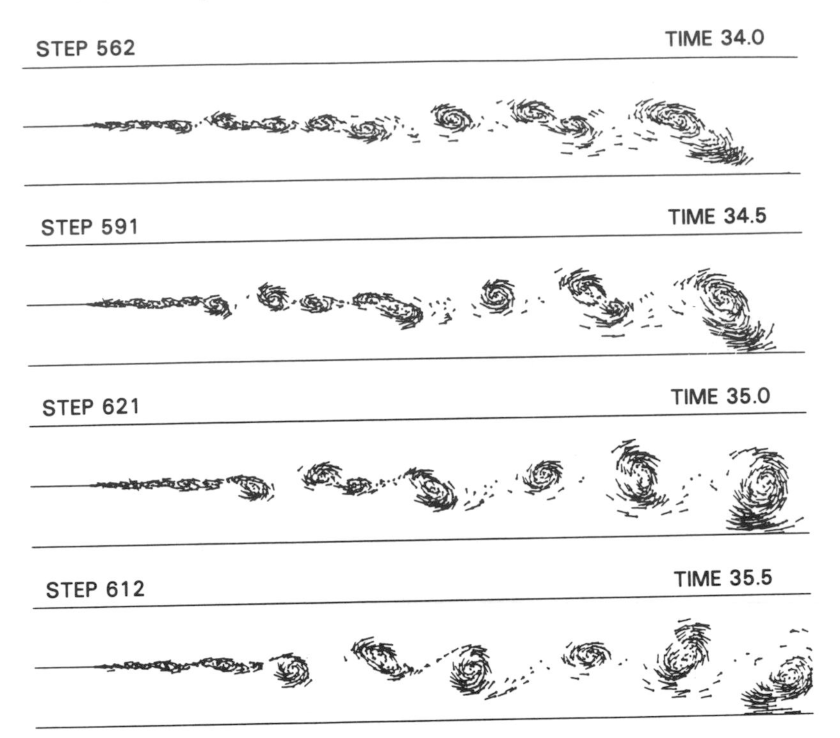

Figure 12–3. Calculation of a spatially evolving two-dimensional shear layer using a vortex-dynamics algorithm (Ghoniem and Ng, 1986a,b).

A Finite-Difference Calculation

Calculations of two- and three-dimensional mixing layers using finite-difference methods have been done by, for example, Mansour et al. (1978), Gatski and Liu (1980), Knight and Murray (1980), Davis and Moore (1985), and Grinstein et al. (1986). In Figure 12–4 we deviate from showing planar calculations and describe a more complicated, compressible, axisymmetric flow calculated by Kailasanath et al. (1986). Here a gas flows out of a long cylindrical inlet into a chamber of larger diameter. The flow leaving this larger chamber is choked, so that it becomes sonic at the exit nozzles. The figure shows a schematic of the chamber and instantaneous streamlines of the flow in the cross-hatched region. Lines originating at the centers of vortices have been drawn between the streamline plots to show the evolution of the vortex structures. The flowfield has an overall repetition cycle of 6000 steps, corresponding to a low-frequency mode driven by acoustic resonances originating in the inlet pipe.

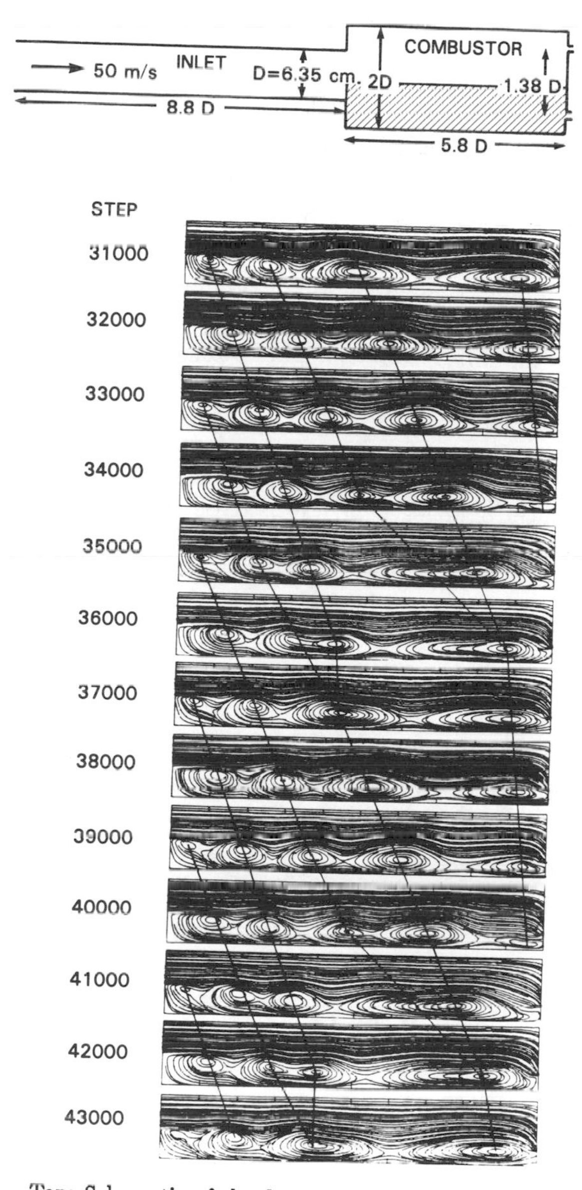

Figure 12–4. Top: Schematic of the flow calculated by Kailasanath et al. (1986). The flow is axisymmetric, with the line of symmetry through the center. Bottom: Calculated instantaneous streamlines in the hatched region of the schematic. Lines through the streamlines connect the centers of vortex structures.

Reactive Flows

The examples given above were for nonreactive flows. Recently, these calculations have been extended to reactive flows. For example, Kailasanath et al. (1986) report finite-difference calculations of chemically reacting flows in the axisymmetric geometry shown in Figure 12–4 (see Section 13–2). Vortex-dynamics methods, used to describe both mixing layers and jets, have been done by, for example, Ghoniem et al. (1981, 1986). Temporally evolving, chemically reacting flows at planar mixing layers have been simulated using spectral methods by McMurtry et al. (1986) and Givi et al. (1987). These calculations, now at the initial stages of development, have the potential of being useful tools for combustion problems.

12–1.2 The Ideal Turbulence Model

We now consider the properties which should be incorporated in an ideal subgrid turbulence model. Understanding the computational consequences of these desired properties can guide us both in extracting information about the nature of real turbulence from detailed calculations and in constructing practical phenomenological models for use in simulations of turbulent flows.

In this discussion, "macroscopic" processes are those that can be resolved on the chosen computational grid. "Microscopic" or "subgrid" phenomena occur on scales too small to be resolved, but they may have macroscopic consequences. Of necessity, computational models of turbulence often assume that subgrid phenomena can be decoupled from the macroscopic flow.

Scaling

Breaking the calculation into macroscopic and subgrid scales is an artifice. In general, the important physics occurs continuously over the whole spectrum of wavenumbers from $k = 0$ to $k = k_{diss}$, the wavenumber above which the spectrum is dominated by viscous damping. The macroscopic and subgrid scale spectra must therefore merge smoothly at k_{cell}, the wavenumber associated with the size of a cell. If k_{cell} is changed, which occurs if numerical resolution is halved or doubled, the predictions of the turbulence model coupled to the macroscopic fluid equations should not change.

Lagrangian Framework

An ideal subgrid model should probably be constructed on a Lagrangian fluid dynamics framework moving with the macroscopic flow. This would ensure that turbulence-induced mixing and molecular mixing are not overshadowed by numerical diffusion. Using a Lagrangian representation would also minimize the possibility of confusing purely local turbulent fluctuations with numerical truncation errors.

Such a three-dimensional Langragian model, or even a good two-dimensional Langrangian model for the types of calculations desired, is still to be perfected. A fallback approach is to formulate the subgrid turbulence phenomenology in an Eulerian representation so that its basic parameters may be convected across a grid with the Lagrangian fluid elements.

Onset and Other Transient Turbulence Phenomena

The turbulence model should be able to predict the onset of turbulence in what is initially laminar flow. Because random unphased fluctuations of the subscale motions can act as infinitesimal triggers of macroscopic instability, a random source term in the macroscopic equations is required wherever there is any subgrid activity. This source term must appear in addition to the averaged diffusion effects discussed below.

The density, temperature, and velocity gradients resolved in the simulation drive macroscopic fluid dynamic instabilities, which in turn initiate subgrid turbulence. Thus these macroscopic quantities determine the energy that is available to drive the subgrid turbulence. Representing density differences that might occur in a time-dependent flow is essential because they are a source of vorticity generation.

Coupling between Macroscopic and Subgrid Regions

The effects of turbulence enter the macroscopic flow equations through phenomenological terms representing averages over the microscopic dynamics. Examples of these terms are eddy viscosity coefficients, diffusivity coefficients, and average chemical heat release terms which would appear as sources in the macroscopic flow equations. Such modified Navier Stokes equations are postulated to include the effects of subgrid turbulence.

Besides providing a definition of such phenomenological terms, the turbulence model must use the information from the large-scale flow to drive the turbulence models correctly. A feedback mechanism is needed to describe energy transfer from the small scales, where mixing and subsequent chemical reactions occur, to the macroscopic flow.

Complicated Reactions and Flows

A turbulence model should be able to account for the effects of chemical kinetics, buoyancy, droplets, condensation, and other processes which may be important in the macroscopic and microscopic scales. Subgrid models seem to be necessary for representing interactions of most physical processes.

If there is a delay as vorticity cascades within the subscales to the short wavelength end of the spectrum, the model should be capable of representing this delay. This allows the representation of bursts and other intermittent phenomena.

Economy

An essentially infinite number of degrees of freedom on the subgrid scale must be represented by a finite number of variables. It is important to keep the number of subgrid variables in each computational cell at a minimum for the model to be generally usable. However, because the number of subgrid variables is finite, any subgrid model is intrinsically incomplete.

12–1.3 Turbulence-Averaged Navier-Stokes Equations

A review of these models has been given by Reynolds (1976). We also recommend the descriptions given in Kuo (1986) and in the recent *Assessment of Current Capabilities and Future Direction in Computational Fluid Dynamics* (National Academy Press, Washington, D.C., 1986). There are a number of statistical approaches to turbulent reacting flows. These are described in the book *Turbulent Reacting Flows* edited by Libby and Williams (1980) and summarized in the review articles by Libby and Williams (1981), Jones and Whitelaw (1982), and Ferziger (1987).

Classical approaches to modeling turbulence are based on a statistical treatment of fluctuations about a stationary or slowly varying flow. The primary variables are broken into two parts: a mean, time-averaged part of the flow, and a fluctuating part representing deviations from this mean. The *Reynolds averages* of the system variables can be defined as

$$f(\mathbf{x},t) \;=\; \overline{f}(\mathbf{x}) + f'(\mathbf{x},t)\;, \qquad (12-1.1)$$

where f indicates, for example, $\mathbf{v}$, P, ρ, n_i, T, or E. The bar over a quantity indicates the time-averaged value,

$$\overline{f} \;\equiv\; \lim_{\Delta t \to \infty} \frac{1}{\Delta t} \int_{t_o}^{t_o+\Delta t} f(t)dt\;. \qquad (12-1.2)$$

The prime indicates the fluctuating part with the condition

$$\overline{f'} \;=\; 0\;. \qquad (12-1.3)$$

When variables written as a mean and fluctuating part, as in Eq. (12–1.1), are substituted in the incompressible, single-component mass and momentum equations, we find for the density equation,

$$\frac{\partial \overline{\rho}}{\partial t} + \nabla \cdot \overline{\rho}\,\overline{\mathbf{v}} = 0 , \qquad (12-1.4)$$

and for the momentum equation

$$\frac{\partial}{\partial t}(\rho\,\overline{\mathbf{v}}) + \nabla \cdot (\rho\,\overline{\mathbf{v}\mathbf{v}}) = -\nabla \overline{P} + \nabla \cdot (\overline{\tau} - \rho\,\overline{\mathbf{v}'\mathbf{v}'}) . \qquad (12-1.5)$$

For incompressible flow, $\rho' = 0$ and $\overline{\rho}$ is constant. These are the *Reynolds-averaged Navier-Stokes* equations. Note that these equations have the same general form as those in Chapter 2, but now there is an additional Reynolds stress term,

$$\rho\,\overline{\mathbf{v}'\mathbf{v}'} = \rho\,\overline{(\mathbf{v} - \overline{\mathbf{v}})\;(\mathbf{v} - \overline{\mathbf{v}})} . \qquad (12-1.6)$$

There is an apparent inconsistency in Eqs. (12–1.4) and (12–1.5) because the quantites $\overline{\rho}$ and $\overline{\mathbf{v}}$ are defined as time averages although they appear with time derivatives. This apparent inconsistency can be justified mathematically when the averaged and fluctuating components of the variable, $\overline{f}$ and f', are both time dependent, but vary on different time scales. Then the time average in Eq. (12–1.2) extends over an interval that is long compared to the fast turbulent variations in f. As discussed earlier, this kind of a separation cannot be made legitimately in turbulent flows.

For compressible flows where $\rho' \neq 0$, there is another term in the density equation

$$\frac{\partial \overline{\rho}}{\partial t} + \nabla \cdot \overline{\rho}\,\overline{\mathbf{v}} + \nabla \cdot \overline{\rho'\mathbf{v}'} = 0 . \qquad (12-1.7)$$

Difficulties associated with modeling extra terms such as this have led to *Favre averaging*, which defines *mass-weighted* quantites such as

$$\tilde{\mathbf{v}} \equiv \frac{\overline{\rho\mathbf{v}}}{\overline{\rho}} . \qquad (12-1.8)$$

The velocity may be written as

$$\mathbf{v}(\mathbf{r},t) \equiv \tilde{\mathbf{v}}(\mathbf{r}) + \mathbf{v}''(\mathbf{r},t) , \qquad (12-1.9)$$

where now $\mathbf{v}''$ is the velocity fluctuation . This formulation simplifies the resulting density equation by eliminating the extra term and giving it the standard form of a sourceless continuity equation. The *Favre-averaged Navier-Stokes equation* for density is

$$\frac{\partial \overline{\rho}}{\partial t} + \nabla \cdot (\overline{\rho}\,\tilde{\mathbf{v}}) = 0 , \qquad (12-1.10)$$

for momentum it is

$$\frac{\partial}{\partial t}(\overline{\rho}\tilde{\mathbf{v}}) + \nabla \cdot (\overline{\rho}\tilde{\mathbf{v}}\tilde{\mathbf{v}}) = -\nabla \overline{P} + \nabla \cdot (\overline{\tau} - \overline{\rho \mathbf{v}'\mathbf{v}'}) , \qquad (12-1.11)$$

and for the energy equation, written in terms of the enthalpy, h, is

$$\begin{aligned} \frac{\partial}{\partial t}(\overline{\rho}\tilde{h}) + \nabla \cdot \overline{\rho}\tilde{h}\tilde{\mathbf{v}} &= \frac{\partial \overline{P}}{\partial t} + \tilde{\mathbf{v}} \cdot \nabla \overline{P} + \overline{\mathbf{v}' \cdot \nabla P} \\ &\quad + \nabla \cdot (-\overline{\mathbf{q}} - \overline{\rho h'\mathbf{v}'}) + \overline{\boldsymbol{\tau} \cdot \nabla \mathbf{v}'} \, . \end{aligned} \qquad (12-1.12)$$

These equations have the same form as the incompressible time-averaged equations, but the Reynolds stresses now include the density variation as well.

To close the set of turbulence-averaged equations requires subgrid models for undefined turbulence quantities, such as the Reynolds stresses and the turbulent heat fluxes in Eqs. (12–1.10) through (12–1.12). This is done with *turbulent scale models.*

Turbulent Scale Models

A turbulence scale model or *turbulence model* provides a description of the connections between the turbulence quantities needed in the averaged equations and other variables or parameters available or easily computable. This relation is usually a combination of algebraic and evolution equations connecting the turbulence parameters with the averaged dependent variables. To close the set of equations, the Reynolds stress is often modeled as a function of a *turbulent velocity scale*, q, a *turbulent viscosity*, ν_T, and a *turbulent length scale*, l, where ν_T is a function of q and l. Such models for the Reynolds stresses are called zero-equation models, one- two- or three-equation models, and stress-equation models (see, for example, Reynolds, 1976).

Zero-equation models solve a partial differential equation such as Eq. (12–1.5) for the mean velocity field, and use a *local algebraic model* to relate the fluctuating quantities to the mean flow. Models of this type work well enough in simple quasi-steady flows for which the adjustable constants have been calibrated. They do not work well for problems in which the mean conditions change abruptly or when a significant region of the flow is affected by turbulent transport. They basically assume that the turbulence has equilibrated with the local mean conditions. These models do not explicitly consider the history of turbulence.

One-equation models add coupled partial differential equations for the turbulence velocity scale. *Two-equation models* add yet another equation for the turbulence length scale. The extra partial differential equations describe the evolution of quantities that are required to calculate the stress tensors in the mean flow equations. These models, which can be described as *scale evolution models*, relate turbulent transport to a combination of local features of the mean flow and one or two scalar parameters of the turbulence. Thus the evolution models take explicit account of the history of the turbulence scales. However, they are still equilibrium models in that they assume that the mean field and turbulence parameters are locally in equilibrium. Evolution equation models generally give a reasonable first-order approximation of the turbulence in simple separated flows or flows with gradual changes in boundary conditions. They fail in flows with strong rotations, density gradients, and chemical energy release.

Stress-equation models, also called *transport evolution models*, solve partial differential equations for all of the components of the turbulent stress tensor. These models, derived from the Navier-Stokes equations themselves, are based on equations for the evolution of the transport terms in the averaged Navier-Stokes equations. In addition, at least one equation is solved for a parameter describing the turbulence length-scale information. Local algebraic models provide the higher-order statistical quantities in these equations. For example, the *Reynolds stress transport* models, in which the Reynolds stresses become dependent variables of the partial differential equations system, are of this type. They are computationally expensive, and using them does not necessarily produce answers substantially better than the simpler algebraic and scalar evolution models. Furthermore, these stress-equation models become even more complex in compressible reactive flows.

The turbulence-averaged equations, whatever their final form, may all be solved by the numerical methods described earlier in this book. They are often solved in situations where the mean flow is nearly steady state. Often the macroscopic flow is rather poorly resolved, assuming that the turbulent viscosity is large enough to mask the unresolved scales. Then physical effects represented by turbulence override the severe effects of numerical diffusion. When defining a mean time-averaged flow at each location and considering fluctuations about that mean, there is some question about what it means to have time-dependent equations. As noted earlier, a multiple time scale approximation is implied, although it is not really correct here.

Models for Turbulence in Reacting Flows

One approach to generating models for turbulent reacting flows is to generalize the various turbulence-averaged, statistical concepts described above. Introducing chemical reactions and heat release means considering the full set of reactive flow equations given in Chapter 2. In addition to the velocity fluctuations, there are now fluctuations in total density, temperature, and individual species densities. Because the set of equations now has terms for correlations of additional variables, more terms must be modeled to close the equations. In the most straightforward models, the higher moments are modeled in a way analogous to that used for nonreacting, constant-density turbulence.

One problem with these approaches is that they ignore intermittency effects that are so obvious in, for example, turbulent flames. The use of *probability density functions*, or PDFs, introduces a formalism that allows modeling these intermittency effects. For example, consider $P(v_i, x, t)$, a PDF defined such that $P(v_i, x, t)dv_i$ is the probability that the ith component of velocity is in the range dv_i at fixed values of $\mathbf{x}$ and t. Also, there are joint PDFs, for example, $P(\rho, v_i, x, t)$, where $P(\rho, v_i)dv_i d\rho$ is the probability that the fluid variables have values in the range dv_i about v_i and $d\rho$ about ρ at fixed $\mathbf{x}$ and t. The probability density function P is normalized so that

$$\int_{-\infty}^{\infty} P(v_i, \mathbf{x}, t)dv_i = 1 \; . \qquad (12-1.13)$$

The value of the turbulence-averaged variable v_i is obtained from

$$\int_{-\infty}^{\infty} v_i P(v_i, \mathbf{x}, t)dv_i = \overline{v}_i(\mathbf{x}, t) \; . \qquad (12-1.14)$$

The assumed forms for PDFs are based on physical grounds or obtained from measurements. For example, a PDF can have delta functions at its extremities, representing the pure fluids that exist on either side of a convoluted interface, and a continuum portion that may be flat, Gaussian, or have some other shape. It is possible to develop evolution equations for PDFs that look like conservation equations. Similar to the stress-transport models, however, this approach requires models of various terms to close the equations. There have been some solutions for a single scalar by finite-difference methods, and more recently there have been Monte Carlo solutions for joint PDFs. A more formal approach which involves derivations of equations for the evolution of PDFs is described by O'Brien (1980). We recommend the recent review article by Pope (1985) for a comprehensive discussion of PDFs and their application to combustion.

The PDF approach remedies some of the problems with the statistical approaches using straightforward Favre averaging. However, one problem with PDF theory is that it is difficult to determine the mass-averaged quantities experimentally and therefore it is sometimes difficult to calibrate the models.

12–1.4 Subgrid Turbulence Models

If the macroscopic flow still has appreciable time variation after the identifiable turbulent fluctuations are averaged, no true steady-state flow exists. Then local models have to be developed that relate the small-scale turbulence to what is happening or has recently happened in the nearby fluid elements. We call such models *subgrid turbulence models*. By definition, they should work in problems with statistically steady states as well as situations in which the macroscopic flow changes rapidly.

The basic idea of subgrid modeling is to represent phenomenologically the macroscopic effects of fluid flow occurring on space scales which are not resolved. This commonly means representing size or length scales which are smaller than a computational cell. This can also mean representing what is happening in a dimension which is not modeled in detail, such as representing the turbulent convective transport in a second or third direction. However, the time averages of macroscopic flow, taken in the statistical turbulence models, are not needed. The recent collection of articles edited by Dwoyer et al. (1985) contains reviews of various approaches to subgrid modeling.

The convective terms in the conservation equations produce a range of scales bounded at the Kolmogorov scale by molecular diffusion. When the Reynolds number is high, we cannot resolve all of the scales numerically and there is some cutoff above the Kolmogorov scale determined by the grid. Different numerical algorithms treat the solution differently at this scale. Some algorithms allow the energy to build up at the small scales, so that the method goes unstable, or at least the results become very nonphysical. For example, Kwak et al. (1975) have shown that if turbulent energy is nonphysically trapped at the small resolved scales rather than cascading down to the Kolmogorov scale, the energy transfer rate from the large scales is too fast. Other algorithms dissipate turbulent energy at the small grid scales into heat. They thus mimic the actual physical process, but at the wrong scale and often by the wrong amount. An important problem in subgrid turbulence modeling is how to remove the energy from the numerically resolved scales at the correct rate.

One approach to subgrid turbulence modeling is to filter the small scales, including a few of the smallest actually resolved. The scales below the cut-off are modeled with a subgrid model. Such a filtering procedure also has the advantage of removing the oscillations caused in many methods by the Gibbs phenomena discussed in Chapter 4. Several kinds of filters have been suggested for nonmonotone algorithms, which do have natural filtering at the small-scale cutoff. These include Gaussian filters (Kwak et al., 1975; Shaanan et al., 1975; Mansour et al., 1978; Moin and Kim, 1982) and the Leonard filter (1974).

It is also important to connect the model of the small scales to the large-scale calculation correctly (Schumann, 1975; Deardorff, 1970). Existing models account for the incoherent feedback from the small scales to the large scales with a diffusionlike term that tends to smooth effects over the short scales. Most subgrid turbulence models developed to date have focussed on representing shorter scales than are resolved in constant density nonreactive turbulence computations. Models for this type of term were first proposed by Smagorinsky (1963) and Lilly (1966). These and other suggestions have been summarized in the reviews by Herring (1979), Ferziger (1977, 1981), and Rogallo and Moin (1984).

Spectral-method calculations must be filtered when the viscosity is too small to prevent nonphysical energy buildup in the small scales. The spectral method itself contains no dissipation at short wavelengths just above the cutoff determined by the series expansion. Spectral models must attach a subgrid model at the designated cutoff, because they do not provide the proper subgrid feedback to the large scales.

Nonlinear monotone algorithms, such as Flux-Corrected Transport (Chapters 8 and 9), are themselves a filter. The monotonicity condition is a physically based filter that prevents nonphysical buildup of short wavelength fluctuations at the grid scale. Such algorithms dissipate the unresolved small-scale energy into heat, and do not allow unphysical feedback to the short, but resolved scales.

To date very little account has been taken of the physically important coherent feedback of the short scales on the large scales. The assumption is made that the retention of subgrid phase information is not important for large-scale dynamics. The issue here is short-wavelength motions with a long-wavelength component. These motions average to zero, but the instantaneous effect of the long-wavelength component can be important to the system and is a source of long-wavelength perturbations. The problem has been discussed by Fox and Lilly (1972) and by Herring (1973). This, unlike the incoherent part of the spectrum discussed above, is not a diffusionlike term, but more like a random fluctuation in the large scales caused by the small scales. Fox and Lilly suggested introducing random fluctuations which react back onto the large scales.

Another approach to subgrid modeling that addresses the problem of feedback from the small spatial scales to the large ones is renormalization group theory, a method originally developed in statistical mechanics. This approach suggests a systematic approach to progressively eliminating the effects of the smallest scales and replacing their mean effect on the larger scales by an effective turbulent viscosity. For a description of this method, we recommend the reviews by McComb (1985) and Herring (1985) and the recent articles by Yakhot and Orszag (1986a,b).

12–1.5 Problems in Modeling Reactive Flow Turbulence

We conclude this section on turbulence with a discussion of physics, not numerics, which extends and amplifies the consideration of fluid dynamics and chemical reactions in Section 2–3. The purpose is to present a coupled, mechanistic picture of turbulent reactive flow to guide future model construction. In particular, we emphasize the ways reactive flow turbulence differs from nonreactive turbulence.

Dynamic Fluid Instabilities and Turbulence Generation

Many hydrodynamic instabilities contribute to the onset of turbulence. At each scale, the quick onset of shorter wavelength fluid instabilities results from variations of the background flow at that scale. When the values of background variables are near threshold, the growth rates are very slow and the unstable wavelengths relatively long. As the background flow evolves into an unstable regime, the maximum growth rates of the most unstable modes increase and instability spreads over a broader band of the spectrum. The turbulent generation of increasingly disordered vorticity is the result of the nonlinear evolution of these instabilities and is an intrinsically transient phenomenon.

Two fluid instabilities commonly discussed in the evolution of turbulence are the *Rayleigh-Taylor* and the *Kelvin-Helmholtz* instabilities. The Rayleigh-Taylor instability is caused by the acceleration of a heavy fluid through a light fluid, and the result is the generation of roughly aligned filamentary vortex pairs of opposite sign as the instability grows. This can occur, for example, in buoyant flows or when sound or shock waves interact with flames or density gradients. Rotationally generated Taylor instabilities also have the same character. The Kelvin-Helmholtz instability occurs at the shear interface between flows of different velocities. It is responsible for the initial formation of coherent structures in splitter-plate and jet experiments. It also tends to fragment existing sheets of vorticity into roughly parallel vortex filaments of the same sign.

The existence of density gradients is one important property of reactive flows that is not generally considered in classical nonreactive turbulence models. Density gradients generate vorticity through interaction with ambient gradients or fluctuations in the pressure field. This is the $\nabla \rho \times \nabla P$ term in the vorticity evolution equation, Eq. (9–1.7). When the mass density is constant, all of the effects potentially depending on the vorticity source term in Eq. (2–2.8) are absent.

In nonreactive turbulence, the vorticity spectrum is driven at macroscopic scales, the intermediate and short wavelengths are populated by cascade due to instabilities and vortex stretching, and the short wavelengths in the spectrum are dissipated viscously. There are also feedback effects of short-wavelength fluctuation on longer scales. In combustion, the localized release of heat in mixed fuel-oxidizer pockets causes strong but transient expansion of the reacting gases. The resulting low density region has scale lengths characteristic of the combustion process and interacts with pressure gradients to generate new vorticity in the flow on these same characteristic scales. Vorticity on these scales is particularly efficient at mixing and can be expected to feed back on itself. The combustion process is in turn enhanced as long as fuel and oxidizer are present. Because the turbulent spectrum seems now to be driven energetically at short wavelengths as well as long, plateaus or even peaks might form in the spectrum altering the usual notions of cascade, scaling, and hence modeling.

There are active, passive, and local roles for turbulent mixing that this fluid expansion activates in chemically reactive flows. These can be understood qualitatively in terms of the local action of a Rayleigh-Taylor instability. The *active role* occurs while a pocket of gas is actually expanding due to the heat released from chemical reactions. A time-varying acceleration accompanying this expansion is felt in the surrounding fluid. An existing region of strong density gradients is Rayleigh-Taylor unstable during this acceleration and vorticity is generated as the energy is being released. A density gradient, whose scale length is comparable to the distance the fluid moves while expanding, exponentiates roughly once from the expansion. The transient expansion induces persistent vortex filaments with a characteristic mixing time. The corresponding induced mixing progresses only about 3% as fast as the driving expansion, but it continues for a long time. The expansion and the velocity accompanying it are intrinsically transient phenomena lasting only a short time.

The *passive role* of expansion is potentially even more important than the active role. In the passive role, the expansion influences turbulent mixing by providing the density gradients, rather than the pressure gradients, which lead to vorticity generation. Consider a fluid vortex rotating at angular frequency Ω. The acceleration of the fluid is $r\Omega^2$ at radius r. Smaller low-density pockets in this vortex, formed by chemical reaction at an earlier time, are unstable to Rayleigh-Taylor modes where the local centrifugal fluid acceleration opposes the density gradient. Because the active generation of vorticity is limited in duration to the time of the expansion phase, it is reasonable to expect that the amount of vorticity generated by the passive interaction of existing density gradients with large-scale vortices is even greater. There are several growth times of the Rayleigh-Taylor mode during a single rotation of the vortex when the mode wavelength is comparable to the vortex radius.

Expansion caused by energy release also plays a *local role* in the interaction between chemical kinetics and fluid dynamics. Density, temperature, and composition changes that occur with chemical reactions, even at constant pressure, can modify the fluid dynamics through diffusive transport processes and changes in the equation of state.

Reactive Interface Dynamics

In reactive turbulence, the microscopic, molecular mixing depends on the motion and contortion of Lagrangian interfaces that originally separate reactive species. We must understand and be able to predict the behavior of these surfaces as they move and stretch with the fluid. As the fuel and oxidizer diffuse into each other at the molecular level, these surfaces continue to lie roughly normal to the strongest species density gradients. Rapid increase of the reactive surface area due to turbulence enhances molecular mixing and potentially speeds reactions. The situation is shown schematically in Figure 12–5. In the upper left-hand panel a cube of fluid 1 cm on a side is shown at time $t = 0$. An interface between reacting species A and B divides the cube in half. Subsequent convolution and stretching of this reactive surface is controlled by the vorticity distribution in the local fluid velocity.

Motion of the fluid in the reactive surface breaks down into a velocity component normal to the surface that can change the integrated area, and components parallel to the surface that cannot. These two situations are shown in Figure 12–6. Both components are important and complement each other in the turbulent mixing process. The component normal to the surface causes interpenetration of parcels of fluid from both sides of the

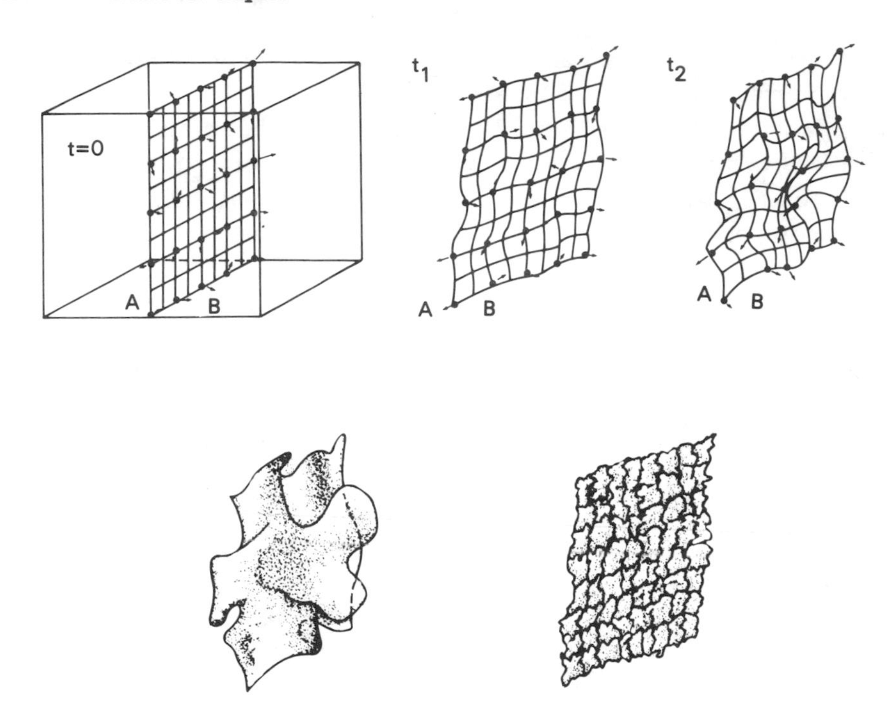

Figure 12–5. Top panels: Three stages during the early deformation of a reactive interface between two fluids A and B. At $t=0$, the surface separating A and B is shown as flat in a three-dimensional volume of fluid. At later times t_1 and t_2, the surface progressively deforms as the result of the Lagrangian motion of surface points. The small arrows indicate the local direction of flow. Lower panels: Two different cases of increase in the area of a reactive interface. When molecular diffusion is important and the turbulence velocity spectrum is enhanced at short wavelengths, the surface is a relatively flat, fluffy surface. When diffusion is small and the velocity spectrum is large, longer wavelength, smoother bulges result.

interface, a process called *entrainment*. When nearby points on the surface separate, so that the surface stretches, the fluid on opposite sides of the surface approaches the surface to keep the flow roughly divergence free. Species and temperature gradients normal to the surface are enhanced. This in turn increases the diffusive interpenetration of the reactants. The surface stretching process is independent of material entrainment. By increasing the actual area of the reactive surface, the bulk reactivity is also increased. Stretching and interpenetration usually occur simultaneously but for clarity have been illustrated separately in the figure.

If the turbulent spectrum is dominated by short wavelengths, an originally smooth surface becomes wrinkled at short wavelengths before larger-scale convolutions have had a chance to grow. This is also illustrated in

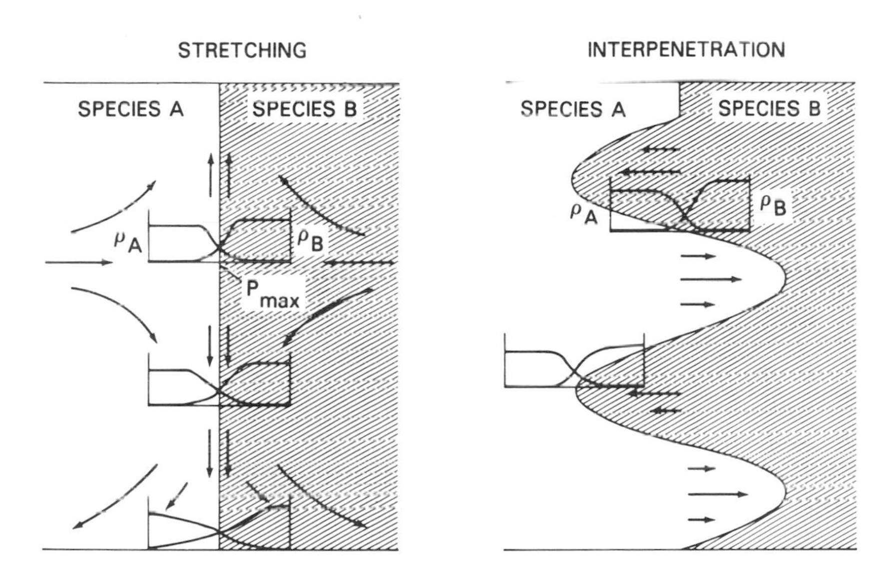

Figure 12–6. Interpenetration of two species A and B as influenced by local reactive-interface dynamics.

Figure 12–5. Because the length scale of the turbulent spectrum is close to molecular mixing lengths, the effective volume in which enhanced mixing occurs is much less than what would be calculated knowing the stretched surface area. Adjacent folds of the surface which approach within the molecular diffusion length of each other tend to merge. The small surface corrugations rapidly get smoothed over with diffusive "fluff" so that one effect of molecular diffusion is to limit the rate of convective growth of the reactive surface area.

In the opposite case, also shown in Figure 12–5, the dominant wavelengths are long compared to the molecular mixing length. The surface is characterized by relatively smooth, long-wavelength bulges which can fill a large volume before any of the mutual interference effects described above can occur. In fact, the area of an ideal Lagrangian surface in an isotropic, homogeneous turbulent velocity field should probably increase in a way that can be represented by a fractal (see, for example, Gouldin, 1987; Mandelbrot, 1983). Molecular mixing now occurs along convolutions which have scales longer than the diffusion length. Eventually, however, overlap of the effective mixed volume occurs as the surface becomes more convoluted. Thus, the molecularly mixed volume is again limited even though the ideal reactive surface area goes to infinity.

Detailed Chemical Kinetics

The third complicating aspect of the reactive-flow mixing problem is the energy released by the chemical reaction. The essentially one-dimensional profiles of fuel, oxidizer, and other reactants perpendicular to the reactive surface change not only from convection and molecular diffusion but also from chemical reactions. Energy release in gaseous flows promotes expansion and, if the flow is sufficiently exothermic, can induce buoyancy. Thus the chemical kinetics feeds back into the hydrodynamic channels.

In some cases the overall reaction rate is governed by the diffusion of hot fuel and oxidizer together through an expanding region of hot products and reactants. In other cases the fuel and oxidizer mix convectively and then molecularly before heating and ignition occur. These latter situations can be ignited by a rapidly moving flame front which travels parallel to the reactive interface rather than perpendicular to it. The situation is complicated significantly by flame stretching, which can extinguish reactions long enough for reactants and hot products to intermix. Here finite-rate chemical kinetics clearly plays a crucial role because it determines the flame speed.

12–1.6 Concluding Remarks

Turbulence is a difficult phenomenon to model and reactive turbulence is perhaps impossibly difficult. However, we must remain undaunted in our attempts to understand and model turbulent phenomena. When a topic is this complex, it is always worthwhile to look for new ideas and potentially useful concepts. Approaches involving fractals, strange attractors, chaos, and cellular automata should all be watched as possible contributors to our understanding turbulent flows. As in computational fluid dynamics, global conservation laws provide useful constraints and bounds that could obviate the need to resolve many of the details of turbulent mixing.

12–2. RADIATION TRANSPORT

The electronic, atomic, and molecular motions associated with internal energy cause materials to emit electromagnetic radiation continuously. This radiation is emitted over a wide spectrum, ranging from radio waves to cosmic rays. It is an important energy-transport mechanism in a number of reactive flow systems, ranging from large-scale fires to small-scale flames. For example, radiation can cause ignition at widely separated points by a phenomenon called *flashover*, in which the radiation from one combustion region heats a distant surface until it spontaneously ignites. Radiation can also be important in engine combustion chambers, where temperatures reach two or three thousand degrees Kelvin. When there is sooting, the soot particles emit and absorb radiation. Radiation from a sooty flame can remove enough heat to markedly change the buoyancy of the hot products. Other applications in which radiant energy transport is important include hypersonic shock layers, rocket propulsion, and dense plasmas such as the sun and nuclear fusion devices.

Emitted radiation depends sensitively on the material temperature and generally becomes more important as the temperature increases. The net radiant energy transferred generally depends on differences of the absolute temperatures of bodies raised to the fourth power, following the Stephan-Boltzmann law. This differs strongly from the energy exchange mechanisms for convection and conduction, which usually depend linearly on the temperature difference.

Another important difference between radiation transport and conduction or convection is that radiative energy, carried by photons, can be transported in a vacuum as well as in a material medium. In convection and conduction, energy is transported by the medium. The mathematical expression of this transport property of radiation gives rise to an integral equation.

This section introduces some concepts that are useful in the most elementary modeling of radiation transport. We first consider the physical basis of radiation transport by describing the quantities transported by the equation of radiative transfer. From this equation we can derive an expression for the quantity $\mathbf{q}_r$, the radiant energy flux, introduced in Eq. (2–1.4) and Eq. (7–0.3). We then discuss several limiting cases of radiation transport. Because these limiting cases are really not general enough for time-dependent numerical simulations of reactive flows, we discuss more useful approaches such as the diffusion approximation and the variable Eddington approximation. For more detailed explanations and derivations, we recommend the books by Zel'dovich and Raizer (1966), Özisik (1973), Rybicki and Lightman (1979), Siegel and Howell (1981), Mihalas and Mihalas (1984), and Kalkofen

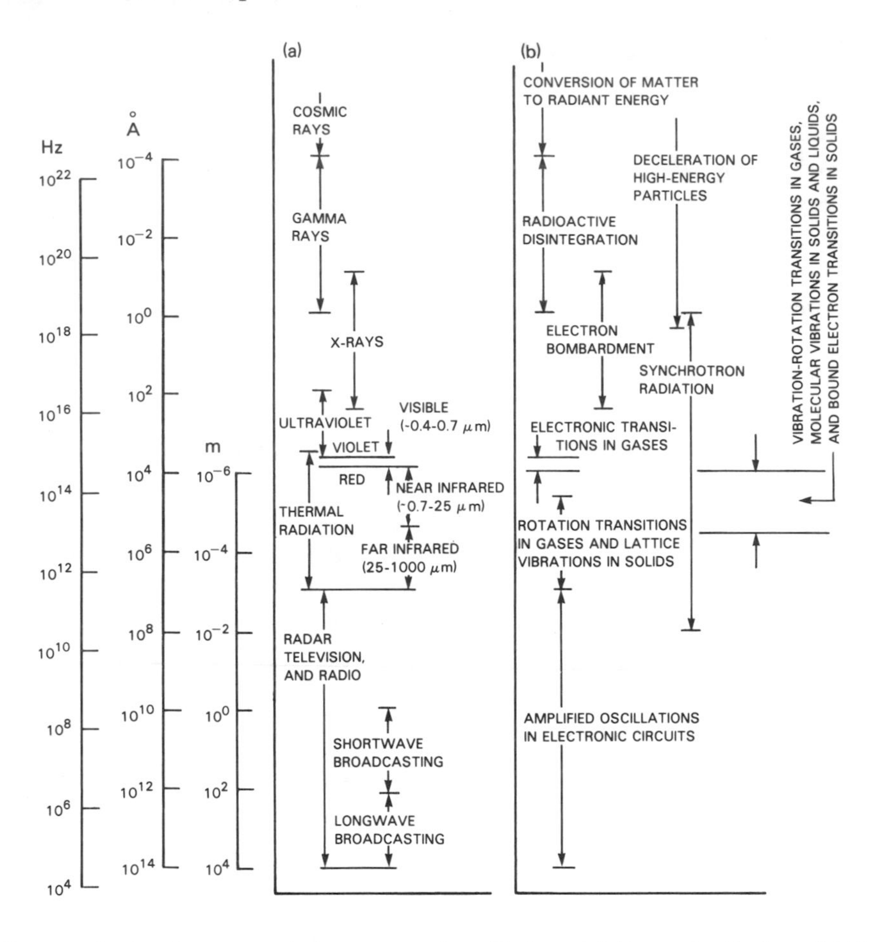

Figure 12–7. Spectrum of electromagnetic radiation. (a) Type of radiation. (b) Production mechanism. (Based on a figure in Siegel and Howell, 1981.)

(1984). This section introduces the material and is a rough guide to simple radiation transport models for reactive flow simulations.

12–2.1 The Physical Basis of Radiation Transport

Thermal radiation is caused by electronic, vibrational, and rotational transitions of the atoms and molecules in gases, by molecular vibration in solids and liquids, bound electron transitions in solids, and by synchrotron radiation. The wavelength regimes of radiation are shown in Figure 12–7.

Radiation from electronic transitions in an atomic system, molecules, ions, or an electron-ion plasma is emitted and absorbed as it is transmitted

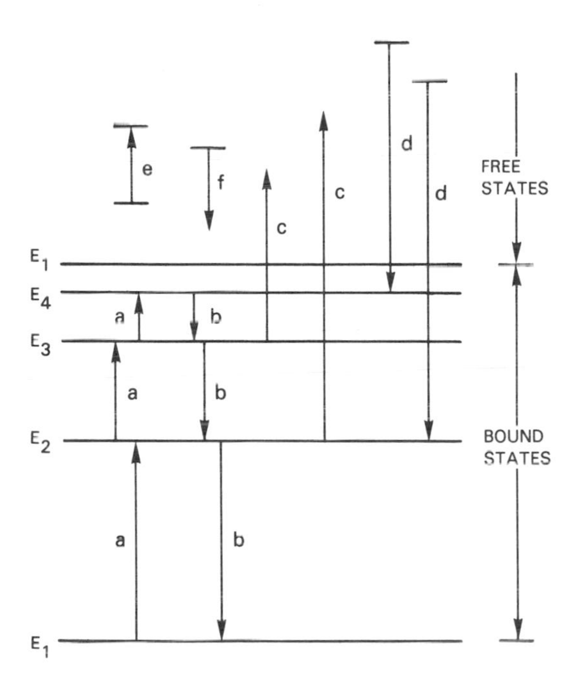

Figure 12–8. Schematic diagram of the energy states and transitions for an atom, ion, or electron. Energy levels are indicated by E_i, i=1,2,...,I, where E_I is the ionization level. a. bound-bound absorption, b. bound-bound emission, c. bound-free absorption, d. bound-free emission, e. free-free absorption, f. free-free emission. (Based on a figure in Siegel and Howell, 1981.)

through a gas. When an atom (or an ion) and a photon interact, the photon can be absorbed, leaving the atom in an excited state. Subsequently, the atom can relax to its original or to another state by emitting another photon. This relaxation may be spontaneous or may be stimulated by the passage of other photons of the same energy. Electronic transitions are generally divided into three types: bound-bound transitions, bound-free transitions, and free-free transitions, as illustrated schematically in Figure 12–8.

In *bound-bound transitions* (b-b), a photon interacting with an atomic system can cause a change from one bound level to another. For example, a photon can be absorbed and a system excited to a higher energy state. A photon can also be emitted, and then the system drops to a lower energy state. Because atomic systems have discrete energy levels, b-b emission produces line spectra. These become band spectra when the system also has vibrational or rotational modes that broaden the transition.

In *bound-free transitions* (b-f), a photon interacting with a bound electron-atom system results in a system which has too much energy for the electron to stay bound, resulting in photoionization. An electron is emitted, and the extra energy becomes kinetic energy of the free electron. The reverse of this process, atomic capture of a free electron, results in the emission of a photon. These kinds of processes produce continuous absorption and emission spectra and the directions of the particles and photons before and after the transition are governed by momentum and energy conservation.

In a *free-free transition* (f-f), a free electron can either emit a photon without losing all of its kinetic energy, and stay free, or it can absorb a photon and its kinetic energy increases. This is called *bremsstrahlung*, and produces continuous emission and absorption spectra.

These basic atomic processes contribute to the absorption and emission of radiation as it passes through a material. The rates of these atomic radiation processes are used in much the same way that chemical reaction rates are used to describe chemical kinetics. The radiation rate data, the emission, absorption, and scattering coefficients, constitute the fundamental input determining emission and absorption spectra, and are input into the equation of radiation transport given below. Determining these input quantities is a large part of the difficulty in solving radiation transport problems.

We are often interested in reactive systems in *local thermal equilibrium*. This means that the radiation field can be described in terms of local state variables, such as composition, temperature, and pressure. The photons are absorbed and reemitted several times as they cross a temperature or density scale length of the radiating fluid. When the radiation is in thermal equilibrium, it is called *thermal radiation*.

12–2.2 The Equation of Radiative Transfer

The fundamental quantity of radiation transport is the *spectral intensity* I_λ, the radiant energy passing through a surface, per unit time, per unit wavelength interval about wavelength λ, per unit surface area normal to the $\Omega = (\theta, \varphi)$ direction, into a unit solid angle centered around Ω. (For example, in cgs units, the spectral intensity has units of erg s^{-1} cm^{-3} $ster^{-1}$.) This is illustrated in Figure 12–9. The spectral intensity depends on material properties of the system, such as temperature, pressure, and the number densities of various species. Below we refer to this quantity as, $I_\lambda(\Omega)$, or simply I_λ, where the subscript λ indicates the wavelength dependence. The explicit dependences on the local properties of the medium are dropped in the discussion below to simplify the presentation.

The *total intensity* in direction Ω is defined as

$$I(\Omega) \equiv \int_0^\infty I_\lambda(\Omega)\, d\lambda \,. \qquad (12-2.1)$$

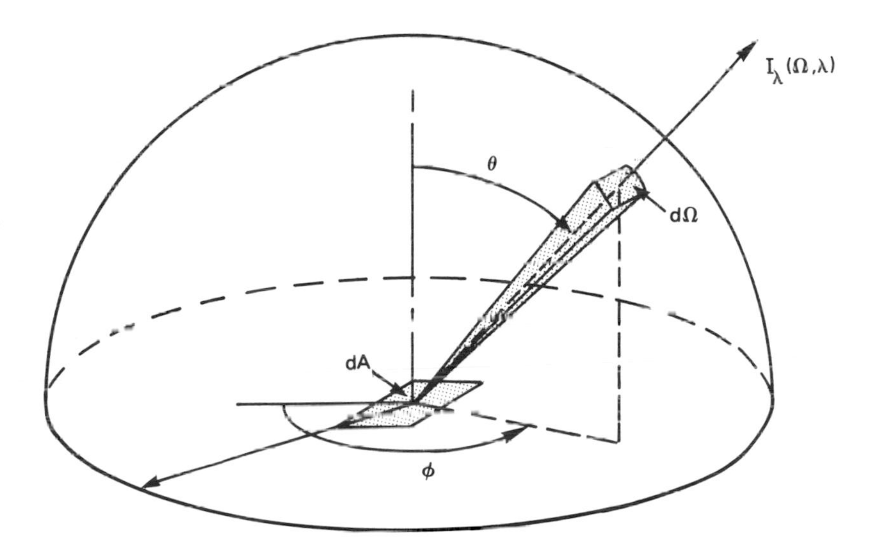

Figure 12–9. Geometry used to define the spectral intensity, I_λ. (Based on a figure in Siegel and Howell, 1981.)

A direction-averaged *mean intensity* is defined as

$$\bar{I}(\lambda) = \frac{1}{4\pi}\int_0^{4\pi} I_\lambda(\Omega)\, d\Omega \,. \qquad (12-2.2)$$

The intensity is a function of the properties of the whole medium globally: how much radiation enters the medium from the walls or outside, how much is absorbed, emitted, or scattered in the medium before it reaches a small volume, and how much is absorbed and emitted in that volume.

The equation of radiative transfer describes the behavior of the spectral intensity I_λ. In general it is a time-dependent conservation equation:

$$\frac{1}{c}\frac{\partial I_\lambda}{\partial t} + \frac{\partial I_\lambda}{\partial s} = W_\lambda \,. \qquad (12-2.3)$$

The independent variable s here measures distance along a ray in direction Ω. The source term, W_λ, has four contributions,

$$W_\lambda \equiv W_{\text{emission}} - W_{\text{absorption}} + W_{\text{scattering in}} - W_{\text{scattering out}} \,. \qquad (12-2.4)$$

We generally require the steady solution of Eq. (12–2.3) because radiation, with c as the characteristic velocity, relaxes toward equilibrium faster than any other process. The radiative time scale for flow problems is much less than the relevant fluid time scales, except for very high energies. The radiation field adjusts itself quickly to changes in the physical conditions at any position in the medium. Thus it is reasonable to ignore explicit time variations in $I_\lambda(\Omega)$ and treat the radiation field as a sequence of quasi-steady states.

The source terms W_λ include:

1. The decrease in I_λ due to absorption of photons along the *ray* parameterized by s, is given by the total absorption,

$$W_{\text{absorption}} = \alpha_\lambda(s) I_\lambda(s) , \qquad (12-2.5)$$

where $\alpha_\lambda(s)$ is the absorption coefficient (units of cm^{-1}). As written in Eq. (12–2.5), $\alpha_\lambda(s)$ contains both absorption and induced emission, each proportional to the spectral intensity in the ray. The total absorption may be positive or negative. When induced emission dominates, α_λ is a gain coefficient describing the potentially exponential growth of the spectral intensity along the ray.

2. The increase in I_λ due to spontaneous emission contributed by the medium along the path is

$$W_{\text{emission}} = \alpha_\lambda(s) I_{\lambda b}(s). \qquad (12-2.6)$$

We have assumed that the radiation is in local thermodynamic equilibrium in Eq. (12–2.6), so that $I_{\lambda b}$ is the Planck function,

$$I_{\lambda b}(T) \equiv \frac{2hc^2}{\lambda^5 \left(\exp(hc/k_b \lambda T) - 1\right)} , \qquad (12-2.7)$$

where h is Planck's constant. The function $I_{\lambda b}$ describes the spectral intensity of *blackbody radiation*, the intensity found in an enclosed vacuum whose walls are at a uniform temperature. The Stephan-Boltzmann law says that the blackbody intensity integrated over all wavelengths is proportional to T^4, that is,

$$I_b(\Omega) = \int_0^\infty I_{\lambda b}(\Omega) d\lambda = \frac{\sigma}{\pi} T^4 , \qquad (12-2.8)$$

where $\sigma \equiv 5.67 \times 10^{-5}$ erg cm^{-2} K^{-4} s^{-1} is the Stephan-Boltzmann constant.

3. The attenuation by scattering out of the ray is

$$W_{\text{scattering out}} = \sigma_\lambda(s) I_\lambda(s) , \qquad (12-2.9)$$

where $\sigma_\lambda(s)$ is the scattering coefficient (units of cm^{-1}). The radiation scattered out of a volume is proportional to the radiation present, similar to absorption. The scattering coefficient, $\sigma_\lambda(T(s), P(s))$, is the reciprocal of the mean free path of a photon of wavelength λ along the path s. The scattering coefficient describes elastic and inelastic scattering and may be isotropic or anisotropic in the directions into which it scatters.

4. The gain of intensity by radiation scattered into the ray along s is

$$W_{\text{scattering in}} = \frac{\sigma_\lambda}{4\pi} \int I_\lambda(s, \Omega_i)\phi(\lambda, \Omega, \Omega_i)\, d\Omega_i \,, \qquad (12-2.10)$$

where the function ϕ is the *phase function*, defined as the scattered intensity in a direction divided by the intensity for isotropic scattering.

The total equation of radiative transfer in the quasi-static limit is

$$\begin{aligned} \frac{dI_\lambda}{ds} = & - a_\lambda I_\lambda(s) - \sigma_\lambda I_\lambda(s) + a_\lambda I_{\lambda b}(s) \\ & + \frac{\sigma_\lambda}{4\pi} \int_0^{4\pi} I_\lambda(s, \Omega_i)\phi(\lambda, \Omega, \Omega_i)\, d\Omega_i \,, \end{aligned} \qquad (12-2.11)$$

a first-order integro-differential equation. This equation shows that the radiation along a path s is attenuated by absorption and scattering, and is enhanced by spontaneous emission, induced emission (included in the net absorption coefficient α_λ), and radiation scattering in from other directions.

Both σ_λ and α_λ are functions of wavelength, the chemical species present, the temperature, and the pressure, but not in general the photon direction. Together, they define the extinction coefficient, $K_\lambda(s)$, where

$$K_\lambda \equiv \alpha_\lambda + \sigma_\lambda . \qquad (12-2.12)$$

The *optical thickness* of a slab of material, $\tau_\lambda(s)$, is a measure of the ability of a given path length s to attenuate radiation,

$$\tau_\lambda(s) \equiv \int_0^s K_\lambda(s')\, ds'. \qquad (12-2.13)$$

Usually the optical thickness is measured relative to characteristic scale lengths, typically gradient lengths, in the medium. When $\tau_\lambda(s) \gg 1$, the material is *optically thick*. The mean penetration distance of each photon is small compared to the characteristic dimensions and scale lengths in the medium. When $\tau_\lambda(s) \ll 1$, the material is *optically thin*. The *opacity* of the medium, a local function of composition, temperature, pressure, and wavelength, is defined by

$$\kappa_\nu \equiv \frac{a_\lambda(s)}{\rho(s)} \qquad (12-2.14)$$

(units of $cm^2\ g^{-1}$).

12–2.3 The Radiant Energy Flux

In Chapter 2 we first introduced the radiant energy flux $\mathbf{q}_r$,

$$\mathbf{q}_r = (q_{r,x}, q_{r,y}, q_{r,z}), \qquad (12-2.15)$$

and it appears again in Eq. (7–0.3). This is the term that couples the radiation transport to the other chemical and physical processes in the conservation equations.

The radiant energy flux crossing an area element dA is a result of the intensities from all directions,

$$\mathbf{q}_r(s) \equiv \int_0^{4\pi} I_\lambda(\Omega, s) \cos\theta \, d\Omega , \qquad (12-2.16)$$

where θ is the angle between each ray direction Ω and the surface normal to dA. Using

$$\begin{aligned} \frac{dI_\lambda}{ds} &= \frac{\partial I_\lambda}{\partial x}\frac{dx}{ds} + \frac{\partial I_\lambda}{\partial y}\frac{dy}{ds} + \frac{\partial I_\lambda}{\partial z}\frac{dz}{ds} \\ &= \frac{\partial I_\lambda}{\partial x}\alpha_x + \frac{\partial I_\lambda}{\partial y}\alpha_y + \frac{\partial I_\lambda}{\partial z}\alpha_z , \end{aligned} \qquad (12-2.17)$$

where α_x, α_y, and α_z, are the direction cosines of Ω,

$$\begin{aligned} \nabla \cdot \mathbf{q}_r = 4\pi \int_0^\infty \Big(& a_\lambda(s) I_{\lambda b} - [a_\lambda(s) + \sigma_\lambda(s)] \overline{I}_\lambda(\lambda) \\ & + \frac{\sigma_\lambda(s)}{4\pi} \int_0^{4\pi} I_\lambda(s, \Omega_i) \, \overline{\phi}(\lambda, \Omega_i) \, d\Omega_i \Big) d\lambda . \end{aligned} \qquad (12-2.18)$$

Here

$$\overline{\phi}(\lambda, \Omega) = \frac{\sigma_\lambda(\Omega)}{\langle\sigma\rangle} \qquad (12-2.19)$$

and

$$\langle\sigma\rangle \equiv \frac{1}{4\pi} \int_0^{4\pi} \sigma_\lambda(\Omega) d\Omega , \qquad (12-2.20)$$

where $\overline{I}_\lambda$ is the mean intensity defined above in Eq. (12–2.2).

Equations (12–2.11) and (12–2.18) look straightforward, but there are two major types of difficulties in solving them. The first is determining the absorption and scattering coefficients, which require a detailed knowledge of the atomic and molecular properties of a medium. The second is the enormous amount of computation required to solve the equations numerically, because each direction must be considered separately. Even assuming local thermodynamic equilibrium, there is also the problem of solving for each I_λ, which couple to the equations for energy and temperature.

12–2.4 Some Important Limiting Cases

First consider the integral of dI_λ/ds from Eq. (12–2.11) to determine $I_\lambda(s)$,

$$\begin{aligned} I_\lambda(s) &= I_\lambda(0)e^{-\int_0^s K_\lambda ds'} \\ &\quad + \int_0^s K_\lambda I_\lambda(s')e^{-\int_{s'}^s K_\lambda ds''} ds' \, , \end{aligned} \qquad (12-2.21)$$

where K_λ, defined in Eq. (12–2.12), can vary along the path. By comparing this to Eq. (12–2.11), we see that $I_\lambda(s)$ depends on the Planck function $I_{\lambda b}$, that is, it depends on the local temperature and on the local scattered radiation. The energy equation must be solved to find the temperature. The radiant energy flux, $\mathbf{q}_r$, used in the energy equation, involves an integral of I_λ over all values of Ω and λ.

Any practical computational approach must avoid solving the full radiative transport equations coupled to the energy equations. Instead, it is necessary to neglect terms, consider the limiting cases of thick or thin radiation, and use an assortment of tricks and approximations to simplify the equations. First consider some of the limiting cases.

The Transparent Gas Approximation

When the opacity is small, the medium is *optically thin*. The exponential attenuation terms in the equation of transfer, Eq. (12–2.11), approach unity and

$$I_\lambda(s) = I_\lambda(0) + \int_0^s K_\lambda I_\lambda(s')ds' \, . \qquad (12-2.22)$$

This equation says that the attentuation along the path s in the medium is very small. Radiation enters at $s = 0$ and keeps essentially the same intensity.

If the extinction coefficient K_λ is large, the optical depth is small, the integral in Eq. (12–2.22) may become small compared to $I_\lambda(0)$. Then

$$I_\lambda(s) = I_\lambda(0) \, , \qquad (12-2.23)$$

which is the *strongly transparent approximation*. The incident intensity is not changed at all as the radiation travels through the medium. A local energy balance is appropriate and much easier to solve than the full transport equations.

The Emission Approximation

In this approximation the medium is optically thin, and not much radiation is entering from the boundaries. Therefore, the spontaneously emitted energy from each small volume of the gas passes through the system without much attenuation. Thus

$$I_\lambda(s) = I_\lambda(0) + \int_0^s a_\lambda(s') I_{\lambda b}(s') ds' . \qquad (12-2.24)$$

Cold Medium with Small Scattering

The radiation emitted and scattered by the medium into the ray along s is negligible compared to that incident from the boundaries or external sources. The local intensity is then the attenuated incident intensity,

$$I_\lambda(s) = I_\lambda(0) e^{-\int_0^s K_\lambda(s' ds')} . \qquad (12-2.25)$$

This limit is important in combustion problems when a cold unignited gas mixture with significant absorbers such as smoke particles is adjacent to a region of strong emission.

12–2.5 More General Approaches

There are other approaches that are somewhat more general, representing more fundamental models. We briefly describe these here before explaining the useful diffusion and variable Eddington approximations.

In the differential approximation, the equations of radiation transfer are approximated by a heirarchy of moment equations. Generally just the first three moments are used, and these have specific physical significance, reminiscent of the equations of fluid dynamics. The first moment represents the radiation energy density, the second moment is the radiative energy flux, and the third moment is the radition stress tensor. One common way of closing these equations reduces to the variable Eddington approximation, discussed in greater detail below.

Approaches to radiation transport using Monte Carlo methods have been developed extensively. Other approaches include replacing the exponential integrals by approximate exponentials, escape probability methods, and the S_n method. We recommend the standard textbooks such as Mihalas and Mihalas (1984) and Siegel and Howell (1981) for in-depth descriptions of these methods.

The work involved in a direct, brute-force solution of the multidimensional equations of radiative transfer is straightforward, but horrendous. First, there is the problem of determining the appropriate coefficients for the various emission and absorption processes. These are usually a function of wavelength, as well as of the local properties of the material. Sometimes they are available in tabular form, but often they must be extrapolated or calculated from very basic considerations. The spectrum is usually divided into wavelength bands and the equations are recast into a collection of equations, each extending over these wavelength bands of width $\Delta\lambda$. Furthermore, space is divided into a finite number of directions so that photons traveling in different directions are treated separately. The general result is a very complex set of coupled finite-difference equations that nominally can be solved by standard methods described in previous chapters of this book. Implicit methods are usually best because of the nonlocal coupling involved.

Because it is not generally practical to try to solve the full set of equations of radiative transfer, some usable approximations are essential. As we have shown, different approximations are valid in different limiting physical conditions. The two most common approximations to radiation transport are the diffusion approximation and the transparent gas approximation because they are relatively easy to compute. In both cases, the radiation intensity is a local instead of a global property of the system.

In many systems the preferential absorption of some frequencies means photons from other spectral lines have longer mean free paths and hence a better chance to escape. In other systems, radiation cooling leaves certain atoms at a different effective temperature than others. In these more complex situations, different parts of the spectrum have to be treated differently. If these problems are to be simulated correctly, some form of multidirectional, multigroup treatment becomes necessary.

Diffusion approximations are the easiest models to use, but require at least some approximate local isotropy of the radiation field. The Eddington approximation uses a diffusion model to consider small, linear anisotropies and two directions of integration. Even the need for separate treatment of distinct wavelength bands in the radiation field is not so impossible computationally if it is valid to assume isotropy in each of the bands.

12–2.6 The Radiation Diffusion Approximation

Like many other physical phenomena, the combination of short photon excursions followed by repeated absorption and reemission results in a random-walk behavior whose macroscopic representation is a diffusion operator to lowest-order. The diffusion approximation is based on the assumption that the optical depth is large enough and the temperature gradients are small enough that the local value of the spectral intensity is a function only of local emission. This leads to a simplified but very useful approximation to the full set of equations.

The primary result was first derived by Rosseland (1936),

$$\mathbf{q}_r = -\frac{4\sigma}{3\alpha_R}\nabla T^4 , \qquad (12-2.26)$$

which comes from relating the energy flux to the local material temperature gradient. This expression gives the diffusive flux of radiation averaged over all wavelengths assuming a local blackbody distribution.

The quantity α_R (cm^{-1}) is the *Rosseland mean absorption coefficient*, an average over the absorption and scattering coefficients α_λ and σ_λ weighted by the blackbody distribution function $I_{\lambda b}(T)$,

$$\alpha_R \equiv \frac{\displaystyle\int_0^\infty \frac{\partial I_{\lambda b}(T)}{\partial T} d\lambda}{\displaystyle\int_0^\infty \frac{\partial I_{\lambda b}(T)}{\partial T} \frac{d\lambda}{\alpha_\lambda + \sigma_\lambda}} , \qquad (12-2.27)$$

where σ is the Stephan-Boltzmann constant defined above. The mean free path of photons corresponding to α_R is the Rosseland mean free path,

$$\lambda_R \equiv \frac{1}{\alpha_R} . \qquad (12-2.28)$$

Once $\mathbf{q}_r$ is determined, the energy conservation equation can be solved by standard numerical methods. The rate of change of the material energy density is given by $-\nabla \cdot \mathbf{q}_r$.

There are many variations of the diffusion approximation, descriptions of which can be found in Siegel and Howell (1981). For example, the Schuster-Schwarzschild approximation (Schuster, 1905; Schwarzschild, 1906) and the Milne-Eddington approximation (Milne, 1930; Eddington, 1959) are both diffusionlike approximations commonly used in astrophysical applications. Both of these approximations assume that there are two directions of propagation and that the intensity in one direction may have a different value than the intensity in another direction. They differ, however, in exactly when in the theory they make the two-direction assumption.

12–2.7 The Variable Eddington Approximation

Consider a situation in which the local radiation temperature T_r deviates significantly from the local material temperature T, but the transfer of energy between the material and the radiation is relatively slow. In this case, each temperature distribution can be essentially Maxwellian but at a different temperature. However, there is always some coupling that tends to equalize the temperatures. In this case, a computationally useful model results by treating the radiation field in terms of a single energy density E_r related to T_r by

$$T_r \equiv \left(\frac{cE_r}{4\sigma}\right)^{1/4} . \qquad (12-2.29)$$

Here σ is the Stephan-Boltzmann constant and c is the speed of light. (The cgs units of E_r are ergs cm^{-3}.) The total energy in the system still has to be conserved, but the fluid energy $E(\mathbf{x},t)$ and the radiation energy E_r interchange locally.

The radiation energy density satisfies a scalar transport equation

$$\frac{1}{c}\frac{\partial E_r}{\partial t} = \nabla \cdot \lambda_R \nabla (f_E E_r) - K_p(E_r - E_{bb}) , \qquad (12-2.30)$$

where scattering is treated in a gradient diffusion approximation using the Rosseland mean free path λ_R. The absorption and emission is governed by the Planck averaged opacity K_p (cm^{-1}) through the second term on the right hand side. The quantity f_E is the dimensionless variable Eddington factor discussed below. The opacity depends on the composition and temperature of the fluid and generally on the wavelength of the photons. Here K_p is the inverse of the mean free path for an optically thin medium, and is given by

$$K_p \equiv \frac{\int_0^\infty K_\lambda I_{\lambda b}(T) d\lambda}{\int_0^\infty I_{\lambda b}(T) d\lambda} . \qquad (12-2.31)$$

In analogy with Eq. (12–2.29), the blackbody energy density appearing in Eq. (12–2.30) is defined as

$$E_{bb} \equiv 4\frac{\sigma T^4}{c} . \qquad (12-2.32)$$

When the actual radiation energy density locally exceeds the blackbody radiation density, E_{bb}, evaluated at the fluid temperature T, the radiation

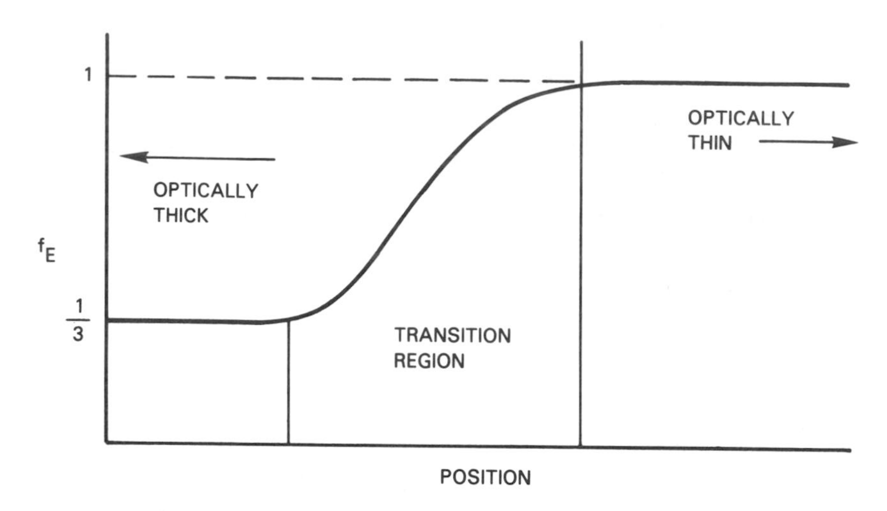

Figure 12–10. Variation of the variable Eddington factor, f_E, from an optically thick to an optically thin region.

energy density decreases in time because the fluid absorbs radiation faster than it is being emitted.

The left hand side of Eq. (12–2.30) is the rate of change of the energy density of the radiation, generally a very small quantity relative to the energy density of the fluid, and generally negligible in Eq. (12–2.30) because the velocity of light is large compared to other speeds in the problem. The variable Eddington factor, f_E, ranges between one for optically thin media and one third for optically thick media, as shown schematically in Figure 12–10. This variable factor gives the model its ability to approximate the local radiation flux in regions between optically thick and optically thin. Generally the variable Eddington factor must be computed from a detailed solution of the equations of radiative transfer. However, for computational simplicity, it is often approximated by a formula determined from the geometry of the problem and the ratio of the radiation flux to the radiation energy density. One example is

$$f_E = \frac{L/3 + \lambda_R}{L + \lambda_R}, \qquad (12-2.33)$$

where L is the radiation field scale length determined from

$$L^{-1} \approx \frac{f_E |\nabla E_r|}{E_r}. \qquad (12-2.34)$$

Since f_E enters the definition of L, a quadratic equation has to be solved to unfold Eqs. (12–2.33) and (12–2.34). This is the ad hoc part of the approximation. The variable Eddington factor, f_E, could also be taken inside the gradient term in Eq. (12–2.34), as used in Eq. (12–2.30). When f_E is taken inside the gradient, the radiation field scale length has to be determined using the values of f_E and ∇E_r from the previous timestep.

When the radiation field has relaxed to a steady state, the diffusionlike term in Eq. (12–2.30) balances the radiation source term and the left hand side of Eq. (12–2.30) vanishes. The radiation field is tightly coupled to the boundary conditions and to the spatial variation of the fluid temperature field through the blackbody energy density. This elliptic equation can be solved by standard techniques as discussed in Chapter 11, although it is expensive in multidimensions.

The fluid dynamic energy density satisfies a similar equation,

$$\frac{\partial E}{\partial t} + \nabla \cdot E\mathbf{v} + \nabla \cdot P\mathbf{v} - \nabla \cdot \lambda_m \nabla T \;=\; cK_p(E_r - E_{bb}) \;, \qquad (12-2.35)$$

where λ_m is the mixture thermal conduction coefficient described in Chapter 2. The velocity of light now occurs in the numerator of the emission and absorption term. This means that even though the radiation energy density is much smaller than the fluid dynamic energy density, the fluid energy can be changed appreciably by the radiation because the speed of light is large.

This coupled system of two energy equations conserves global energy, as we can see by multiplying Eq. (12–2.30) by c and adding it to Eq. (12–2.35) Also, the variation of fluid dynamic energy density, and hence temperature, feeds back into Eq. (12–2.30) through the temperature variation of the blackbody energy density $E_{bb}(T)$ and the Rosseland mean $\lambda_R(T)$, and therefore the system can be stiff. This potential stiffness of the coupled equations complicates their numerical solution, although the techniques described in previous chapters are adequate, at least in principle, to deal with this.

The variable Eddington approximation reduces to the diffusion model when E_{bb} and E_r are nearly equal. In this case the elliptic term in Eq. (12–2.30) can be written with $E_{bb}(T)$ in place of E_r. When the result is substituted for the right hand side of Eq. (12–2.35), we find

$$\frac{\partial E}{\partial t} + \nabla \cdot E\mathbf{v} + \nabla \cdot P\mathbf{v} - \nabla \cdot \lambda_m \nabla T \;=\; \nabla \cdot \lambda_R \nabla \left(\frac{4}{3}\sigma T^4\right) \;. \qquad (12-2.36)$$

By comparing Eq. (12–2.36) with the Rosseland diffusion flux given in Eq. (12–2.26), we can see that the variable Eddington model reduces to the diffusion model in the optically thick limit, $f_E = 1/3$.

In the optically thin limit, it is sometimes necessary to put a limit on the radiation flux. The limit ensures that the amount of radiation flux is bounded by its free-streaming value, cE_r. Physically the flux cannot exceed this value, but it can in a diffusion model, as discussed in Chapter 7. The breakdown in this case comes about because the system is in a regime in which the diffusion approximation is not valid.

The computational problem becomes more severe, however, when the spectral intensity varies markedly with direction. This occurs, for example, where a medium goes from optically thick to optically thin. At the boundary of a dense radiating region, the radiation distribution changes from isotropic to essentially unidirectional streaming away from the source. To describe this phenomena accurately is currently a challenge. Very often the variable Eddington method is fully consistent with the level of approximation of the problem. When it is not, the geometric complexity again requires a full nonlocal treatment of the radiation field. There is really no completely satisfying yet practical way to solve the full nonlocal radiation transport problem.

12–3. MULTIPHASE FLOWS

Multiphase flows occur in a wide range of problems, including the behavior of suspended grain dust or coal dust, droplets and sprays, propellant burning, charring, soot, and smoke formation, slurries, bubbles in liquids, rain, and sedimentation. There are distinguishing characteristics in all of these problems which keep any particular multiphase modeling approach or approximation from being generally applicable. The result is that many disjoint modeling communities have formed that use their own specific formulations and approximations.

Treating multiphase systems computationally with anything resembling the detail with which we treat gas phase systems is an essentially impossible job. As a result, many levels and types of computational models have been explored to simplify the multiphase problem. One approach is to simplify detailed equations based on well understood assumptions about the system. Just writing down a completely consistent set of equations, however, is very difficult. Another approach is to graft important physical properties of the system onto simpler fluid dynamic models in an ad hoc manner using approximate phenomenologies. This approach can miss important interaction terms. In general, the decision about what to treat from first principles and what to represent through a simplifying phenomenology is often not obvious.

There are a number of excellent reviews of the basic equations and approximations used in multiphase flows. Ishii (1975) and Soo (1967, 1984) have organized the evolving knowledge about multiphase flows and presented a set of multicomponent fluid equations. Faeth (1983) has written a comprehensive review of the state of understanding and modeling sprays and related multiphase media. We also recommend the overview of multiphase flow problems given in the text by Kuo (1986). Smoot (1984) has described modeling coal combustion; Khalil (1982) described models of furnaces and combustors; and Ohlemiller (1985) described models of smoldering combustion. The proceedings of a NATO conference on multiphase systems edited by Alper (1983) discusses many practical problem areas. The article by Stewart and Wendroff (1984) describes two-phase flow with a discussion of numerical methods.

The material presented in this section is *not* a complete review of the modeling of multiphase flows. The subject is simply too diverse and too complicated, and should be the subject of an entire book. Instead, we present an overview with a classification and discussion of models showing how the methods described in the previous chapters can be used for multiphase flows. The multiphase reactive flow equations are composed of terms that resemble those described earlier. However, the equations now contain new coupling terms between the phases and new phenomenologies added to describe specifically multiphase processes.

12–3.1 Interactions among Phases

There are four major types of differences that occur in the properties of coexisting phases:

1. *Chemical Differences.* Because chemical reactions between different phases change the relative numbers of molecules of each phase at each point and time, it is often necessary to treat the phases and species individually. The complexity introduced by including multispecies equations can be enormous. We have dealt with this problem already for multicomponent gas phase flows where the coupling is written in terms of a balance of production and loss terms.
2. *Thermal Differences.* The different phases can have different temperatures even though they may occupy the same macroscopic fluid element. This occurs because velocity equilibration can be much faster than temperature equilibration. If each phase or component must be described by a different temperature, the number of equations to be solved increases. When there is a temperature difference among the phases, coupling terms are needed to describe energy transfer. Sometimes solving the temperature coupling equations is difficult. These terms are generally proportional to $(T_i - T_j)$, where the indices i and j refer to different phases or components of the composite medium.

3. *Dynamic Differences.* Sometimes it is necessary to describe each phase by its own average velocity field. This introduces new coupling terms proportional to $(\mathbf{v}_i - \mathbf{v}_j)$. This, again, is a problem addressed to some extent when we considered different species velocities in Chapters 2 and 7. There can also be phenomena arising from differential accelerations on drops or particles in a fluid. The *effective mass* of the particle appears larger than the actual mass because the background fluid must be displaced when the particle or droplet accelerates through it.
4. *Spatial Separations.* Sometimes we need to consider the spatial extent of the different phases. With different phases in the same region, we cannot always assume that they completely interpentrate and occupy the same volume. This was not a problem in multicomponent gases, which interpenetrate fully. When the spatial separation between phases is not completely resolved in the model, it might be necessary to consider what fraction of the volume is occupied by each material. A much more fundamental but less general approach is to resolve the spatial separation and consider the material interfaces, as discussed in Chapter 11.

To formulate a multiphase flow problem, it is necessary to consider the type and extent of interactions among the phases. It is helpful to know ahead of time whether dynamic equilibrium and temperature equilibrium are valid assumptions. To help determine this, evaluate the level of *collisionality* between the phases. When the flow is *fully collisional*, all the phases move together and a single macroscopic velocity field is adequate to describe the motions. This simplification, called a single-fluid model, is used frequently even when it is not strictly valid because treating only one flow field greatly reduces the computational cost. The next level of approximation considers small differential velocities of the individual phases about the mean flow. The differential velocities allow more complete treatment of the interpenetration without the need to treat a number of separate convection velocities.

The use of local thermodynamic equilibrium for each of the collisionless interpenetrating components is attractive but may not be justified. In collisionless or nearly collisionless systems, using a continuum momentum model is open to some question. In nearly collisionless situations, the approximation made is called a *multifluid approximation*. Each phase has its own velocity field. The fact that there are no collisions also means that there is seldom local thermodynamic equilibrium. Particle models as opposed to fluid models are often necessary to represent the full complexity of a truly collisionless medium.

Difficulties invariably arise in trying to deal with situations between extremes. The interaction between grains of dust and a surrounding gas could be appreciable and the motions of both materials could be strongly correlated. However, full coupling cannot generally be assumed. When the coupling is appreciable, the equations describing momentum conservation of each phase are stiff in the mathematical sense. The time constants for decay of the differential velocities may be much shorter than the timesteps, and yet the velocity differences can be important.

Single-Fluid Models

At the crudest level, the different phases are considered a single fluid whose properties are composites of those of the various phases present. All phases of the flow are assumed to be closely coupled and thus move with a single velocity. Flows containing very small droplets or particles can often be treated this way. Such a model improperly describes rain, however, because the relative motion between air and water droplets is important. Such a model might be used to describe a detonation propagating through a homogeneous mixture of materials.

When phases are in thermal and dynamic equilibrium, a single temperature and a single-fluid velocity are adequate. The separate phases behave rather like separate chemical species. The assumption is made that the exact locations of individual particles in a macroscopic volume are not important. Averaging this information out leaves a locally homogeneous, low-resolution model.

Increasing the complexity of the problem to include temperature differences between the phases is often an adequate generalization. Two-temperature and multitemperature models are more complicated than single-temperature models, but they cover a wide range of conditions in multiphase flows. This generalization leads to coupling terms between the two energy equations of the form $(T_1 - T_2)$.

Two-Fluid Models

The simplest extensions of the single-fluid model assume that the different phases move at the same mean velocity. The difference in velocity between the phases is often small enough that the different phases can be assumed to travel at the same velocity as far as bulk convection is concerned. However, the difference may be enough to be a significant source term in differential compression or expansion or to describe settling of heavier particles. It also might be a significant perturbation to the momentum in an accelerating flow in which the different phases do not equilibrate momentum as fast as the background flow properties change.

In solving the convection equations for the various phases, we can use a difference between the momentum equations for each phase to describe the different motions. For example, including two momentum equations might allow us to calculate the way gravity pulls sand out of a sandstorm. A common assumption is that all material of a given phase behaves in a similar way. For example, all of the particles in the solid phase behave as one type or size of particle. However, in the presence of a gravitational field, there is always a difference in velocity as the heavier particles do not tend to settle out of the gas or liquid at the terminal velocity. If the fluid is moving upward, the particles won't move upward quite as quickly as the surrounding gas. In a swirling flow, the heavier phases migrate to the outside.

The equations describing this kind of flow are usually cast in terms of the drift velocities of the constituents relative to the mean flow. Often the velocity differences need to be estimated only crudely. Fully self-consistent calculations of two-fluid models have to be used in cases where the difference in velocity can be comparable to the velocity itself.

The momentum coupling terms that appear in the momentum equations also have corresponding terms that must appear in the energy equations. When temperature differences are allowed, the partitioning of energy between the phases has to be determined. As with separate chemical species, all possible transport terms in all the equations interact, generating additional unknown quantities for which calibrated phenomenologies have to be developed.

Multifluid Models

In multiphase flows, such as dust in air and coal fragments in water or air, particles of different sizes travel at different velocities and react faster or slower to changes in the flow. Each volume of fluid is characterized by a distribution of particle or droplet sizes and characteristics. Models for multifluid, multiphase flows exist in the sense that it is possible to write down a set of equations to describe the system. In practice, solving these equations is computationally expensive and more difficult than solving the most complicated single- and two-fluid models.

With a broad range of particle sizes, many different characteristic behaviors are all possible at the same time. The smaller particles tend to acquire the fluid velocity very quickly and the larger particles accelerate more slowly. Thus the relative concentration of large and small particles changes in time whenever the background flow turns or accelerates.

Sometimes it is attractive to treat a random selection of the typical Lagrangian particles or droplets in a continuum background flow to represent the effects of the complete distribution. These hybrid models contain more physical effects than fully continuum approximations, but become very expensive because it is difficult to model enough particles to have statistically smooth, essentially continuous interactions with the fluid equations.

12–3.2 Equations for Multiphase Flows

Multiphase flow is intrinsically a nonequilibrium process. Generally we assume that each phase is in local thermodynamic equilibrium, but that the different phases are not necessarily in equilibrium with each other. We then need to specify the rates of transfer of mass, momentum, and energy among the phases in order to close the set of equations.

Sets of multiphase flow equations have been derived from continuum mechanics constraints on conservation of mass, momentum, and energy. A major point of continuing confusion is the lack of a unique set of equations and assumptions to fit every situation. We now describe several mathematical models in order to illustrate the general form and show some of the complexities.

Equations of Stewart and Wendroff

We begin with a generalization of the two-fluid equations presented by Stewart and Wendroff (1984). A set of fluid equations can be written for each phase, labeled here by i,

$$\frac{\partial \alpha_i \rho_i}{\partial t} + \nabla \cdot \alpha_i \rho_i \mathbf{v}_i = \Gamma_i \qquad (12-3.1)$$

$$\frac{\partial (\alpha_i \rho_i \mathbf{v}_i)}{\partial t} + \nabla \cdot (\alpha_i \rho_i \mathbf{v}_i \mathbf{v}_i) + \nabla(\alpha_i P_i) = \mathbf{m}_i - \mathbf{f}_i \qquad (12-3.2)$$

$$\frac{\partial \alpha_i E_i}{\partial t} + \nabla \cdot (\alpha_i E_i \mathbf{v}_i) + P_i \frac{\partial \alpha_i}{\partial t} + P_i \nabla \cdot (\alpha_i \mathbf{v}_i) + \nabla \cdot \mathbf{q}_i = L_i \, . \qquad (12-3.3)$$

These equations are very similar to those given for multicomponent gas flows in Chapter 2. The $\{\Gamma_i\}$, $\{\mathbf{m}_i\}$, $\{\mathbf{f}_i\}$, and $\{L_i\}$ are coupling terms among the phases. The $\{\Gamma_i\}$ are mass transfer terms representing the rate of production of phase i. The effect of mass transfer on the momentum equation is given by $\mathbf{m}_i$, which is proportional to Γ_i. The effects of external forces comes into the momentum equation through the term $\{\mathbf{f}_i\}$. For example, frictional drag is described by

$$\mathbf{f}_i = \lambda_{ij}(\mathbf{v}_i - \mathbf{v}_j) \; . \tag{12-3.4}$$

The energy coupling terms, $\{L_i\}$ have the form

$$L_i = c_i\Gamma_i + \sum_j \lambda_{ij}(\mathbf{v}_i - \mathbf{v}_j)^2 + \sum_j \kappa_{ij}(T_i - T_j) \; , \tag{12-3.5}$$

where λ_{ij} and κ_{ij} are coupling coefficients.

Besides the new coupling terms, there is a new variable here, α_i. The physical meaning of this quantity and any equations describing its behavior depend on the specific problem. Stewart and Wendroff were particularly interested in stratified flows or bubbly flows. For a stratified flow, α_i could be a height above a surface. For a bubbly flow, α_i would be a volume fraction. The mixture quantities are then

$$\rho_m = \sum_i \alpha_i \rho_i \tag{12-3.6}$$

$$\rho_m \mathbf{v}_m = \sum_i \alpha_i \rho_i \mathbf{v}_i \; , \tag{12-3.7}$$

Note that the scalar pressure P is often assumed equal in the various phases, although the vicosities μ_i that would appear in a pressure tensor $\mathbf{P}_i$ may not be. If surface tension is present, the pressures of the different phases are not equal, as indicated by the P_i in Eqs. (12–3.1) – (12–3.3). Pressure equilibrium and conservation constraints help to define the values of various input coefficients, but most of the basic physics of the problem lies in the choice of coupling parameters.

The Formulation by Soo

Soo (1967, 1984) has formulated a mathematical model based on a set of equations of motion for mean quantities that represent the density, momentum, and energy, ρ_m, $\rho_m \mathbf{v}_m$, and E_m, of the mixture, where the subscript m indicates a mixture quantity. These global equations look very similar to the equations for density, momentum, and energy given in Chapter 2,

$$\frac{\partial \rho_m}{\partial t} = -\nabla \cdot \rho_m \mathbf{v}_m \tag{12-3.8}$$

$$\frac{\partial \rho_m \mathbf{v}_m}{\partial t} = -\nabla \cdot (\rho_m \mathbf{v}_m \mathbf{v}_m) - \nabla \cdot \mathbf{P}_m + \mathbf{f}_m \tag{12-3.9}$$

$$\frac{\partial E_m}{\partial t} = -\nabla \cdot E_m \mathbf{v}_m - \nabla \cdot (\mathbf{v}_m \cdot \mathbf{P}_m) - \nabla \cdot \mathbf{q}_m + f_{Em} \; . \tag{12-3.10}$$

Here $\mathbf{f}_m$ and f_{Em} represent the effects of external forces. There are constraints on various variables, such as density, momentum, and specific heats,

$$\rho_m = \sum_i \rho_i \tag{12-3.11}$$

$$\rho_m \mathbf{v}_m = \sum_i \rho_i \mathbf{v}_i \tag{12-3.12}$$

$$\rho_m c_{vm} = \sum_i \rho_i c_{vi} , \tag{12-3.13}$$

that we can use to find a set of equations for $\frac{\partial(\rho_i \mathbf{v}_i)}{\partial t}$ and $\frac{\partial E_i}{\partial t}$, as well as quantities such as μ_m and T_m. The form of these equations is slightly different from that given by Stewart and Wendroff. The momentum equations now contain terms such as $(\mathbf{v}_i - \mathbf{v}_m)$ in addition to $(\mathbf{v}_i - \mathbf{v}_j)$. In its general philosophy, however, this approach is an extension of the multicomponent approach to gases we described in Chapter 2.

Equations for a Gas-Particulate Mixture by Kuo

Kuo (1986) presented a set of two-fluid equations for describing combustion of solid particles burning in a hot, flowing gas confined in a duct. These equations illustrate some of the necessary complications that arise in attempting to describe a multiphase flow system realistically. Kuo derives a set of conservation equations for mass, momentum, and energy by accounting for fluxes of gas and particulates as they move into and out of a volume. The system is assumed to be composed of gas and particulate phases, both of whose behavior can be treated in the fluid approximation. In the equations below we have omitted the interaction of the fluids with the confining walls, included in the more general equations in Kuo (1986) and Kuo et al. (1976). The reader is referred to these works for a more complete model and its applications.

The fractional porosity, ϕ, of the material is defined as

$$\phi \equiv 1 - \frac{n_p m_p}{\rho_p} = \frac{\text{volume of void}}{\text{total volume}} , \tag{12-3.14}$$

where n_p is the number density of the particles, m_p is the mass of each particle, and ρ_p is the mass density of particles. In addition, we define k_s as

$$k_s = S_b n_p = 4\pi r_p^2 \frac{3(1-\phi)}{4\pi r_p^3} = \frac{3(1-\phi)}{r_p} , \tag{12-3.15}$$

related to S_b, the burning surface of the spherical particle whose radius is r_p.

Using these definitions, the one-dimensional fluid equations for the density of the gas and particulates, respectively, are:

$$\frac{\partial(\rho\phi)}{\partial t} + \frac{\partial(\rho\phi v_g)}{\partial x} = k_s \rho_p v_b \qquad (12-3.16a)$$

$$\frac{\partial}{\partial t}\left[\rho_p(1-\phi)\right] + \frac{\partial}{\partial x}\left[\rho_p(1-\phi)v_p\right] = -k_s \rho_p v_b \,. \qquad (12-3.16b)$$

Here v_b is the burning rate, the rate at which a particle shrinks, and v_g is the velocity of the gas.

The equations for momentum conservation for the gas and the particles, respectively, are:

$$\begin{aligned}\frac{\partial(\rho\phi v_g)}{\partial t} + \frac{\partial(\rho\phi v_g^2)}{\partial x} &= -\phi\frac{\partial P}{\partial x} + k_s\rho_p v_b v_p \\ &\quad - k_s D_v + \frac{\partial}{\partial x}(\tau_{xx}\phi)\end{aligned} \qquad (12-3.17a)$$

$$\begin{aligned}\frac{\partial}{\partial t}\left[\rho_p(1-\phi)v_p\right] + \frac{\partial}{\partial x}\left[\rho_p(1-\phi)v_p^2\right] &= \frac{\partial}{\partial x}\left[\tau_p(1-\phi)\right] - k_s\rho_p v_b v_p \\ &\quad + k_s D_t \,.\end{aligned} \qquad (12-3.17b)$$

Here D_v is the drag due to the relative velocity betwen the gas and the particles and $D_t = D_v + D_p$ is the total drag, where D_p is the drag due to the porosity gradient. Also τ_p is the intragranular stress and τ_{xx} is the normal stress in the x-direction. The viscous stress is the pressure tensor minus the scalar pressure.

The equations for energy conservation for the gas and the particles, respectively, are:

$$\begin{aligned}&\frac{\partial}{\partial t}(\rho\phi E) + \frac{\partial}{\partial x}(\rho\phi E v_g) + \frac{\partial}{\partial x}P v_g \phi \\ &\quad = k_s\rho_p v_b\left(h_f + \frac{v_p^2}{2}\right) - k_s\overline{h}_t(T - T_{ps}) \\ &\quad\quad - k_s D_v v_p - P\frac{\partial\phi}{\partial t} \\ &\quad\quad - \frac{\partial}{\partial x}(q\phi) + \frac{\partial}{\partial x}(\tau_{xx} v_g \phi)\end{aligned} \qquad (12-3.18a)$$

$$\frac{\partial T_p}{\partial t} + \frac{\partial(T_p v_p)}{\partial x} = \frac{\alpha_p}{r}\frac{\partial^2(rT_p)}{\partial r^2} \,. \qquad (12-3.18b)$$

Here T_{ps} is the particle surface temperature, T_p is the particle temperature, h_f is the enthalpy of the gas at the flame temperature, h_t is the total heat transfer coefficient between the gas and the particle, q is the heat flux, and α_p is the thermal diffusivity, defined as $\lambda/(\rho c_p)$.

For a dense gas, Kuo suggests using the Noble-Abel equation of state which has the form

$$P\left(\frac{1}{\rho} - b\right) = RT \; . \qquad (12-3.19)$$

The solid particles are assumed to have constant density,

$$\rho_p = \text{constant} \; , \qquad (12-3.20)$$

implying that the particle phase is incompressible, and hence cannot be used to treat problems at high pressures or with shocks.

To complete the problem definition, there is still quite a bit of information that must be supplied. For example, we need to know the constitutive law for the intergranular stress, τ_p, the drag correlation for D_v, and the burning rate law to give v_b. Sometimes these can be obtained from the literature, as sometimes they have to be estimated or guessed. As an example of the application of these equations, we refer to the work by Chen et al. (1981) who applied them to interior ballistics problems.

12–3.3 Spray and Droplet Models

Spray dispersion and combustion in hot, turbulent, oxidizing flows is an example of a complex multiphase flow. Recent reviews have been written by Law (1982), Sirignano (1983), and Faeth (1983).

Faeth (1983) identifies three categories of spray models: *locally homogeneous flow* (LHF) *models*, *separated flow* (SF) *models*, and *drop life-history* (DLH) *models*. In the LHF models, the gas and liquid phases are assumed to be in dynamic and thermodynamic equilibrium, so that the phases have the same velocity and temperature. This limiting case only accurately represents a spray consisting of very small drops. In the SF models, the effects of finite rates of transport between the phases are considered. This is really a very broad category of models, ranging from those in which particulates are treated as Lagrangian particles in a continuum background to the case of full multiphase, multifluid continuum models such as those based on the types of equations shown above. The third type of model, the DLH models, focuses on individual droplets and attempts to calculate their behavior in various environments in some detail. These DLH models, for example, provide input to the phenomenologies in LHF and SF models by either calibrating input terms or by summing over an ensemble of droplets to estimate the average mass, momentum, and energy exchange.

The LHF models differ from the homogeneous gas phase reactive flow equations in Chapter 2 primarily in the interpretation of the composite equations of state and the ordinary differential equations describing reaction kinetics. In addition, there are differences because the equation of state is now a composite of the equations of state of the various materials, and it is not necessarily as simple as in Chapter 2. The phases all overlap in the same macroscopic volume even though they occupy distinct regions with interfaces on the microscopic level. Thus particles and droplets whose volumes are small behave as if they contribute essentially no pressure to the gas in which they are suspended. Particles entrained in a locally homogeneous flow can weigh it down appreciably, however, and contribute significantly to its heat capacity. Further, the local reaction equations describe evaporation and condensation as phenomenologies as well as volumetric and surface-catalyzed chemical reactions.

Faeth identifies several kinds of SF models:

1. *Particle Source in Cell Model* or *Discrete Droplet Model.* In discrete droplet models, a finite number of particle groups are used to represent the entire spray. The motion of these representative parts of the spray are tracked using a Lagrangian formulation to advance the droplet groups. An Eulerian formulation is used to solve the governing equations for the gas phase in which the droplet groups are embedded. The effects of droplets on the gas phase is accounted for by introducing appropriate source and sink terms in the gas phase equations.
2. *Continuous Droplet Model.* Here a distribution function, $f_j(\mathbf{r}, \mathbf{x}, \mathbf{v}, t)$, is used to evaluate the statistical distribution of the droplet temperature, concentration, and so on. The quantity f_j is the number of particles of chemical composition j in the size range $d\mathbf{r}$ about $\mathbf{r}$, in the position range $d\mathbf{x}$ about $\mathbf{x}$, with velocities in the range $d\mathbf{v}$ about $\mathbf{v}$ at time t. The transport equations for $\{f_j\}$ are solved in conjunction with the gas conservation equations to give the properties of the spray. Again, the effects of the droplets on the gas phase are included through source and sink terms in the gas phase equations.
3. *Continuum-Formulation Model.* The motion of both the droplets and the gas are treated as interpenetrating continua. A continuum formulation, as illustrated above, is used to model the problem. The equations for the two phases now look similar even though the unresolved microscopic behavior can be very different.

These models are described in some detail in Faeth (1983) and summarized again in Kuo (1986). Although they provide useful insights and occasionally quantitative estimates, their capabilities for a priori prediction are limited. The reason for this is that the number and complications of the equations in even the conceptually crudest models precludes easy or accurate solution. In the complicated set of equations derived by Kuo and given above, there are a number of essentially unknown input parameters requiring calibrated, time-dependent phenomenologies. Substantial development and experience with spray models is required before such multiphase models become reliable design tools or even relatively easy to use.

Generally it is not possible to simulate simultaneously all the important aspects of the microscopic and macroscopic physics of interest. It is possible to treat either the macroscopic multifluid equations accurately or treat the dynamics of a few particles accurately. One extreme is to treat the background flow in some detail, but treat the particle interaction with the flow phenomenologically. The other extreme is to treat one or two particles in detail in an assumed background flow field. The first case has already been discussed. The second type of simulation, which includes the DLH models, can be very difficult because of the level of resolution required.

The differences inherent in the calculations that use locally homogeneous flow or separated flow models and drop life-history models are related to the problems discussed in Chapter 11 on tracking interfaces. In the case of LHF or SF models, as in the case of tracking an interface, phenomenologies are introduced to represent many of the details of the unresolved interactions. In the case of a DLH model with a resolved interface, the basic purpose of the calculation is to simulate the interface accurately and describe the events at the interface from basic principles.

Recently there have been new computational approaches to drop life histories. In the approach of Patnaik (1986) and Patnaik et al. (1986), and Sirignano (1986), the droplets are treated as receding spheres that are responsible for the transfer of mass due to evaporation, momentum due to frictional drag, and heat through thermal conduction to a background gas. Figure 12–11 shows the results of a typical calculation in which two dodecane droplets are modeled using two overlapping nonorthogonal grids. The background air is hot, 1000 K, and the drops are initially at room temperature. The background air moves from left to right. At the time shown in the figure, fuel vapor from the first drop has reached the second drop. However, because the second drop is outside the wake of the first drop, there is minimal hydrodynamic interaction. The second drop sees modified ambient conditions only. The droplets are treated as a surface interacting

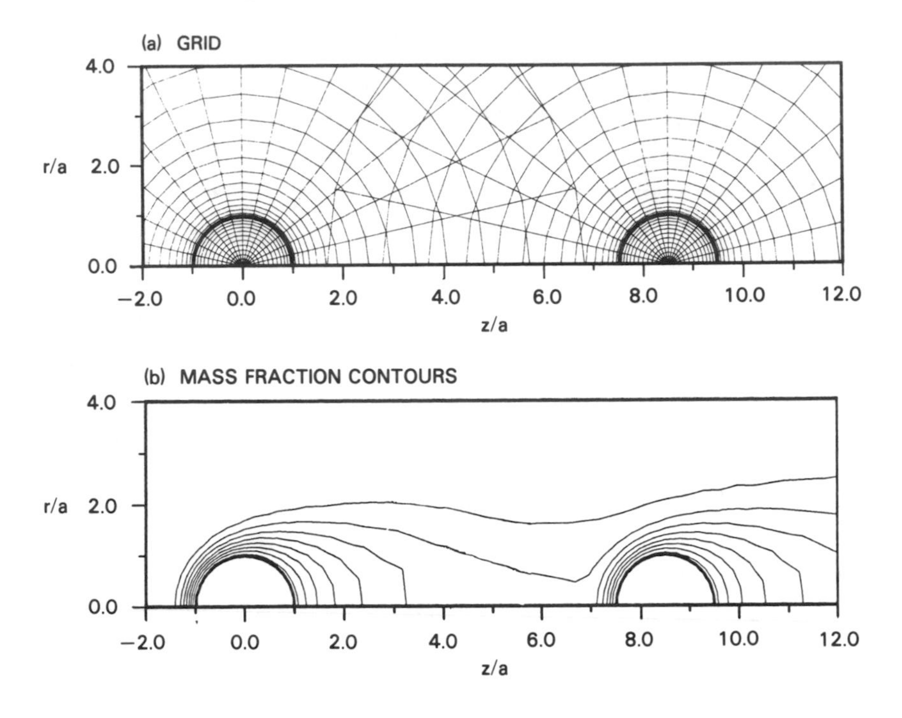

Figure 12–11. (a) Grid and (b) mass fraction contours from the two-droplet computation by Patnaik (1986). The droplet radius is a, and mass fraction coutours are spaced 0.1 apart.

with the air. Patnaik and Sirignano use an ADI method for the convection and an SOR method for the pressure. The calculation is first-order accurate in time, and second-order accurate in space.

Contrasting calculations by Fyfe et al. (1987) are described in Section 13–4. These examples show the application of a Lagrangian approach on a restructuring triangular grid to study the distortion and breakup of the droplets due to collisions and shear. This modeling approach emphasizes the dynamic behavior of the droplet, calculating the internal and external flows. A potentially important calculation would involve merging the physical models in the Patnaik and Sirignano simulations with those in the Fyfe et al. simulations.

12–3.4 A Model of Deflagration-to-Detonation Transition

In porous explosives or propellants, combustion often originates as a subsonic burning *deflagration* that can later accelerate to a detonation. The fundamental experimental observation of this was made by Griffiths and Groocock (1960) using columns of the granular explosive HMX, where they observed a detonation occurring in a region far removed from the location where combustion first began. It was later observed that the explosive was compressed appreciably in the region ahead of the slow combustion wave (Bernecker et al., 1981; Sandusky, 1983).

The phenomenon they observed, *deflagration-to-detonation transition*, or *DDT*, is a complex combustion process that has four regimes: (1) burning dominated by thermal conduction, (2) burning with convection, (3) compression and shock formation, and (4) detonation. Here we use the work of Baer and Nunziato (1986) and Baer et al. (1986a,b) as an example of a multiphase compressible flow model used to study DDT in a porous bed of the explosive, CP, which is $C_2H_{15}N_{10}CoCl_2O_6$. When this explosive is ignited thermally, it is highly reactive and a detonation occurs quickly.

The formulation of the model of this system considers two compressible continuous phases, the solid and the hot product gases, that are *not* in thermodynamic equilibrium. Therefore each phase has its own pressure, density, temperature, and entropy, and the volume fractions of each phase are treated as independent variables. The basic equations for density, momentum, and energy are:

$$\frac{\partial}{\partial t}(\phi_a \rho_a) + \frac{\partial}{\partial x}(\phi_a \rho_a v_a) = c_a^+ \qquad (12-3.21)$$

$$\frac{\partial}{\partial t}(\phi_a \rho_a v_a) + \frac{\partial}{\partial x}(\phi_a \rho_a v_a^2 + \phi_a P_a) = m_a^+ \qquad (12-3.22)$$

$$\frac{\partial}{\partial t}(\phi_a \rho_a E_a) + \frac{\partial}{\partial x}((\phi_a \rho_a E_a + \phi_a P_a) v_a + q_a) = e_a^+ \, . \qquad (12-3.23)$$

Here the subscript a indicates a particular phase, denoted separately by g for gas or s for solid, ϕ_a is the volume fraction of species a, q_a is the thermal flux, and e_a is the internal energy. The total energy of each phase, E_a is given by

$$E_a = e_a + \frac{v_a^2}{2} \, . \qquad (12-3.24)$$

All phase-interaction quantities are denoted with a superscript $+$ and are constrained to maintain conservation between the two phases.

Because the phases are not in equilibrium, a different equation of state must be used for each phase to calcualte P_s, e_s, P_g, and T_g. The phase-interaction terms were taken as

$$c_s^+ = -c_g^+ = \alpha P_g^n \,, \tag{12-3.25}$$

$$m_s^+ = -m_g^+ = P_g \frac{\partial \phi_s}{\partial x} - f + c_s^+ v_s \,, \tag{12-3.26}$$

$$\begin{aligned} e_s^+ &= -e_g^+ \\ &= \left(P_g \frac{\partial \phi_s}{\partial x} - f\right) v_s - (P_s - \beta_s) \left(\frac{\partial \phi_s}{\partial t} + v_s \frac{\partial \phi_s}{\partial x} - \frac{c_s^+}{\rho_s}\right) \\ &\quad - h(T_s - T_g) + E_s c_s^+ \,, \end{aligned} \tag{12-3.27}$$

where f is a drag factor and β_s is the intragranular stress. The heat flux for each phase a is given by

$$q_a = -k_a \frac{\partial T_a}{\partial x} \,. \tag{12-3.28}$$

The equation for the solid volume fraction includes the effects of pressure differences and combustion,

$$\frac{\partial \phi_s}{\partial t} + v_s \frac{\partial \phi_s}{\partial x} = \frac{\phi_s \phi_g}{\mu_c} (P_s - P_g - \beta_s) + \frac{c_s^+}{\rho_s} \,, \tag{12-3.29}$$

which is constrained by

$$\phi_g = 1 - \phi_s \,. \tag{12-3.30}$$

In this description, an experimentally determined intragranular stress β_s is used and the material viscosity, μ_c, is related to changes of volume induced by pressure differences.

To complete the model, a description of the combustion process is necessary that specifies the change in the amount of solid material due to burning. The burn law postulated was:

$$c_s^+ = \frac{-6 \phi_s \rho_s}{d_p} F(\phi_g^o) \frac{P_g}{P_g^{\mathrm{ref}}} \,, \tag{12-3.31}$$

where F is determined as a function of ϕ_g^o and burning is not allowed until the mixture reaches an ignition temperature defined by a volume fraction-weighted temperature.

Baer et al. (1986) give more detailed explanations, tables of constants, expressions for equations of state, and describe how various parameters in these equations are derived from experimental data. Using this set of equations, they carried out computations for the ignition, burning, and subsequent transition to detonation of CP. The one-dimensional calculations were done using a method of lines approach (see, for example, Hyman, 1979), and with a one-dimensional Flux-Corrected Transport module (Chapter 9). The two-dimensional calculations used Flux-Corrected Transport for convection, coupled to other algorithms for the additional physical processes by timestep-splitting methods. Extensive tests were reported in Gross and Baer (1986).

Figure 12–12 shows the evolution of ϕ_s and the gas and solid phase pressures in a one-dimensional simulation. Flame spread is initially caused by convection of the hot gas that penetrates into the unreacted material. A rapid increase in pressure causes the formation of a "compaction wave" that impedes gas flow. Figure 12–12a shows the development of the compaction wave during combustion. Flame spread compresses and heats both phases. The combustion wave continues to accelerate and forms a shock front that eventually overtakes the compaction wave to produce the detonation. Figure 12–12b shows that the solid-phase pressure runs ahead of the gas during the final stages of transition to detonation.

Figure 12–13 shows a comparison for the calculations to experiments for three values of ϕ_s^o, corresponding to initial densities of 1.2, 1.4, and 1.6 g/cm^3. The trajectory of the burn front was taken from image-enhanced photographic streak records and compared to the location of the burn front as predicted by the calculations. The flame spread from deflagration to detonation is predicted fairly well. Two distinct modes of flame spread, a convective mode and a compressive mode, were noticed, each characterized with a rapid transition to the next mode of burning.

Figure 12–14 shows an extension of this work to two dimensions (Baer et al., 1986b). Contours of total pressure and solid volume fraction are shown for ignition of a cylindrical flame front in a rectangular bed of CP. Similar to the one-dimensional calculations, a fast deflagration wave develops, driven by convective heat transfer. Different pressure waves in the gas and solid phase are formed that produce a cylindrical compaction wave ahead of the reaction front. A rapid increase in gas pressure combined with high drag forces cause a shock to form. The shock accelerates to a detonation before the combustion wave interacts with the rigid boundaries. When the reacting wave reaches these boundries, a more complicated shock structure results consisting of a Mach stem, a reflected wave, and an inwardly curved reaction and shock front.

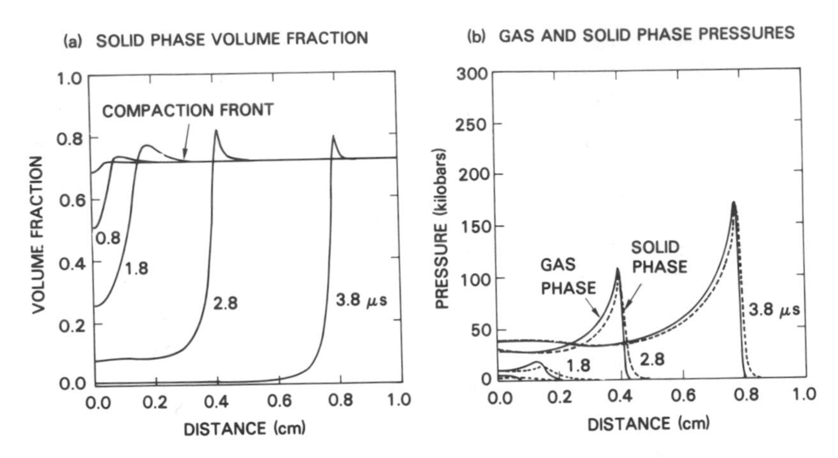

Figure 12–12. Calculated temporal variation of (a) the solid volume fraction and (b) the gas and solid pressure during combustion of a porous explosive (Baer et al., 1986b).

Figure 12–13. Comparison of calculations (solid lines) and experimental data (points) for the temporal variation of the combustion front for the solid explosive CP for three values of the initial density: (a) 1.2 g/cm^3, (b) 1.4 g/cm^3, (c) 1.6 g/cm^3.

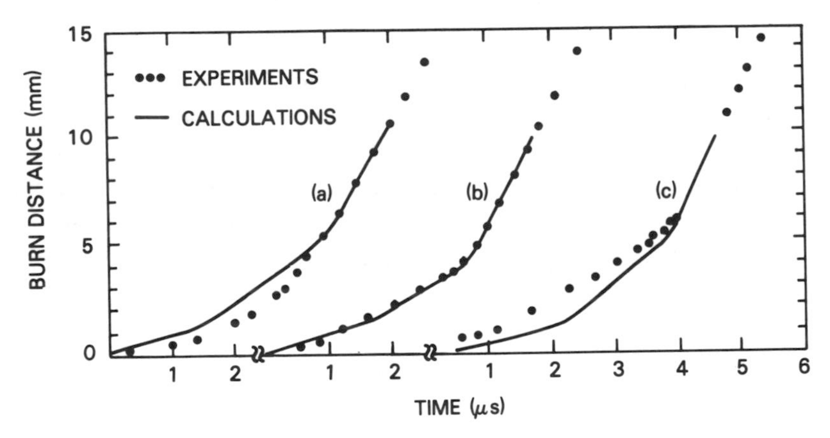

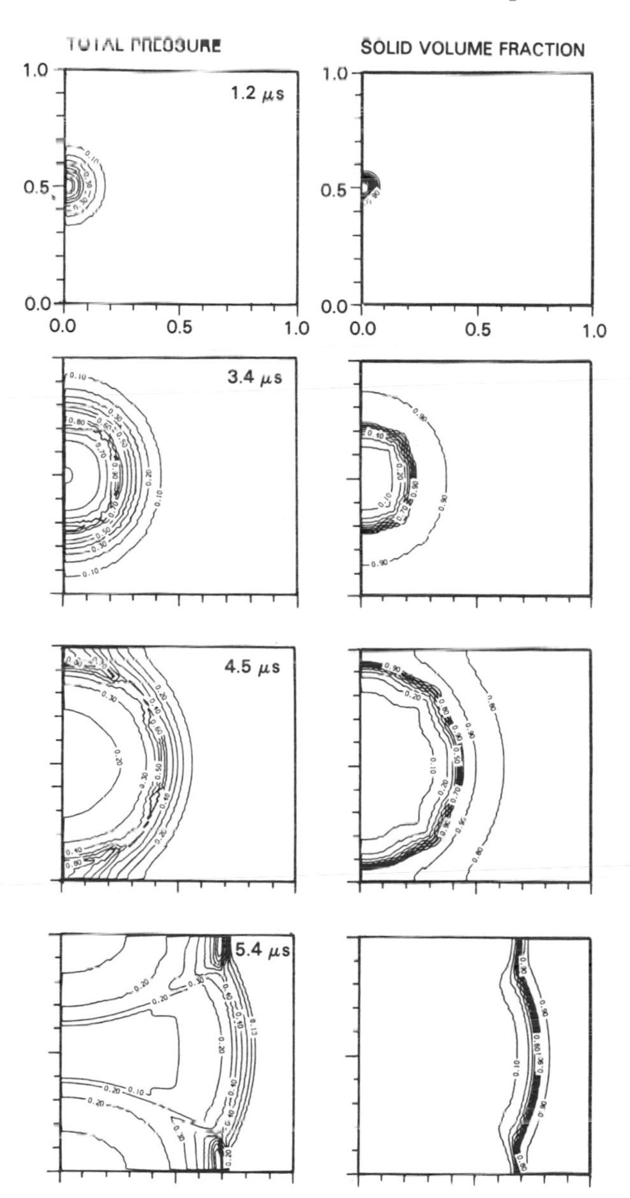

Figure 12–14. Contours of total pressure and solid volume fraction during combustion in a confined rectangular granular bed of the explosive CP (Baer et al., 1986b).

12–3.5 Even More Basic Models

Molecular dynamics and Monte Carlo methods have not been emphasized in this book. These methods, however, are a potentially valuable way to simulate, from first principles, the kinds of multiphase processes that occur in coagulation or at the interfaces between different phases. They can be used to develop and calibrate the phenomenologies needed in macroscopic SF or LHF models.

Such calculations are at the particle level and do not use the continuum approximations. It is necessary to specify the interactions between different types of particles, usually in terms of a force law, and also the initial conditions of the particles such as mass, momentum, and energy state. Then a great many particles are allowed to interact and their collective behavior is studied. For combustion problems, we imagine using these methods to study particle coagulation, surface reactions, or detonations through crystal lattices. In the first and second cases, we would need to specify the rules for particles sticking together and reactions between them. In the second and third cases, we can use the finite-particle properties of the model to set up a model of a crystal lattice, and then investigate how waves and particles propagate through and interact with the lattice.

These particle approaches provide valuable information needed in the more macroscopic models, but are difficult and expensive, and therefore not to be undertaken lightly. New methods for speeding up such calculations are currently being developed but they are unlikely to supplant the macroscopic fluid multiphase models.

12–3.6 Modeling Multiphase Flows

The sets of equations presented above for modeling multiphase flows are at least as difficult to solve as the set of equations introduced in Chapter 2 for a gas mixture. Extra classes of behavior in multiphase flows introduce many assumptions into the equations that can only be tested by comparing the computed results with carefully designed and controlled experiments.

In the estimation of chemical reactions, surface reactions are important and their behavior can differ markedly from volumetric estimates of the same rates. Subgrid models must be developed to account for the propagation of reaction fronts through volumes where diffusion effects on unresolved spatial scales govern the macroscopic behavior of the system. The presence of material and thermodynamic interfaces moving with the flow means that boundary layers and particle scales have to be considered and sometimes resolved. Because these problems are so complex, they cannot be simulated in the detail possible with more idealized reactive flows. For example, fewer scales are represented in detail and fewer individual species can be included because the models require more degrees of freedom to describe and update the state of the composite medium. A greater range of scale sizes must be represented by phenomenologies rather than simulated by detailed models.

Multiphase flows sometimes behave very differently on the macroscopic scale than an apparently similar homogeneous system. Counterintuitive behavior arises from the interactions of the various local, diffusive, convective, and oscillatory phenomena in the flow. This is expected because the equations are more complex, allowing new modes of reponse through possible new interactions.

For example, consider a locally homogeneous dust of relatively inert particles suspended in a combustible gas. A shock propagating through the pure gas may be able to ignite it, but the same shock cannot necessarily ignite the multiphase suspension. The bulk ignition process is strongly influenced by the particles which cool the gas after the shock has passed. If the particles are large, the gas might ignite because the particulate surface area is relatively small. If the same mass of dust is made up of smaller particles, the dust might quench the chemical reaction before the chemical induction time has elapsed. When the dust mass is too small, the gas again ignites.

Acoustic waves in multiphase media also show interesting effects. Because gases are easily compressed but liquids and particles are not, collective behavior, sometimes called *slow sound*, can arise from the extra degrees of freedom associated with the differential compressiblity of the two media. A froth of bubbles in liquid can be very much denser than air, for example, and yet have essentially the same compressibility. The gas in the bubbles and between the droplets compresses as the result of a pressure rise in almost the same way as if the liquid phase were absent, but the effective mass density of the composite medium is much heavier than air. Thus sound waves in the multiphase system travel much slower than they would in air.

Despite the variety of new phenomena that are possible, the equations for multiphase flow generally contain terms that are of the same generic type we have already considered. Because of this, we can use the methods discussed previously in this book to solve the sets of equations. These include, in particular, the convection algorithms discussed in Chapters 8, 9, and 10, the diffusion algorithms in Chapter 7, the chemistry algorithms in Chapter 5, and the various gridding, representation, and boundary and interface information given in Chapters 6 and 11.

References

Acton, E., 1976, The Modelling of Large Eddies in a Two-Dimensional Shear Layer, *J. Fluid Mech.* 76: 561–592.

Alper, E., Ed., 1983, *Mass Transfer with Chemical Reaction in Multiphase Systems*, Volumes I and II, NATO ASI Series E: Applied Sciences — No. 72 and 73, Nijhoff, Boston.

Aref, H., and E.D. Siggia, 1980, Vortex Dynamics of the Two-Dimensional Turbulent Shear Layer, *J. Fluid Mech.* 100: 705–737.

Ashurst, W.T., 1979, Numerical Simulation of Turbulent Mixing Layers via Vortex Dynamics, in F. Durst, B.E. Launder, F.W. Schmit, AND J.H. Whitelaw, eds., *Turbulent Shear Flows I*, Springer-Verlag, New York, 402–413.

Baer, M.R., and J.W. Nunziato, 1986, to appear in *Int. J. Multiphase Flow.*

Baer, M.R., R.J. Gross, J.W. Nunziato, and E.A. Igel, 1986a, An Experimental and Theoretical Study of Deflagration-to-Detonation Transition (DDT) in the Granular Explosive, CP, *Comb. Flame* 65: 15–30.

Baer, M.R., R.E. Benner, R.J. Gross, and J.W. Nunziato, 1986b, Modeling and Computation of Deflagration-to-Detonation Transition (DDT) in Reactive Granular Materials, *Lect. Appl. Math.* 24: 479–498.

Bernecker, R., H. Sandusky, and A. Clairmont, Jr., 1981, Deflagration-to-Detonation Transition Studies of Porous Explosive Charges in Plastic Tubes, *Proceedings of the Seventh Symposium (International) on Detonation*, pp. 119–138, Report No. NSWC-MP-82-334, Naval Surface Weapons Center, Silver Spring, MD, [AD-A126 667].

Chen, D.Y., V. Yang, and K.K. Kuo, 1981, Boundary Condition Specification for Mobile Granular Propellant Bed Combustion Processes, *AIAA J.* 19: 1429–1437.

Couet, B., and A. Leonard, 1981, Mixing Layer Simulation by an Improved Three-Dimensional Vortex-in-Cell Algorithm, in W.C. Reynolds and R.W. MacCormack, eds., *Proceedings of the Seventh International Conference on Numerical Methods in Fluid Dynamics*, Springer-Verlag, New York, 125–131.

Davis, R.W., and E.F. Moore, 1985, A Numerical Study of Vortex Merging in Mixing Layers, *Phys. Fluids* 28: 1626–1635.

Deardorff, J.W., 1970, A Numerical Study of Three-Dimensional Turbulent Channel Flow at Large Reynolds Numbers, *J. Fluid Mech.* 41: 453–480.

Dwoyer, D.L., M.Y. Hussaini, and R.G. Voigt, Eds., 1985, *Theoretical Approaches to Turbulence*, Springer, New York.

Eddington, A.S., 1959, *The Internal Constitution of the Stars*, Dover, New York.

Faeth, G.M., 1983, Evaporation and Combustion of Sprays, *Prog. Energy Comb. Sci.* 9: 1–76.

Ferziger, J.H., 1977, Large-Eddy Numerical Simulations of Turbulent Flows, *AIAA J.* 15: 1261–1267.

Ferziger, J.H., 1981, Higher-Level Simulations of Turbulent Flows, Report No. TF-16, Department of Mechanical Engineering, Stanford University, Palo Alto, CA.

Ferziger, J.H., 1987, Simulation of Incompressible Turbulent Flows, *J. Comp. Phys.* 69: 1–48.

Fox, D.G., and D.K. Lilly, 1972, Numerical Simulation of Turbulent Flows, *Rev. Geophys. Space Phys.* 10: 51–72.

Fyfe, D.E., E.S. Oran, and M.J.Fritts, 1987, Surface Tension and Viscosity with Lagrangian Hydrodynamics on a Triangular Mesh, *J. Comp. Phys.*, to appear.

Gardner, J.H., and S.E. Bodner, 1986, High-Efficiency Targets for High-Gain Inertial Confinement Fusion, *Phys. Fluids* 29: 2672–2678.

Gatski, T.B., and J.T.C Liu, 1980, On the Interactions between Large-Scale Structure and Fine-Grained Turbulence in a Free Shear Flow, III. A Numerical Solution, *Philos. Trans R. Soc. London* A 293: 473–509.

Ghoniem, A.F., A.J. Chorin, and A.K. Oppenheim, 1981, Numerical Modeling of Turbulent Combustion in Premixed Gases, *Eighteenth Symposium (International) on Combustion*, The Combustion Institute, Pittsburgh, PA, 1375–1383.

Ghoniem, A.F., D.Y. Chen, and A.K. Oppenheim, 1986, Formation and Inflammation of a Turbulent Jet, *AIAA J.* 24: 224–229.

Ghoniem, A.F., and K.K. Ng, 1986a, Effect of Harmonic Modulation on Rates of Entrainment in a Confined Shear Layer, AIAA Paper AIAA-86-0056, AIAA, NY.

Ghoniem, A.F., and K.K. Ng, 1986b, Numerical Study of the Dynamics of Forced Shear Layer, *Phys. Fluids*, to appear.

Givi, P., W.-H. Jou, and R.W. Metcalfe, 1987, Flame Extinction in a Temporally Developing Mixing Layer, *Proceedings of the Twenty-First Symposium (International) on Combustion*, The Combustion Institute, Pittsburgh, PA, to appear.

Gouldin, F.C., 1987, An Application of Fractals to Modeling Premixed Turbulent Flames, *Proceedings of the Twenty-First Symposium (International) on Combustion*, The Combustion Institute, Pittsburgh, PA, to appear.

Griffiths, N., and J.M. Groocock, 1960, Burning to Detonation of Solid Explosives, *J. Chem. Soc. London* 814: 4154–4162.

Grinstein, F.F., E.S. Oran, and J.P. Boris, 1986, Numerical Simulations of Asymmetric Mixing in Planar Shear Flows, *J. Fluid Mech.* 165: 201–220.

Gross, R.J., and M.R. Baer, 1986, *A Study of Numerical Solution Methods for Two-Phase Flows*, SAND-84-1633, Sandia National Laboratories, Albuquerque, NM [DE86012265/XAB].

Herring, J.R., 1973, Statistical Turbulence Theory and Turbulence Phenomenology, *Proceedings of the Langley Working Conference on Free Turbulent Shear Flows*, NASA SP 321, Langley Research Center, VA, 41-66.

Herring, J.R., 1979, Subgrid Scale Modeling — An Introduction and Overview, in F. Durst, B.E. Launder, F.W. Schmit, and J.H. Whitelaw, eds., *Turbulent Shear Flows I*, Springer-Verlag, New York, 347–352.

Herring, J.R., 1985, An Introduction and Overview of Various Theoretical Approaches to Turbulence, in D.L. Dwoyer, M.Y. Hussaini, and R.G. Voigt, eds., *Theoretical Approaches to Turbulence*, Springer, New York, 73–90.

Hyman, J.M., 1979, *A Method of Lines Approach to the Numerical Solution of Conservation Laws*, Los Alamos Report 79-837, Los Alamos National Laboratories, Los Alamos, NM [LA-UR-79-837].

Ishii, M., 1975, *Thermo-Fluid Dynamic Theory of Two-Phase Flows*, Eyrolles, Paris.

Jones, W.P., and J.H. Whitelaw, 1982, Calculation Methods for Reacting Turbulent Flows: A Review, *Comb. Flame* 48: 1–26.

Kailasanath, K., J. Gardner, J. Boris, and E. Oran, 1986, Acoustic-Vortex Interactions in an Idealized Ramjet Combustor, *Proceedings of the 22nd JANNAF Combustion Meeting*, Chemical Propulsion Information Agency, Johns Hopkins University, Applied Physics Laboratory, Laurel, MD.

Kalkofen, W., Ed., 1984, *Methods in Radiative Transfer*, Cambridge University Press, Cambridge, New York.

Khalil, E.E., 1982, *Modelling of Furnaces and Combustors*, Abacus Press, Kent, England.

Knight, D.D., and B.T. Murray, 1980, Theoretical Investigation of Interaction and Coalescence of Large Scale Structures in the Turbulent Mixing Layer, in J. Jininez, ed., *The Role of Coherent Structures in Modelling Turbulence and Mixing*, Springer-Verlag, New York, 62–91.

Kuo, K.K., 1986, *Principles of Combustion*, Wiley, New York.

Kuo, K.K., J.H. Koo, T.R. Davis, and G.R. Coates, 1976, Transient Combustion in Gas-Permeable Propellants, *Acta Astro.* 3: 573–591.

Kwak, D., W.C. Reynolds, and J.H. Ferziger, 1975, *Three-Dimensional Time Dependent Computation of Turbulent Flow*, Report No. SU-TF-5, Department of Mechanical Engineering, Stanford University, Palo Alto, CA [NASA-CR-143347] [N75-30477/4].

Law, C.K., 1982, Recent Advances in Droplet Vaporization and Combustion, *Prog. Ener. Comb. Sci.* 8: 171–202.

Leonard, A., 1974, Energy Cascade in Large-Eddy Simulations of Turbulent Fluid Flows, *Advances in Geophysics*, 18A: 237–248, Academic, New York.

Libby, P.A., and F.A. Williams, Eds., 1980, *Turbulent Reacting Flows*, Springer-Verlag, New York.

Libby, P.A., and F.A. Williams, 1981, Some Implications of Recent Theoretical Studies in Turbulent Combustion, *AIAA J.* 19: 261–274.

Lilly, D.K., 1966, *On the Application of the Eddy Viscosity Concept in the Inertial Subrange of Turbulence*, NCAR Manuscript 123, National Center for Atmospheric Research, Boulder, CO.

Mandelbrot, B.B., 1983, *The Fractal Geometry of Nature*, Freeman, San Francisco.

Mansour, N.N., J.H. Ferziger, and W.C. Reynolds, 1978, *Large-Eddy Simulation of a Turbulent Mixing Layer*, Report TF-11, Department of Mechanical Engineering, Stanford University, Stanford, CA [NASA-CR-156575] [N78-22027/4].

Metcalfe, R.W., S.A. Orszag, M.E. Brached, and J.R. Riley, 1987, Secondary Instability of a Temporally Growing Mixing Layer, *J. Fluid Mech.*, to appear.

McComb, W.D., 1985, Renormalization Group Methods Applied to the Numerical Simulation of Fluid Turbulence, in D.L. Dwoyer, M.Y. Hussaini, and R.G. Voigt, eds., *Theoretical Approaches to Turbulence*, Springer, New York, 187–208.

McMurtry, P.A., W.-H. Jou, J.J. Riley, and R.W. Metcalfe, 1986, Direct Numerical Simulations of a Reacting Mixing Layer with Chemical Heat Release, *AIAA J.* 24: 962–970.

Mihalas, D., and B.W. Mihalas, 1984, *Foundations of Radiation Hydrodynamics*, Oxford University Press, New York.

Milne, F.A., 1930, Thermodynamics of the Stars, *Handbuch der Astrophysik* 3: 65–255, Springer, Berlin.

Moin, P., and J. Kim, 1982, Numerical Investigation of Turbulent Channel Flow, *J. Fluid Mech.* 118: 341–377.

O'Brien, E.E., 1980, The Probability Density Function (pdf) Approach to Reacting Turbulent Flows, in P.A. Libby and F.A. Williams, eds., *Turbulent Reacting Flows*, Springer-Verlag, New York, 185–218.

Ohlemiller, J.T., 1985, Modelling of Smoldering Combustion Propagation, *Prog. Energy Comb. Sci.* 11: 277–310. Also U.S. National Bureau of Standards Report NBSIR-84/2895, 1984 [PB84-236389].

Özisik, M.N., 1973, *Radiative Transfer and Interactions with Conduction and Convection*, Wiley, New York.

Patnaik, G., 1986, *A Numerical Solution of Droplet Vaporization with Convection*, Ph.D. Thesis, Carnegie-Mellon University, Pittsburgh, PA, University Microfilm, Ann Arbor, MI [AAD86-16517].

Patnaik, G., W.A. Sirignano, H.A. Dwyer, and B.R. Sanders, 1986, A Numerical Technique for the Solution of a Vaporizing Fuel Droplet, in J.R. Bowen, J.-C. Leyer, and R.I. Soloukhin, eds., *Dynamics of Reactive Systems, II. Modeling and Heterogeneous Combustion*, *Prog. Astro. Aero.* 105: 253–266, AIAA, New York.

Pope, S.B., 1985, PDF Methods for Turbulent Reactive Flows, *Prog. Energy Comb. Sci.* 11: 119–192.

Reynolds, W.C., 1976, Computation of Turbulent Flows, *Ann. Rev. Fluid Mech.* 8: 183–208.

Riley, J.J., and R.W. Metcalfe, 1980, Direct Numerical Simulation of a Perturbed, Turbulent Mixing Layer, AIAA Paper No. 80-0274, AIAA, New York.

Rogallo, R.S., and P. Moin, 1984, Numerical Simulation of Turbulent Flows, *Ann. Rev. Fluid Mech.* 16: 99–137.

Rosseland, S., 1936, *Theoretical Astrophysics; Atomic Theory and the Analysis of Stellar Atmospheres and Envelopes*, Clarendon Press, Oxford.

Rybicki, G.B., and A.P. Lightman, 1979, *Radiative Processes in Astrophysics*, Wiley, New York.

Sandusky, H.W., 1983, Compressive Ignition and Burning in Porous Beds of Energetic Materials, *Proceedings of the 3rd JANNAF Propulsion Systems Hazards Meeting*, 1: 249–257, Chemical Propulsion Information Agency Publication (CPIA), Laurel, MD.

Schumann, U., 1975, Subgrid Scale Model for Finite Difference Simulations of Turbulent Flows in Plane Channels and Annuli, *J. Comp. Phys.* 18: 376–404.

Schuster, A., Radiation through a Foggy Atmosphere, 1905, *Astrophys. J.* 122: 488–497.

Schwarzschild, K., Equilibrium of the Sun's Atmosphere, 1906, *Ges. Wiss.* Gottingen, *Nachr., Math-Phys. Klasse* 1: 41–53.

Shaanan, S., J.H. Ferziger, and W.C. Reynolds, 1975, *Numerical Simulation of Turbulence in the Presence of Shear*, Rep. TF-6, Department of Mechanical Engineering, Stanford University, Palo Alto, CA [N75-30476/6].

Siegel, R., and J.R. Howell, 1981, *Thermal Radiation Heat Transfer*, Hemisphere Publishing Corporation, New York.

Sirignano, W.A., 1983, Fuel Droplet Vaporization and Spray Combustion, *Prog. Energy Comb. Sci.* 9: 291–322.

Smagorinsky, J., 1963, General Circulation Experiments with the Primitive Equations. I. The Basic Experiment, *Mon. Weather Rev.* 91: 99–164.

Smoot, L.D., 1984, Modeling of Coal-Combustion Processes, *Prog. Energy Comb. Sci.* 10: 229–272.

Soo, S.L., 1967, *Fluid Dynamics of Multiphase Systems*, Blaisdell, Waltham, MA.

Soo, S.L., 1984, Development of Dynamics of Multiphase Flow, *Int. J. Sci. Eng.* 1: 13–29.

Stewart, H.B., and B. Wendroff, 1984, Two-Phase Flow: Models and Methods, *J. Comp. Phys.* 56: 363–409.

Yakhot, V., and S.A. Orszag, 1986a, Renormalization Group Analysis of Turbulence, I. Basic Theory, *J. Sci. Comput.* 1: 3–51.

Yakhot, V., and S.A. Orszag, 1986b, Renormalization-Group Analysis of Turbulence, *Phys. Rev. Lett.* 57: 1722–1724,

Zel'dovich, Ya. B., and Yu. P. Raizer, 1966, *Physics of Shock Waves and High-Temperature Hydrodynamic Phenomena*, Academic, New York.

Chapter 13

MODELS FOR SHOCKS, FLAMES, AND DETONATIONS

This chapter discusses complete reactive flow models constructed by combining algorithms described earlier in the book. The structure of a typical computer program is described with emphasis on the way algorithms representing individual processes are coupled together. Although we use some of the specific models we have constructed as examples, the information presented about coupling processes together is generic. Examples of both reacting and nonreacting, subsonic and supersonic flows in both Eulerian and Lagrangian representations are given to illustrate what can be done with each type of model.

The computer programs described are not "black boxes." They are ad hoc assemblies of relatively independent software modules, each solving a particular type of term in the conservation equations. These models are constructed, reconstructed, and changed frequently by the users. This approach to building complex computer programs stresses flexibility in the algorithms and techniques. The specific modules are the enduring elements of the effort, rather than completely integrated simulation models. Because of this modular structure, algorithms for specific types of terms can be updated and new physical processes can be added with minimal interference in the structure of the program.

13–1. ONE-DIMENSIONAL SUPERSONIC REACTIVE FLOWS

In this section, a group of numerical models, collectively called RSHOCK are used to illustrate the construction of programs for solving shock and detonation problems with detailed models of chemical kinetics. These are the simplest models described in this chapter because they couple only two interacting physcial processes, convection and chemical reactions, with an ideal gas or an empirically determined equation of state. Here we emphasize how the models for the separate processes are coupled together and the problems encountered in doing this. The individual algorithms used for these processes were described in previous chapters.

13–1.1 The Structure of RSHOCK Models

RSHOCK models are one-dimensional, time-dependent, compressible, explicit and Eulerian, originally designed for gas-phase flows. These models have been used for a variety of applications, including descriptions of flows in shock tubes, studies of the shock-to-detonation transition, and studies of chemical-acoustic coupling. Some of the applications are described in Oran et al. (1979, 1981, 1982a), Oran and Boris (1981, 1982), and Kailasanath and Oran (1983). Multidimensional versions, described in the next section, have been used for studies of the structure of propagating detonations (Oran et al., 1982b; Kailasanath et al., 1985b), for shock structure studies (Book et al., 1981), and calculations of shocks through density discontinuities (Picone et al., 1984). A condensed-phase RSHOCK model was developed by Guirguis and Oran (1983), and used for studies of detonations in liquid nitromethane (Guirguis et al., 1986, 1987). Other versions that include a radiation transport model and separate ion and electron temperature equations have been used for modeling the solar corona (see, for example, Mariska et al., 1982; Oran et al., 1982; Karpen et al., 1982, 1984), inertial confinement fusion (see, for example, Gardner and Bodner, 1981, 1984), and the interaction of ion beams with materials (DeVore et al., 1984). In this chapter we emphasize the combustion applications.

RSHOCK combines optimized modules for the different physical processes through timestep splitting. These algorithms include:

- FCT (Flux-Corrected Transport) for solving continuity equations (Boris and Book, 1976), as discussed in Chapters 8 and 9.
- Adaptive-gridding techniques using an automatic sliding rezone discussed in Chapter 6. The implementation with FCT was described in Chapter 8.
- CHEMEQ, for solving ordinary differential equations, as discussed in Chapter 5.
- A polynomial representation of the enthalpies as a function of temperature used to represent the equation of state, as discussed in Chapter 5.
- Timestep-splitting techniques to couple the various physical and chemical processes together, as introduced in Chapter 4.

A major advantage of timestep splitting is being able to use a monotone, positivity-preserving algorithm for convection. It is crucial to have minimal numerical diffusion for calculations that run many thousands of timesteps. Maintaining steep gradients is important when chemical reactions and energy release are coupled to convection.

At the end of a convective transport step, truncation errors usually cause some inconsistency among the calculated quantities. For example, the sum of the densities of the individual species in a cell, each of which has been convected separately by FCT, do not necessarily equal the value obtained from convecting the total density. This may occur even though the global conservation laws are individually satisfied. It is important that the values be locally consistent. When the disparity between the total density and the sum of the species densities is noticeable, the species densities should be renormalized to be consistent with the total mass density. The new pressure and temperature are calculated from the conserved total energy and the species densities using the equation of state. The new velocity is calculated from the conserved total mass density and momentum.

Using FCT, the computational grid can be changed each timestep. This allows the cells to stretch and compress, ensuring resolution where it is needed. The movement of the cell boundaries is controlled by predetermined criteria based on the existence of local gradients. The idea is to cluster fine cells at the gradients and allow them to move with moving gradients. Various numerical indicators are used to locate and track gradients. For example, a gradient in OH density indicates the initiation of chemical reactions. A gradient in pressure could indicate a shock or detonation front.

The hybrid algorithm CHEMEQ, discussed in detail in Chapter 5, is used to solve the coupled, nonlinear, ordinary differential equations describing chemical interactions. In RSHOCK models, we use a special machine-dependent version vectorized for parallel processing over the separate computational cells as well as the separate chemical species. An important feature of CHEMEQ is that it is self-starting. It can be restarted at the beginning of each full timestep Δt with no penalties in start-up time. When it is restarted during a timestep, it uses only the most current values of the temperature and the species densities so there is no penalty in computer storage. It is relatively inexpensive because there are no matrix inversions, a very important feature when it is necessary to integrate the chemical equations at thousands of computational cells.

CHEMEQ determines its own timestep internally. When this is much smaller than Δt required for the convection, the chemistry integration is subcycled internally until the full convection timestep is reached. Accuracy is controlled by predetermined convergence parameters which change the stiffness criterion, the timestep, and thus effectively the degree of conservation. This timestep-splitting and subcycling procedure assumes that the chemistry can be decoupled from the other processes for the length of a convection timestep Δt. The adequacy of this assumption is tested by running the program with two different values of Δt and comparing answers.

In each of the timestep subcycles associated with the chemistry integration, the temperature can vary. Thus, in addition to solving the species equations, it is necessary to solve an algebraic equation for the temperature. RSHOCK models invert the algebraic power series equations of state by a Newton-Raphson method, as described in Chapter 5.

There are some situations when a convenient form for the equation of state does not exist. This can occur, for example, at high temperature where there might not be any tabulated enthalpies of individual species. This is the case for atmospheric chemistry and laser chemistry problems, that often involve fluids at more than one temperature. In these cases, another ordinary differential equation must be introduced to describe the behavior of the temperature (see, for example, Lampe, et al., 1984). Such versions of RSHOCK are not discussed here.

Even using these tricks to save time, the chemistry integration is the most expensive part of the simulation. To date, techniques to reduce the cost of integrating ordinary differential equations are hampered by large increases in required storage. Table lookup routines, for example, could be used instead of continually evaluating chemical rates with costly exponentials. These tables, however, require extensive computer memory even for modest chemical kinetics systems.

Most methods that integrate stiff ordinary differential equations are not perfectly conservative. When the number of atoms in the system is counted before the chemistry integrator is called and this same number is counted afterwards, there is a discrepancy. This nonphysical discrepancy indicates the accuracy of the integration method. However, the discrepancy must be controlled so it may be necessary to tighten the convergence criteria.

13–1.2 Timestep Splitting

An overview of the sequence of computations in RSHOCK models is given in Table 13–1. The entire problem is broken down into a number of *substeps*, each for a different physical process. In a complete timestep, Δt, it is necessary to

1. Evaluate Δt, based on the Courant condition (Chapters 8 and 9). If the temperature change due to release of chemical energy in the previous timestep is greater than a specified amount, reduce Δt by some fraction even if the convection algorithms are stable.
2. Determine how much energy is deposited at each location in the computational domain from outside the system in Δt. This is where a model of an ignition source such as a laser, spark, or hot wall, can be added. Such processes change the temperature or chemical composition at specified

locations. If the energy deposition is very strong or very fast, it may be necessary to reduce Δt.

3. Evaluate the effects of convective transport. This substep may also involve moving the grid. During this substep, the densities, velocities, energies, and pressures can change. To find the pressures and temperatures, the polynomial equations for the enthalpies are used.
4. Integrate the chemical reactions over Δt. This usually requires internal subcycling at some of the grid points. This should be done automatically in the routines that integrate the ordinary differential equations.
5. Update all of the variables at the end of the complete timestep, making sure that the densities, energies, pressures, and temperatures are consistent with the independent variables known at that time.

The terms describing convection are solved for the *entire* time interval, Δt. Values of the variables are then updated. The starting values for the chemistry integration are those most recent updated values from convection. The final values for the completed timestep are those at the end of the chemical reaction substep.

Timestep splitting becomes more accurate as the timestep is made smaller, and it is exact in the limit. In a large simulation, the timestep should be just small enough that the results are acceptably accurate. The most expensive process computationally should limit the accuracy. If this is not the case, the computational cost can be reduced further without reducing the overall accuracy of the coupled simulation by decreasing the accuracy of the most expensive part.

There are ways of estimating the appropriate timestep for each physical process, but there is no sure way to determine the appropriate timestep for the whole system when several processes are coupled. The nature of the interactions often allows the overall timestep to be larger than the minimum required by one of the contributing processes. This is the case in RSHOCK, where the stiff chemical reaction rates often require a timestep much shorter than the convection timestep. We have used the convection criterion to determine Δt and then subcycle the chemistry within this timestep. Situations might arise with slow flows, slow chemistry, but with fast sound waves for which the procedure should be reversed.

Other cases can be imagined in which the overall timestep should be smaller than estimates based on individual processes. Nonlinear coupling between processes can require a smaller timestep, as when two stiff reactions compete on about the same time scale. Usually reducing the timestep by a factor of two solves this problem.

Table 13–1. Sequence of Computations in RSHOCK Models

Initialize Variables

* Determine Δt, so that $t^n = t^o + \Delta t$

1. External Energy Deposition

    Problem dependent, for example,

    $T(x) \rightarrow T^{(1)}(x)$, $n_i(x) \rightarrow n_i^{(1)}(x)$, etc.

    (Variables now labeled with superscript 1.)

2. Convective Transport

    a. Move grid as needed to maintain resolution

    (Variables now labeled with superscript 1′.)

    Convect variables 1′ from t to $t+\Delta t$

    $\rho^{(1')}(x) \rightarrow \rho^{(2)}(x), \quad v^{(1')}(x) \rightarrow v^{(2)}(x)$

    $E^{(1')}(x) \rightarrow E^{(2)}(x), \quad P^{(1')}(x) \rightarrow P^{(2)}(x)$

    $T^{(1')}(x) \rightarrow T^{(2)}(x), \quad \{n_i^{(1')}(x)\} \rightarrow \{n_i^{(2)}(x)\}$

    $\gamma^{(1')}(x) \rightarrow \gamma^{(2)}(x)$

    (Variables now labeled with superscript 2.)

3. Chemical Reactions

    Take variables labeled 2 from t to $t+\Delta t$

    $\{n_i^{(2)}(x)\} \rightarrow \{n_i^{(3)}(x)\}, \quad T^{(2)}(x) \rightarrow T^{(3)}(x)$

    $P^{(2)}(x) \rightarrow P^{(3)}(x), \quad \gamma^{(2)}(x) \rightarrow \gamma^{(3)}(x)$

    $\rho^{(2)}(x) = \rho^{(3)}(x), \quad v^{(2)}(x) = v^{(3)}(x)$

    $E^{(2)}(x) = E^{(3)}(x)$

    (Variables now labeled with superscript 3.)

4. Update Variables

    $\rho^{(3)}(x) = \rho(x), \quad v^{(3)}(x) = v(x)$

    $E^{(3)}(x) = E(x), \quad T^{(3)}(x) = T(x)$

    $P^{(3)}(x) = P(x), \quad \{n_i^{(3)}(x)\} = \{n_i(x)\}$

    $\gamma^{(3)}(x) = \gamma(x)$

Start New Timestep (go to * above).

Ideally, the order of applying the convection and chemistry algorithms should not matter. Invariance to the order of the substeps in timestep splitting is also a good test of accuracy and convergence. The advantage of performing convection first is that the grid is generally modified during the convection step. This means that at the end of the timestep, the chemistry variables can be calculated and updated on the same grid as the fluid variables.

A problem arises in coupling chemical kinetic energy release to fluid dynamics when there are nonphysical oscillations in the fluid variables. Overshoots or undershoots in primary calculated quantities often lead to corresponding errors in the derived temperature. Because the chemical reaction rates have an exponential form, the nonphysical overshoots can cause the release of energy much sooner than it should be released. This, in turn, feeds back into the convection, and the result can be runaway chemical reactions and nonphysical flame or detonation velocities. The convection portion of a simulation with chemical energy release must be more accurate than is usually required in pure convection calculations.

The accuracy of fluid variables in reactive regions must be monitored even when a fourth- or higher-order monotone method is used. It is generally not a good idea to disrupt the fluid calculation by permanently modifying extreme values of energy, momentum, or density that might occur during a timestep. It is better to limit the derived temperature used by the chemical kinetics. One way to do this is to pass the chemistry routines a temperature that is a weighted average of the locally calculated temperature at each computational cell and the surrounding temperature values. For example,

$$T_{\text{chem},i} = \frac{T_{i-1} + 2T_i + T_{i+1}}{4} \,. \tag{13-1.1}$$

This trick is usually adequate to restrict the temperature from above. It is also useful in both chemistry and convection to limit the temperature from below. For example, consider using $T_{\text{min}} = 250$ K. Another trick is to convect the fluid entropy as well as energy with the chemical species concentrations. In this way the temperature can be prevented from dropping below the adiabatically compressed values which are determined from the local density and entropy values.

In contrast to the accuracy criteria for convection, which have to be tightened when there is energy release, we have found that convergence requirements for the chemical kinetics can be *relaxed* somewhat in reactive flows. A convergence criterion for the ordinary differential equations of $\epsilon = 0.001$ (Chapter 5) gives acceptable results when the equations are integrated alone. When they are coupled to convection, however, allowing expansion and rarefaction in the densities, temperatures, and pressures, the answers are acceptable with $\epsilon = 0.01$. This has been tested repeatedly for both flame and detonation calculations. Some of the advantages gained from reducing ϵ come from the nature of CHEMEQ, which works best when the system is slightly out of equilibrium. Fluid dynamics, with rarefactions, compressions, and fluctuations, tends to guarantee at least small deviations from chemical equilibrium.

13–1.3 Calculations with Reactions and Shocks

The examples described below are taken from a series of calculations of shocks, reaction waves, and detonations in mixtures of hydrogen and oxygen. These examples show typical calculations that can be done with RSHOCK models.

Reflected Shock Wave Calculations

Simulations of reflected shock experiments have been performed to study the process of shock ignition of detonations (Oran et al., 1982a; Oran and Boris, 1982). As an example, consider a homogeneous mixture of hydrogen, oxygen, and argon in the ratio 2:1:7, which is initially at atmospheric temperature and pressure. A shock wave moves to the left through this system (Figure 13–1a), raises the temperature and pressure, and induces a uniform subsonic flow to the left behind the shock. The temperature behind this primary shock is not high enough to cause rapid chemical reactions.

Eventually the shock reflects from a rigid wall, and the reflected shock moves to the right back through the previously shocked material (Figure 13–1b). The gas behind the reflected shock is heated further by the reflected shock, and the fluid behind the reflected shock is essentially brought to rest. This temperature is now high enough to speed up the chemical reactions substantially. Ignition occurs near the wall where the doubly shocked gas has been heated the longest. After a time, a reaction wave develops at the wall and propagates through the gas (Figure 13–1c).

Figure 13–1 also shows the location of the variably spaced moving grid used in the problem. There are two fine-zoned regions. At the beginning of the calculation, one of these regions surrounds the incident shock and moves with the shock to the left. The other fine-gridded region sits at the reflecting wall. After a time, the incident shock reflects off the wall, and moves back to the right taking one fine-gridded region with it. Eventually the energy-releasing chemical reactions are initiated at the wall. When this happens, the second fine-gridded region resolves the development of the reaction wave, and moves with it as it propagates to the right. If the reaction wave catches up with the shock front, the two fine regions merge and now resolve the motion of the detonation wave. The movement of the region around the shock is keyed to a pressure gradient. The movement of the region at the wall is keyed to the change in OH number density.

These calculations used a detailed chemical reaction mechanism for hydrogen-oxygen combustion including eight reacting species, H_2, O_2,H, O, OH, HO_2, H_2O_2, H_2O, and the diluent Ar. About fifty elementary

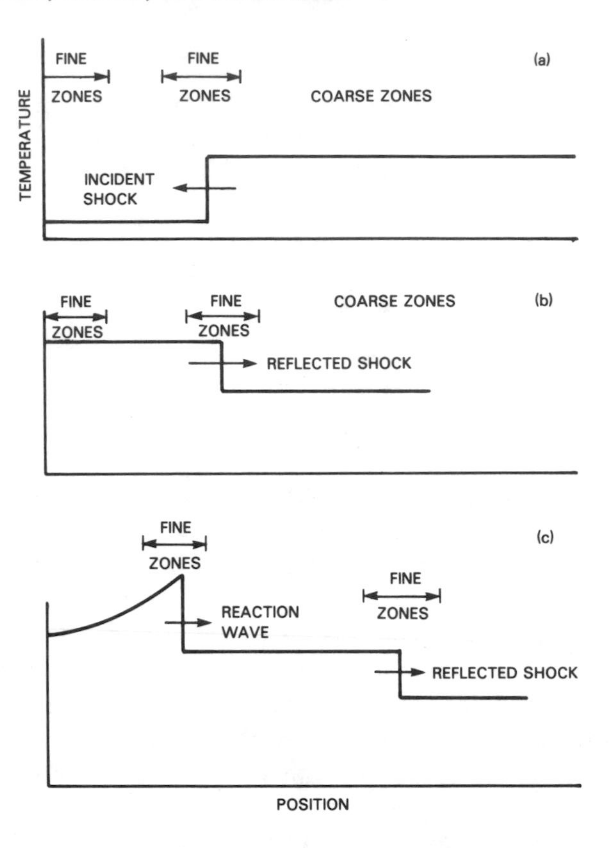

Figure 13–1. Schematic of the reflected shock tube problem. (a) Incident shock moves toward the wall. (b) Reflected shock moves back through the system. (c) Reaction wave moves from the wall in the shocked mixture.

chemical reaction rates were used (Oran et al., 1982a). Typical calculations use 150–200 computational cells, many of which resolve steep gradients.

Figure 13–2 shows the positions of the shock front, the reaction wave, the merged detonation wave, and the contact discontinuity formed as the reflected shock and reaction waves merge, as a function of time. The reaction wave begins slowly and then accelerates to a detonation. After it merges with the reflected shock front, it decelerates relative to the laboratory coordinates because the incoming shocked flow has no longer been hot for an appreciable time before the reaction wave arrives. The contact discontinuity formed when the reaction wave and shock waves merge moves forward more slowly, at the fluid velocity. Figure 13–3 shows temperature and pressure profiles of the reaction wave before it merges with the reflected shock. The reaction wave looks like a detonation propagating in a shocked mixture.

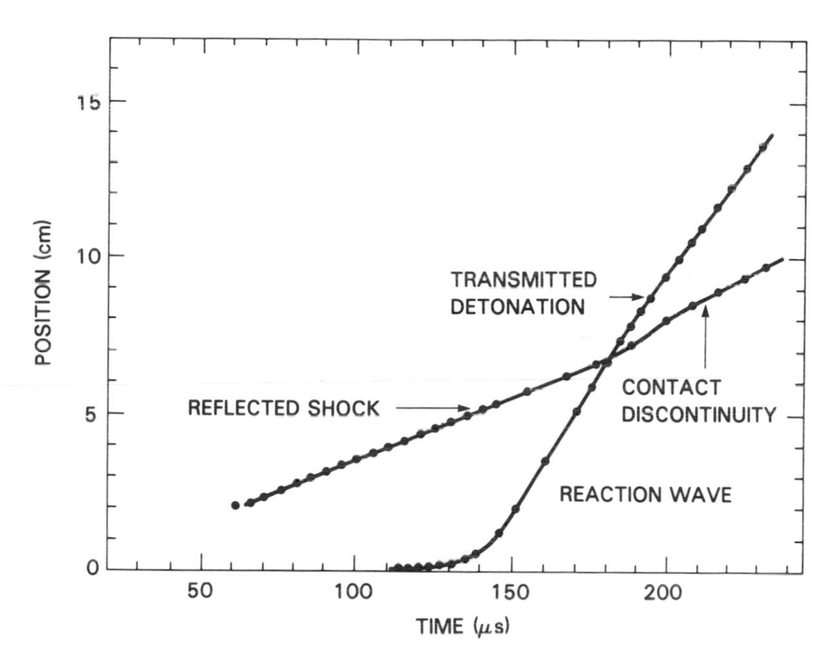

Figure 13–2. Position versus time of the several important waves in the reflected shock simulation (Oran et al., 1982a).

Hot-Spot Ignition of Detonations

To study ignition behind a shock wave, an RSHOCK model was configured to simulate a shock tube in which a diaphragm bursts and a propagating shock is generated (Kailasanath and Oran, 1983). In these simulations, the diaphragm initially separates a region containing a combustible mixture of hydrogen and oxygen from a region of high-pressure helium gas. The simulation begins without a diaphragm in place, but with discontinuities in the fluid variables that model the diaphragm. A shock is generated that propagates through the mixture of hydrogen and oxygen. The temperature and pressure behind this primary shock are high enough to ignite the gas. Some time after the shock has started propagating, a reaction wave begins near the contact surface. As with the reflected shock calculation described above, the reaction wave propagates forward in the shocked material and finally merges with the shock front from the bursting diaphragm, causing transition to detonation.

Figure 13–4 shows the general shape of the temperature profile after the diaphragm has burst but before ignition. As in the reflected shock tube problem, a combination of variably spaced and moving grids were used to

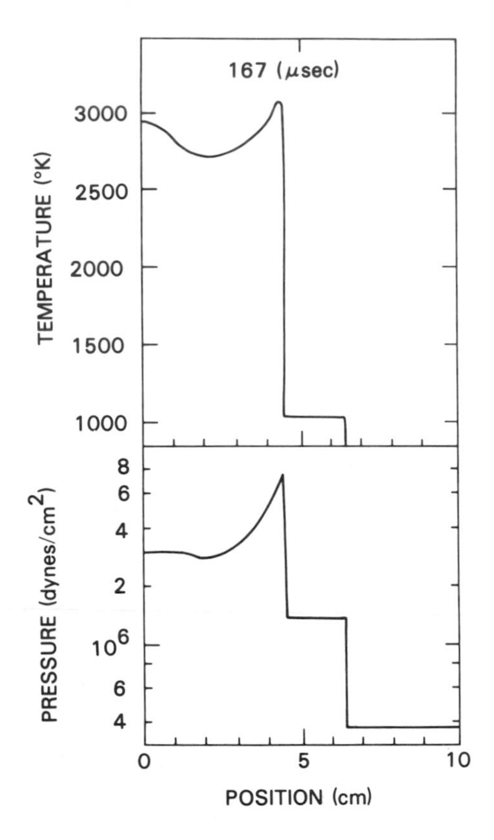

Figure 13–3. Reaction wave propagating through a shocked region in a reflected shock calculation (Oran et al., 1982a).

resolve the contact discontinuity, the reaction wave, and the shock front. A finely gridded region was placed at the contact discontinuity and kept there throughout the course of the calculation. The movement of this finely gridded region was keyed to the velocity of the contact discontinuity, that is, the fluid velocity. The grid is approximately Lagrangian in this region. A second finely gridded region was held near the contact discontinuity in the shocked region until a reaction wave started. This second region was then allowed to move with the reaction wave into the shocked gas. A third finely gridded region moved with the shock front, and eventually merged with the second region. This complicated gridding exercise produced extremely well resolved calculations.

One set of these calculations was performed for a mixture of hydrogen and oxygen in a temperature and pressure range characteristic of "weak

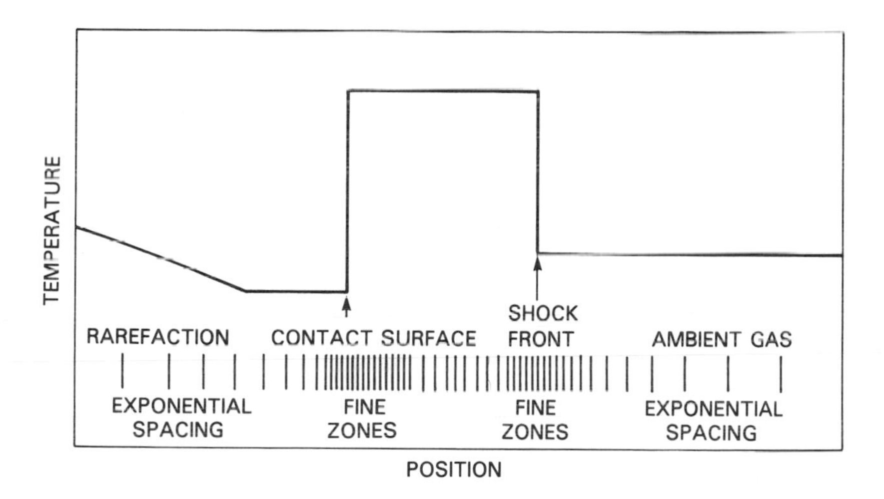

Figure 13–4. Schematic showing an idealized temperature profile in a bursting diaphragm shock problem. Regions of coarse and fine gridding used in the calculation are marked.

ignition." In this regime, the chemical induction time is very sensitive to small changes in temperature and pressure, thus the ignition process is not as laminar as expected. Because relatively small fluctuations in the background quantities can initiate or speed up chemical reactions, hot spots and ignition occur at locations behind the shock front that are not very close to the contact discontinuity. The shock initially travels at a nearly constant velocity, so the temperature and pressure of the shocked mixture are raised to a near-constant value. Chemical reactions in the heated gas first occur relatively near the contact surface where the temperature has been high for the longest time. These reactions generate rather small pressure disturbances which travel forward at a velocity which is the sum of the sonic and fluid velocities behind the shock. When they reach the shock, these pressure disturbances accelerate the shock slightly, giving another temperature increase behind the shock. This small increase in temperature increases the rate of energy release. This process is shown in the sequence of temperature profiles in Figure 13–5.

The irregular structure that results behind the shock wave prior to ignition can been seen in Figure 13–6, which shows distinct peaks in OH mole fraction and a temperature profile generally increasing toward the shock front. We see three reacting centers. The temperature in the one nearest the shock front eventually increases enough to cause a reaction wave. Figure 13–7 shows the subsequent evolution of the hot spot and the transition to

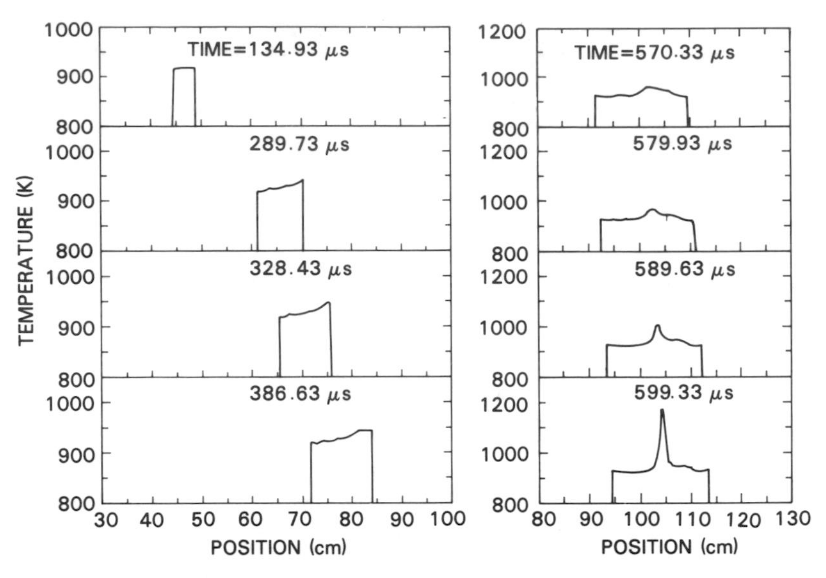

Figure 13–5. Calculated temperature profiles in the bursting diaphragm shock problem (Kailasanath and Oran, 1983).

Figure 13–6. Profiles of temperature and OH mole fraction in the early stages of a bursting diaphragm shock calculation (Kailasanath and Oran, 1983).

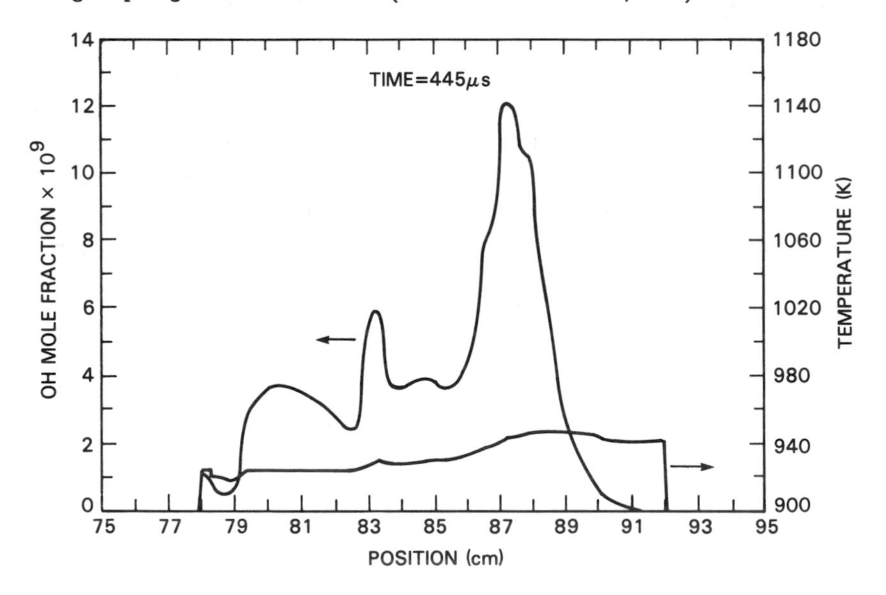

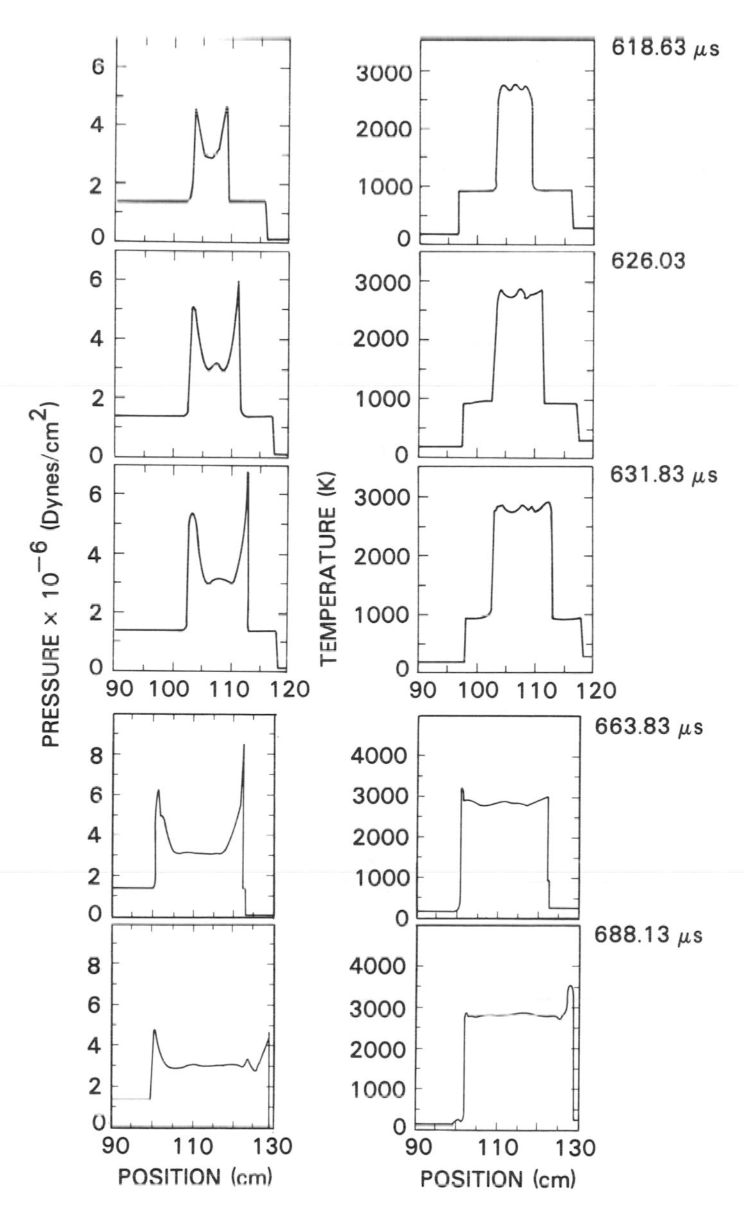

Figure 13-7. Temperature and pressure profiles in the later stages of the incident shock calculation (Kailasanath and Oran, 1983).

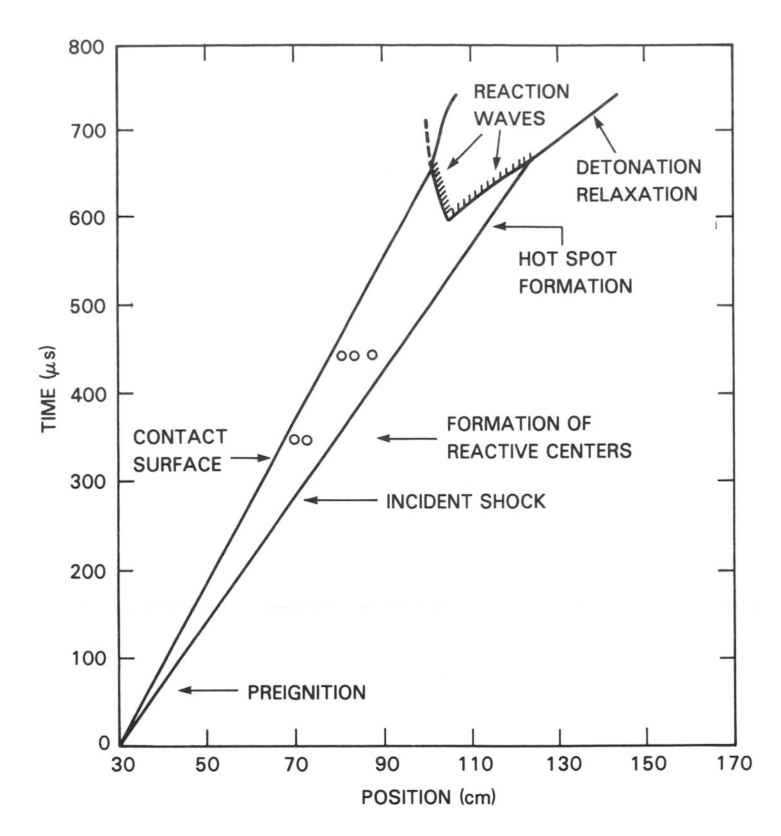

Figure 13–8. Time history of the important waves in the system in the bursting diaphragm shock calculation (Kailasanath and Oran, 1983). A circle represents the time and location of the formation of a reactive center.

detonation. Figure 13–8 summarizes the calculation and shows the location in time and space of the formation of reactive centers.

13–2. MULTIDIMENSIONAL COMPRESSIBLE REACTIVE FLOWS

FAST2D is a generic name for time-dependent, two-dimensional, compressible. Eulerian programs that use direction splitting and timestep splitting with the Flux-Corrected Transport algorithm. The direction-splitting approach was described in Section 9–2 in the context of an axisymmetric $r - z$ geometry. The conversion to other geometries is straightforward using switches in the FCT subroutines. The one-dimensional models described above used detailed models of chemical processes. In the two-dimensional models described here, the chemical reaction and energy release models are simplified to reduce the calculation cost.

Applications to subsonic flows include calculations of vortex shedding and asymmetric mixing in shear layers by Grinstein et al. (1986, 1987) and acoustic-vortex interactions in a ramjet combustor configuration by Kailasanath et al. (1985a, 1986, 1987). Applications to supersonic flows include studies of the initiation and structure of propagating detonations by Oran et al. (1982b), Picone and Boris (1983), Kailasanath et al. (1985b), and Guirguis et al. (1986, 1987), studies of shock-bubble interactions by Picone et al. (1984, 1986), interior ballistics by Edwards (1986) and Boris et al. (1983).

FAST2D, and the three-dimensional version, FAST3D, have been used for plasma, ionospheric, astrophysical, and solar physics studies. For example, Zalesak et al. (1982, 1985) used similar models for simulating equatorial spread-F and the evolution of barium clouds injected into the Earth's ionosphere, and Lyon et al. (1981, 1986) modeled the dynamics of the Earth's magnetotail and simulated a geomagnetic substorm. Chevalier and Gardner (1974) studied supernova explosions in the galactic disk. Dahlburg et al. (1986) studied two-dimensional condensational instabilities in the solar corona. Emery et al. (1982a,b) used similar models to simulate high-density fully-ionized plasmas generated by high-intensity lasers.

The models we describe below include the effects of gravity, thermal conduction, and have a phenomenological model for energy release to represent chemical reactions. The generic FAST2D description below assumes that the reader has already read Section 13–1 describing the equivalent one-dimensional models.

13–2.1 Structure and Coupling in FAST2D

FAST2D models are constructed by linking together a number of optimized routines for different physical processes through timestep-splitting methods. These algorithms include:

- Flux-Corrected Transport, which in this model is applied with direction splitting in two orthogonal dimensions.
- Boundary conditions representing inflow, outflow for subsonic, choked, and supersonic flows, and closed boundary conditions.
- Adaptive gridding methods, to move the grid from timestep to timestep according to specified physical criteria or to cluster the grid where higher resolution is needed.
- An implicit thermal conduction module for problems in which convection does not dominate the energy transport.

Table 13–2. Outline of FAST2D

Initialize Variables: $t^n = t^o + \Delta t$
* Determine Δt
  1. External Energy Deposition: Problem dependent.
  2. Grid Adjustment: Move grid as needed to maintain resolution
  3. Convection in first direction: As described in Section 9–2.
  4. Convection in second direction: As described in Section 9–2.
  5. Gravity: As described in text.
  6. Chemical Energy Change:
     Effects of energy release and species conversion.
  7. Update Variables

Start New Timestep (go to * above).

- The induction parameter model, to simulate chemical reactions and energy release.
- Equations of state.
- A gravity term appearing as an external force.
- Timestep-splitting techniques to couple the various physical and chemical processes together, as introduced in Chapter 4.

An overview of the order of operations in FAST2D is given in Table 13–2.

Strong Thermal Conduction

Using an explicit time integration algorithm generally restricts the timestep to a fraction of the acoustic transit time across the smallest cell. In normal reactive flows this usually means that physical diffusion and thermal conduction can be integrated accurately using a simple explicit algorithm. There are, however, situations in which the sound speed is not the fastest characteristic velocity in the system and yet we need to integrate the model for a number of acoustic transit times. FAST2D models have been applied to plasma systems where the high temperature partially or fully strips the atoms of their electrons. When the thermal conduction is dominated by the motions of the free electrons which are much faster than the sound speed, an implicit treatment of the thermal conduction is needed. In addition, the thermal conduction varies as $T^{5/2}$ and thus is even more nonlinear than the $T^{3/2}$ dependence considered in Chapter 7. Algorithms for nonlinear thermal conduction and for multidimensional diffusion are described in some detail in Sections 7–3 and 7–4.

Emery et al. (1982a) and Dahlburg et al. (1986) consider the two-dimensional dynamics and stability of laser-driven ablation layers. In these problems, the temperature varies by tens of electron volts over a distance of a few microns. Figure 11–7 showed mass density contours for a thin slab of plastic heated by a high-intensity laser shining on the slab from the right. The layer develops a rippled surface from the Rayleigh-Taylor instability, and the shape of this surface differs markedly depending on the wavelength of the ripple. In these simulations, the heat conduction equation was integrated implicitly in the direction of the incident laser. The y-components were treated explicitly. This simplification over having to treat two directions implicitly worked because the steepest temperature gradients were in the x-direction and because the cell sizes in the y-direction are larger by a factor of five or ten than the smallest cell sizes in the x-direction.

The Equation of State

Although it is expensive to incorporate the correct equations of state in multidimensional calculations, sometimes it is necessary to do this to get quantitatively correct answers. For the gas-phase problem, a complete equation of state usually requires some kind of iteration to determine consistent variables, and this adds to the overall cost. Multidimensional calculations that include the full equations of state for the explosive CP were shown in Section 12–3. Below we show calculations that include a complicated equation of state for liquid nitromethane (Guirguis et al., 1986, 1987).

For the gas phase, it is convenient to evaluate the equation of state using power-series formulas for species enthalpies (Chapter 5). When this is too expensive, it is sometimes possible to use a simpler approximation. If there are no chemical reactions or they are particularly simple, a single table lookup model giving $P(\rho, E)$ might be adequate. The temperature can be determined from the ideal gas law. When different species are being formed and consumed through chemical reactions, this approach does not work because the enthalpy functions, and hence the equations of state, differ from one species to another. When the chemical species exist in different proportions in different locations, the equation of state depends not only on mass and internal energy density but also on the individual species densities.

By assuming that each species has a separate value for the enthalpy h_i, and heat of formation h_{oi}, the temperature or energy release can be calculated from the change in number densities of the species. In turn, this can be used to calculate the pressure change and hence the new pressure.

Chemical Kinetics

A detailed chemistry model can be included in FAST2D, as described for RSHOCK. However, we seldom do this. Integrating the full chemical kinetics is usually the most expensive part of a calculation, and only occasionally affordable in multidimensions. In addition, the complete detailed chemical reaction scheme is often not known well enough to make this worthwhile. Because of this, the chemical reactions and energy release are often approximated by simplified models.

One useful chemistry phenomenology is the *induction parameter model*, for numerical simulations by Oran et al. (1982b). The form described in Guirguis et al. (1986, 1987) considers a two-step model for ignition of a combustible mixture. The first step is a chemical induction period in which

$$\text{Step 1}: \quad Fuel + Oxidizer \rightarrow Radicals \; . \tag{13-2.2}$$

In the second step,

$$\text{Step 2}: \quad Fuel + Oxidizer + Radicals \rightarrow Products \; . \tag{13-2.3}$$

When the concentration of radicals is low, they are produced by Step 1 much faster than they are consumed by Step 2. This is a period in which the number of radicals increases rapidly but their number is still relatively small. There is little temperature change and the amount of products formed is negligible. The rate of Step 1 can be expressed in terms of an induction time $\tau_o(T, \rho)$. One way of finding the input data $\tau_o(T, \rho)$ is to integrate a full set of detailed chemical equations; another way is to extract it from experimental data. The induction period is complete when appreciable products from Step 2 begin to appear.

Let f denote the fraction of the induction time elapsed at time t, then

$$\frac{df}{dt} = \frac{1}{\tau_o(T, \rho)} \; , \tag{13-2.4}$$

where $f(\mathbf{x}, 0) = 0$. The rate of Step 2 is kept at zero until f exceeds unity. This formula provides a convenient way of averaging over changing temperature and density during the chemical induction period. Equation (13–2.4) reduces automatically to the correct induction delay when the density and temperature are constant.

Once the induction delay has elapsed, the fuel and oxidizer are consumed according to

$$\frac{d}{dt} N_{\text{fuel}} = -N_{\text{fuel}} \, A \exp(-E/RT) \tag{13-2.5}$$

where N_{fuel} is the density of fuel. In this model, f can be interpreted as the fraction of the critical radical concentration. When this model is used in a moving system, d/dt in Eqs. (13–2.4) and (13–2.5) denotes the derivative following the fluid flow. The quantity f must be convected with the fluid. Several examples of detonation simulations using this model are described below.

Incorporating the Effects of Gravity

There are two different ways to include the effects of gravity. The first uses the idea of a *gravitational potential energy*. Then the energy conservation equation contains an expanded definition of the energy density,

$$E^* \equiv \frac{1}{2}\rho v^2 + \epsilon + \phi(\mathbf{x},t) . \qquad (13-2.6)$$

For example, the gravitational potential can be written as $\phi = \rho g h$ in a planar geometry with constant gravity. More generally, the gravitational potential is defined as

$$\phi(\mathbf{x},t) \equiv -\rho(\mathbf{x},t)\int_{\mathbf{x}_o}^{\mathbf{x}} \mathbf{g}(\mathbf{x})\cdot d\mathbf{x} , \qquad (13-2.7)$$

where $\mathbf{g}(\mathbf{x})$ is the vector gravitational field and $\mathbf{x}_o$ is a reference location where the gravitational potential is taken to be zero.

The energy conservation equation is written just as before with the new definition of the energy density, E^*,

$$\frac{\partial E^*}{\partial t} = -\nabla\cdot E^*\mathbf{v} - \nabla\cdot P\mathbf{v} . \qquad (13-2.8)$$

The form of the equation has not changed, but the definition of the total conserved energy has. There is also a force $\rho\mathbf{g}$ which must be added to the momentum equation.

A problem arises with this approach through errors accumulating from taking differences of large numbers. When the system covers many gravitational scale heights, the gravitational potential term in the energy equation can exceed the internal and kinetic energies by large factors. It then becomes difficult to find the temperature accurately.

A second approach is to integrate the equation for the internal plus kinetic energy without the potential energy term included in Eq. (13–2.6). Large numbers of scale heights can be represented because the calculations are entirely local. However, the momentum change and the kinetic energy change from the gravitational acceleration must be consistent. One way to ensure this is to timestep split the gravitational terms in the momentum and energy equations and treat them together consistently. Let ρ^o, v^o, and E^o be the mass density, velocity, and energy density after one timestep calculated without gravity. The updated values of these quantities due to gravitational effects, ρ^n, v^n, and E^n, can then be determined from

$$\rho^n = \rho^o \; , \qquad (13-2.9)$$

$$\rho^o(\mathbf{v}^n - \mathbf{v}^o) = \rho^o g \Delta t \; , \qquad (13-2.10)$$

$$E^n - E^o = \frac{1}{2}\rho^o(\mathbf{v}^n \cdot \mathbf{v}^n - \mathbf{v}^o \cdot \mathbf{v}^o) \; . \qquad (13-2.11)$$

The temperature and pressure are not directly affected by gravity.

Timestep Splitting in Two Dimensions

When direction splitting is used to reduce a multidimensional problem to a sequence of one-dimensional problems, the immediate question is "which direction should be integrated first?" Whichever choice is made, errors from asymmetries in the solution procedure cannot be avoided. When these errors become important, there are ways to eliminate them. One way is to calculate the convection twice, alternating directions, and take an average. That is, denote the beginning of the convective transport substep as "old" values. Then,

1. Starting with the old values, integrate in the x-direction and then in the y-direction. Do not update the system variables, but save the new ones.
2. Starting with the old values, integrate in the y-direction and then in the x-direction. Do not update the system variables, but save the new ones.
3. Average the two results and update the system variables with the averaged values.

This approach is adequate, but is more expensive. Another approach is to use the fully multidimensional FCT versions described in Chapter 9.

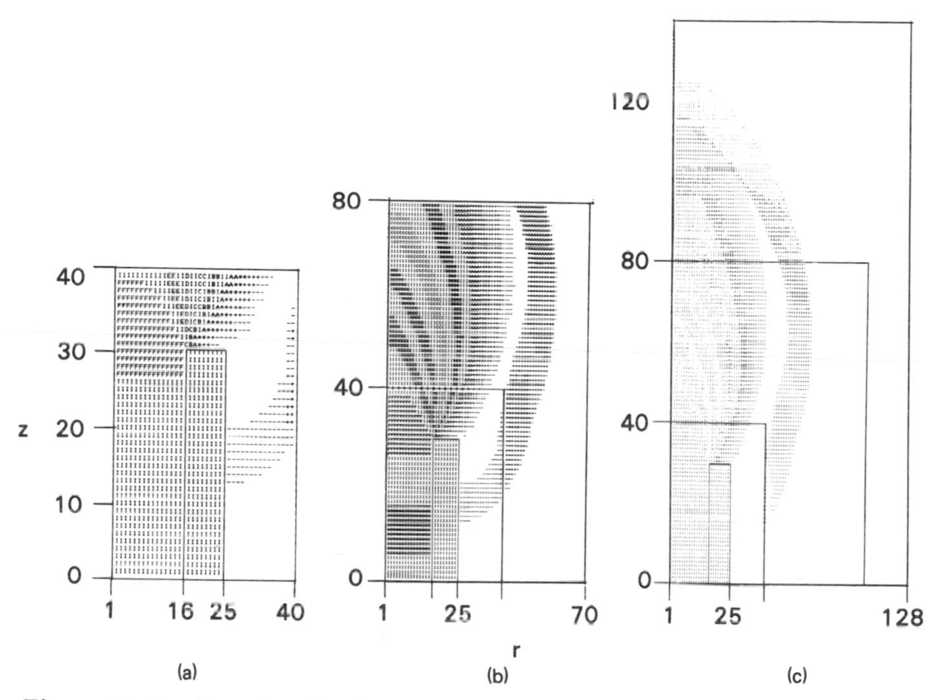

Figure 13–9. Test of outflow boundary conditions in a calculation of a shock leaving a barrel: (a) 40 × 40 grid; (b) 80 × 80 grid; (c) 150 × 300 grid. The computational cell size is the same in each calculation (Boris et al., 1983).

13–2.2 Some Examples

Tests of Boundary Conditions and Convergence

Much work has been done to test boundary conditions and convergence in FAST2D models. Boris et al. (1983) described tests of boundary conditions in both supersonic and subsonic flows, and Grinstein et al. (1986) described boundary-condition tests and convergence tests in subsonic and choked flows. In addition, Kailasanath et al. (1986, 1987) described convergence tests in subsonic flows, and Guirguis et al. (1986) and Edwards (1986) have reported convergence tests in supersonic flows. Here we describe several tests of particular interest to reactive flows.

Figure 13–9 from Boris et al. (1983) shows calculations of a shock exiting a solid, finite-thickness cylinder and then expanding. This could, for example, model a muzzle blast. The calculation on the left used an evenly spaced 40 × 40 grid with the axis of symmetry on the left side and outflow boundary conditions (Section 11–1) on the top and right side. The center

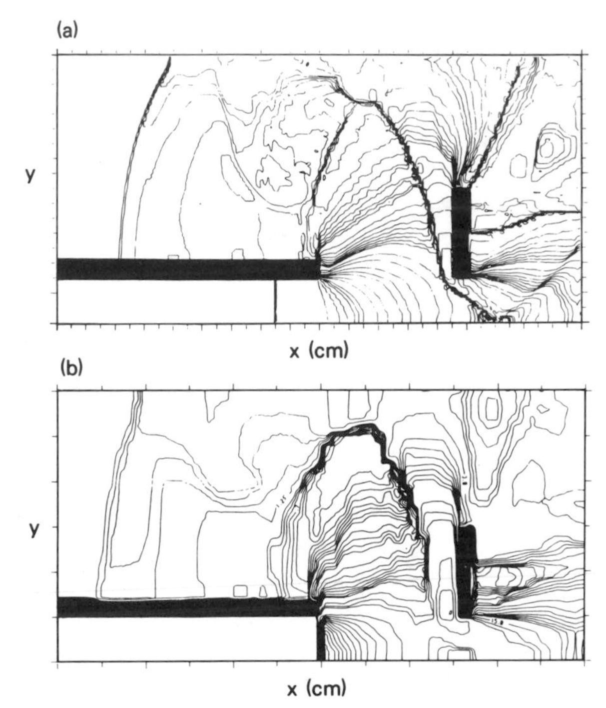

Figure 13–10. Pressure contours for a muzzle blast calculation. Calculation (a) has three times the resolution of calculation (b). (Figure is courtesy of G. Edwards, Fluid Gravity, Ltd, Surrey, England.)

panel shows exactly the same problem on an 80×80 grid, where the lower left 40×40 cells are exactly the same as in the 40×40 calculation. The panel on the right shows the same calculation, this time using an even larger region and a 150×300 grid. All three calculations are shown at timestep 200.

In each case the approximate outflow boundary conditions used on the outer edges of the smaller grid, described in Table 11–3, can be tested for a time against the "exact" solution on the larger grid. The larger-grid solution includes all of the expanding primary shock that has not yet reached the boundary in the larger grid. Therefore any boundary condition errors in the larger-grid solution cannot yet influence the solution. The three calculations have essentially the same values along the top and right hand side walls. This is remarkable since any numerical reflections from the top boundary of the 40×40 simulation have had time to pass down and back up the length of the system by step 200.

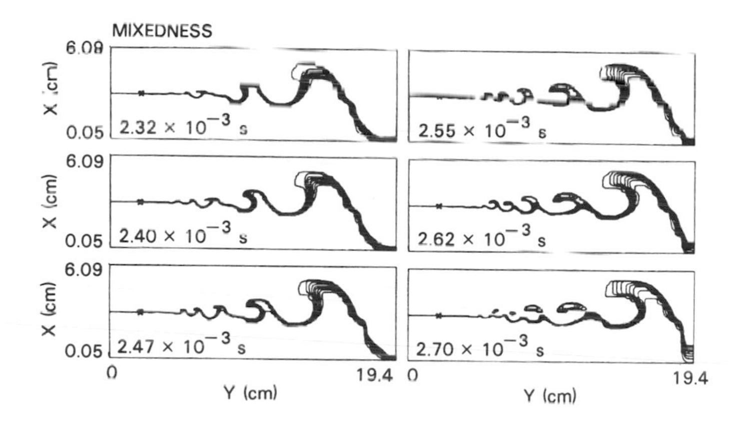

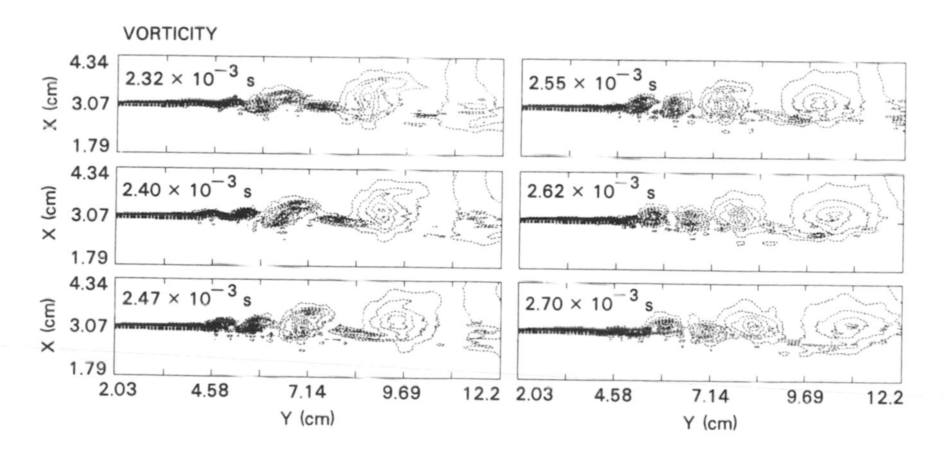

Figure 13–11. Contours of vorticity and mixedness (see text) for a calculation of a planar mixing layer (Grinstein et al., 1986). Both gas streams consist of air, with the fast stream on the bottom.

Edwards (1986) used a very similar geometry to study muzzle blasts. An example of his resolution convergence tests is shown in Figure 13–10. The axis of symmetry is now at the bottom of each panel and an additional "blast shield" is included as an impermeable region in the middle right. The lower panel shows pressure contours from a calculation using 60×35 cells. The upper panel is from a calculation with three times the spatial resolution. In this case the more resolved calculation not only has sharper shocks, but also shows additional structure.

Subsonic Flows

Mixing and subsequent combustion in shear-layer geometries are important in practical applications. Section 12–1.1 highlighted calculations that showed development of vorticity in these mixing layers. Figure 13–11 (Grinstein et al., 1986) shows vorticity contours and contours of a mixedness parameter, R, for a calculation in which the fast flow enters from the bottom left and a slower flow enters from the top left. The mixedness parameter R is defined as

$$R \equiv \frac{N_{\mathrm{fast}}}{N_{\mathrm{fast}} + N_{\mathrm{slow}}} , \qquad (13-2.12)$$

where N_{fast} is the density of the fast flow and N_{slow} is the density of the slow flow. The grid for this calculation was shown in Figure 6–9.

Kailasanath et al. (1985a, 1986, 1987) used a version of FAST2D to study the interactions between vortex structures and the acoustics in a ramjet combustion chamber. Nonreactive calculations in confined, axisymmetric geometries were shown in Figure 12–4, and the grid was described in Figure 9–1. Figure 12–4 showed a subsonic, slightly preheated, but nonreacting, mixture of hydrogen and air entering into a combustor. The mixture evolved to a quasi-periodic pattern as vortices were shed, merged, and exited through the nozzle. The temperature before ignition was essentially uniform.

In Figure 13–12, an originally unreacting flow containing hydrogen and oxygen is "ignited" in a small region near the chamber inlet at step 160,000. Chemical reactions and heat release now occur according to an induction parameter model. The instantaneous streamlines and temperature contours in Figure 13–12 show the flame front moving down the chamber. A quasi-steady combustion pattern is eventually set up with large-amplitude recurrent oscillations. The flame front is located on the temperature contours both by the location of the high-temperature contours and the dark lines caused by the closely spaced contours. In time, the rollup in the shear layer causes the reaction front to curve downward and engulf the cold mixture which subsequently burns. As the reaction moves downstream, a new vortex forms near the step between timesteps 175,000 and 180,000. This mixes the burned gases with the incoming mixture and acts as an ignition source. The flow field undergoes a cycle of roughly 25,000 timesteps, or 3.463 ms.

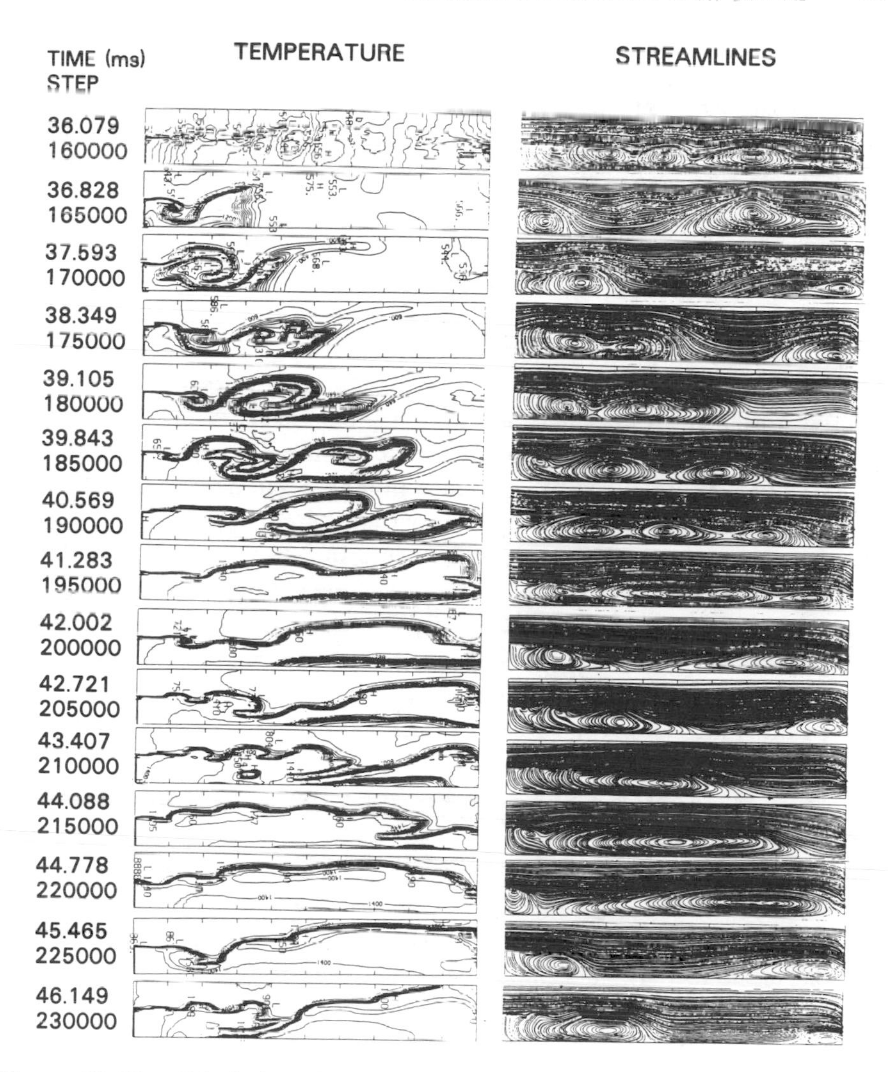

Figure 13–12. Calculation of a reacting flow initiated in a cold flow similar to that shown in Figure 12–4. Instantaneous streamlines and temperature contours are shown for an axisymmetric combustor (Kailasanath et al., 1986a).

Propagating Detonations

Models of the FAST2D type are useful for studying the multidimensional dynamics of explosions and detonations. Of particular interest is the multidimensional structure of detonation fronts, as described briefly in Chapter 2. The detonation front, shown schematically in Figure 2–2, is made up of intersecting shock waves. The intersections, called *triple points*, move and trace out diamondlike structures called *detonation cells*. The size and regularity of the patterns of detonation cells depends on the particular material and on the initial temperature and pressure.

Figure 6–9 showed contours of density and temperature and a schematic of the grid used by Kailasanath et al. (1985) for a calculation of a detonation propagating in a channel filled with hydrogen and oxygen diluted with argon. These simulations were performed in planar geometry using the sliding Eulerian rezone (Chapter 6) to keep the fine zones at the detonation front throughout a calculation. From contours such as these, it is possible to trace out the pattern of triple points to see the formation of detonation cells. Figure 13–13 shows a sequence of density contours from such a calculation. The lines with arrows indicate the direction of motion of the triple points. Figure 13–14 show a sequence of these traces for calculations in channels of three different sizes. The largest channel, 9 cm, shows a characteristic detonation cell pattern.

Figure 13–15 shows contours of the extent of reaction and the temperature for a detonation propagating in the same mixture in a very narrow channel. The dark region on the right is cold, unreacted material ahead of the advancing detonation. The light region to the left is fully reacted material in which energy release is completed. A peculiar feature of these calculations is the formation of unburned gas pockets at the detonation fronts, first noted in simulations by Picone, Oran, and Boris and reported in Oran et al. (1982). These pockets may be cut off as transverse waves hit the wall or by other transverse waves. Note also that the temperature in these pockets is cooler than the surrounding temperature. The unburned pockets subsequently burn and release energy, but their formation takes energy out of the propagating detonation. These pockets are a potential mechanism for detonation failure.

Recently Guirguis et al. (1986, 1987) studied detonations in liquid nitromethane including the correct equation of state and an induction parameter model based on experimental data. The simulations were initiated by perturbing a steady one-dimensional detonation with a pocket of reacting material in the center of the induction zone, as shown in step 200 in Figure 13–16. When the shock wave generated by the pocket reaches the

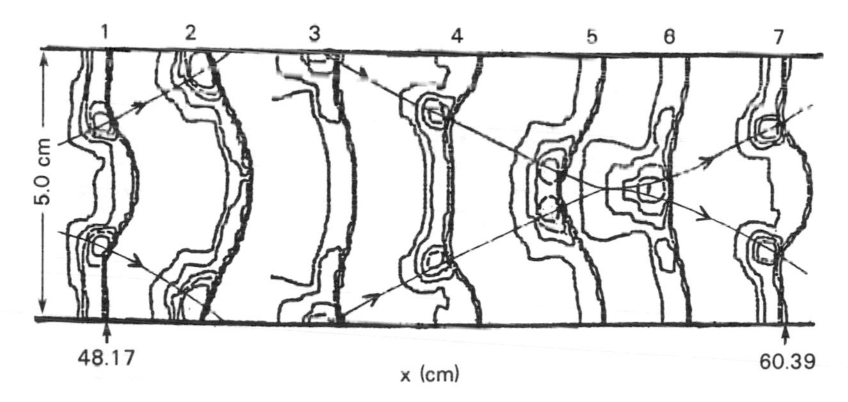

Figure 13–13. Density contours from seven timesteps in the simulation of a detonation propagating in a 60% argon diluted hydrogen-oxygen mixture in a 5 cm wide channel (Kailasanath et al., 1985). The lines with arrows indicate the direction of the triple points.

Figure 13–14. Calculated paths of triple points for the same mixture as in Figure 13–13 (Kailasanath et al., 1985). (a) 5 cm wide channel; (b) 7 cm; (c) 9 cm.

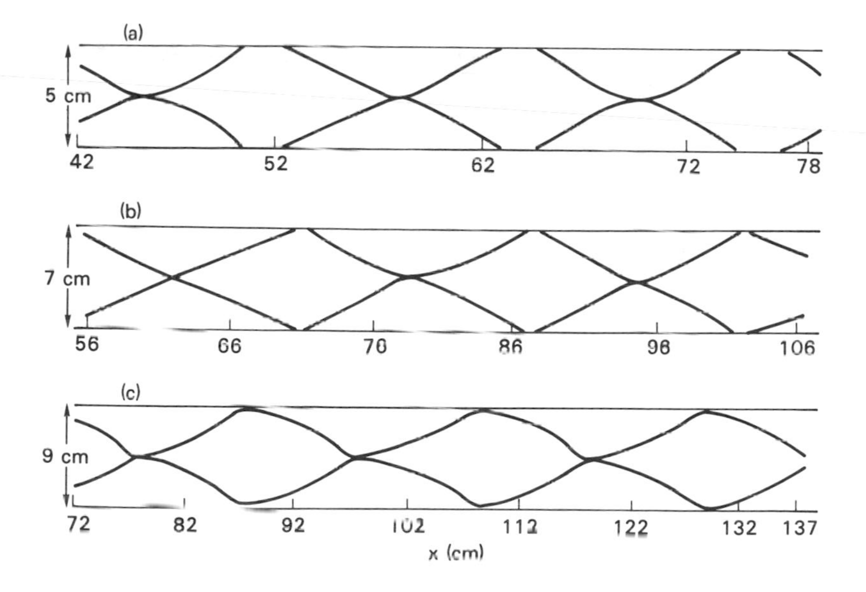

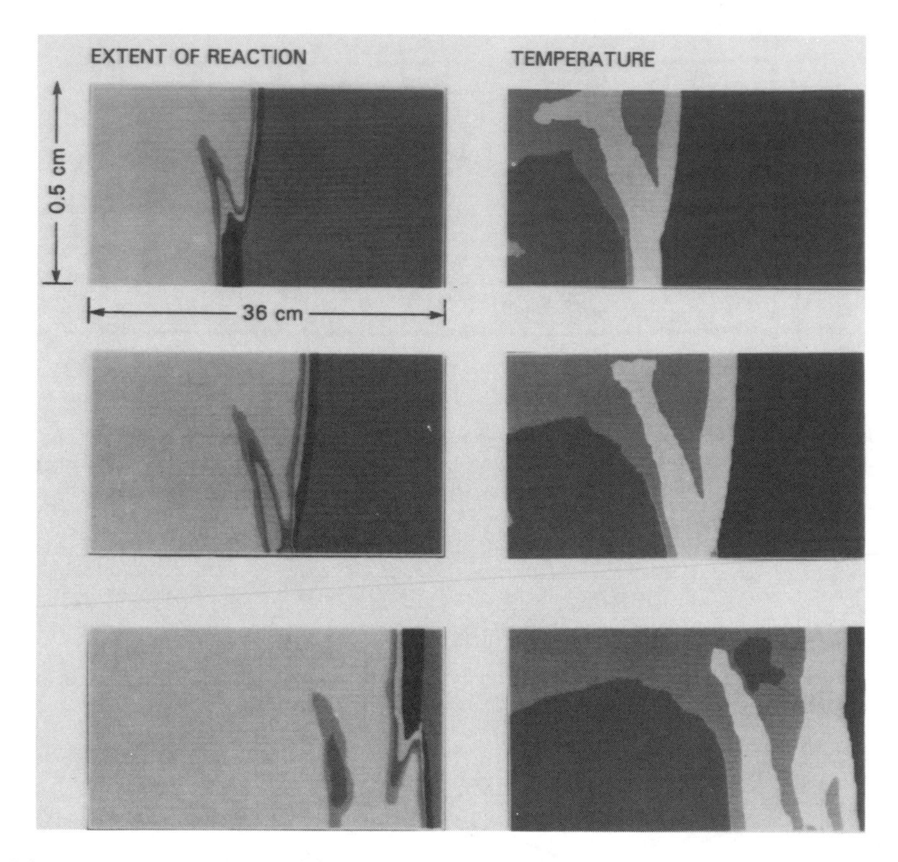

Figure 13–15. Extent of reaction and temperature contours from three steps in a calculation of a detonation propagating in a 60% argon diluted hydrogen-oxygen mixture in a very narrow channel. The detonation propagates from left to right. (Figure is based on material presented in Oran et al., 1982.)

detonation front, two transverse waves are formed, as illustrated by pressure contours at steps 400, 500, and 600. This figure also shows the details of the interaction between the shock wave generated by the energy pocket and the rear wall. After a few collisions with the channel walls, the transverse waves establish the structure typical of the detonation front.

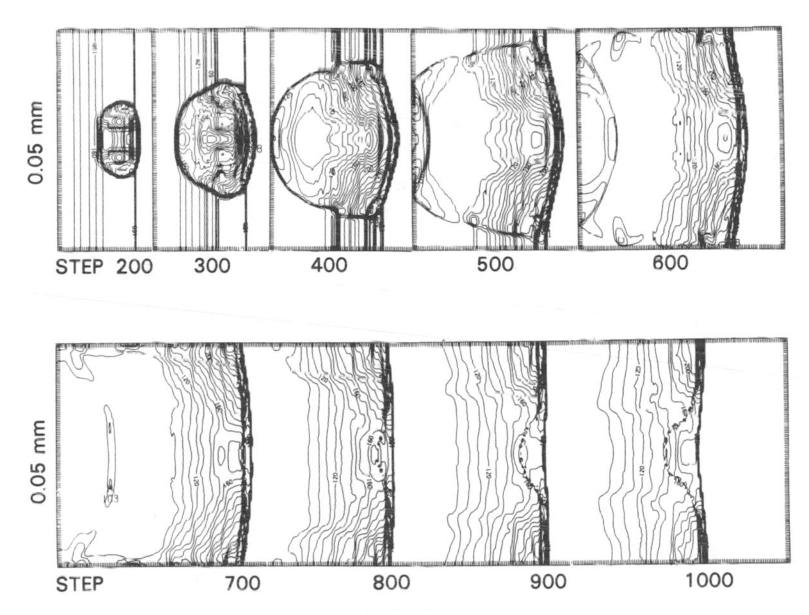

Figure 13–16. Pressure contours of a typical initiation of the multidimensional structure in a liquid nitromethane detonation (Guirguis et al., 1986).

Figure 13–17 shows a later sequence of pressure contours behind the advancing detonation front. The curious feature of these contours is that the structure is very irregular compared to the hydrogen-oxygen detonation diluted in argon. In fact, detonations cells are generally very irregular. Guirguis et al. have shown that by changing the temperature dependence of the induction time, the structure can be made very regular, as shown in Figure 13–18. These figure are from a series of studies done to isolate the factors affecting the regularity of the detonation cell structure.

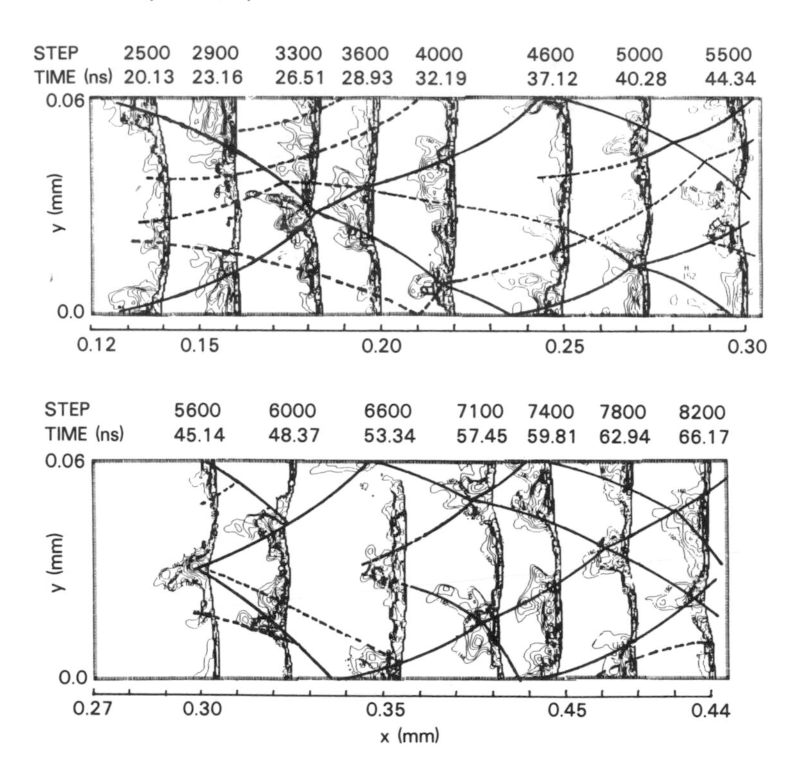

Figure 13–17. Pressure contours of a detonation propagating in liquid nitromethane (Guirguis et al., 1986). The heavy solid lines are the paths of the main triple points. The dashed lines are the paths of secondary triple points.

Figure 13–18. Pressure contours for a detonation propagating in liquid nitromethane (Guirguis et al., 1986). By adjusting the induction time, the weak triple points in Figure 13–17 disappear.

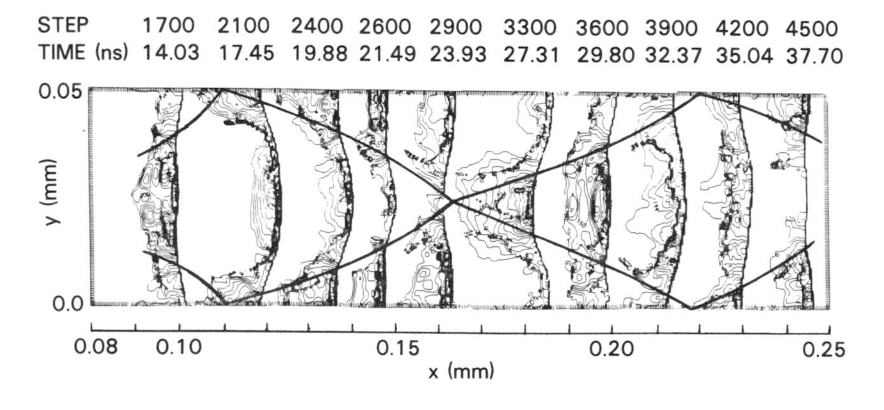

13–3. ONE DIMENSIONAL SUBSONIC REACTIVE FLOWS

Convective transport, thermal conduction, molecular diffusion, chemical kinetics and energy release, radiation transport, and perhaps other heat or material sources and sinks are all important in a gas-phase flame. For a review of the physics and chemistry of flames, we recommend Lewis and von Elbe (1961), Gaydon and Wolfhard (1979), Strehlow (1984), and Williams (1985).

This this section describes FLAME1D, a set of one-dimensional, time-dependent, Lagrangian models for subsonic reactive flows. They have been used for a variety of laminar flame studies, including calculations of flame ignition (Oran and Boris, 1981), minimum ignition energies and quench volumes (Kailasanath et al., 1982a), and burning velocities (Kailasanath et al., 1982b). Comparisons of calculations with both analytical results and experimental data have validated the models. A review of FLAME1D and some of its applications is given by Kailasanath et al. (1982c). The book, *Numerical Methods in Laminar Flame Propagation* edited by Warnatz and Peters (1982) summarizes the features of a number of steady-state and time-dependent one-dimensional flame models, including FLAME1D.

If the diffusive transport processes were added to RSHOCK models, they could in principle be used to describe flames as well as shocks and detonations. However, this is not the best approach for several reasons. First, an RSHOCK model would require more resolution for a flame calculation than a FLAME1D model because FLAME1D models are Lagrangian and have no numerical diffusion in the convective transport step. Second, the timestep in RSHOCK models is determined by the Courant condition, based on the cell size and the sound speed of the material. The implicit method in FLAME1D allows longer timesteps. Thus a considerable amount of computer time is saved when the characteristic flow speeds are subsonic and the cells are small.

13–3.1 Structure of FLAME1D Models

FLAME1D combines the following algorithms:

- ADINC, an implicit, Lagrangian method for solving the convective parts of the conservation equations (Boris, 1979), discussed in Chapter 10.
- Lagrangian rezoning to split and merge cells to maintain accuracy in selected regions, discussed in Chapters 6 and 10.
- DFLUX, an algorithm for determining diffusion fluxes (Jones and Boris, 1981), discussed in Chapter 7.

- SPRED, conservatively differenced algorithms for solving a diffusion term, discussed in Chapter 7.
- CHEMEQ, for ordinary differential equations (Young, 1977, 1979), discussed in Chapter 5.
- A polynomial representation of the enthalpies as a function of temperature for the equation of state, discussed in Section 5.
- Timestep-splitting techniques, somewhat modified from those used for the supersonic flows described above.

Table 13–3 outlines the structure of these programs. The first obvious feature is that there are more physical processes included in FLAME1D than in RSHOCK or FAST2D. Thermal conduction, ordinary diffusion, and thermal diffusion are included in addition to convection, chemical reactions, and heat release. These three diffusion processes are the bare minimum needed for quantitative description of laminar flames. Other versions, including viscosity and radiation transport, are not discussed here.

The term representing thermal conduction is conservatively differenced and solved explicitly. The timesteps required by the thermal conductivity or the species diffusion may be unacceptably small when a light, relatively fast-moving species such as hydrogen is involved. If the whole program were run at the timesteps these fast diffusion processes require, the cost would be exorbitant. The required diffusion timestep may even be smaller than the timestep required by the Courant condition. To get around this, we have used subcycling to determine $\{v_{di}\}$, similar to that applied in the chemistry integration stage of RSHOCK and FAST2D.

The overall timestep, Δt, can no longer be chosen by considering only the convection timestep imposed by ADINC. This timestep is generally much too large to accurately resolve the other physical processes. The only real requirement for ADINC is that the cell interfaces do not cross. The timestep must be chosen by considering the stability and accuracy requirements of all processes included in the model, and taking the largest step permitted. This does not mean using timesteps as small as the stiff ordinary differential equations require chemistry can still be subcycled as discussed in Section 13–1. In principle, many of these processes, including turbulence and radiation transport models, can be calculated implicitly or subcycled to allow larger timesteps.

A flame front propagates at a velocity determined by the various convective and diffusive transport processes. This "interface" does not move at the fluid velocity. As the flame moves through the unburned gas mixture, expansion occurs at the flame front because energy is released. Thus a Lagrangian grid tends to spread out just when it should be most resolved. The

Table 13–3. Outline of FLAME1D

Initialize Variables.

* Determine Δt

    Considers all physical processes involved.

Rezone to Improve Resolution

    Add and delete cells.

1. External Energy Deposition

    Varies depending on the problem.

    All variables now labeled with superscript 1.

    Calculate ΔP_1.

2. Chemical Reactions

    Integrate from t to $t+\Delta t$.

    Only update $\{n_i(x)\}$.

    Calculate ΔP_2.

Evaluate Diffusive Transport Coefficients

    $\lambda_m(x)$, $\{D_{ij}(x)\}$, $D_i^T(x)$.

3. Thermal Conduction

    Integrate from t to $t+\Delta t$.

    Calculate ΔP_3.

    Do not update any variables.

4. Ordinary Diffusion and Thermal Diffusion

    Integrate from t to $t+\Delta t$.

    Only update $\{n_i(x)\}$.

    Calculate ΔP_4.

5. Convective Transport

    Calculate the input to ADINC using new pressure,

    $$P^n = P^o + \sum_i \Delta P_i \quad .$$

    Integrate from t to $t+\Delta t$.

    Update all variables: $\rho(x)$, $P(x)$, $T(x)$, $E(x)$, $\{n_i(x)\}$, $\gamma(x)$.

Start New Timestep (go to * above).

answer is to use a rezone procedure that splits or merges cells as required. Adding or removing cells in the regions of a steep temperature gradient requires interpolating values to obtain smaller cells or combining values to get larger cells. Unless the interpolation is very good, the process is diffusive. Rezoning really works only in regions where not much is happening and it is in fact not needed yet. The solution to this problem is to rezone somewhat ahead of and behind the flame front. That is, split cells ahead of the flame front and merge them well behind it. This is straightforward because the direction of flame propagation is known.

Closed boundary conditions are simulated by requiring that the velocity at the wall is zero. In this case, the flame propagates in a confined chamber and the pressure increases in time. To study unconfined flames, we need an algorithm for an open boundary. One approach is to increase the size of the computational cells as they get further from the flame. Thus the computational domain becomes very large without a corresponding increase in computer storage. This cell stretching should be gradual to limit inaccuracies which arise as a result of the varying cell size. Adaptive gridding is important because the resolution around the front must increase to maintain accuracy as the flame front moves toward the large cells.

A second approach moves the open boundary to keep the pressure essentially constant. The movement of the last cell boundary on the open end of the system allows the pressure in each cell at the end of a timestep to relax adiabatically to the pressure before the timestep. This is done by calculating the change in each cell volume V_c using

$$V_c(t + \Delta t) = V_c(t) \left(\frac{P^n}{P^o} \right)^{1/\gamma} , \qquad (13-3.1)$$

where P^n is the pressure calculated before the convection step, as discussed below. The quantity P^o is the pressure at the beginning of the timestep. Changes in the volume of all the cells cause the location of the open boundary to change. The location of the open boundary and the fluid velocity in the last cell (which is also the velocity of the open boundary) are then used as input conditions to ADINC.

13–3.2 Timestep Splitting

The order of the substeps, as shown in the table, is important. The rezoning, the timestep determination, and the energy deposition should be carried out in that order. In coupling all of these effects, convection should be carried out last. The diffusive transport processes and chemistry can be carried out in any order.

The timestep splitting used here appears similar to that used in RSHOCK, but it is quite different in a fundamental way. The major differences arise when both chemical kinetics and sound waves are stiff. The FLAME1D approach allows expansion and diffusive transport to relieve the pressure from the flame region on time scales characteristic of the sound speed. Thus the pressure stays effectively constant if the timestep is enough larger than the sound transit time across a cell. After each substep of the split timestep, a change of pressure is calculated that would occur due to that particular process alone. These provisional pressure changes are added to the old pressure before the convection stage to obtain a new pressure to use in ADINC.

The first process is external energy deposition. Any change in $\{n_i\}$ that this process causes in Δt is used to update $\{n_i\}$. In addition, any change in pressure that this process causes is saved in ΔP_1. The other variables are not changed as a result of this process.

At the end of the chemistry substep, the heat released is converted to an effective pressure change at constant volume because the cell volume has been held fixed during the chemical kinetics calculation. The chemical energy released and the temperature and pressure this implies can be calculated from the heats of formation and the changes in the number densities of the various species, as described in Chapter 5. Except for the $\{n_i\}$, the variables are not updated at this stage. The temperature is held constant during the chemistry substep.

Finally, the convective transport step is used to update all of the variables. It is necessary to make sure that all of the variables are fully consistent with each other by renormalizing the individual species densities, as discussed in Section 13–1. The ratios of species densities to use are those derived in the chemistry stage of the timestep.

13–3.3 Several Examples

The examples in this section are taken from studies of laminar, premixed hydrogen-oxygen flames diluted with nitrogen. The specific chemical reaction scheme is the same one used in the RSHOCK examples of Section 13–1. The diffusive transport coefficients and thermophysical properties were described in Chapter 7.

The FLAME1D models were used to study ignition in spherical geometry. Figure 13–19, taken from one of these calculations, shows the evolution of a temperature profile after ignition of a flame. The flame was initiated by a localized hot spot specified as a Gaussian temperature profile centered at the origin. A predetermined amount of energy, 4 mJ, was deposited linearly over a period of 10^{-4} seconds near the origin, in a deposition radius of 0.1 cm. The calculation was initially set up with 50 computational cells, and Lagrangian split-merge rezoning was allowed to occur as needed. An open boundary simulated a very large system. The steep rise in temperature behind the flame front shows the effects of initial conditions. In Figure 13–20 we show species densities for the last profile of Figure 13–19 and also show a blowup of the region around the flame front.

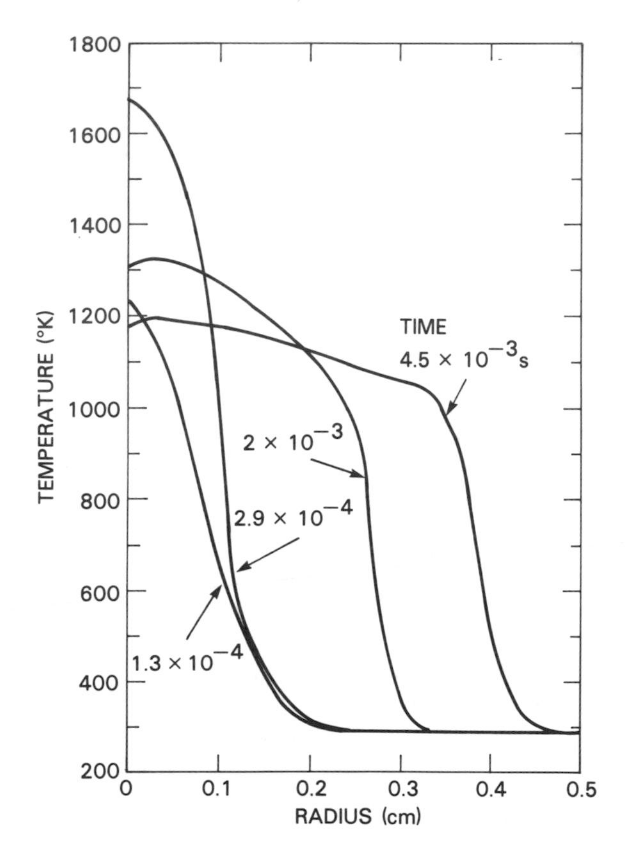

Figure 13–19. Temperature profiles at a sequence of times shortly after the spherical ignition of a mixture of hydrogen, oxygen, and nitrogen in the ratio 2:1:10. The mixture was ignited by depositing 4.0 mJ of energy inside a radius of 0.1 cm in 10^{-4} s (Kailasanath et al., 1982).

Even after the end of the period of energy deposition, the central temperature increases due to the continuing heat release from combustion. Eventually the temperature near the center decreases and the temperature away from the center increases due to diffusive transport. By 4.5 ms, the temperature distribution approaches that of a propagating flame. When the deposited energy was reduced to 3 mJ, the temperature distribution did not develop into a profile associated with a steadily propagating flame. In the 3 mJ case, the gradient at the flame front decreases in time and the flame dies away. Thus the 3 mJ of energy deposited is below the *minimum ignition energy*.

By repeating the computations with varying amounts of energy deposition, a minimum ignition energy for a particular radius can be obtained. A

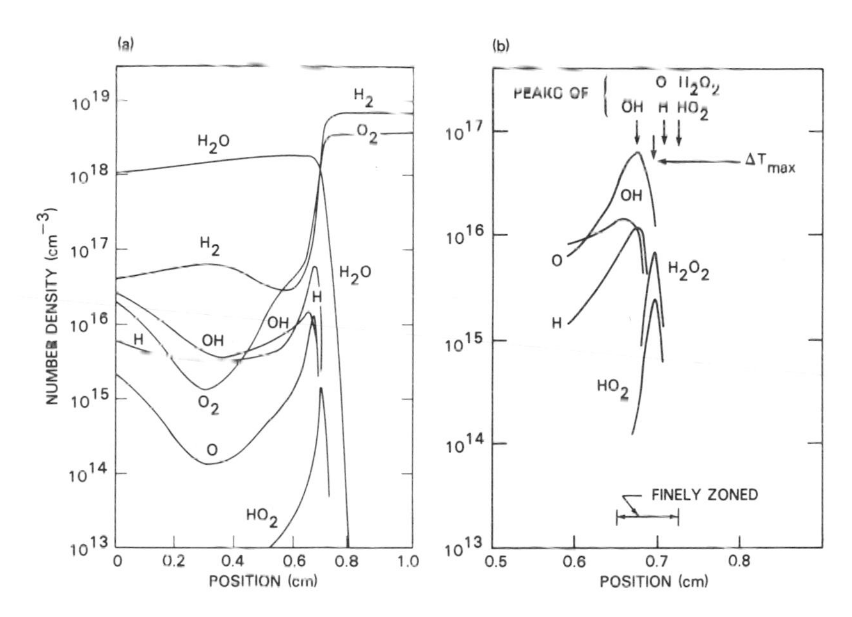

Figure 13–20. (a) Species profiles in a propagating flame in a mixture of hydrogen and oxygen. (b) Details of the flame front (Oran and Boris, 1981).

summary of many calculations with different values of the radius of energy deposition are shown in Figure 13–21. For example, a propagating flame resulted when 3.8 mJ was deposited in a sphere with a radius of 0.1 cm. However, when the same amount of energy was deposited in a sphere of smaller radius, the rate of heat liberation was insufficient to compensate for the rate of diffusive heat loss, and consequently there was no ignition. This radius, 0.1 cm, is the "quench radius" for this particular mixture. For larger radii (larger than 0.11 cm), the minimum ignition energy increases rapidly with increasing radii. Therefore the absolute minimum ignition energy is about 3.7 mJ for the system under study. These observations are in qualitative agreement with the spark ignition experiments described in Lewis and von Elbe (1961, pp. 323, 267). Quantitative comparisons are difficult because the mixture composition and the energy-deposition time were different. Furthermore, we do not know how to account for losses to the electrodes.

When an ignition kernel becomes a propagating steady flame, it is possible to calculate laminar flame velocities and burning velocities. If enough energy is deposited at the origin of a sphere and the computations are carried out for long enough, the temperature distribution attains a typical flame profile. To determine the burning velocity and flame speed, we first need

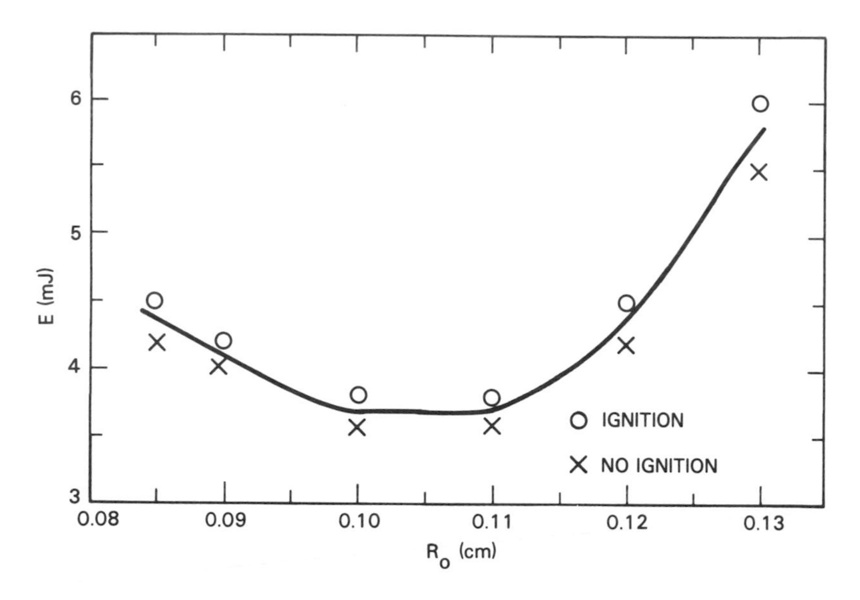

Figure 13–21. Minimum ignition energy as a function of radius and energy deposition for the same mixture described in Figure 13–19 (Kailasanath et al., 1982).

criteria for determining the location of the flame front and the appropriate value of the flame velocity to use. The burning velocity, v_b, can then be written

$$v_b = v_f - v_{\text{fluid}} , \qquad (13-3.2)$$

where v_f is the velocity at which the flame moves, and v_{fluid} is the fluid velocity in the unburned mixture ahead of the flame.

These criteria can sometimes be ambiguous, as shown by Figure 13–22 showing the fluid velocities and the temperature profiles in a planar and spherical flame propagating in a mixture with the same temperature, pressure, and stoichiometry. For the planar flame in Figure 13–22a, or any thin flame, there is not much ambiguity since the fluid velocity ahead of the flame is constant. However, when the flame has a finite thickness and is curved, as shown for a spherical flame in Figure 13–22b, the criteria are ambiguous. The fluid velocity reaches a maximum within the flame, and then decreases ahead of the flame due to spherical expansion effects. We generally use two reference fluid velocities to calculate the burning velocity. The first is the maximum fluid velocity in the system, v_{max}, and the second is velocity of the unburned gases corresponding to the first location ahead of the flame with a temperature of 300 K. The lower estimate v_b is obtained as $v_f - v_{\text{max}}$, and

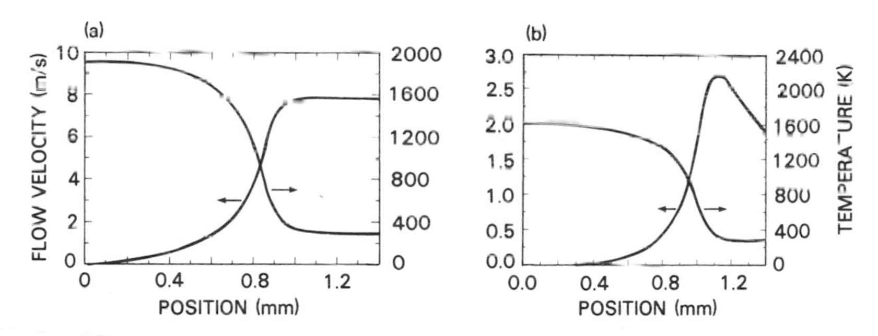

Figure 13–22. Flow velocity and temperature profile in a mixture of hydrogen, oxygen, and nitrogen in the ration 2:1:4 for (a) a planar flame and (b) a spherical flame (Kailasanath and Oran, 1986).

Figure 13–23. Flame velocity, v_f, maximum fluid velocity, v_{max}, and velocity of the fluid at 300 K, v_{300}, as a function of time in spherically propagating flames in mixtures of hydrogen, oxygen, and nitrogen in the ratios (a) 2:1:4, (b) 2:1:7, and (c) 2:1:13 (Kailasanath and Oran, 1986).

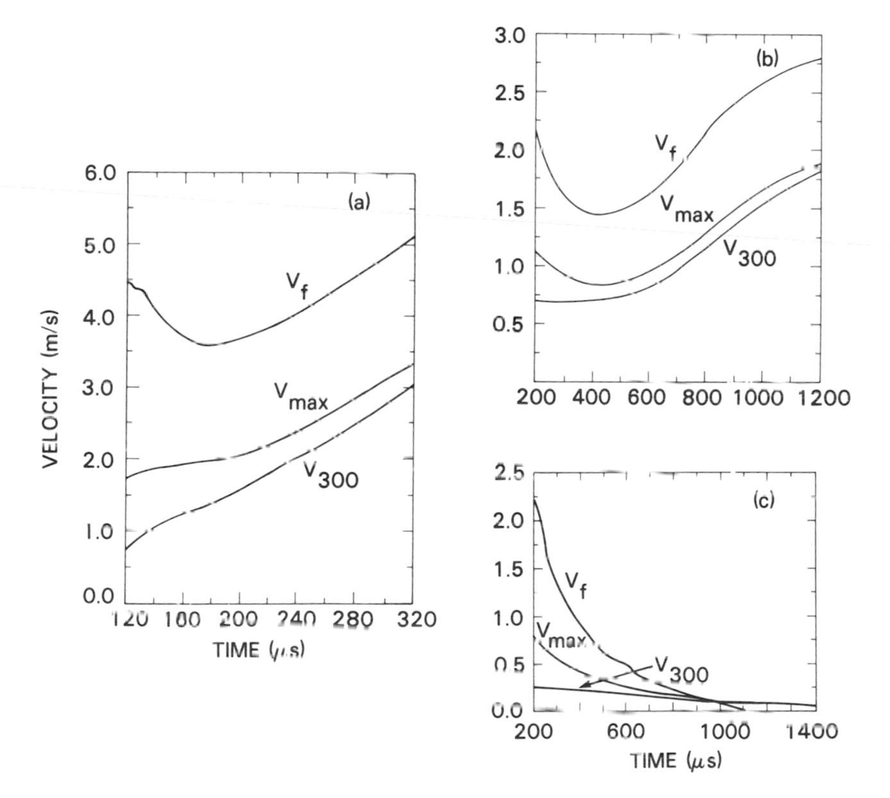

the upper estimate is from $v_f - v_{300}$. For planar flames, these two estimates are identical.

Simulations have been carried out with FLAME1D to study the effects of curvature and dilution on hydrogen-oxygen flames. Several examples are shown in Figure 13–23. These figures are taken from a series of calculations in spherical geometry for which the dilution with nitrogen was steadily increased. As the dilution is increased, the burning velocity decreases. Finally, in Figure 13–23c, no steady propagating flame is achieved. Similar results have been found for planar flames.

13–4. A TWO-DIMENSIONAL LAGRANGIAN MODEL

The use of Lagrangian methods in numerical simulations has generally been restricted to "well-behaved" flows. Shear, fluid separation, or even large amplitude motions produce severe grid distortion. These distortions arise because grid points move far enough that their nearest neighbors change in the course of a calculation. When differential operators are calculated on a highly distorted mesh, the approximations usually become inaccurate. Attempting to maintain accuracy through regridding and interpolating physical quantities onto a grid with a better structure introduces numerical diffusion into the calculation. Further, the difficulties that arise in standard Lagrangian methods often outweigh the benefits of essentially no new numerical diffusion.

Here we present more detailed information about the structure and use of the Lagrangian program SPLISH, which was described in Chapter 10. This method avoids grid tangling by using a dynamically restructuring grid of triangles. The method was first described in some detail by Fritts and Boris (1979). It has been applied to a number of different types of problems, including calculations of nonlinear waves (Fritts, 1976a; Fritts and Boris, 1977), Kelvin-Helmholtz instabilities (Fritts, 1976b), Rayleigh-Taylor instabilities (Emery et al., 1982b), Couette flows and Taylor vortex flows (Emery et al., 1981), and most recently to the flows in and around droplets (Fritts et al., 1984; Fyfe et al., 1987). Here the structure of the SPLISH computer program is described and then the surface tension algorithm is presented. Section 10–2 is a necessary preamble to both of these discussions.

13–4.1 The Structure of SPLISH

The various parts of the SPLISH program are more tightly coupled than the loose assembly of algorithms and modules described previously in this chapter. There are five major parts of SPLISH:

- Initializing variables, such as vertex positions, velocities, and pressures.
- Calculating the evolution of the fluid dynamics, described in Chapter 10, and outlined below.
- Calculating the surface tension forces.
- Readjusting the vertex connectivity, inserting and deleting vertices to obtain an optimal grid.
- Preparing tabular and graphical output.

Table 13–4 shows the order of operations and serves as a general outline for the discussion. Note that variables are defined either as triangle-centered quantities, sometime called triangle quantities, denoted by subscript t, or as vertex quantities, denoted by subscript v.

The first step in the calculation is determining the timestep, Δt. Currently Δt is determined using the criterion that no vertex should pass through any of the opposite triangle sides by the end of a timestep. This condition is calculated from the vertex velocities, $\{\mathbf{v}_v\}$. Because the algorithm is implicit, a timestep greater than that by the Courant limit can be used. A larger timestep might even be acceptable for simple translational flows where the vertices remain in the same general relationship to each other during one timestep, and no grid restructuring is necessary.

The next step is the fluid dynamics component of the calculation. The first operation is to integrate the velocities forward a half timestep. Surface tension forces are incorporated in the calculation of ∇P_v to find half-step values of the triangle-centered velocity $\mathbf{v}_t^{n-\frac{1}{2}}$ according to

$$\mathbf{v}_t^{n-\frac{1}{2}} = \mathbf{v}_t^{n-1} + \frac{1}{\rho}\frac{\Delta t}{2}\left[\nabla^{n-1} P_v^{n-1} - \mu(\nabla^{n-1}\cdot\nabla^{n-1})\mathbf{v}_t^{n-1} + \mathbf{f}_t\right] . \tag{13 – 4.1}$$

Here $\mathbf{f}$ is an external force. Note that the ∇ operator depends on the grid vertex locations, and therefore changes at each timestep. Also, taking the gradient of a vertex quantity, such as P_v produces a triangle-centered quantity, such as $(\nabla P_v)_t$. We generally write this as ∇P_v and drop subscript t. Values of the new velocities and pressures, $\mathbf{v}_t^n$ and P_v^n, are estimated. The initial iteration sets $\mathbf{v}_t^n$ equal to $\mathbf{v}_t^{n-\frac{1}{2}}$ and P_v^n equal to P_v^{n-1}, the value of the pressure at the beginning of the timestep. The estimated values of the new velocity and pressure used to start the iteration are called $\mathbf{v}_t^{(0),n}$ and $P_t^{(0),n}$, respectively.

Table 13–4. Outline of SPLISH

Initialize Grid and Variables.
* Determine Δt
    Based on grid-movement criteria.
    $t^n = t^{n-1} + \Delta t$

1. Fluid Dynamics
    a. Advance Velocities a Half Step.
        Calculate $\mathbf{v}_t^{n-\frac{1}{2}}$
    b. Estimate $\mathbf{v}_t^n$ and P_v^n.
    c. Iterate to find $\mathbf{x}_v^n$, $\mathbf{v}_t^n$, P_v^n.
        Compute the new positions, $\mathbf{x}_v^n$.
        Apply residual algorithm: ensure mass conservation.
        Apply rotator: ensure conservation of circulation.
        Generate new surface tension forces.
        Update P_v^n
        Update $\mathbf{v}_t^n$
        End loop if no change in $\mathbf{x}_v^n$, $\mathbf{v}_t^n$ and P_v^n.
2. Grid Restructuring
    a. Reconnect to optimize the Poisson solver.
    b. Iterate to optimize the grid.
        Delete small triangles.
        Add vertices as needed.
        Delete very small lines as needed.
        Add vertices on interfaces if needed.
        Reconnect to optimize Poisson solver.
        Iterate until there is no change in grid.

Start New Timestep (go to * above).

The iteration then determines progressively better values of the new vertex locations, velocities, and pressures. To calculate new vertex locations, first calculate $\{\mathbf{v}_v^{(0),n}\}$ by interpolating from the estimated new triangle-centered values. Then calculate a half-step vertex-centered velocity, $\{\mathbf{v}_v^{(k),n-\frac{1}{2}}\}$, from

$$\mathbf{v}_v^{(k),n-\frac{1}{2}} = \frac{\mathbf{v}_v^{(k),n} + \mathbf{v}_v^{n-1}}{2} , \qquad (13-4.2)$$

where the $\{\mathbf{v}_v^{n-1}\}$ are the values at the beginning of the timestep. The superscript (k) indicates the kth iterate. The new vertex positions are calculated from

$$\mathbf{x}_v^{(k),n} = \mathbf{x}_v^{n-1} + \Delta t\, \mathbf{v}_v^{(k),n-\frac{1}{2}} . \qquad (13-4.3)$$

Next, the residual algorithm and the rotator algorithm described in Chapter 10 are applied.

A straight triangle side would not actually remain straight during one timestep in a general convective flow field. The straight sides through the vertex positions that form new triangles give a slightly different value for the mass about a vertex than the correct curved sides of the translated old triangles would give. The residual algorithm adjusts the vertices so that the new straight triangle sides give the same vertex mass that would occur in the real flow.

The rotator algorithm changes the halfstep velocities to ensure that circulation is exactly conserved numerically. As the grid rotates, the triangle velocities should also rotate. Specifically, conservation of circulation is guaranteed by solving

$$\nabla^{(k),n} \times \tilde{\mathbf{v}}_t^{(k),n-\frac{1}{2}} = \nabla^{n-1} \times \mathbf{v}_t^{n-\frac{1}{2}} \qquad (13-4.4)$$

for $\tilde{\mathbf{v}}_t^{(k),n-\frac{1}{2}}$.

Surface tension forces are calculated and these are used to calculate a new $\{P_v^{(k),n}\}$ from

$$\nabla^{(k),n} \cdot \frac{1}{\rho} \nabla^{(k),n} P_v^{(k),n} = \frac{2}{\Delta t} \nabla^{(k),n} \cdot \tilde{\mathbf{v}}_t^{(k),n-\frac{1}{2}} . \qquad (13-4.5)$$

Finally the new velocities are calculated from

$$\mathbf{v}_t^{(k),n} = \tilde{\mathbf{v}}_t^{(k),n-\frac{1}{2}} + \frac{1}{\rho}\frac{\Delta t}{2} \left[\nabla^{(k),n} P_v^{(k),n} - \mu \left(\nabla^{(k),n} \cdot \nabla^{(k),n} \right) \mathbf{v}_t^{(k-1),n} + \mathbf{f}_t \right] \qquad (13-4.6)$$

which uses the latest estimate of the new velocities on the right hand side.

The total iteration described in item (1c) of Table 13–4 is continued until there is no significant change in calculated new velocities and pressures from one iteration to the next. Typically about three iterations are required.

The final major section of the program concerns the triangular gridding. After the fluid dynamics calculation, the vertices are reconnected, new vertices are added, and unnecessary vertices are deleted to optimize the local structure of the grid. One design goal is to keep the triangles close to equilateral. The result is a diagonally dominant matrix when the Poisson equation is solved, which means that the solutions converge much more quickly. This regridding section of the program also involves an iteration which proceeds until there is no change in the grid.

13–4.2 Surface Tension

The surface tension at an interface between two materials depends on the curvature of the interface. The conventional mathematical representation of surface tension is cast into a finite-difference form by fitting vertices to the material interface using a parametric function. Differentiating this function then provides an estimate of the local curvature. Once the interface curvature is known, the surface tension force is evaluated and used to accelerate the interface vertices.

The traditional prescription for the application of forces to interface vertices fails in SPLISH for two reasons. First, the interface vertices would be accelerated directly by surface tension forces evaluated on the vertices. Since velocities are centered on triangles in SPLISH, the velocity field responds to these accelerations a half timestep late unless a secondary calculation is made. As a result, the pressure within the droplet would be inconsistent with the pressure found from the surface tension formula. Second, since the pressure gradient forces and surface tension forces are not calculated in the same manner, numerical inconsistencies result which can grow with each timestep.

Both of these problems are eliminated when a surface tension potential is used to generate the forces. The surface tension force is determined as the gradient of a potential present only at the interfaces. With this method, the pressure gradient forces are calculated in the same manner and on the same grid as the forces derived from the surface tension potential. Therefore the surface tension potential and the pressure are dynamically similar. The pressure drop across the interface must exactly cancel the surface tension forces.

The algorithms for calculating the surface tension are straightforward. The surface tension forces are included through Laplace's formula for the pressure jump across an interface (Landau and Lifshitz, 1959),

$$P_i - P_o = \sigma / R , \qquad (14-4.7)$$

where P_i is the pressure just "inside" the interface, P_o is the pressure just "outside" the interface, σ is the surface tension coefficient associated with the two media which define the interface, and R is the radius of curvature of the interface in the two-dimensional plane. The radius of curvature is positive at points on the interface where the droplet surface is convex (a circle is convex everywhere) and negative when the surface is concave. These pressure jumps are included in the Poisson equation for the pressure. The average pressure, $(P_i + P_o)/2$, is computed at the interface vertices. From the average pressure and the pressure jump, SPLISH computes a pressure gradient centered on triangles, both inside and outside the surface. This pressure gradient is used in the momentum equation.

The radius of curvature is computed from a parametric cubic spline to interpolate between the interface vertices. The parametric interpolant used in SPLISH generates the twice-differentiable periodic spline interpolants, $\mathbf{r}(s) = (x(s), y(s))$ (deBoor, 1978). This spline is also used for regridding. When the regridding algorithm needs to bisect a triangle bordering the interface between the two materials, a new vertex is added on the spline interpolant between the vertices rather than along the straight triangle side connecting the vertices.

13–4.3 Flows in and around Droplets

The calculations described below are taken from Fritts et al. (1984) and Fyfe et al. (1987). The surface tension and viscosity algorithms described above were originally developed for the study of droplet distortion and breakup due to shear in the background flow and for droplet-droplet collisions.

Flow about a Droplet

The first problem described is a simulation of an incompressible, inviscid flow about a droplet only twice as dense as the background fluid. The gridding routines put the droplet at the center of the computational grid, and allow the resolution inside and outside the droplet to be variable. Additional routines initialize the background flow by a pressure pulse at the left boundary for the first half timestep, or linearly increase the flow speed from zero if the initial perturbation would cause transients that are too large. For all later times the left and right boundary conditions are periodic. Rigid wall boundary conditions are imposed at the top and bottom of the computational region. The problem, therefore, is really an infinite row of cylindrical droplets in an impulsively started flow moving from left to right.

The droplet was divided initially into 28 triangles in a total system of 552 triangles, as shown in the first frame of Figure 13–24. Only the triangle vertices are shown within the droplet itself, although the vertices are connected outside the droplet. This relatively coarse resolution is adequate to demonstrate the algorithm.

Figure 13–24 shows the evolution of the droplet and the triangular grid. Early in the calculation, a recirculation zone forms behind the droplet, compressing it in the direction parallel to the flow. The internal flow is driven

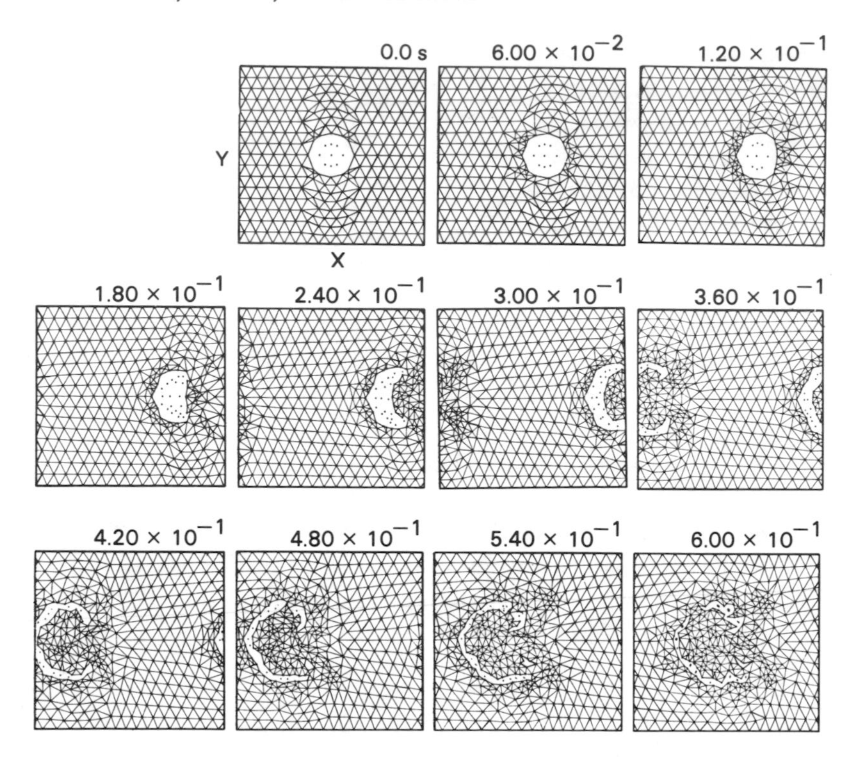

Figure 13–24. Simulation of the distortion and breakup of a low-density droplet with no viscosity or surface tension forces (Fritts et al., 1984). Background flow is from left to right. Boundary conditions in x-direction are periodic. Points inside the droplet indicate locations of triangle vertices.

by the compression set up between the front and rear stagnation points and by the high shear that extends around the top and bottom of the droplet and recirculation zone. The bulges formed at the top and bottom of the distorted droplet are convected around the recirculation zone by the shear flow that is maximum in these regions. Eventually, as seen in the figure, the droplet stretches into a thin layer coating the recirculation zone. The thin, elongated droplet then shatters into smaller pieces, first at the rear of the droplet and later in the more laminar flow toward the front.

A study of the movement of the individual triangle vertices shows that the flow is fairly smooth at all times, despite the distorted shape of the droplet. Subtracting out the mean flow leaves a large stationary double vortex. In spherical geometry, this recirculation zone forms a vortex ring, and the thinned droplet coating the ring fragments in both the radial and azimuthal directions.

An obvious feature of this calculation is its lack of symmetry about the center line at later times, even though the initial conditions are symmetric. This asymmetry results from a combination of two features in the algorithm. First, there is a minimum resolution allowed in the calculation, determined by specifying the minimum length of a triangle side that is kept in the calculation. If a triangle develops a shorter side, it is eliminated by the regridding procedure. Second, the regridding is a local operation, not global. Regridding at one location is independent of regridding some distance away. The result is that triangles are added and deleted in slightly different ways at different locations on the mesh, restricted only by conditions on a triangle side. Regridding at one place changes the overall mesh. Regridding at the supposed symmetrical points can occur somewhat differently depending on the order of operations. However, there is less apparent asymmetry as the grid becomes finer, which indicates convergence.

An important quantity monitored in these simulations is the mass of the droplet. If the timestep is too large, or if there is insufficient resolution, the mass changes during the calculation. When the mass deviates beyond a fixed tolerance, the calculation is invalid and must be redone with finer resolution.

Next consider forced fluid flow due to a fast air stream about an initially stationary kerosene droplet. Table 13–5 gives the physical parameters of the droplet and the background flow field. A total of 309 vertices were used initially with 12 vertices at the droplet interface. The initial flow is a steady-state potential flow about a periodic series of cylinders. As in the previous example, the boundary conditions on the sides are periodic, and the upper and lower boundary conditions are reflecting walls. Initially, the droplet is perfectly circular and at rest in the background flow, providing a smooth start for the calculation. This initialization describes a situation in which the flow velocity is ramped up to its final value before any significant structure develops in the flow, and before the droplet has acquired a substantial velocity. These are relatively high-speed flows, and the Reynolds number is fairly high. The large density difference between the kerosene droplet and the air makes the calculation more difficult and more expensive than the 2:1 calculation shown in Figure 13–24. Because the Poisson equation has a $1/\rho$ factor in the coefficient matrix, the matrix equation is less well conditioned and requires more iterations for its solution.

Figure 13–25 shows the evolution of pathlines for a portion of the calculation. A pathline is constructed by storing the five most recent locations of each vertex, and putting an asterisk on the most recent one to represent an arrowhead. This figure shows both the internal and external flow fields.

Table 13–5. Parameters of Droplet Calculations

density of kerosene	0.82 g/cc
density of air	0.0013 g/cc
surface tension (STP)	30 dynes/cm
viscosity of kerosene	1.8 centipoise
viscosity of air	0.018 centipoise
air velocity	100 and 120 m/s
initial droplet velocity	0.0 m/s
droplet radius	125 microns
Reynolds number	≈ 2000

The fourth panel at 3.5×10^{-6} s shows a pair of counter-rotating vortices, the first clear indication of the development of the recirculation region. The recirculation zone continues to develop throughout the calculation, although at times the vortex pair is not as evident in the pathlines due to the deletion and addition of vertices which interrupt the continuity of the pathlines. By the last panel, another pair of vortices is forming near the droplet, and the original pair has been shed.

Distortions in the face of the droplet are evident by the seventh panel. These distortions occur because the curvature has increased and the streamlines in the external flow are condensed by the approaching wake. The internal velocities are small compared to the external flow rates and therefore cannot be distinguished as pathlines. Indication of the (small) internal recirculation is visible by comparing internal vertex positions at various timesteps.

Figure 13–26 shows the grid at times in the calculation corresponding to those in Figure 13–25. During the course of the calculation, considerable vertex addition and deletion has occured. Vertex addition, however, is most noticeable in the wake of the droplet and around its interface. There were 300 vertices at the beginning of the calculation, but there are 450 at the end.

As seen in Figure 13–26, the computational grid needs further refinement because the perturbations cannot be resolved by the limits set on minimum triangle size originally chosen for the calculation. A sign that the calculation is under resolved is that one of the crests of the surface wave is spanned by a single triangle, a situation which allows no communication between that surface fluid and the interior of the droplet.

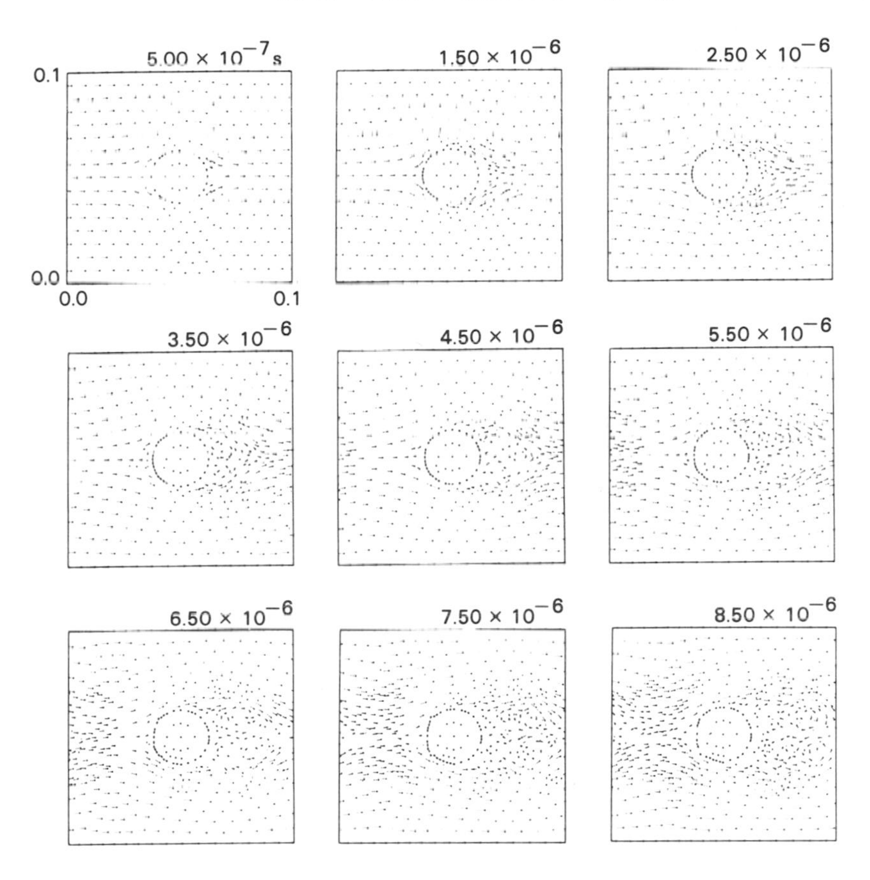

Figure 13–25. Pathlines from a calculation of a flow around a kerosene droplet that includes surface tension and viscosity (Fritts et al., 1984). The background flow velocity is 120 m/s. $Re \approx 2000$.

A Droplet in a Shear Flow

An important problem in atomization is how a droplet breaks up due to shear forces. To investigate this computationally, we have simulated droplets in a shear flow. The initial flow is $v_x = (y - y_d)G$ for points outside the droplet and $v_x = 0$ for points inside the droplet. The parameter y_d is the y-coordinate of the center of the drop and G gives the magnitude of the shear. The y-component of the velocity is initially zero everywhere.

Results of shear on a kerosene droplet in hot air are shown in the three panels in Figure 13–27 for $G = 5 \times 10^3$ s^{-1}. We used 0.013 g/cm^{3}as the density for hot air. The remaining physical parameters are the same as in Table 13–5. The droplet is initially round, but soon becomes elongated

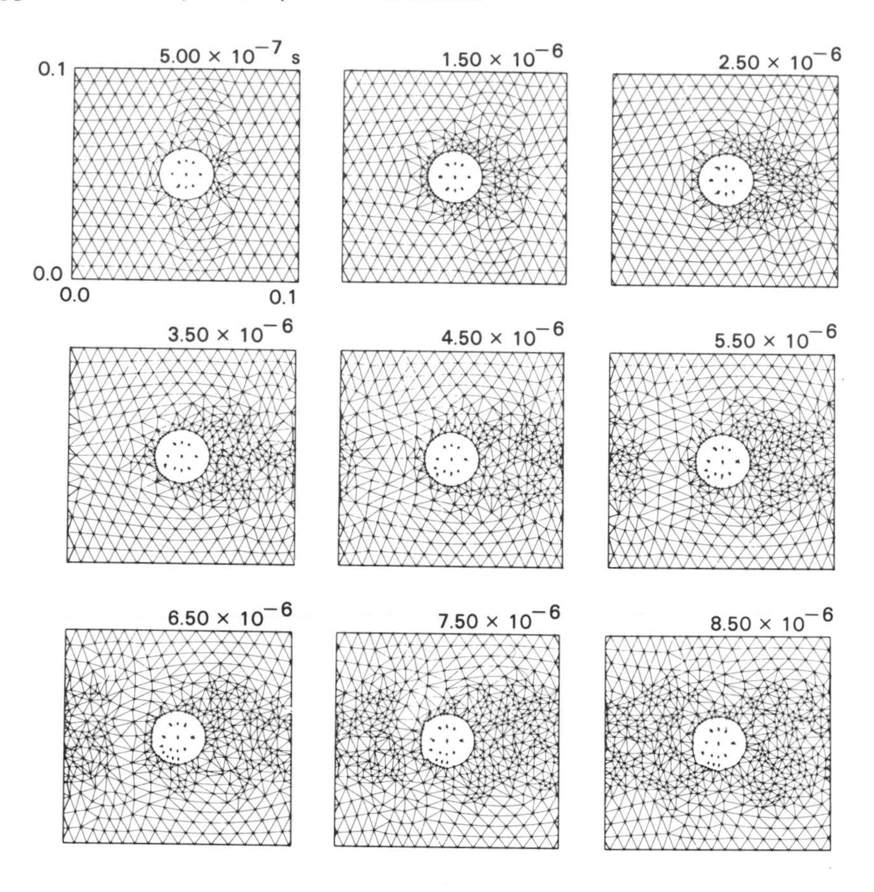

Figure 13–26. Grid corresponding to pathlines shown in Figure 13–25 (Fritts et al., 1984).

in the direction of the shear. At a later time, it has become even more elongated. In the interval between these two figures, very small droplets have been drawn off the large drop, but their size was so small that they were subsequently deleted from the calculations. In the last panel in Figure 13–27, small droplets move off of both sides. The small droplets sometimes seem to move counter to the flow of the main shear layer because a recirculation zone forms at the upper left and the lower right of the large droplet.

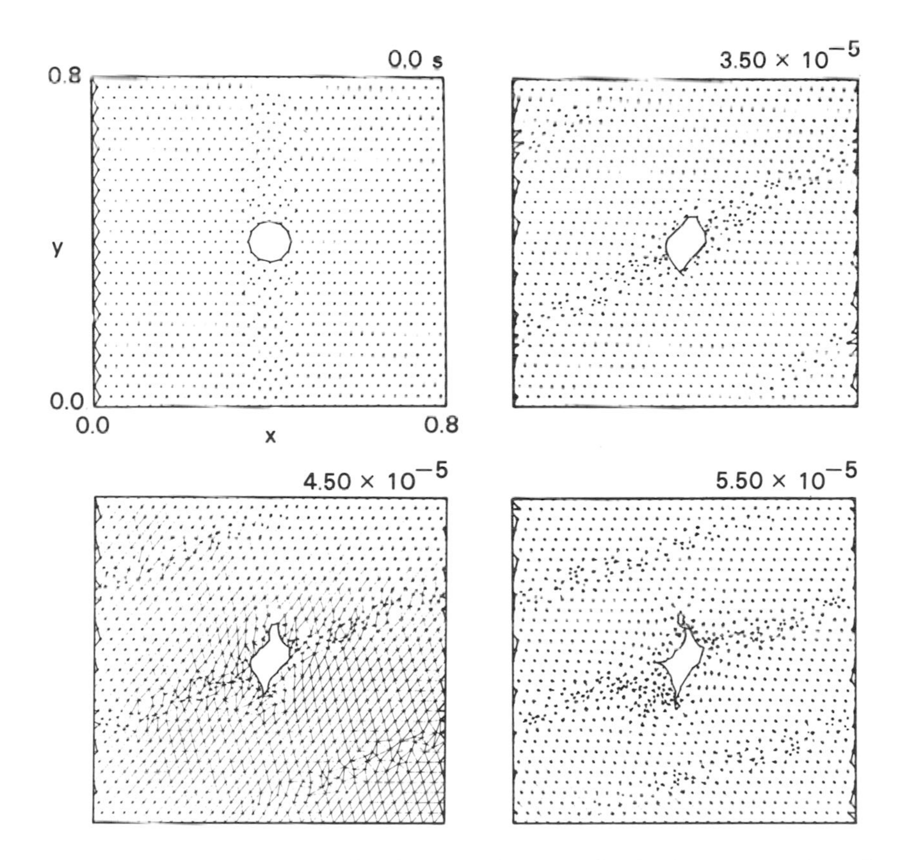

Figure 13–27. Distortion and breakup of a kerosene droplet in a strong shear flow. (Figure is courtesy of D.E. Fyfe.)

13–5. RULES OF THUMB

We summarize some of the important lessons learned in the course of constructing and using complex numerical models for simulating reactive flows.

1. Specifying the Mathematical and Computational Model

   - Before starting work on a simulation model, decide as clearly as possible what the objectives of the study are. It is important to pose scientific questions in terms that can be answered effectively by the simulation.
   - Look for a new approach if large improvement is needed in an already resonably optimized method.
   - Recognize and work within the constraints of the existing simulation environment. This involves understanding the limitations of the avail-

able methods, computer, personal abilities, and the time available to get the job done.

- A physical phenomenon cannot be simulated if the important controlling processes are not incorporated in the computational model and resolved in the simulation.
- Choose a representation that allows some generalization for future applications. If the simulation model cannot be adapted to more realistic boundary conditions or generalized to higher dimensionality, it may not be flexible enough even though it has other advantages.
- Do not try to program complex, multiprocess reactive flow simulations in dimensionless units. Dimensionless units work well for small, idealized, single-process programs. When programs in dimensionless units are extended to complex reactive flows, errors are inevitable. It is hard to debug a complicated program when all the numbers it generates, right or wrong, are of order unity. The dimensionless analysis can *always* be performed on the dimensional results.

2. Constructing the Numerical Simulation Model

- Three rules of thumb should be kept in mind about resolution:
  - Adequate resolution in the region of steep gradients is important and may be crucial.
  - Profiles should not change more than 10–15% per timestep.
  - Fast, stiff phenomena must either be in a slowly varying dynamic equilibrium in which two effects nearly cancel, or they must be resolved with small timesteps.
- Modest improvements in accuracy and resolution arise from big improvements in computer power. Do not expect slightly larger computers to bring a large change in capability.
- Three rules should also be kept in mind when constructing computational algorithms to represent the various processes in reactive flow calculations:
  - Algorithms must keep mass densities and energy densities positive where the real physical quantities are positive. Negative temperatures and densities usually give nonsensical results in chemical reaction mechanisms designed around positive definite quantities.
  - Energy-conserving algorithms are crucial in shock and flame calculations.
  - Convection algorithms must minimize both numerical diffusion and nonphysical fluctuations.

- Keep the computer program clean. Discard unused or erroneous statements rather than keeping them in the program. Confusion results from not doing this, and time is wasted each time the program is modified. Document within the program what the models do. Do not document what the program does not do by "commenting out" unused statements.
- Change the model slowly and systematically so there is always progress from one known situation to another. Test and verify the modifications before adding more changes. Making too many changes at one time can waste an inordinate amount of time debugging.
- If a program or module is obsolete, throw it out and start over.
- Build known physical properties into numerical algorithms as constraints on the solutions. Basic properties such as symmetry, conservation, positivity, causality, and reversibility should be enforced where appropriate.

3. Using the Numerical Simulation Model

- Justification for placing trust in complex simulations is based on the extent to which the model is tested and the results compared to theoretical and experimental cases where the answers are known. Therefore, it is worth the time and effort to do the job right.
- It always helps to document simulations as they are being performed. The explanations should be complete enough to be an immediate reminder of what you were trying to accomplish and what you learned.
- Make sure that the timestep is just small enough for the results to be acceptably accurate. The practical stability of algorithms generally requires a timestep that is a factor of two smaller than the most stringent mathematical stability conditions for the processes and interactions treated.
- The most expensive process computationally should be the one that limits the accuracy. If this is not true, the cost can be reduced without reducing the overall accuracy of the simulation by decreasing the excess accuracy of the most expensive part.
- Printing out the results of a reactive flow calculation to four, five, or even six significant figures is usually misleading because the basic input data are not even known to 10% accuracy.
- Ten half-hour calculations are generally worth much more than one five-hour calculation.
- Calculations tend to take roughly an hour on any computer system. This is a psychological result rather than a constraint of hardware performance or software speed. Shorter calculations have less resolution or

are computed for fewer timesteps. Longer calculations delay the other users and run the risk of wasting too much computer time if something is wrong.

- Personnel costs are usually 3 or 4 times as much as computing costs. This should help determine how time is spent.
- There are always conflicts between what you want to do and what you can afford. For best results, it is necessary to understand the trade-offs and respect the limitations of the model and the computer system.
- Maintain a healthy skepticism. Nothing works until it is well tested, and probably not even then.

References

Book, D., J. Boris, A. Kuhl, M. Picone, E. Oran, and S. Zalesak, 1981, Simulation of Complex Shock Reflections from Wedges in Inert and Reactive Mixtures, in W.C. Reynolds and R.W. MacCormack, eds., *Proceedings of the 7th International Conference on Numerical Methods in Fluid Dynamics*, Springer, New York, 84–90.

Boris, J.P., 1979, *ADINC: An Implicit Lagrangian Hydrodynamics Code*, NRL Memorandum Report 4022, Naval Research Laboratory, Washington, DC.

Boris, J.P., and D.L. Book, 1976, Solution of the Continuity Equation by the Method of Flux-Corrected Transport, *Methods in Computational Physics*, 16: 85–129.

Boris, J.P., E.S. Oran, M.J. Fritts, and C. Oswald, 1983, *Time Dependent, Compressible Simulations of Shear Flows: Tests of Outflow Boundary Conditions*, NRL Memorandum Report 5249, Naval Research Laboratory, Washington, DC.

Chevalier, R.A., and J.H. Gardner, 1974, The Evolution of Supernova Remnants, II. Models of an Explosion in a Plane-Stretched Medium, *Astrophys. J.* 192: 457–463.

Dahlburg, R.B., C.R. DeVore, J.M. Picone, J.T. Mariska, and J.T. Karpen, 1986, Nonlinear Evolution of Radiation-Driven Thermally Unstable Fluids, *Astrophys. J.*, to appear.

deBoor, C., 1978, *A Practical Guide to Splines*, Springer, New York, 316–322.

DeVore, C.R., J.H Gardner, J.P. Boris, and D. Mosher, 1984, Hydrodynamic Simulations of Light Ion Beam – Matter Interactions: Ablative Acceleration of Thin Foils, *Laser and Particle Beams*, 2: 227–243.

Edwards, D.G., 1986, private communication, Fluid Gravity (Eng.) Ltd., Surrey, England.

Emery, M.H., M.J. Fritts, and R.C. Shockley, 1981, *Lagrangian Simulation of Taylor-Couette Flow*, NRL Memorandum Report 4569, Naval Research Laboratory, Washington, DC.

Emery, M.H., J.H. Gardner, and J.P. Boris, 1982a, Nonlinear Aspects of Hydrodynamic Instabilities in Laser Ablation, *Appl. Phys. Lett.* 41: 808-810.

Emery, M.H., S.E. Bodner, J.P. Boris, D.G. Colombant, A.L. Cooper, M.J. Fritts, and M.J. Herbst, 1982b, *Stability and Symmetry in Inertial Confinement Fusion*, NRL Memorandum Report 4947, Naval Research Laboratory, Washington, DC.

Fritts, M.J., 1976a, *A Numerical Study of Free-Surface Waves*, SAIC Report SAI-76-528-WA, Science Applications International, Inc., McLean, VA.

Fritts, M.J., 1976b, *Lagrangian Simulations of the Kelvin-Helmholtz Instability*, SAIC Report SAI-76-632-WA, Science Applications International, Inc., McLean, VA.

Fritts, M.J., and J.P. Boris, 1977, Transient Free Surface Hydrodynamics, in J.V. Wehausen and N. Salvesen, eds., *Second International Conference on Numerical Ship Hydrodynamics*, pp. University Extension Publications, Berkeley, 319–328.

Fritts, M.J., and J.P. Boris, 1979, The Lagrangian Solution of Transient Problems in Hydrodynamics Using a Triangular Mesh, *J. Comp. Phys.* 31: 173–215.

Fritts, M.J., D.E. Fyfe, and E.S. Oran, 1984, *Numerical Simulations of Fuel Droplet Flows Using a Lagrangian Triangular Mesh*, NRL Memorandum Report 5408, Naval Research Laboratory, Washington, DC.

Fyfe, D.E., E.S. Oran, and M.J.Fritts, 1987, Surface Tension and Viscosity with Lagrangian Hydrodynamics on a Triangular Mesh, *J. Comp. Phys.*, to appear.

Gardner, J.H., and S.E. Bodner, 1981, Wavelength Scaling for Reactor-Size Laser-Fusion Targets, *Phys. Rev. Lett.* 47: 1137–1140.

Gardner, J.H., and S.E. Bodner, 1984, High Efficiency Targets for High-Gain Inertial Confinement Fusion, *Phys. Fluids* 29: 2672–2678.

Gaydon, A.G., and H.G. Wolfhard, 1979, *Flames*, Chapman and Hall, London.

Grinstein, F.F., E.S. Oran, and J.P. Boris, 1986, Numerical Simulations of Asymmetric Mixing in Planar Shear Flows, *J. Fluid Mech.* 165: 201–220.

Grinstein, F.F., E.S. Oran, and J.P. Boris, 1987, Direct Numerical Simulation of Axisymmetric Jets, *AIAA J.* 25: 92–97.

Guirguis, R., and E. Oran, 1983, *Reactive Shock Phenomena in Condensed Materials: Formulation of the Problem and Method of Solution*, NRL Memorandum Report 5228, Naval Research Laboratory, Washington, DC.

Guirguis, R., E.S. Oran, and K. Kailasanath, 1986, Numerical Simulations of the Cellular Structure of Detonations in Liquid Nitromethane — Regularity of the Cell Structure, *Comb. Flame* 65: 339–366.

Guirguis, R., E.S. Oran, and K. Kailasanath, 1987, The Effect of Energy Release on the Regularity of Detonation Cells in Liquid Nitromethane, *Proceedings of the 21th Symposium (International) on Combustion*, The Combustion Institute, Pittsburgh, PA, to appear.

Jones, W.W., and J.P. Boris, 1981, An Algorithm for Multispecies Diffusion Fluxes, *Comp. Chem.* 5: 139–146.

Kailasanath, K., and E.S. Oran, 1983, Ignition of Flamelets behind Incident Shock Waves and the Transition to Detonation, *Comb. Sci. Tech.* 34: 345–362.

Kailasanath, K., and E.S. Oran, 1986, Effects of Curvature and Dilution on Unsteady, Premixed Flame Propagation, in Dynamics of Reactive Systesm, I. Flames and Configurations, *Prog. Aero. Astro.* 105: 167–179, AIAA, New York.

Kailasanath, K., E.S. Oran, and J.P. Boris, 1982a, A Theoretical Study of the Ignition of Premixed Gases, *Comb. Flame* 47: 193–190.

Kailasanath, K., and E.S. Oran, 1986, Effects of Curvature and Dilution on Unsteady Flame Propagation, in J.R. Bowen, J.-C. Leyer, and R.I. Soloukhin, eds., *Dyanmics of Reactive Systems, I. Flames and Configurations, Prog. Astro. Aero.* 105: 167–179.

Kailasanath, K., E.S. Oran, and J.P. Boris, 1982b, Time-Dependent Simulation of Flames in Hydrogen-Oxygen-Nitrogen Mixtures, in J. Warnatz and N. Peters, eds., *Proceedings of the GAMM-Workshop on Numerical Methdos in Laminar Flame Propagation*, Vieweg, Wiesbaden, 152–166.

Kailasanath, K., E.S. Oran, and J.P. Boris, 1982c, *A One-Dimensional Time-Dependent Model for Flame Initiation, Propagation and Quenching*, NRL Memorandum Report 4910, Naval Research Laboratory, Washington, DC.

Kailasanath, K., J. Gardner, J. Boris, and E. Oran, 1985a, Acoustic-Vortex Interactions in an Idealized Ramjet Combustor, *Proceedings of the 22nd JANNAF Combustion Meeting*, Chemical Propulsion Information Agency (CPIA), Johns Hopkins University, Applied Physics Laboratory, Laurel, MD.

Kailasanath, K., E.S. Oran, and J.P. Boris, 1985b, Determination of Detonation Cell Size and the Role of Transverse Waves in Two-Dimensional Detonations, *Comb. Flame* 61: 199–209.

Kailasanath, K., J.H. Gardner, J.P. Boris, and E.S. Oran, 1986, Interactions between Acoustics and Vortex Structures in a Central Dump Combustor, Paper No. AIAA-86-1609, AIAA, New York.

Kailasanath, K., J. Gardner, J. Boris, and E. Oran, 1987, Numerical Simulations of Acoustic-Vortex Interactions in a Central-Dump Ramjet Combustor, *J. Prop. Power*, to appear.

Karpen, J.T., E.S. Oran, J.P. Boris, and G.T. Breuckner, 1982, The Dynamics of Accelerating Coronal Bullets, *Astrophys. J.* 261: 375–386.

Karpen, J.T., E.S. Oran, and J.P. Boris, 1984, Detailed Studies of the Dynamics and Energetics of Coronal Bullets, *Astrophys. J.* 287: 396–403.

Lampe, M., W. Ali, G. Joyce, B. Hui, J.M. Picone, R. Hubbard, R. Fernsler, and S. Slinker, 1984, *Beam Propagation Studies at NRL, July 1983 — June 1984, Volume 2*, NRL Memorandum Report 5412, Naval Research Laboratory, Washington, DC.

Landau, L.D., and E.M. Lifshitz, 1959, *Fluid Mechanics*, Pergammon, New York, 230–234.

Lewis, B., and G. von Elbe, 1961, *Combustion, Flames and Explosions of Gases*, Academic Press, New York.

Lyon, J.G., S.H. Brecht, J.D. Huba, J.A. Fedder, and P.J. Palmadesso, 1981, Computer Simulation of a Geomagnetic Substorm, *Phys. Rev. Lett.* 46: 1038–1041.

Lyon, J.G., J.A. Fedder, and J.D. Huba, 1986, The Effect of Different Resistivity Models on Magnetotail Dynamics, *J. Geophys. Res.* 91: 8057–8064.

Mariska, J.T., J.P. Boris, E.S. Oran, G.A. Doschek, and T.R. Young, 1982, Solar Transition Region Response to Variations in the Heating Rate, *Astrophys. J.* 255: 783–796.

Oran, E., and J.P. Boris, 1981, Theoretical and Computational Approach to Modeling Flame Ignition, in J.R. Bowen and A.K. Oppenheim, eds., *Combustion in Reactive Systems, Prog. Astro. Aero.*, 76: 154–171, AIAA, New York.

Oran, E.S., and J.P. Boris, 1982, Weak and Strong Ignition: II. Sensitivity of the Hydrogen-Oxygen System, *Comb. Flame* 48: 149–161.

Oran, E., T.R. Young, and J.P. Boris, 1979, Application of Time-Dependent Numerical Methods to the Description of Reactive Shocks, *Proceedings of the 17th Symposium (International) on Combustion*, The Combustion Institute, Pittsburgh, PA, 43–54.

Oran, E., J.P. Boris, T.R. Young, and J.M. Picone, 1981, Numerical Simulations of Detonations in Hydrogen-Air and Methane-Air Mixtures, *Proceedings of the 18th Symposium (International) on Combustion*, The Combustion Institute, Pittsburgh, PA, 1641–1649.

Oran, E., T.R. Young, J.P. Boris, and A. Cohen, 1982a, Weak and Strong Ignition: I. Numerical Simulations of Shock Tube Experiments, *Comb. Flame* 48: 135–148.

Oran, E., T.R. Young, J.P. Boris, J.M. Picone, and D.H. Edwards, 1982b, A Study of Detonation Structure: The Formation of Unreacted Gas Pickets, *Proceedings of the 19th Symposium (International) on Combustion*, The Combustion Institute, Pittsburgh, PA, 573–582.

Oran, E.S., J.T. Mariska, and J.P. Boris, 1982c, The Condensational Instability in the Solar Transition Region and Corona, *Astrophys. J.* 254: 349–360.

Picone, J.M., and J.P. Boris, 1983, Vorticity Generation by Asymmetric Energy Deposition in a Gaseous Medium, *Phys. Fluids* 26: 365–382.

Picone, J.M., E.S. Oran, J.P. Boris, and T.R. Young, Jr., 1984, Theory of Vorticity Generation by Shock Wave and Flame Interactions, in J.R. Bowen, N. Manson, A.K. Oppenheim, and R.I. Soloukhin, eds., *Dynamics of Shock Waves, Explosions, and Detonations, Prog. Aero. and Astro.* 94: 429–448.

Picone, J.M., J.P. Boris, E.S. Oran, and R. Ahearne, 1986, Rotational Motion Generated by Shock Propagation through a Non-Uniform Gas, in D. Bershader and R. Hanson, eds., *Proceedings of the 15th International Symposium on Shock Waves and Shock Tubes*, Stanford University Press, Palo Alto, CA, 523–529.

Strehlow, R.A., 1984, *Combustion Fundamentals*, McGraw-Hill, New York.

Warnatz, J., and N. Peters, Eds., 1982, *Proceedings of the GAMM-Workshop on Numerical Methdos in Laminar Flame Propagation*, Vieweg, Wiesbaden.

Williams, F.A., 1985, *Combustion Theory*, Second Edition, Benjamin Cummings, Menlo Park, CA.

Young, T.R., and J.P. Boris, 1977, A Numerical Technique for Solving Stiff Ordinary Differential Equations Associated with the Chemical Kinetics of Reactive-Flow Problems, *J. Phys. Chem.* 81: 2424–2427.

Young, T.R., 1979, *CHEMEQ — Subroutine for Solving Stiff Ordinary Differential Equations*, NRL Memorandum Report 4091, Naval Research Laboratory, Washington, DC [AD AO83545].

Zalesak, S.T., S.L. Ossakow, and P.K. Chaturvedi, 1982, Non-linear Equatorial Spread-F: The Effect of Neutral Winds and Background Pedersen Conductivity, *J. Geophys. Res.* 87: 151–166.

Zalesak, S.T., P.K. Chaturvedi, S.L. Ossakow, and J.A. Fedder, 1985, Finite Temperature Effects on the Evolution of Ionospheric Barium Clouds in the Presence of a Conducting Background Ionosphere 1. The Simplest Case: Incompressible Background Ionosphere, Equipotential Magnetic Field Lines, and an Altitude-Invariant Neutral Wind, *J. Geophys. Res.* 90: 4299–4310.

Index

B

D

G

H

P

U